GERMANIUM-BASED TECHNOLOGIES

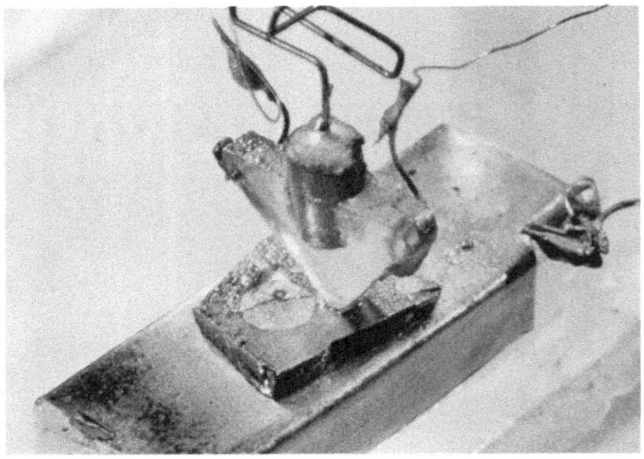

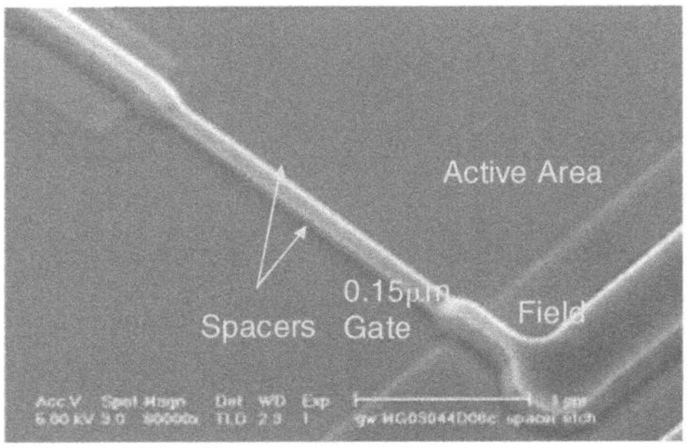
Active Area
Spacers
0.15μm
Gate
Field

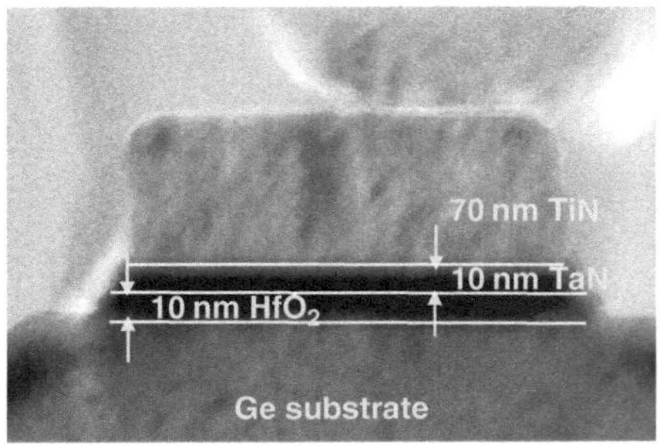
70 nm TiN
10 nm TaN
10 nm HfO_2
Ge substrate

GERMANIUM-BASED TECHNOLOGIES

From Materials to Devices

Cor Claeys

and

Eddy Simoen

ELSEVIER

Amsterdam • Boston • Heidelberg • London • New York • Oxford
Paris • San Diego • San Francisco • Singapore • Sydney • Tokyo

Elsevier
The Boulevard, Langford Lane, Kidlington, Oxford OX5 1GB, UK
Radarweg 29, PO Box 211, 1000 AE Amsterdam, The Netherlands

First edition 2007

British Library Cataloguing in Publication Data
Germanium-based technologies: from materials to devices
1. Germanium – Industrial application 2. Germanium
I. Claeys, Cor L. II. Simoen, E. (Eddy)
661'.0684

Library of Congress Catalog Number: 2006939883

ISBN-13: 978-0-08-044953-1
ISBN-10: 0-08-044953-0

For information on all elsevier publications
visit our web site at books.elsevier.com

Typeset by Charon Tec Ltd (A Macmillan Company), Chennai, India
www.charontec.com

Transferred to Digital Print 2008

Printed and bound in Great Britainby CPI Antony Rowe, Eastbourne

Contents

Editors

Cor Claeys
IMEC
Kapeldreef 75
B-3001 Leuven
Belgium
claeys@imec.be

Also at E.E. Department
KU Leuven
Kasteelpark Arenberg 10
B-3001 Leuven
Belgium

Eddy Simoen
IMEC
Kapeldreef 75
B-3001 Leuven
Belgium
simoen@imec.be

Contributors

Alessandra Satta
IMEC
Kapeldreef 75
B-3001 Leuven
Belgium
Alessandra.Satta@imec.be

Anthony R Peaker
UMIST
Sackville Street Building
Manchester M60 1QD
UK
peaker@manchester.ac.uk

Antoon Theuwis
Umicore
Watertorenstraat 33
B-2250 Olen
Belgium
antoon.theuwis@umicore.com

Arne Nylandsted Larsen
University of Aarhus
Institute of Physics and Astronomy
DK-8000 Aarhus
Denmark
anl@phys.au.dk

Ben Depuydt
Umicore
Watertorenstraat 33
B-2250 Olen
Belgium
ben.depuydt@umicore.com

Brice De Jaeger
IMEC
Kapeldreef 75
B-3001 Leuven
Belgium
Brice.DeJaeger@imec.be

Carl Quaeyhaegens
Umicore: Watertorenstraat 33
B-2250 olen,
Belgium,
carl.quaeyhaegens@umicore.com

Chi On Chui
UCLA
7440D Boelter Hall
Los Angeles, 90095-1594
chui@ee.ucla.edu

Chrystel Deguet
CEA-LETI
Rude des Martyrs 17
F-38054 Grenoble
France
Chrystel.deguet@cea.fr

Cor Claeys
IMEC
Kapeldreef 75
B-3001 Leuven
Belgium
claeys@imec.be

David Esseni
University of Udine
DEIGM, Via delle Scienze 208
I-33100 Udine
Italy
esseni@uniud.It

Eddy Simoen
IMEC
Kapeldreef 75
B-3001 Leuven
Belgium
Eddy.Simoen@imec.be

Enrico Sangiorgi
University of Bologna
ARCES-DEIS
Bologna
Italy
esangiorgi@deis.unibo.it

Erich Kasper
University of Stuttgart
Pfaffenwaldring 47
D-70569 Stuttgart
Germany
kasper@iht.uni-stuttgart.de

Fabrice Letertre
SOITEC
Parc Technologique de Fontaines 38
F-190 Bernin
France
fabrice.letertre@soitec.fr

Igor Romandic
Umicore
Watertorenstraat 33
B-2250 Olen
Belgium
Igor.romandic@umicore.be

Jan Vanhellemont
Ghent University
Krijgslaan 281-S1
B-9000 Gent
Belgium
jan.vanhellemont@telenet.be

José Coutinho
Department of Physics
University of Aveiro
3810-193 Aveiro
Portugal
coutinho@fis.ua.pt

Krishna C. Saraswat
Stanford University
Center for Integrated System
Stanford, CA
USA
saraswat@cis.stanford.edu

Marc De Jonghe
Umicore
Watertorenstraat 33
B-2250 Olen
Belgium
Marc.dejonghe@umicore.be

Marc Heyns
IMEC
Kapeldreef 75
B-3001 Leuven
Belgium
Marc.Heyns@imec.be

Marc Meuris
IMEC
Kapeldreef 75
B-3001 Leuven
Belgium
Marc.Meuris@imec.be

Matty Caymax
IMEC
Kapeldreef 75
B-3001 Leuven
Belgium
Matty.caymax@imec.be

Michel Houssa
IMEC
Kapeldreef 75
B-3001 Leuven
Belgium
Michel.Houssa@imec.be

Paul Clauws
Ghent University
Krijgslaan 281-S1
B-9000 Gent
Belgium
paul.clauws@UGent.be

Pierpaolo Palestri
University of Udine
DEIGM, Via delle Scienze 208
I-33100 Udine
Italy
palestri@uniud.it

Robert Jones
University of Exeter
Stocker Road
Exeter
UK
jones@excc.ex.ac.uk

Takeshi Akatsu
SOITEC
Parc Technologique de Fontaines 38
F-190 Bernin
France
takeshi.akatsu@soitec.fr

Vladimir Markevich
University of Manchester
Sackville Street Building
Manchester M60 1QD
UK
V.Markevich@manchester.ac.uk

Walter De Baets
Umicore
Watertorenstraat 33
B-2250 Olen
Belgium
Walter.debaets@umicore.be

List of Acronyms

AFM	Atomic force microscopy
ALD	Atomic layer deposition
APD	Anti-phase domain
AR-XPS	Angle-resolved XPS
BOX	Buried oxide
BTE	Boltzmann transport equation
BZ	Brillouin zone
CCD	Charge-coupled device
CHF	Cyclic HF clean
CMOS	Complementary metal-oxide–semiconductor
CMP	Chemical–mechanical polishing
COP	Crystal originated particles
CRSS	Critical resolved shear stress
C–V	Capacitance–voltage
CVD	Chemical vapor deposition
Cz	Czochralski
DBR	Dielectric Bragg reflector
DCS	Device coordinate system
DD	Double donor
DEG	2D electron gas
DFT	Density functional theory
DG FET	Double gate field-effect transistor
DHG	2D hole gas
DI	Deionized water
DIBL	Drain-induced barrier lowering
DLTS	Deep level transient spectroscopy
DoS	Density-of-states
ECS	Ellipsoid coordinate system
EL	Electroluminescence
EMA	Effective mass approximation

EOR	End of range
EOT	Equivalent oxide thickness
EPR	Electronic paramagnetic resonance
EPD	Etch pit density
ESR	Electron spin resonance
FA	Furnace anneal
FET	Field-effect transistor
FTIR	Fourier transform infrared spectroscopy
GOI	Germanium-on-insulator
G–V	Conductance–voltage
HBT	Hetero junction bipolar transistor
HDD	Highly doped drain
HEMT	High electron mobility transistor
HF	Hartree Fock
HP-Ge	High-purity Ge
HVEM	High-voltage transmission electron microscopy
IC	Integrated circuit
IMFP	Inelastic mean free-path
IPA	Isopropyl alcohol
IR	Infrared
ITRS	International technology roadmap for semiconductors
LCBB	Linear combination of bulk bands
LED	Light emitting diode
LPCVD	Low-pressure chemical vapor deposition
LPDs	Light point defects
LSE	Latex sphere equivalent
LTO	Low-temperature oxide
LVM	Local vibrational mode
MBE	Molecular beam epitaxy
MC	Monte Carlo
MEIS	Medium energy ion scattering
MHAH	Multiple hydrogen annealing for hetero-epitaxy
MIS	Metal–insulator–semiconductor
MOCVD	Metal–organic-chemical vapor deposition
MODFET	Modulation-doped-field-effect transistor
MOIC	Micro-/optoelectronics integrated circuit
MOSFET	Metal-oxide–semiconductor field-effect transistor
MQW	Multiple quantum well
MSMC	Multi subband Monte Carlo
NDR	Negative differential resistance
NEGF	Non-equilibrium Green function
NIEL	Non-ionizing energy loss
NIR	Near infrared

OEIC	Optoelectronic integrated circuit
OTD	Oxygen thermal donor
PACS	Perturbed angular correlation spectroscopy
PDA	Post deposition anneal
PECVD	Plasma-enhanced chemical vapor deposition
PGA	Post gate anneal
PL	Photoluminescence
PSG	Phosphosilicate glass
PTIS	Photothermal ionisation spectroscopy
PVD	Physical vapor deposition
QCSE	Quantum confined Stark effect
QD	Quantum dot
QE	Quantum efficiency
RBS	Rutherford backscattering spectroscopy
RCE	Resonant cavity enhanced
RF	Radio frequency
RTA	Rapid thermal annealing
RTN	Rapid thermal nitridation
RTO	Rapid thermal oxidation
RTP	Rapid thermal processing
SAM	Scanning acoustic microscopy
SCE	Short channel effect
SE	Spectroscopic ellipsometry
SIMS	Secondary ion mass spectrometry
SL	Superlattice
SLS	Strained layer superlattice
SME	Surfactant mediated epi
SOI	Silicon-on-insulator
SP	Smooth pits
SPER	Solid phase epitaxial regrowth
SRB	Strain relaxed buffer
SRH	Shockley–Read–Hall
SRIM	Stopping and range of ions in matter
SRP	Spreading resistance probe
SR-PES	Synchrotron radiation photoemission spectroscopy
SSD	Solid-source diffusion
sSi	Strained silicon
SSL	Solid solubility limitation
STI	Shallow trench isolation
TB	Tight binding
TD	Thermal donor
TDD	Threading dislocation density

TED	Transient-enhanced diffusion
TEM	Transmission electron microscopy
TLM	Transmission line measurement
TXRF	Total X-ray fluorescence
UHV	Ultra high vacuum
UHV-CVD	Ultra high vacuum chemical vapor deposition
ULSI	Ultra large scale integration
UVO	Ultraviolet ozone oxidation
VGF	Vertical gradient freeze
VLSI	Very large scale integration
VPD-ICPMS	Vapor phase decomposition inductively coupled plasma mass spectrometry
XPS	X-ray photoemission spectra
XTEM	Cross-section transmission electron microscopy

List of Symbols

a	Lattice parameter (Ge = 0.565 nm)
A_j	Junction area (cm^2)
C_{HF}	High-frequency capacitance ($F\,cm^{-2}$)
C_L	Concentration of an impurity in the melt (liquid phase), relative to the atomic density
C_{LOAD}	Load capacitance at the output node
C_{ox}	Oxide capacitance density ($F\,cm^{-2}$)
C_{pn}	Junction capacitance
C_S	Concentration of an impurity in the solid, relative to the atomic density
C_V^{eq}	Equilibrium concentration of vacancies, relative to the atomic density
C_V^{eq0}	Equilibrium concentration of neutral vacancies, relative to the atomic density
$C_V^{eq0,+}$	Equilibrium concentration of positively charged vacancies, relative to the atomic density
$C_V^{eq0,-}$	Equilibrium concentration of negatively charged vacancies, relative to the atomic density
$C_V^{eq0,i}$	Equilibrium concentration of neutral vacancies in intrinsic Ge, relative to the atomic density
$C_V^{eq0+,i}$	Equilibrium concentration of positively charged vacancies in intrinsic Ge, relative to the atomic density
$C_V^{eq-,i}$	Equilibrium concentration of negatively charged vacancies in intrinsic Ge, relative to the atomic density
D	Diffusion coefficient ($cm^2 s^{-1}$)
D_{Ge}	Self-diffusion coefficient in Ge ($cm^2 s^{-1}$)
D_{Ge}^+	Self-diffusion coefficient for positively charged vacancies ($cm^2 s^{-1}$)
D_{Ge}^-	Self-diffusion coefficient for negatively charged vacancies ($cm^2 s^{-1}$)
D_I	Diffusion coefficient of interstitials ($cm^2 s^{-1}$)
D_{it}	Interface state density ($cm^{-2} eV^{-1}$)
D_n	Diffusion coefficient of electrons ($cm^2 s^{-1}$)
D_p	Diffusion coefficient of holes ($cm^2 s^{-1}$)
D_V	Diffusion coefficient of vacancies ($cm^2 s^{-1}$)
D_V^0	Self-diffusion coefficient for neutral vacancies ($cm^2 s^{-1}$)
D_0	Pre-exponential factor diffusion coefficient ($cm^2 s^{-1}$)
D_{0I}	Pre-exponential factor interstitial-mediated diffusion coefficient ($cm^2 s^{-1}$)

D_{0V}	Pre-exponential factor vacancy-mediated diffusion coefficient ($cm^2 s^{-1}$)
E_A	Ionization energy of a shallow acceptor (eV)
E_C	Conduction band minimum (eV)
E_D	Ionization energy of a shallow donor (eV)
E_{eff}	Effective electric field ($V cm^{-1}$)
E_F	Fermi energy (eV)
E_f	Formation energy of a native point defect (eV)
E_G	Band gap (eV)
E_i	Intrinsic Fermi level (eV)
E_T	Activation energy of a trap or deep level center (eV)
E_V	Maximum of the valence band (eV)
F	Electric field (V/m)
f_c	Calibration factor for the interstitial oxygen concentration in Ge
f_V	Correlation factor for vacancy-mediated diffusion
G	Temperature gradient at the melt/solid interface
G_M	MOSFET transconductance
H	Activation enthalpy (eV)
H_f	Enthalpy of formation of a native point defect (eV)
H_I	Activation enthalpy for the interstitial-mediated diffusion coefficient (eV)
H_m	Enthalpy of migration of a point defect (eV)
H_{fV}	Enthalpy of formation of a vacancy (eV)
H_{mV}	Enthalpy of migration of a vacancy (eV)
H_V	Activation enthalpy for the vacancy-mediated diffusion coefficient (eV)
I	Self-interstitial
I_{Dsat}	Saturation drive current
I_{gen}	Generation current
I_{gsP}	Peripheral surface generation current
I_{ph}	Photocurrent
I_{pho}	Saturated photocurrent
I_{pn}	Current through a p–n junction
I_{pn0}	Junction saturation current
I_{rsP}	Peripheral surface recombination current
J_A	Area current density ($A cm^{-2}$)
J_C	Corner current density ($A corner^{-1}$)
J_P	Perimeter current density ($A cm^{-1}$)
k	Boltzmann's constant
k_d	Distribution or segregation coefficient
k_{dm}	Distribution or segregation coefficient of an impurity at the melt temperature
L_{gate}	MOSFET channel length (cm)
L_n	diffusion length of electrons
L_p	Diffusion length of holes
m	Diode ideality factor
M	Mass of an atom
n	Free electron concentration (cm^{-3})
n_{disl}	Dislocation density (cm^{-2})
n_i	Intrinsic carrier concentration (cm^{-3})

N_a	Atomic density (4.413×10^{22} cm^{-3} for Ge)
N_A	Acceptor concentration (cm^{-3})
N_C	Density of states in the conduction band (cm^{-3})
N_D	Donor concentration (cm^{-3})
N_{it}	Interface state density (cm^{-2})
N_{st}	Surface density of interface traps (cm^{-2})
N_T	Bulk trap or deep level concentration (cm^{-3})
N_V	Density of states in the valence band (cm^{-3})
p	Free hole concentration in cm^{-3}
P	Pressure in Pa
P_j	Junction perimeter (cm)
q	Electron charge in absolute value (1.60218×10^{-19} C)
Q_{inv}	Inversion charge inside the MOSFET channel
R_s	Series resistance
s	Strain
S_{fV}	Entropy of vacancy formation (eV K^{-1})
S_g	Surface generation velocity (cm s^{-1})
S_{mV}	Entropy of vacancy migration (eV K^{-1})
S_r	Surface recombination velocity (cm s^{-1})
S_0	Equilibrium solid solubility of an impurity (cm^{-3})
S_{0max}	Maximum solid solubility of an impurity (cm^{-3})
t	Time (s)
t_A	Anneal time
$t_{high\text{-}\kappa}$	Thickness of the high-κ layer (cm)
t_{ox}	Gate oxide thickness (cm)
T	Temperature in K or °C
T_h	Hole temperature
T_m	Melt temperature of the host lattice
v	Pulling speed of the Czochralski crystal
V	Vacancy
V_{bi}	Built-in potential
V_{DD}	Supply voltage
V_{FB}	Flat band voltage (V)
v_{fV}	Vacancy formation contribution to the activation volume for self-diffusion (cm^3)
v_{Ge}	Activation volume of self-diffusion in germanium (cm^3)
v_{inj}	Thermal injection velocity at the source-channel barrier
V_{pn}	Bias over a p–n junction
v_{src}	Carrier injection velocity at the source-channel barrier
V_T	Threshold voltage (V)
v_{th}	Thermal velocity of carriers
W	Depletion region
x_d	Depth depletion region
x_j	Junction depth (cm)
ε_o	Permittivity of free space (8.85×10^{-12} F m^{-1})
ε_s	Permittivity of the semiconductor (F cm^{-1})
$\Phi_{CNL,x}$	Charge neutrality level of Ge

Φ_{n}	Schottky barrier height
χ_{Ge}	Electron affinity of Ge
χ_{MO_x}	Electron affinity of Mo_x
λ_{d}	Debye length
μ_{eff}	Effective carrier mobility ($\mathrm{cm}^2\mathrm{Vs}^{-1}$)
ν_{V}	Jump frequency into an adjacent vacancy (s^{-1})
ρ_{s}	Sheet resistivity (Ω/sq)
σ_{n}	Capture cross section for electrons (cm^2)
σ_{p}	Capture cross section for holes (cm^2)
σ_{s}	Capture cross section of surface states (cm^2)
τ	Time constant (s)
τ_{g}	Generation time constant (s)
τ_{r}	Recombination time constant (s)

Introduction

Cor Claeys and Eddy Simoen

1. Introduction

Although predicted by Mendeleev, the element Ge of the IVa column of the Periodic Table was only discovered in 1886 by Winkler [1], some 120 years ago and 66 years later than silicon. It is interesting to remark that Winklers PhD thesis was on Si and Si compounds. While Ge was the key substrate material used during the early days of the semiconductor transistor and integrated circuit developments, it was later completely overruled by silicon. One can clearly state that the road to silicon was paved with germanium [2]. The first section is giving a historical perspective on germanium and discusses some key milestones in the Ge technology. A second section deals with the opportunities and challenges for Ge to be used as a so-called advanced substrate for ultra-large scale integration (ULSI) applications beyond the 32 nm node, a topic which will extensively be addressed in different chapters of the book. It is, however, important to remark that this book is not only focusing on IC technology but will also discuss alternative types of devices such as e.g. detectors and solar cells. The last section outlines the content of the different chapters and gives some general aspects for making optimal use of this book.

2. Historical Perspective and Milestones

The invention of the semiconductor transistor happened accidentally by Shockley, Bardeen and Brattain [3], while on the search for the field-effect device. In the late forties, germanium was the material of choice, as it was available with the best crystalline quality achievable at that time [4]. The first point contact transistor fabricated in 1948 by researchers

Germanium-Based Technologies: From Materials to Devices. C. Claeys and E. Simoen (Editors)

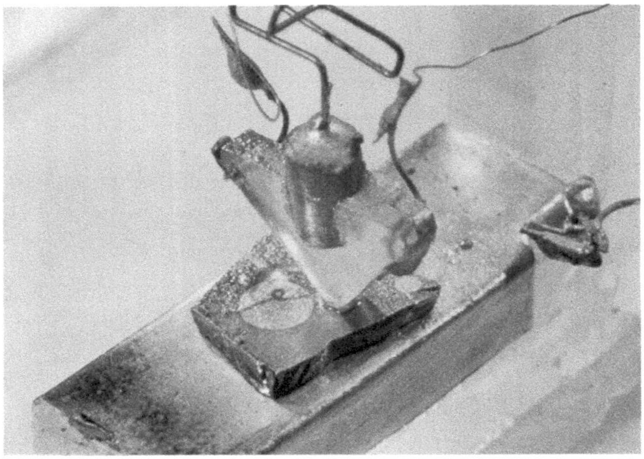

Figure 1. The first point contact transistor fabricated by scientists from Bell Labs in 1948.

from Bell Labs is shown in Figure 1. The Ge slab, into which a plastic wedge presses two strips of gold foil, is about half a centimeter long. Needless to say that scientifically and technologically this discovery is one of the most important ones of the 20th century. It paved the way to what is currently the largest industry worldwide, namely, the Microelectronics Industry, with for 2005 a semiconductors turn over of 228 billion US$, and expecting to reach 365 billion US$ by 2010. By that time the semiconductor content of electronic systems will be about 25%. Microelectronics products have enabled the Internet revolution, provide for global communication and are expected to play an important role in future healthcare, automotives and our daily life. A recent historical overview on germanium from its discovery to SiGe devices has been written by Haller [5].

For more than a decade, till the late fifties, early sixties, Ge has been the workhorse for the development of bipolar transistors, while the progress in field-effect devices was less prosperous, due to the unfavorable surface properties of Ge. This is mainly related to the low quality of the oxide (GeO_2) [6], providing insufficient passivation of the surface. Eventually, this, in combination with the far better behavior of SiO_2 and the introduction of the planar technology on silicon wafers in the early sixties [7] turned out to be a showstopper for the use of Ge as a microelectronics material. Silicon took over for a long time and has been the workhorse of the microelectronics industry ever since. To put this in perspective, the annual consumption of silicon in 2004 is around 8.000 M US$, while for Ge, it is only a small fraction. Germanium is utilized nowadays in some niche markets, like epitaxial substrates for III–V solar cells for satellite applications or high-resistivity material for the fabrication of nuclear-radiation detectors.

During the pioneering years of semiconductor electronics, a vast amount of exploration of the basic device and material physics was carried out on germanium. This has led to the formulation of several important theories and physical principles which are still of prime importance today. One of the famous examples is the McWhorter theory for 1/f noise [8], which states that the fluctuations in the charge transport through a device are related to tunneling in so-called slow oxide traps. With some refinements, this theory is still used nowadays to explain the 1/f noise in state-of-the-art deep submicron silicon metal-oxide-semi conductor field effect transistors (MOSFETs) [9]. The state-of-the-art deep submicron Ge devices presently under investigation will allow rechecking some of the early assumptions.

The first evidence of plasticity at higher temperatures in a covalent cubic face cubic centered (fcc) diamond lattice semiconductor was reported on Ge by Gallagher in 1952 [10]. At room temperature, on the other hand, Group IV semiconductors are brittle materials, due to the high Peierls energy of the dislocations. Further studies on deformed Ge led to new insights in the mechanical properties of crystalline materials: it was for example observed that the stress–strain characteristic showed a yield behavior not found for fcc metals [11], which could only be explained by the dynamics of dislocations in the material. Germanium has shown to be also a thankful medium for dislocation studies by transmission electron microscopy (TEM), as there was no risk for plastic flow (i.e. dislocation glide or multiplication) at room temperature during sample preparation. In this way, useful fundamental insights were revealed which were also valuable for other materials. Germanium was one of the first materials were the Frank–Read dislocation sources [12] were actually revealed [13].

Most of the interesting properties of semiconductor materials stem from point defects and impurities which provide for the local resistivity, the doping type and density. Doping is normally achieved by Group III (p-type) or Group V (n-type) impurities, which are characterized by a high maximum solubility (S_{max}) up to a few times $10^{20}\,cm^{-3}$ (Figure 2) [14]. Other impurities, like the transition metals, have a much lower S_{max}, so that they are less useful as shallow-level dopants. They generally give rise to deep levels in the band gap, acting as recombination or generation centers, which can be successfully described by the Shockley–Read–Hall theory [15]. In other words, metal impurities in Group IV semiconductors are efficient lifetime killers and may be centers of leakage current generation. For these and other reasons, both shallow and deep level impurities have been heavily investigated in the fifties [16,17]. One of the main properties is the diffusivity (D), which gives a measure of the ease of mass transport under a concentration gradient in the host lattice. Some of the basic diffusion models in semiconductors were developed for Ge. The Frank–Turnbull model [18] originally proposed to explain the diffusion of copper in germanium provides still the best picture for metal-diffusion in Ge [19] and other semiconductors [20]. It is based on the exchange between a rather immobile substitutional fraction, determining the solubility and a mobile interstitial fraction, whereby exchange is through a germanium vacancy (V). In general, it has been established that dopant diffusion in Ge takes place through a vacancy mechanism [21].

Besides electrically active impurities, there exist also inactive or neutral ones, which do not give rise to a level in the band gap. Classical examples are Group IV elements in another Group IV semiconductor: a silicon atom can perfectly replace a germanium atom on a lattice site, except for the mechanical strain associated with it. Ge and Si are perfectly miscible over the complete concentration range and for a sufficiently large atomic fraction, the related strain will change the band gap of the material. In other words, by finetuning the concentration of Ge in Si (or vice versa) one can perform band-gap engineering, which has found many applications in recent years [22]. Besides the built-in band offsets, one can

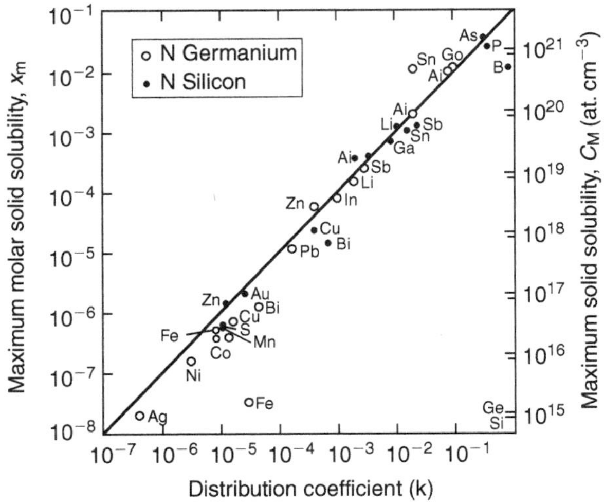

Figure 2. Maximum molar solid solubility (x_M) versus distribution coefficient at the melting point (K) for impurities in Ge and Si. The ordinate scales on the right-hand side of the plot gives maximum solubilities in at. cm^{-3} (after Fischler [14]).

also benefit from the increase in mobility when going to the Ge-rich end of the SiGe compounds.

Other neutral impurities that may be present in as-grown material are oxygen and hydrogen, since they can be introduced via the atmosphere or a quartz crucible. It was soon discovered that interstitial oxygen itself, sitting in a bond-centered position between two Ge atoms (an embedded GeO_2 molecule) is only infrared active (vibrational mode at 11.7 μm) [23]. However, upon thermal annealing in the range 300–500°C, it has been found that a donor activity is associated with oxygen clusters. Originally, this was assigned to a GeO_4 complex, as the oxygen thermal donor (OTD) concentration in the first instance scaled proportionally with $[O_i]^4$, with $[O_i]$ the initial interstitial oxygen concentration. It was later discovered that a whole family of related double donor centers exists, having the same core structure, but a different amount of oxygen atoms attached to it [24]. It is accepted by now that the ionization energy of the OTDs becomes less deep for a higher number of oxygen atoms in the cluster, although there is still no complete understanding of the defect structure that enables to explain all experimental features of the OTDs. It soon turned out that oxygen behaved similar in Czochralski silicon, where it is technologically far more relevant. This explains the huge research efforts devoted to the study of oxygen and its precipitation behavior in silicon, leading to concepts like intrinsic gettering (IG) [25], denuded-zone formation, new thermal donors [26] and so on.

Hydrogen is another neutral impurity, which has gained strong technological relevance in the past two decades [27]. Initial studies on hydrogen focused on its diffusivity in germanium and silicon, like in the classical work by van Wieringen and Warmoltz [28]. Much later, it was realized that besides a passivation action, hydrogen can also induce electrical activation of otherwise neutral impurities like C, O and Si. This was for the first time demonstrated in high-purity germanium (HP-Ge), which is grown under a hydrogen atmosphere [29] and was the start of the lively field of H-related defect research in semiconductors. It has among others led to some new applications of "hydrogen engineering" like the so-called Smart-Cut® technology for layer exfoliation [30].

As the original developments of semiconductor electronics was military inspired, the concern of radiation damage was there from the beginning. Initial radiation defect studies have mainly been performed on germanium, leading to the basic concepts of Frenkel pair (vacancy – V; interstitial – I) generation, followed by subsequent secondary reactions between "V's and I's" and other impurities in the material [31]. A bye-product of this research was the invention of neutron transmutation doping [32].

Besides more fundamental and materials oriented research also some device principles were demonstrated first (or in parallel) on germanium. An interesting example is the Esaki diode [33], consisting of a high–high p^{++}–n^{++} junction. In forward operation, a region of negative differential resistance (NDR) occurs, which can be used in principle as a resonant tunneling diode. NDR was also found in metal doped Ge resistors [34,35] and junctions [36]. Another field that was pioneered on germanium is cryogenic electronics. Already in 1959, McWhorter and Rediker proposed the Cryosar [37], a component for low-temperature computing and this, three decades before the Si-based ETA computer operating at 77 K [38]. In fact, the operation principle of this early cryogenic device is very similar to a recently proposed Impact-ionization MOS (I-MOS) which has been demonstrated on Ge at 77 K [39]. Finally, it should be remarked that the first characterization of a high-κ MOS on germanium was already performed in 1956 [40]: Montgomery and Brown used a 0.5 mm plate of strontium titanate with a κ of 300 to investigate the field-effect.

3. Ge as a Novel ULSI Substrate: Opportunities and Challenges

Until the 90 nm CMOS technology node, device scaling could be achieved by the dimensional scaling of the feature size (device length L_{gate}; gate oxide thickness t_{ox}, depletion width W) and to some extent, the supply voltage. However, the physical thickness of the gate oxide has become smaller than the limit for electron tunneling (~3 nm) so that the gate leakage current density in the channel off-state becomes unacceptably high. In order to allow for low-voltage, low-power operation of 65 nm circuits and below, the classical SiO_2 has to be replaced by a so-called high-κ dielectric. Candidates are HfO_2, ZrO_2, LaO_2, etc. [41]. The use of a material with a higher dielectric constant (·) relaxes the thickness constraints by a factor $\kappa/3.9$, so that a thicker gate oxide can be implemented, resulting in a lower gate leakage current density. Unfortunately, the use of high-κ gate oxides is confronted with some serious problems. One is that the quality with respect to oxide and interface charges for state-of-the-art deposition by metal–organic chemical vapor deposition (MOCVD) or atomic layer deposition (ALD) is significantly lower compared with SiO_2 [41]. This leads to threshold voltage instabilities, current hysteresis and a rather high and asymmetric flat-band voltage shift between n- and p-type devices. Perhaps the largest consequence is that scattering at these charges limits the high-field mobility of the transistor so that the advantage of having a high-κ gate is overcompensated by a lower drive current. A second problem is the hard to avoid growth of an interfacial layer between the silicon substrate and the high-κ material during high-κ deposition, which lowers the effective gate capacitance density and again destroys part of the benefits to be expected from it.

For these reasons, interest has risen for the so-called high-mobility substrates [22], like strained silicon, strained SiGe and last but not least germanium. It offers a 4 times higher hole and a 3 times higher electron mobility at low-fields. Moreover, the instability of GeO_2 – in the past an enormous handicap – now turns into a major advantage. Indeed, it has been demonstrated that for the same deposition conditions, the interfacial oxide thickness is significantly lower in the case of a Ge substrate compared with silicon [42]. As will be seen in detail in Chapters 8 and 11, submicron p-MOS transistors can be processed on a Ge substrate, yielding improved performance compared with silicon counterparts.

In spite of these encouraging results, there are still some major obstacles to be overcome in the development of Ge FETs. To name a few, there is the problem of the passivation of the Ge-oxide interface, which seems difficult to achieve by classical means like the hydrogen passivation of dangling bonds. A second problem is the high leakage current of Ge p–n junctions which may dominate the off-state leakage of a transistor. This is related to the relatively small band gap of Ge ($\sim$0.67 eV at room temperature), yielding an intrinsic carrier concentration $n_i \sim 10^{13}\,cm^{-3}$ at 300 K. Perhaps the largest drawback will be material availability. It is clear that if the microelectronics industry decides to switch over to Ge, there will be not sufficient 200 and 300 mm wafers and suppliers to cover the market needs. This can be resolved by employing only a thin layer of Ge, for example in a germanium-on-insulator (GOI) scheme. This could also be a template for the development of future hybrid solutions based on combining Ge and III–V compounds on a Si substrate, which will be touched upon in Chapter 13.

4. Outline of the Book

As the title of the book suggests, it is aimed to cover all issues related to crystal growth, grown-in and processing-induced defects, different material aspects, processing related issues up to state-of-the-art ULSI devices. While Chapters 1–7 are to be considered more introductory, covering more fundamental materials, defects and device aspects, Chapters 8 till 12 are more applications and processing oriented. Finally, Chapter 13 wraps up and gives an outlook on long-term prospects of Ge as a material for microelectronics. The content of the different chapters is briefly addressed.

Chapter 1: Germanium materials

This chapter focuses on the material aspects and discusses more in detail the Ge fabrication techniques. Besides the manufacturing of Czochralski bulk Ge substrates, attention is also given to the possible approaches for making GOI materials. The present status and future outlook for 200 and 300 mm wafers is reviewed.

Chapter 2: Grown-in defects in germanium

Similar as for silicon, a good knowledge and control of the grown-in defects are key for achieving optimal device performance. In contrast to Si, little is known about the possible gettering approaches that can be applied for Ge. The type of harmful defects is depending on the application that is envisaged. Different vacancy and interstitial related defects are studied in function of the crystal growth parameters. Concerning the extrinsic point defects a review is given about oxygen, hydrogen, nitrogen, carbon, silicon and neutral impurities. The dislocation formation is also addressed.

Chapter 3: Diffusion and solubility of dopants in germanium

A good insight in the diffusion and solubility behavior of intrinsic and extrinsic point defects is essential for a better understanding of their behavior and the way in which that these point defects can be controlled. Besides a review of the older data published in the literature, some recent insights in the material are critically discussed. Self-diffusion, Group IV diffusion and the diffusion of Groups III and V dopants are analyzed. The chapter discusses some basic concepts which are important for several of the other chapters.

Chapter 4: Oxygen in germanium

Not only in silicon but also in germanium extensive studies have been devoted to the different aspects of oxygen in the material: interstitial concentration, detection techniques, precipitation behavior and the formation of oxygen related thermal donors. In contrast to Si, the Czochralski growth technique only leads to moderate oxygen levels. A critical analysis is given of the experimental data reported in the literature.

Chapter 5: Metals in germanium

The metallic contaminants in Ge have a strong impact on the electrical properties of the devices such as leakage current and carrier lifetime. This chapter gives an in-depth review on the knowledge related to a large variety of metals in Ge such as Cu, Ag, Au, Pt, Ni, Co, Mn and Fe. The chemical trends in the properties of these metals and their impact on the optical properties and the carrier lifetime are discussed.

Chapter 6: Ab-initio modeling of defects in germanium

The experimental study of the impact of point defects on optical and electrical properties is very time consuming so that much attention has been given to the modeling of these defects. This chapter reviews the *ab-initio* modeling activities and outlines the differences between first-principles quantum mechanical methods and more approximate techniques. Several of the simulated results are compared with experimental data in order to validate the potential of the simulation techniques.

Chapter 7: Radiation performance of Ge technologies

The focus is on the identification of the radiation-induced defects, their annealing behavior and their impact on the electrical properties. The different defects and defect reactions are critically reviewed. From a device performance viewpoint distinction has to be made between transient and long-term effects resulting from both the ionization and displacement damage. A special form of radiation damage is caused by ion implantation. Wherever relevant a comparison is made with the well-known radiation damage in silicon.

Chapter 8: Electrical performance of Ge devices

The electrical performance of basic structures such as MOS capacitors and pn diodes are reviewed. This allows achieving an insight in the performance parameters and obtaining a better understanding of the basic mechanisms involved. A good control of the performance of these building blocks is essential for optimizing the transistor performance. Some basic measurement techniques are explained in detail.

Chapter 9: Device modeling

The present understanding of Ge device modeling is critically discussed and illustrated with different examples. The semi-classical modeling approaches are based on the Boltzmann transport equation. Also drift-diffusion approaches and Monte–Carlo techniques are applied.

Chapter 10: Nanoscale germanium MOS dielectrics and junctions

The fabrication of integrated circuits and state-of-the-art deep submicron transistors can only be achieved if high quality MOS capacitors and shallow junctions can be made in a reproducible manner. The chapter first reviews the extensive work on going in order to deposit high-κ dielectrics on Ge substrates. Different gate dielectrics fabricated with a variety of processing techniques are experimentally evaluated. The second part of the chapter deals with the formation of high quality shallow junctions using either solid state dopant diffusion or ion implantation. Essential are the activation of the dopants and the avoidance of process-induced defects. The chapter ends with a discussion of metal germanide contacts.

Chapter 11: Advanced Ge devices

The worldwide activities concerning high mobility Ge MOS devices, strained epitaxial Ge channels, Schottky Ge MOSFETs and Ge nanowires are briefly reviewed in order to obtain a clear view on what is present-day achievable and to give an insight in the remaining challenges.

Chapter 12: Alternative Ge devices

This chapter presents an overview of the large variety of potential Ge device applications outside the CMOS field. This includes detectors, solar cells, bolometers, etc.

Chapter 13: Trends and outlook

A brief summary is given of some material and device issues which have not been addressed in previous chapters. Based on the present state-of-the-art Ge device processing an outlook is given of the future type of Ge-based devices that have a potential and could lead to a breakthrough.

As such, the book will be useful both to materials scientists, device physicists and process engineers, who have a good background in silicon and want to broaden their scope to a related semiconductor material. The book also contains a large amount of basic materials information for scientists who have a general or more specific interest in germanium.

References

[1] H.C.A. Winkler, *J. Prak. Chemie* **34** (1886) 177.

[2] M. Riordan, Proc. Semiconductor Silicon, H.R. Huff, U. Gösele and H. Tsuya Editors, *Electrochem. Soc.* **98–01** (1998) 135.

[3] J. Bardeen and W.H. Brattain, *Phys. Rev.* **75** (1949) 1208.

[4] G.K. Teal, M. Sparks and E. Buehler, *Proc. I.R.E.* **40** (1952) 906.

[5] E. Haller, *Proc. E-MRS Symp.* "Ge-based devices from materials to devices", *Mat. Sci. Semicond. Proc.* **9** (2006) 408.

[6] K. Prabhakaran and T. Ogino, *Surf. Sci.* **325** (1995) 263.

[7] J.S. Kilby, *IEEE Trans. Electron Dev.* **23** (1976) 648.

[8] A.L. McWhorter, 1/f noise and germanium surface properties, In: *Semiconductor Surface Physics*, edited by R.H. Kingston, University of Pennsylvania Press, Philadelphia, PA (1957) 207.

[9] E. Simoen and C. Claeys, *Solid-State Electron.* **43** (1999) 865.

[10] C.J. Gallagher, *Phys. Rev.* **88** (1952) 721.

[11] H. Alexander and P. Haasen, *Solid State Phys.* **22** (1968) 27158.

[12] F.C. Frank and W.T. Read, *Phys. Rev.* **79** (1950) 722.

[13] W.W. Tyler and W.C. Dash, *J. Appl. Phys.* **28** (1957) 121.

[14] S. Fischler, *J. Appl. Phys.* **33** (1962) 1615.

[15] W. Shockley and W.T. Read Jr., *Phys. Rev.* **87** (1952) 835.

[16] W. Dunlap Jr., *Prog. Semicond.* **2** (1957) 167.

[17] E.M. Conwell, *Proc. I.R.E.* **46** (1958) 1281.

[18] F.C. Frank and D. Turnbull, *Phys. Rev.* **104** (1956) 617.

[19] H. Bracht, N.A. Stolwijk and H. Mehrer, *Phys. Rev.* **B43** (1991) 14465.

[20] R.N. Hall and J.H. Racette, *J. Appl. Phys.* **35** (1964) 369.

[21] A. Seeger and K.P. Chik, *Phys. Stat. Sol.* **19** (1968) 455.

[22] M. Lee, E.A. Fitzgerald, M.T. Bulsara, M.T. Currie and A. Lochtefeld, *J. Appl. Phys.* **97** (2005) 011101.

[23] C.S. Fuller, W. Kaiser and C.D. Thurmond, *J. Phys., Chem. Solids* **17** (1961) 301.

[24] P. Clauws, *Mat. Sci. Eng.* **B36** (1996) 213.

[25] T.Y. Tan, E.E. Gardner and W.K. Tice, *Appl. Phys. Lett.* **30** (1977) 175.

[26] G. Pensl, M. Schulz, K. Hölzlein, W. Bergholz and J.L. Hutchison, *Appl. Phys.* **A48** (1989) 49.

[27] S.J. Pearton, J.W. Corbett and T.S. Shi, *Appl. Phys.* **A43** (1987) 153.

[28] A. van Wieringen and N. Warmoltz, *Physica* **23** (1956) 849.

[29] E.E. Haller. W.L. Hansen and F.S. Goulding, *Adv. Phys.* **30** (1981) 93.

[30] M. Bruel, *IEE Electron. Lett.* **31** (2) (1995) 1201.

[31] H.M. James and K. Lark-Horovitz, *Z. physik. Chem.* **198** (1951) 107.

[32] C. Claeys and E. Simoen, *Radiation Effects in Advanced Semiconductor Materials and Devices*, Springer Verlag (2002).

[33] L. Esaki, *Phys. Rev.* **109** (1958) 603.

[34] R.G. Pratt and B.K. Ridley, *J. Phys. Chem. Solids* **26** (1965) 11 & 21.

[35] P.H. Wisbey and B.K. Ridley, *J. Phys. C: Solid State Phys.* **3** (1970) 211.

[36] W.W. Tyler, *Phys. Rev.* **96** (1954) 236.

[37] A.L. McWhorter and R.H. Rediker, *Proc. I.R.E.* **47** (1959) 1207.

[38] D.M. Carlson, D.C. Sullivan, R.E. Bach and D.R. Resnick, *IEEE Trans. Electron Dev.* **36** (1989) 1404.

[39] K. Gopalakrishnan, R. Woo, C. Jungemann, P.B. Griffin and J.D. Plummer, *IEEE Trans. Electron Dev.* **52** (2005) 77.

[40] H.C. Montgomery and W.L. Brown, *Phys. Rev.* **103** (1956) 865.

[41] G.D. Wilk, R.M. Wallace and J.M. Anthony, *J. Appl. Phys.* **89** (2001) 5243.

[42] K. Kita, K. Kyuno and A. Toriumi, *Appl. Phys. Lett.* **85** (2004) 52.

CHAPTER 1

Germanium Materials

Ben Depuydt, Marc De Jonghe, Walter De Baets, Igor Romandic, Antoon Theuwis, Carl Quaeyhaegens, Chrystel Deguet, Takeshi Akatsu and Fabrice Letertre

1.1. Introduction

Nowadays, germanium and some derived chemicals (GeO_2 and $GeCl_4$) are key materials for a wide variety of applications. The main applications of bulk single crystalline germanium are lenses and windows for infrared (IR) optics (thermal vision), detectors for gamma radiation and substrates for III–V based opto-electronic devices (see Sections 1.2 and 1.3).

Over the last two decades, germanium has regained a lot of interest as a semiconductor material for opto-electronic and electronic applications. Germanium and gallium arsenide show only a slight lattice mismatch, therefore germanium fulfils one of the main criteria to be considered as a substrate for epitaxial III–V growth.

Additionally, Ge substrates offer certain advantages over GaAs substrates: high crystallographic perfection, high mechanical strength and germanium is recyclable. These factors have led to the wide use of Ge wafers as a substrate for GaAs/Ge solar cells for telecommunication satellites [1], and they also make Ge a viable competitor for GaAs related devices different from solar cells. The feasibility has been shown of using Ge instead of GaAs for the fabrication of magnetoresistive sensors, high electron mobility transistors (HEMTs) [2,3], light-emitting diodes (LEDs) [3] and laser diodes [4].

Recently germanium regained a lot of attention in the semiconductor industry since it has attractive properties that can provide solutions for some of the major roadblocks that silicon

Germanium-Based Technologies: From Materials to Devices. C. Claeys and E. Simoen (Editors)

technology currently faces in the development of advanced nanoscale transistor structures. The continuation of scaling down the transistor dimensions, driven by the electronics industry's striving for higher performance at lower cost per function and reflected in Moore's law, is facing the physical limits of the conventional Si/SiO_2-based materials system. A viable option for achieving a higher transistor drive current at adequate off-current is the use of new channel materials.

The high mobility of charge carriers in germanium (two times higher than Si for electrons, four times for holes) makes this material ideally suited as channel material for the formation of high-speed circuits where it can potentially provide improved performance even compared to advanced strained silicon layers. A further discussion on this topic is given in Chapter 8.

The high absorption coefficient of germanium compared to silicon in the wavelength range 800–1550 nm, combined with its high carrier mobilities make the material a viable candidate for integration of optical detectors and modulators on complementary-metal-oxide-semiconductor (CMOS) circuits for optical interconnection [5,6]. Also for this application germanium can pave the way for integration of III–V optical devices with Si-based electronics, since high-quality epitaxial GaAs can be grown directly on Ge.

Today, single crystals grown by the Czochralski technique provide the closest to perfect germanium material, and are indispensable for a variety of applications (see Sections 1.2.2 and 1.2.3). Moreover, it is the material of choice for fundamental scientific research on its physical material parameters (see Chapters 2–6). The use of bulk germanium wafers for the above mentioned new applications in CMOS transistors and optical interconnects is restricted due to problems related to the high density of germanium and hence the high weight of bulk wafers, to their fragility compared to silicon and to the relatively high price of germanium. Therefore, alternative solutions are being developed: epitaxial growth of germanium on a silicon substrate, and the transfer of a germanium layer from a bulk substrate onto an oxidized silicon wafer (germanium-on-insulator or GOI, covered in Section 1.3).

1.2. Bulk Wafer Manufacturing

1.2.1. Germanium raw materials: supply and production flow sheet

1.2.1.1. Supply

Germanium is widely dispersed and has been estimated at 6.7 parts per million (ppm) in the earth's crust [7]. As a matter of fact there are no real germanium minerals, but only minerals who contain important amounts of germanium. These minerals, which are listed in Table 1.1, have been mined in the past in Africa (Namibia, Congo), Germany and Latin America (Bolivia). A photograph of the mineral renierite is given in Figure 1.1. Some other minerals exist, like stottite ($FeGe(OH)_6$), shauerteite ($Ca_3Ge(SO_4)_2(OH)_6$-$3H_2O$) and sriarite ($Cu_2(Fe,Zn)GeS_4$), but have no industrial exploitation to date.

Table 1.1. Industrially exploited germanium-containing minerals.

Mineral	Typical Ge content (%)
Argyrodite ($4Ag_2S$-GeS_2)	6–7
Germanite ($7CuS$-FeS-GeS_2)	8.7
Canfeldite (Ag_8SnS_6)	1.8
Renierite (complex of Cu,Fe,Ge and As)	7.8

Figure 1.1. The mineral renierite (picture from Johari, Umba Kyamitala).

The two main sources for the industrial production of germanium today are zinc ores and coal. The contribution of these sources varies over time and is largely dictated by the economics of coal usage in power stations. The main source of germanium in the former Soviet Union and still today in Russia and in the Ukraine, is coal from the Sakhalin Island in Eastern Russia, with smaller quantities from coking coal plants in Ukraine. Recently, other germanium bearing coal deposits in Primorski (Eastern Russia) and in the Krasnojarsk region (Central Russia) are under investigation. In China, coal from Mongolia and Lincang contributes substantially to the germanium output of China.

In the past germanium has been recovered from coal in the UK and Eastern Germany. Due to the availability of flue dust and ashes, many germanium producers have studied the possibility of recovering germanium from these stacks. However, many of these projects struggle with the economics as well as the presence of many other, heavy and undesirable metallic elements. At present the contribution of coal in the worldwide supply of germanium can be estimated in the range of 20–30%.

China (all producers together) can be considered as the main supplier of germanium, however, TeckCominco in Canada is the number one producer and contributes about 25% of the market needs of germanium from its zinc ores. Traditional zinc sources for germanium like the Gordonsville mine (US-Pasminco), Nanisivic (Canada), Asturiana de Zinc (Spain) and

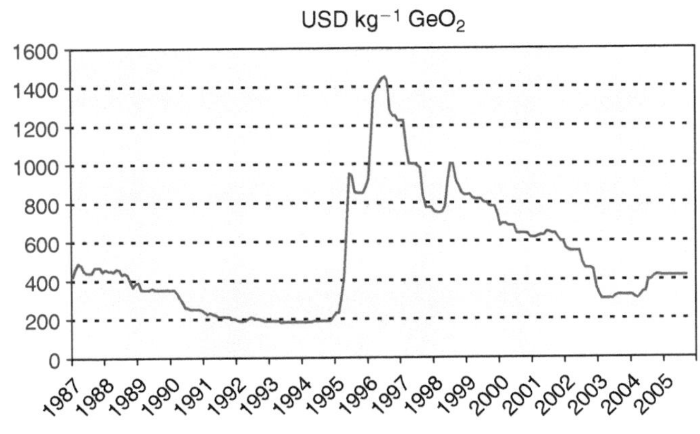

Figure 1.2. Evolution of the price of germanium (as contained in GeO_2).

others have ceased their operations in recent years. In the late nineties, OMG (Outokumpu Mooney Group) invested in a smelter in Lubumbashi (Congo-Zaire) to recover mainly Cu, Co and Zn from the "Big Hill". It is estimated that this slag dump contains about 3000 m of germanium. The annual output can be estimated at 5–10 m of Ge contained.

Today a number of smaller projects are under study. Just to name a few; War Eagle mining is studying the possibility to recover germanium from an old zinc mine in Mexico, Tres Marias, that ceased its operations in the early nineties. Another example is the project from Zincox to recover germanium from an old slag dump at Tsumeb (Namibia). However, the feasibility of these and similar projects suffer from the current "low" germanium prices (Figure 1.2).

Many producers have developed over the years recycling processes to recover germanium from their internal scrap generation, from end of life or broken germanium lenses and from wastes generated by the optical fiber industry ($GeCl_4$ is used as a dopant to the core of the optical fiber). Today the contribution of recycling of germanium bearing wastes (excluding internal recycling) can be estimated at 30%. This percentage is largely dependent on the price of germanium, as most of these recycling processes are costly.

Other substantial contributors of elemental Ge to the market are the governmental stockpiles, storing strategic metals and materials in their warehouses. Due to the end of the cold war the US and the former Soviet Union have decided to release these strategic metals over time. This year for instance the DNSC (Defense National Stockpile Center) in the US has a plan to release 8000 kg of germanium. Equally in Russia and Ukraine germanium units are sold on a regular basis from their strategic stocks.

1.2.1.2. Production flow sheet

Because there are no real germanium minerals the overall production process is always divided into two steps: production of germanium concentrate and the production of germanium. As described in the previous section the two main sources are zinc ores and coal. From the

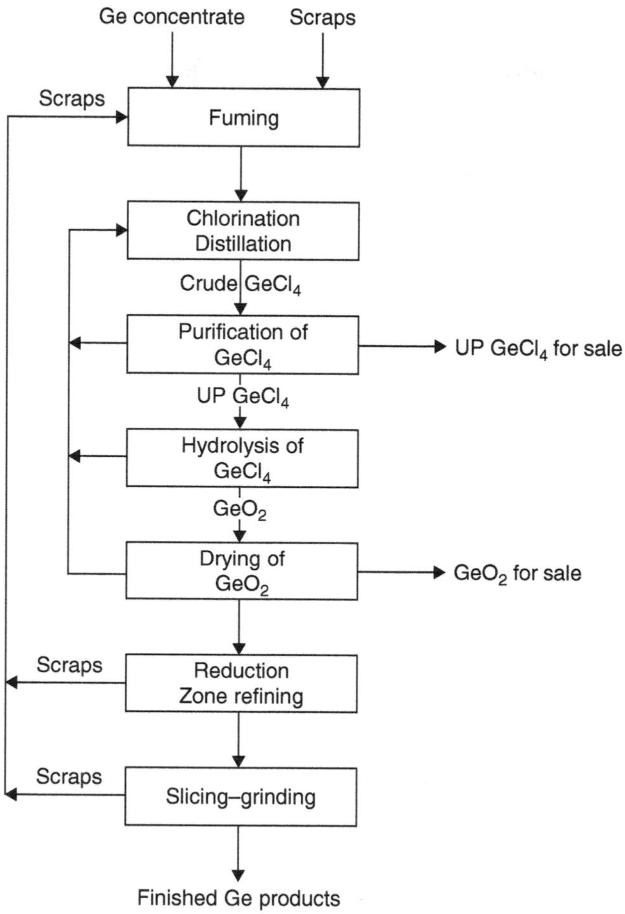

Figure 1.3. A possible production flow starting form a Ge concentrate.

first one the germanium is concentrated either through a pyrometallurgical or hydrometallurgical process. The most common process is the pyrometallurgical one. The germanium is volatized as GeO or GeS. From coal, germanium volatizes as GeO and is concentrated in the fly ashes. In both processes concentrations of 1 to 6% Ge are reached. Another important source today is the recycling of Ge coming from the optical fiber industry like scrubber solutions and glass scraps. Except for glass scrap, direct chlorination of these products is possible but with low yields. The obvious process is pyrometallurgical.

A possible production flow starting from a Ge concentrate is shown in Figure 1.3. Ge is fumed in a smelter at a temperature of about 1200°C.

The Ge concentrates in the fumes and concentrations of 5–30% Ge are obtained. The fumes are chlorinated in concentrated hydrochlorine acid (6 M) and the following reaction takes place:

$$GeO_2 + 4HCl \rightarrow GeCl_4 + 2H_2O \quad (1.1)$$

Because the reaction occurs at the boiling point of $GeCl_4$ (83°C) the $GeCl_4$ vaporizes. A crude $GeCl_4$ is obtained. A next distillation step is necessary to obtain ultra pure $GeCl_4$.

To obtain GeO_2 the ultra pure $GeCl_4$ is hydrolyzed with high-purity water:

$$GeCl_4 + 2H_2O \rightarrow GeO_2 + 4HCl \quad (1.2)$$

Subsequent filtration and appropriate drying yields various grades of GeO_2. Electronic grade GeO_2 is reduced in a hydrogen atmosphere. Care has to be taken to avoid generation of volatile GeO. Therefore, the temperature must be kept below 700°C during the reduction process.

1.2.2. Germanium crystal growth

1.2.2.1. Introduction and specific features of Czochralski Ge crystal growth

While germanium single crystals can be grown by various crystal growth techniques, the most widely used method is the Czochralski (Cz) crystal pulling technique. Germanium was the material on which the Czochralski pulling method was first pioneered by Teal and Little at Bell Labs in the 1950s [8]. Other Ge crystal growth techniques include horizontal or vertical Bridgman systems, and the related vertical gradient-freeze (VGF) method [9].

Polycrystalline material of usable IR optical-grade quality can also be obtained by other directional solidification systems, e.g. by the Stockbarger method or the heat exchange technique. Single crystals of germanium cannot commercially be grown by the floating zone technique which has been so successfully applied to silicon, because the ratio of the melt surface tension to the density is too small to support a floating zone of more than 1 cm diameter. Recent developments in Ge single crystal growth include the use of a relatively new crystal growth technique called axial heat processing (AHP), which requires further optimization before it could be used commercially [10], experiments with detached growth of Ge crystals [11–14], and gradient-freeze Ge crystal growth in micro-gravity conditions [15,16].

This review further concentrates on the Czochralski method only. The main features of Cz crystal growth are illustrated in Plate 1. For a detailed description of the Czochralski crystal pulling technique see Ref. [17]. Whereas the production scale of silicon crystals by the Czochralski method is nowadays vastly bigger than that of germanium, germanium has always been the favorite material in basic studies of semiconductor crystal pulling. The reason for this is its relatively low melting point (937°C), its negligibly low vapor pressure at the melting temperature, its low toxicity and only moderate reactivity of molten germanium with graphite and quartz crucible materials.

The general design features of a Czochralski puller for germanium growth are close to those for silicon. In fact Cz pullers developed and optimized for the growth of Si crystals can be readily used for the growth of Ge crystals, after some mechanical adjustments (more robust hardware design due to a higher density of Ge) and hot zone modifications. Automatic diameter control can be done by means of a weight sensing system, or by camera monitoring of the meniscus (even though the meniscus ring is less bright than in the Si case, due to the much lower melting temperature of germanium and its low liquid emissivity ((0.2) compared to the solid (0.55)). For the crystal pulling, either a shaft or a chain or a wire can be used. The disadvantages of a shaft system are the elevated height of the puller, and vacuum sealing issues. A chain or wire system is more prone to vibrations and orbiting of the crystal in the melt.

The pulling ambient can be any inert gas, or H_2, or vacuum. Pulling under vacuum is not suitable for large-diameter crystals, given the large amount of latent heat of fusion to be removed ($0.0025\,J\,kg^{-1}$ for Ge) and the relatively low thermal conduction in the solid germanium (only $25.14\,W\,mK^{-1}$, compared to $147\,W\,mK^{-1}$ for Si). Typically two pressure ranges can be applied, first slightly above atmospheric pressure, and second, 10–30 mbar. Crucibles can generally be made of graphite (isostatic pressed quality). For certain material quality grades, however, other crucible materials are required (see further). Heating is generally done with a graphite picket fence type resistor heater, using three phase AC or DC. For the growth of high-purity germanium (HP-Ge), which requires radio-frequency induction heating, a quartz tube can be used for thermal shielding, and a high-purity graphite element as RF susceptor.

Starting material for the Ge crystal growth has to be of high purity, particularly with respect to electrically active impurities. For the growth of IR-grade and electronic-grade germanium single crystals, the concentration of electrically active impurities in the starting materials must not exceed $10^{13}\,cm^{-3}$. Polycrystalline zone-refined Ge bars and scraps from single crystals can be used for this purpose. For the growth of HP-Ge crystals for γ-ray detector applications, the starting material has to be of impurity content less than $10^{11}\,cm^{-3}$, which is achieved by zone refining of Ge with single crystal seeding. Prior to insertion into the crucible, the material is etch-cleaned, rinsed and dried thoroughly in clean-room conditions. Doping elements can be added to the starting material in elemental form or in solid solution in Ge. The latter option is favorable in the case that low doping levels have to be achieved, e.g. for detector crystals. Typical charge size ranges from 10 kg for detector crystals, 30–200 kg for electronic-grade crystals, up to 300 kg for IR optical-grade crystals.

The seed is usually a small cylinder cut from a single crystal of Ge, the cylinder axis being along the crystallographic direction of the crystal to be pulled. The seed is etched before use, to remove any surface damage and contamination. The seed is lowered down close to the melt surface and allowed to get in thermal equilibrium with the melt, while it is already rotating. The seed is then dipped into the melt, and if necessary the temperature is lowered slightly until a meniscus can be supported by the seed crystal. The continuation of the growth process depends strongly on the specified quality (regarding both crystallographic structure and impurity content) of the germanium crystal to be grown, which in turn depends on the application for which the crystal is intended.

1.2.2.2. Ge single crystals for IR optics

The foremost requirements for IR optical-grade Ge crystals is high transparency for IR radiation, especially in the 3–5 and 8–12 μm wavelength bands (most intense parts of the

Figure 1.4. IR optical-grade germanium single crystal (top view). The {111} facets on the crown in a 6-fold symmetry can be easily seen.

spectrum emitted by blackbody radiation), optical uniformity and isotropy (i.e. the homogeneity of the refractive index must be very low (typically $\Delta n < 10^{-4}$)), and finally the inherent birefringence must be minimized ($<1\,\mu m\,cm^{-1}$). To be highly transparent, the IR Ge crystals must be n-type doped with a donor concentration between 4×10^{13} and $1 \times 10^{15}\,cm^{-3}$, corresponding to a resistivity range 5–40 Ω cm. Figure 1.4 shows an IR optical-grade Ge crystal, clearly illustrating the {111} facets on the crown in a sixfold symmetry.

Typical absorption coefficient for such IR Ge crystals will be less than $0.02\,cm^{-1}$ at room temperature, but will quickly rise when temperature increases due to typical extrinsic absorption (free carrier absorption). Both high transparency and optical uniformity and isotropy require the minimization of mosaicity and low-angle grain boundaries [12]. The optically detrimental effect of low-angle grain boundaries is the least for crystals grown in the <111> direction. Furthermore, to ensure low birefringence, the residual stresses in the crystal must be minimized [18]. Residual stresses are the result of the plastic strain (through the formation, movement and multiplication of dislocations) which relieves the thermal stresses exceeding the critical resolved shear stress (CRSS) during cooling down of the crystal above the crystal/melt interface [13]. To minimize these thermal stresses, this interface should be kept as flat as possible during the growth, corresponding with an almost vanishing radial temperature gradient. These conditions can be accomplished by the use of appropriate heat shields and/or after-heaters above the melt surface.

1.2.2.3. HP-Ge crystals for radiation detectors

The HP-Ge crystals have an electrically active impurity concentration of only 10^{9}–$10^{10}\,cm^{-3}$, depending on detector size. The larger the detector, the lower the impurity concentration must be in order to create a space-charge region throughout the whole volume of the crystal at a feasible reverse bias voltage. HP-Ge crystals are usually grown from fused silica

Table 1.2. Some mechanical and thermal properties of germanium and silicon at 300 K.

Property	Unit	Germanium	Silicon
[100] Young's modulus	$[10^{11}\,\text{dyne cm}^{-2}]$	10.3	13.0
[100] Poisson ratio		0.26	0.28
Bulk modulus	$[10^{11}\,\text{dyne cm}^{-2}]$	7.5	9.8
Shear modulus	$[10^{11}\,\text{dyne cm}^{-2}]$	4.1	5.2
Hardness	[Mohs]	6	7
Knoop hardness	$[\text{kg mm}^{-2}]$	780	1150
Density	$[\text{g cm}^{-3}]$	5.323	2.329
C_{11}	$[10^{11}\,\text{dyne cm}^{-2}]$	12.60	16.60
C_{12}	$[10^{11}\,\text{dyne cm}^{-2}]$	4.40	6.40
C_{44}	$[10^{11}\,\text{dyne cm}^{-2}]$	6.77	7.96
Thermal conductivity	$[\text{W cm}^{-1}\,{}^{\circ}\text{C}^{-1}]$	0.58	1.30
Thermal diffusivity	$[\text{cm}^{-2}\,\text{s}^{-1}]$	0.36	0.80
Thermal expansion, linear	$[10^{-6}\,{}^{\circ}\text{C}^{-1}]$	5.9	2.6

After Ioffe [22].

crucibles in hydrogen ambient, up to 100 mm diameter size. Controllable dislocation content of HP-Ge crystals with dislocation density between 10^2 and $10^4\,\text{cm}^{-2}$ must be achieved throughout the crystal length, and represents an additional growth challenge. Dislocation-free HP-Ge crystals contain too high concentrations of the divacancy-hydrogen complex (V_2H) deep band gap level, which acts as charge trapping center [19–21]. If the dislocation density exceeds $10^4\,\text{cm}^{-2}$, the dislocations themselves start acting as charge trapping centers [19]. HP-Ge crystals are therefore grown along the <100> crystal direction, making it possible to achieve the specified dislocation content.

1.2.2.4. Dislocation-free Ge crystals

The growth of dislocation-free Ge single crystals by the Czochralski method is physically possible because dislocation generation in the crystal is thermodynamically not favorable as it increases the Gibbs free energy in the crystal. Too high thermal stresses in the material are the main reason for dislocation generation, and should be minimized during the growth process. On the other hand, the mechanical and thermal properties of germanium are less favorable for (dislocation-free) Cz crystal pulling compared to those of silicon, as illustrated in Table 1.2.

The combination of larger weight of the Ge crystal, lower mechanical strength, lower thermal conductivity and diffusivity and higher thermal expansion makes larger seed and neck diameters necessary for similar crystal lengths and increases the generation of thermal stresses in the material during the crystal growth process, as compared to Si. CRSS, the threshold stress value for the generation of dislocation in the material, is also much lower in Ge than in Si, which in turn increases the probability of dislocation nucleation and multiplication. Furthermore, studies of mechanical properties of dislocations in germanium and their relation with material strength and plastic flow revealed much higher dislocation mobility in germanium than in silicon [23].

When the seed crystal is dipped into the (temperature stabilized) germanium melt, dislocations are generated in the seed crystal because of the high thermal stresses induced by the temperature shock. These dislocations can be out-grown by the so-called "Dash necking" technique, if the seed crystal is reduced in diameter and fast pulled for a few centimeters [24]. Dislocations in diamond structure crystals propagate preferably in {111} planes (main glide planes), which are oblique to the growth axis for crystals pulled along <100> or <111> direction, allowing the dislocations to glide out and terminate at the crystal surface. The remaining thermal stresses during the necking phase must be kept low enough not to allow the movement of the existing dislocations and generation of new ones. Furthermore, no new dislocations should be formed at the interface.

After the dislocation-free crystal structure is achieved, the crystal diameter can be carefully increased to the desired one, and the crystal body further grown. The necessary conditions for successful dislocation-free growth of a Ge crystal are very good thermal stability of the Cz puller, high purity of the starting charge material and optimized growth recipe and growth parameters. Once achieved, the dislocation-free growth is further relatively stable if the above-mentioned conditions are fulfilled. The reason for this is the high energy needed to generate a first dislocation in the crystal. If the shear stresses along the principal glide planes do not *at any point* exceed the value of CRSS σ_{cr} for the nucleation of dislocations or for the growth of very small dislocation loops, the crystal will remain macroscopically dislocation free. The possible causes for the generation of dislocations during the crystal growth are too high thermal stresses caused by large crystal/melt interface deflection (concave or convex), too high thermal gradients along peripheral {111} facets, thermal shocks, melt vibrations and mechanical shocks, impurity micro-segregation due to constitutional supercooling and inclusion of solid particles or gas bubbles at the crystal interface or surface. At the end of the pulling process, when the crystal has to be withdrawn from the residual melt, the thermal shock at this separation can lead to stress-generated dislocations running back up into the solid all over the still plastic temperature range (above 500°C), and consequently to yield losses. To avoid this, the end of the crystal is grown tapered in a cone to a very small or point contact diameter withdrawal.

The majority of dislocation-free Ge crystals produced today are highly doped 4 in. (100 mm) diameter crystals for the production of Ge substrates, used for the metal-organic chemical vapor deposition (MOCVD) growth of high-efficiency GaAs solar cells for space applications. Typically n-type crystals are doped with Sb to concentrations of $2 \times 10^{17}\,cm^{-3}$ (resistivity 20 mΩ cm), or with P to concentrations of up to $2 \times 10^{18}\,cm^{-3}$ (resistivity 4 mΩ cm), while p-type crystals are typically doped with Ga up to $2 \times 10^{19}\,cm^{-3}$ (resistivity 3 mΩ cm). The choice of doping elements depends on their segregation behavior, solid solubility in germanium and available purity.

Development and optimization of larger-diameter dislocation-free Ge crystals has been realized by Umicore EOM in recent years, triggered by the renewed interest for the use of germanium in advanced nano-electronic applications. The growth of 200 mm dislocation-free Ge crystals has already been optimized, with 80 cm useful crystal length for the production of 200 mm Ge substrates. Both moderately doped n-type and p-type 200 mm Ge crystals can be routinely pulled. Average resistivity dispersion measured across p-type

200 mm diameter Ge crystal is typically below 5%; 300 mm dislocation-free Ge crystals have also already been produced; however, the optimization of the quality and productivity of these crystals still needs to be realized.

1.2.2.5. Modeling of Ge crystal growth

Finite element computer modeling of germanium Cz growth processes for various material grades and crystal diameters provides continuous support for the development of new pulling processes and improvement and optimization of Ge crystal quality. The modeling results also provide valuable insights in different physical phenomena occurring during the Cz growth process, as well as drastically reduce the number of growth experiments for achieving the desired crystal properties.

The quality of the modeling results strongly depends on the correct values of material parameters for all materials used inside the Cz puller. The temperature dependence of many of these parameters is not very well known for the whole temperature range of the growth process (300–1210 K), therefore the modeling results have to be continuously validated by means of *in-situ* temperature measurements and/or comparison of calculated and measured crystal properties. The material parameters and different numerical models for heat and mass transport, gas and melt convection, etc. can then be fine-tuned based on the experimental results.

The temperature distribution in a Cz puller during the growth of a 200 mm Ge crystal has been simulated and is illustrated in Plate 2.

Three-dimensional unsteady simulations of the Cz growth of 4 in. (100 mm) dislocation-free Ge crystals revealed that the crystal pulling process is relatively unstable at the beginning of the body growth because of large-scale low-frequency periodic fluctuations of the melt temperature under the crystal/melt interface and variations of the crystallization rate across the interface. These findings were experimentally validated by observed fluctuations in the measured crystal/melt interface shape having the same time constant as the temperature fluctuations. These phenomena are caused by the strong turbulent mixing in the Ge melt under the grown crystal. The growth process reaches a much more steady state after about one-third of the body length with the virtual disappearance of the temperature oscillations and interface shape fluctuations.

Furthermore, the simulation results for the Cz growth of a 4 in. (100 mm) Ge crystal are essential for improvements in hot zone design (developments of heat shields, use of new materials in the hot zone), energy conservation through the use of new thermal insulation materials, and growth process optimization. The calculated axial temperature profiles and gradients inside the crystal are also used as input for the simulations of point defects distributions and clustering in the growing dislocation-free Ge crystals [26]. An illustration of the comparison between simulated and calculated temperature distributions is shown in Figure 1.5 for the case of the variation of the melt/crystal interface deflection during the growth of a 200 mm dislocation-free Ge crystal.

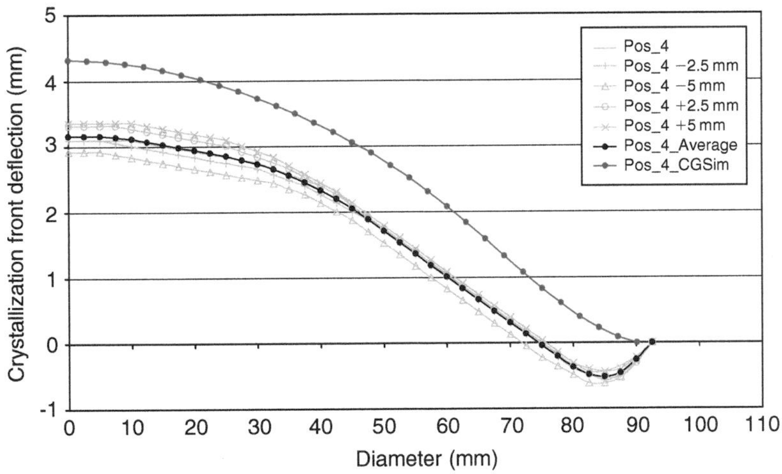

Figure 1.5. Comparison of calculated melt/crystal interface deflection (red line) with the experimental interface deflections (blue and black lines) as measured with Nomarski Optical Microscope, for an 200 mm dislocation-free Ge crystal. Even though the general shape of the actual interface is reproduced by the modelling, improvements in the melt convection model should ensure better quantitative agreement.

The main goal for thermal modeling of IR optical-grade Ge crystals is the minimization of thermal stresses during and immediately after the growth process, allowing for better optical homogeneity and lower birefringence of the IR Ge crystals. The calculated von Mises invariant stress distributions can be compared to the real distributions of residual stresses in the growth crystal via IR optical polarimetry and interferometry measurements on polished crystal slices [18].

1.2.3. Germanium wafer manufacturing

1.2.3.1. Introduction

Nowadays, germanium finds a use in a wide variety of applications as a direct result of both its unique material characteristics and the advanced state of Ge refining and Ge crystal growth. Here, the focus will be on Ge bulk wafer products. Germanium wafers are predominantly used as a substrate for III–V solar cells [27–31]. The energy supply of satellites is frequently and increasingly provided by high-efficiency III–V solar cells. In the past, these solar cells were produced by epitaxial growth of III–V layers on GaAs bulk substrates. However, the excellent match between the lattice characteristics of Ge and GaAs (crystal structure, lattice parameter and thermal expansion coefficient, see Table 1.3) has resulted in the substitution of GaAs substrates by Ge substrates without compromising the cell efficiency. From Table 1.3, it is clear that Ge wafers are mechanically stronger as compared to GaAs wafers. As such, thinner substrates can be used resulting in lower-weight solar

Table 1.3. Some relevant material properties of Si, Ge and GaAs.

Material property	Si	Ge	GaAs
Crystal structure[a]	Diamond	Diamond	Zincblende
Cleavage plane[a]	<111>	<111>	<100>
Density ($g\,cm^{-3}$)[a]	2.33	5.32	5.32
Band gap (eV)[a]	1.12	0.66	1.43
Melting point T_m (°C)[a]	1412	937	1240
Thermal conductivity ($W\,cm^{-1}C^{-1}$)[a]	1.30	0.58	0.55
Linear thermal expansion, linear ($°C^{-1}$)[a]	2.6×10^{-6}	5.9×10^{-6}	5.73×10^{-6}
CRSS at T_m (MPa)[b]	4–8	1	0.033
CRSS at $0.7 \times T_m$ (MPa)[b]	110	7	1.9 ($0.65 \times T_m$)
Tensile strength (MPa)	700–7000	40–95	–
Knoop surface hardness ($kg\,mm^{-2}$)[a]	1150	780	750
Mohs's hardness[a]	7	6	4–5
Young's modulus (GPa) [100][a]	130–190	103	86
Bulk modulus (GPa)[a]	98	75	75
Shear modulus (GPa)[a]	52	41	33

[a] Data taken from Ioffe [22].
[b] Chernov [33].
[c] Lane [32].
T_{mp}: melting point.

cells and thus reduced satellite launching cost. Furthermore, the low band gap of Ge allows manufacturing multi-junction solar cells with an extra p–n junction in the Ge substrate yielding an increase of the overall cell efficiency.

Other advantages are the defect properties of Ge, i.e. Ge wafers can be produced completely free from dislocations, and the available wafer diameter. Figure 1.6 illustrates that Ge wafers up to 300 mm in diameter can be produced.

Aside from the use of germanium for high-efficiency solar cells, Ge is also used as a substrate for LED production and as a high-mobility substrate for micro-electronics applications. The reader is referred to the following chapters for more information on the latter applications.

1.2.3.2. Wafer preparation: general remarks

The process of transforming an as-grown crystal into thin wafers with damage free, mirror-like, clean surfaces typically requires a combination of purely mechanical machining steps such as cutting and grinding, chemical steps (wet etching, cleaning) and mixed process steps such as chemical–mechanical polishing (CMP) where an interplay between chemical and mechanical removal mechanisms exists [32,34]. In the following sections some differences between the mechanical properties of Si, Ge and GaAs are discussed.

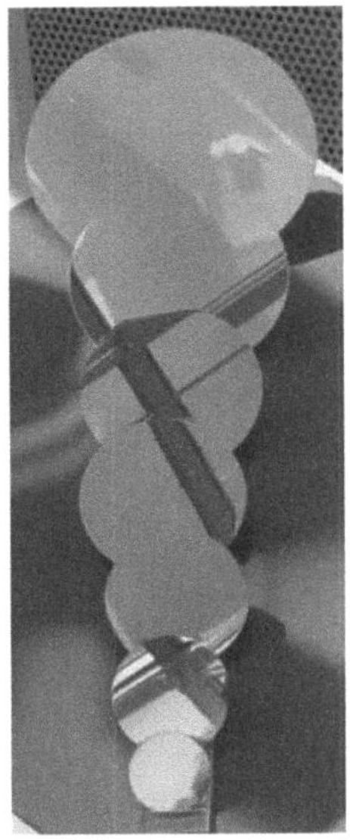

Figure 1.6. Photograph of dislocation-free Ge wafers with diameters ranging from 2 in. (50 mm) to 300 mm.

Mechanical properties

When comparing the mechanical properties of Ge to those of Si or GaAs, it is clear that the mechanical properties of Ge are intermediate between those of Si and GaAs (see Table 1.3).

Tensile strength
The tensile strength of Ge is substantially lower than the one of Si. Given the fact that the density and thus the wafer weight (for a fixed wafer thickness) of Ge is approximately twice that of Si, the sensitivity for wafer breakage increases accordingly to the sequence Si < Ge.

Material deformation
Another material parameter listed in Table 1.3 is Young's modulus. This property is a measure of the resistance of a solid to deformation when mechanical forces are applied to it. Given the large density of Ge as compared to Si, Ge wafers will be more prone to gravitational forces as compared to Si, resulting into higher bow/warp values for Ge.

Hardness
Given the lower hardness of germanium, Ge wafers will more easily be scratched by particles, etc. during processing as compared to Si wafers.

Critical resolved shear stress
The shear stress, i.e. stress resulting from thermal and/or mechanical forces applied, is referred to as the CRSS when it is higher than the threshold for dislocation movement. Ge is more sensitive to the presence of shear stress due to its lower thermal conductivity/higher thermal expansion coefficient and its increased density as compared to Si. If the dislocation threshold

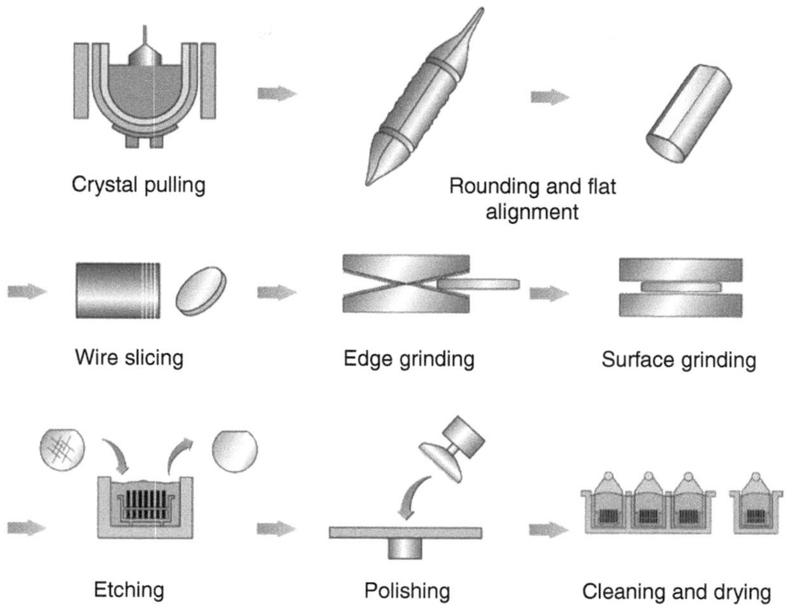

Figure 1.7. Schematic flow of the Ge wafer production process.

is exceeded, the stress will result into the occurrence of slip in the solid. This is to be avoided as the presence of a vast amount of dislocations/slip will substantially complicate further wafer processing and compromises device performance and reliability. Table 1.3 clearly shows that Ge has a CRSS value intermediate between the CRSS values of Si and GaAs. Therefore, the sensitivity of Ge to slip formation is also intermediate between Si and GaAs.

1.2.3.3. Wafer preparation: process steps

Figure 1.7 shows the typical sequence of processing steps used to produce Ge wafers. A detailed discussion of wafer preparation process steps is presented in Refs [32,34] for Si and InP, respectively.

Crystal pulling

The germanium crystals are grown either by Czochralski crystal pulling or by the VGF method. Reference is made to Section 1.2.2 for more information. In this section only the Cz-grown crystals are considered.

During the crystal pulling, wafer properties such as the conductivity type, the wafer resistivity and the main defect properties are fixed. The crystals are grown according to a specific crystallographic orientation which mostly but not always corresponds to the final wafer orientation.

Preferably, the parameters of the crystal growth process are tuned to produce cylindrically round crystal as this shape minimizes further material loss. The diameter of the crystals typically is chosen to be somewhat larger than the final wafer diameter. The diameter control during crystal growth is insufficiently accurate to meet wafer diameter specifications. Wafer processes such as crystal rounding followed by edge grinding are therefore used to determine the final wafer diameter. The crystal length is chosen to be a multiple of the typical maximum length of the work piece that can be processed by a multi-wire saw (see below).

Cropping

After crystal growth the conical ends of the germanium crystal with a smaller than required diameter (crown and taper), are removed using OD (outer diameter) sawing or band sawing. The removed end pieces are subsequently used for assessing crystal properties such as the conductivity type, the crystal conductivity, the structural integrity (dislocations and other defects) or the chemical purity. As will be discussed further, Ge ingots are mostly cut using the multi-wire slicing method. State-of-the-art multi-wire saws can handle work pieces that have a maximum length of 40–50 cm, depending on the machine manufacturer. Therefore, the crystals have to be cut into multiples of 40–50 cm. Given the specific mechanic properties of Ge, care has to be taken when selecting the machine parameters to avoid excessive damage.

Crystal rounding

Crystal rounding is used to produce an ingot with a fixed diameter. The diameter after the crystal rounding has to be large enough to allow for some material removal during the edge grinding. Ge crystal rounding is mostly realized using center grinding. Center pieces are glued to the top and bottom section of the cropped crystal. The ensemble is mounted in a grinding machine, similar to a lathe. A grinding wheel is slowly translated along the crystal length while quickly rotating (see Figure 1.8). The Ge crystal rotates slowly around its axis and Ge is removed in several passes until the desired ingot diameter is attained. The single-pass removal rate, the number of machining passes, the type of grinding wheel, the rotation rates of the wheel and ingot and the coolant flow have to be optimized to achieve a good throughput without compromising the surface quality of the ingot.

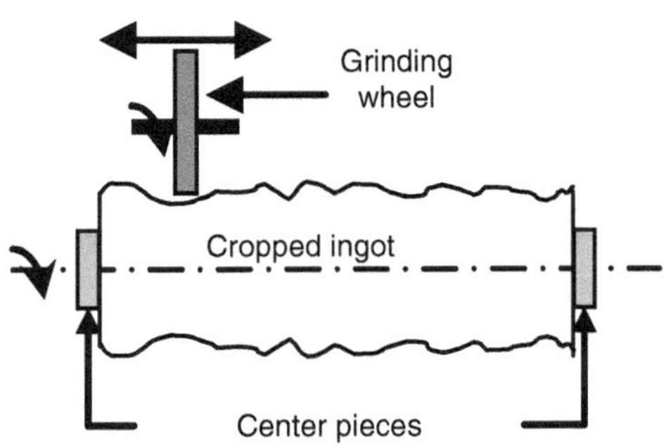

Figure 1.8. Schematic illustration of the crystal rounding process.

Flat/notch grinding

Similar to the case of Si, Ge wafers usually carry some kind of physical references to the crystallographic orientation of the crystal and the dopant type. The reference is realized by accurately grinding one or two flat sections (wafer diameter up to 150 mm) or a notched section (200 mm and 300 mm Ge wafers) on the ingot along a crystallographic direction. The flat section that identifies the crystallographic orientation is referred to as the primary flat, while the other one is called the secondary flat. In the case of 100 mm <100> oriented wafers, used as substrates for III–V solar cells, the primary flat is located on a <100> direction. Large-diameter wafers (200 mm and 300 mm) with an <100> orientation usually have a single notch that refers to the <110> direction. The crystallographic orientation of the ingot before the flat/notch grinding is done by X-ray goniometry. After the orientation, the ingot is transferred to the grinder, immobilized and the flat/notch is ground to the desired dimension. Here also, care has to be taken to minimize damage due to the fixed abrasive machining.

Wire slicing

Historically, OD and later inner diameter (ID) slicing were used to cut Si and Ge ingots into thin wafers. An OD saw is a cut-off machine that is fitted with a metal blade that holds a fixed abrasive. The blade rotates at high speed and cuts the ingot in one pass while the ingot is clamped. Due to the large kerf loss resulting from the thick OD slicing blade, ID slicing rapidly took over in the sixties. In this method, a large stainless steel blade with a central hole of which the inner edge is coated with a fixed abrasive, is tensioned in the machine and rotates at high speed. The ingot is firmly mounted on a graphite or epoxy holder, a so-called beam, using glue and is positioned on an adjustable holder on the machine. The latter allows fine-tuning of the cutting plane and thus the resulting wafer orientation in two dimensions. Both horizontal and vertical setups are commercially available. In the horizontal setup, the ingot is positioned horizontally inside the hole in the ID blade and either the ingot or the ID blade is moved vertically to produce the cut. Alternatively, the mounted ingot can be positioned vertically in the hole of the horizontally suspended ID blade. This cutting method consumes less crystal material as compared to the OD slicing method as the ID blade is much thinner. Reference [32] gives an excellent discussion of the OD and ID slicing methods.

Nowadays, multi-wire slicing is used to cut Ge ingots into wafers. The ingot, glued to a beam and previously accurately oriented using X-ray goniometry, is pushed very slowly through a web of moving thin steel wires on which a slurry of small abrasive particles (SiC) and a liquid such as mineral oil or ethylene glycol is being dispersed. Cutting proceeds by the same physical mechanism as lapping: the abrasive particles are transported by the liquid that sticks to the wires, pushed against the Ge work piece by the wires and abrade it.

State-of-the-art multi-wire slicing machines can cut ingots up to 300 mm diameter and a length of 40–50 cm in one pass and produce hundreds of wafers in one cut. The thickness of the as cut wafers is determined by the distance between the wires, the thickness of the wires and the abrasive size. The kerf loss resulting from multi-wire sawing is substantially lower as compared to ID sawing due to the very thin wire used and the small size of the abrasive particles. Also, the force exercised by the metal wire web on the work piece is lower as compared to the force applied on the ingot during OD slicing, resulting into a smaller amount of slicing damage. Implementing multi-wire slicing requires a lot of fine-tuning and optimization. The main parameters are slurry composition, slurry temperature, slurry flow rate, the tension, the velocity of the wire web, the feed rate of the work piece and the alignment of the pulleys that guide the wire web.

Laser marking

Laser marking can be used to identify Ge wafers with a code. This code typically contains a first series of alphanumeric characters that refers to the crystal growth run and a second series that identifies the specific wafer. Marked wafers allow for (improved) traceability both at the wafer and cell manufacturer site.

Edge grinding

After wire slicing, the edge of the wafers is damaged due to crystal grinding and slicing processes. Further processing such edge-damaged wafers would result in yield losses due to increased chipping and breakage during further wafer preparation processes (contact with cassettes, etc.) and customer processes. Therefore, the edge of the wafers is profiled using fully automated cassette-to-cassette edge grinding equipment. The wafer is held on a slowly rotating vacuum chuck while a rotating grooved wheel with fixed abrasives inside the groove moves along the wafer edge and the flat and grinds the desired profile (see Figure 1.7).

Surface grinding

Surface grinding is used to flatten wafers after the slicing and to remove the thickness variations due to the slicing. Historically, single-side or double-side lapping was used for this process. Nowadays, surface grinding is more commonly used. A rotating vacuum chuck is used to hold a wafer and germanium is removed by a rotating grinding wheel with fixed abrasives. After one side is grinded, the wafer is flipped over and the other side is grinded. By carefully adjusting the position of the grinding wheel versus the chuck, both sides of the wafer can be surface ground achieving a wafer total thickness variation of less than 1 μm. After the surface grinding, the surface of the wafer shows a spiral pattern of grinding marks. Fully automated cassette-to-cassette wafer single-side and double-side surface grinders are commercially available today.

Etching

The mechanical processing steps that were used to transform the as-grown ingot into ground wafers inflict damage to both the surface and the edge of the germanium wafers. This frequently results in wafer warpage. Wet etching is used to relieve the Ge wafers from their strained state.

In the fifties and sixties, a lot of research work on the electrochemistry and wet etching of Ge was done and published. Reference is made to [35–38] and references therein for details. Here only two types of etching mixtures that are frequently used for the wet etching of Ge will be mentioned.

HNO_3–HF mixtures

A wide variety of etching compositions exist that contain HNO_3 and HF as basic ingredients, HNO_3 being a strong oxidizing agent and HF being a strong complexing chemical. Aside from these two components, the etching mixtures may contain a catalyst (Br_2, I_2, KI, dissolved Ge), a weak acid (oxalic acid, acetic acid, tartaric acid) or metal salts ($PdCl_2$, $AgNO_3$, $Hg(NO_3)_2$, $Cu(NO_3)_2$, $Pb(NO_3)_2$). The behavior of the etching mixtures depends strongly upon the exact composition and both polishing etching and defect-selective etching can be realized by choosing the right composition and temperature. It should be noted that all mixtures that contain HNO_3 are not straightforward to control. The etch rate shows an induction period due to the fact that not HNO_3 but a decomposition product such as NO is the actual etching agent.

H_2O_2–HF mixtures

Here H_2O_2 replaces HNO_3 as the oxidizer and these mixtures are much better controllable as an induction period is absent. The etch rate is sensitive to the crystallographic orientation in all cases.

Polishing

After the etching process, strain is relieved from the wafers but the surface of the wafers is not yet damage free. A final processing step is needed to render the surface defect free and mirror-like without compromising the wafer flatness and thickness variation too much. This is realized by CMP of the Ge wafers. Somewhat simplified, CMP is a process step in which a work piece such as a semiconductor substrate, mounted in/on a carrier, is pressed against a rotating platen or belt, covered by a polishing pad. A polishing fluid is dispensed on the platen while down pressure is exercised on the carrier. The polishing fluid typically contains dispensed abrasive particles (SiO_2, Al_2O_3, . . .), some stabilizing/dispersing chemicals and sometimes an etching agent. Due to the combined effect of the chemical corrosion of the work piece, chemical dissolution and the mechanical action of the abrasive particles on the work piece, material is removed from the work piece in a controlled way. Consequently, this type of process step is generally referred to as CMP. By carefully selecting the parameter values such as down pressure, spindle speed, the type of polishing pad and carrier, the flow rate and composition of the slurry, the surface of the work piece is rendered flat and smooth. State-of-the-art polishing equipment allows one to manufacture substrates that are very smooth and flat. Both fully automatically single-side and double-side polishing machines equipped with *in-situ* metrology and cleaning tools are commercially available today. Due to the complex interactions between the work piece, the chemicals, the pad, the carrier and the abrasive particles, many different parameters exercise a profound influence on the removal rate and the final roughness and flatness of the work piece. Consequently, when one has to implement or optimize a CMP process step in practice, one generally encounters difficulties in selecting the appropriate consumables from the wide product range commercially available and establishing the "winning" combination of parameter values. Using an optimized CMP process, the surface roughness (as measured by atomic force microscopy (AFM)) of the Ge wafers can easily be brought to the sub-nanometer level.

Cleaning, drying and inspection

The cleaning processes are key process steps in the fabrication of the Ge substrates (or any substrates). Basically, two purposes are served. Firstly, traces of polishing fluid from the preceding process step have to be removed from the Ge surface. Secondly, the metal surface concentration, particulate surface concentration and the concentration of other relevant impurities have to be brought to an acceptable level. When this has been achieved, the substrates are ready to be dried, inspected and packed. Two main complications arise when cleaning Ge with wet chemical baths.

Firstly, when comparing typical etch rates of Ge and Si in, for example a typical Si cleaning chemistry, it becomes clear that Ge has a much higher reactivity toward etching agents when compared to Si. Ge even dissolves in pure oxygen-containing water although very slow! This fact complicates the use of wet chemical solutions for cleaning purposes. The solutions have to be carefully chosen and their performance has to be optimized to minimize surface etching.

Secondly, when Ge is exposed to an ambient environment, a native oxide layer will be present on the surface. The latter is apparently composed of both tetravalent germanium oxide (GeO_2) and some sub oxides GeO_x. The tetravalent oxide is soluble in water whereas the sub-oxide is not. From an industrial perspective, a wet chemical treatment that would leave a stable, oxide-free and passivated surface would be highly desirable. Such a wet treatment is, however, not available at present (see [39–41]).

Drying of Ge wafers after the cleaning step can be done using spin rinse drying or IPA vapor drying.

1.2.3.4. Germanium recycling

As already mentioned in one of the previous sections, economically exploitable germanium sources are rather scarce [42]. As such, germanium is an expensive material as compared to silicon, for example. In order to produce germanium wafers in a cost-efficient way, germanium has to be recycled from all relevant waste streams. The nature of these streams depends strongly on the wafering operation considered.

Aside from the cost benefits already mentioned, the ever increasing EH&S (Environment, Safety and Health) awareness of the materials processing industry (and other industries as well) further consists a strong driver for recycling waste streams and minimizing emissions to the environment.

1.3. GOI Substrates

GOI wafers constitute a materials platform which combines the advantages of SOI (silicon-on-insulator) with the advantages of Ge over Si with respect to charge carrier mobilities (relevant for application as CMOS channel material), optical band gap and absorption coefficient (relevant for high-bandwidth optical interconnects), and lattice match with GaAs (relevant for integration of III–V-based opto-electronics on a Si-based electronics).

Several techniques have been or are being developed for the realization of GOI structures, such as layer transfer techniques (Section 1.3.2), the Ge condensation method [43], Ge epitaxial growth on SOI, liquid phase epitaxy [44] and bond and grind/etch back (Section 1.3.1).

1.3.1. Back-grind SOI

Si CMOS technology can be expected to remain, at least for one more decade, the workhorse of the micro-electronics industry. The need to further increase the relative information capacity (bits s^{-1}) in our society will act as pull for the convergence of electronics and photonics and will result in the need for integration of electronic and optical functionalities in single devices to allow for low cost and high performance. However, the band gap of silicon does not allow for adequate light detection at the optimized and standardized frequency bands of the telecommunication industry.

Devices which integrate electronic and optical functionalities can be developed on a new material platform: GOI which recently became available in diameters up to 200 mm. The band gap of Ge is ideal for optical communications and near-IR imaging, the high carrier mobility in Ge allows for short detector transit times. Additionally Ge is almost lattice matched to GaAs, which allows for the heteroepitaxial growth and integration of solid light sources.

Some simple inter-chip, intra-chip and near-IR integration schemes are shown in Figure 1.9. All integration schemes of Figure 1.9 show the presence of a dielectric layer between

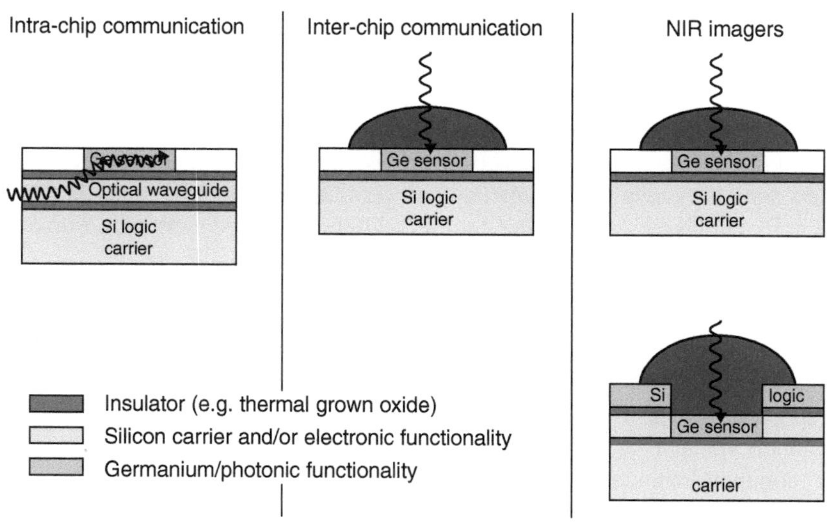

Figure 1.9. Simple integration schemes using the GOI engineered substrate platform.

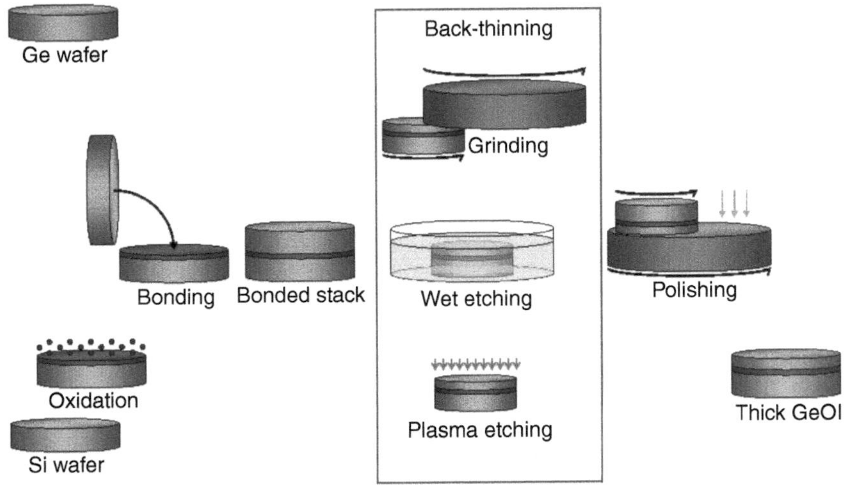

Figure 1.10. Thick GOI manufacturing techniques.

Table 1.4. Some thick GOI engineered wafer specifications, best in class.

Wafer diameter	Up to 200 mm
Device/handle layer	Si and Ge
Device layer thickness	>1.5 μm for 50–100 mm, ⩾2.0 μm for +150 mm
Device layer TTV	<1.0 μm for 50–150 mm
Buried oxide (BOX) thickness	0.1 to >2 μm (1.0 μm ± 5% standard)
Handle wafer thickness	>350 μm for 50–150 mm
Handle wafer TTV	<1.0 μm for 50–150 mm
Handle wafer back surface	Polished or lapped and etched
Voids	C-SAM inspection

germanium and silicon. The dielectric layer can be thermally grown oxide on silicon, its presence is however not always a prerequisite. The targeted applications require a functional Ge layer thicknesses in excess of 1 μm which can more easily be achieved using a bond and grind back technique compared to layer transfer techniques such as the patented Smart Cut™ and NanoCleave™. Per definition GOI engineered substrates with a Ge layer thickness in excess of 1 μm are grouped in the category thick GOI.

Recent developments in the field of wafer fabrication and the progress made with wafer bonding equipment created the foundations to allow thick GOI engineered substrates to become available. Figure 1.10 gives the material flow for the manufacturing of thick GOI. After the fabrication of the bonded wafer pair, several Ge wafer thinning techniques can be used to reduce the thickness of the Ge wafer hundreds of micron thick into the micron range, besides the most common techniques of back grinding, wet or plasma etching techniques can be used.

Table 1.4 gives an overview of the best in class specifications for thick GOI wafers, scanning acoustic microscopy (SAM) is the technique of choice to determine bond integrity.

1.3.2. GOI substrates by layer transfer

1.3.2.1. Donor wafers

The most commonly used technique for the realization of a thin GOI structure is the Smart Cut™ technology which was first developed on Si [45]. The method is versatile, has been adapted and applied to different germanium donor wafers (100 mm and 200 mm Ge bulk wafers called "bulk Ge" [46,47], and 200 mm Ge epitaxially grown on silicon wafers called "epitaxy Ge" [48]).

For the epitaxy Ge approach, the lattice mismatch between Ge and Si has to be absorbed in the epitaxy layer, in order to minimize the defectivity in the Ge layer. By using low temperature seed layer growth and high temperature thick layer growth followed by a thermal cycling, pure Ge epiwafers on Si(001) have been obtained with a threading dislocation density of $6 \times 10^6 \, cm^{-2}$ for a 2.5 μm thick Ge epilayer [49].

On the other hand, for needs of higher crystal quality, Ge bulk wafers should be used as donor. Czochralski-grown Ge dislocations-free wafers are commercially available (Sections 1.2.2 and 1.2.3). The authors used bulk germanium wafers with different doping levels (n-type or p-type), orientations ((100) on axis, 6° off axis toward [111]) and different diameters (from 100 up to 200 mm).

According to the final applications, there are advantages and drawbacks for each donor wafer. For example for high-performance MOS field effect transistor (MOSFET) applications, Ge would be potentially integrated only for the 32 nm technology node or beyond, thus 300 mm or larger diameters will be necessary. The "epitaxy Ge" is not limited by such future diameters as it is grown on the mature Si substrates, so could be chosen. If the quality in terms of dislocation density is sensitive to the application, for instance for photonic applications, then a bulk approach would be preferred. For these reasons the two donor wafers (bulk and epitaxy) have been developed in parallel.

1.3.2.2. GOI realization

This paragraph presents the realization of GOI substrates using the Smart Cut™ technology and discusses the characterization results using both substrates. The process flow of the Smart Cut™ technology adapted to germanium wafer is described in Figure 1.11. First we developed a HF-based Ge cleaning process with low Ge consumption and a high particle removal rate. The metallic contamination levels do not exceed $5 \times 10^{10} \, at \, cm^{-2}$ as measured using the VPD ICPMS technique, and no degradation of the surface roughness was observed. Then, a plasma enhanced chemical vapor deposition (PECVD) silicon oxide was deposited for a few hundred nanometers on the Ge donor wafer to form a part of the final buried oxide, followed by densification at a temperature around 600°C. Then Ge wafers were ion-implanted with H^+, with dose and energy in the mid-$10^{16} \, at \, cm^{-2}$ range and 50–100 keV, respectively. Then the wafer was cleaned and hydrophilic bonded to thermally oxidized Si handling substrates. Final steps such as thermal treatment and polishing process were

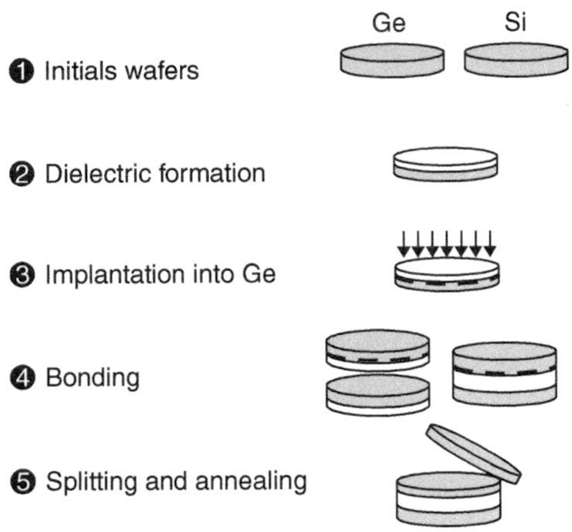

Figure 1.11. Smart Cut™ technology applied to Germanium wafers.

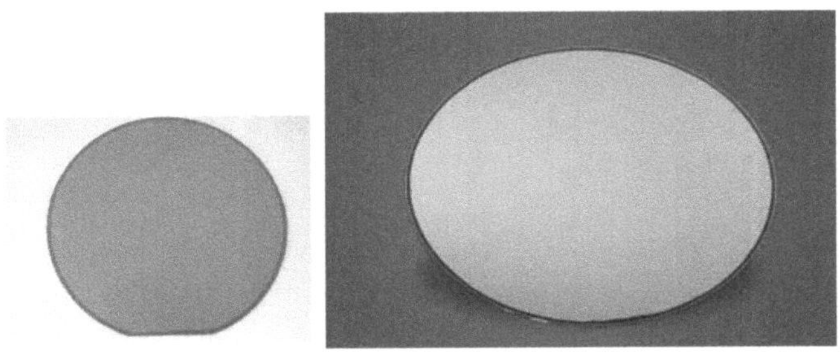

Figure 1.12. Final GOI wafers 100 and 200 mm in diameter (the donor wafers are bulk Ge wafers).

conducted. In the case of a Ge bulk donor wafer, the bonded Si/Ge wafers go through a thermal stress at elevated temperature which increases with diameter, due to the difference in thermal expansion coefficients (CTE) of Si and Ge (Table 1.2). Therefore, the ion implantation and layer transfer conditions were optimized for 100 mm as well as 200 mm. GOI substrates from 100 up to 200 mm have been demonstrated (Figure 1.12) with a top germanium layer from 200 down to 40 mm to meet a variety of applications such as GaAs epitaxial growth, partially depleted and fully depleted MOSFET fabrication. The Smart Cut™ technology also provides some opportunities for the re-use of resource-limited bulk Ge donor wafers.

1.3.2.3. Characterization of GOI substrates

Typically, the transferred Ge thickness depends on the implantation energy. The final Ge thickness measured by spectroscopic ellipsometry (SE), was reached using an SOI-like CMP process, using a slurry based on colloidal silica. Typically the average thickness on the final substrate is about 200 nm with a standard deviation equal to 5%. Roughness values of

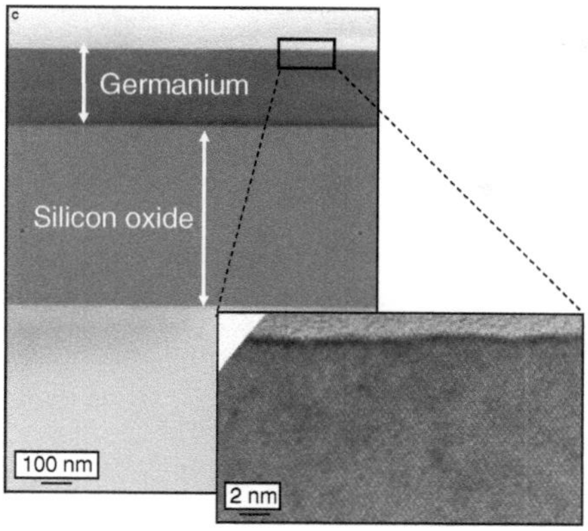

Figure 1.13. Cross-sectional conventional TEM images of a GOI substrate.

0.2 nm RMS and 2 nm in Z range were obtained by AFM for 5 μm × 5 μm scans, which is comparable to or even better than those of the Ge bulk donor wafers.

Cross-sectional transmission electron microscopy (TEM) images of the Ge thickness of 200 nm, and the buried oxide thickness of 350 nm are shown in Figure 1.13. The bonding interface between thermal silicon oxide and deposited oxide can be seen. Even using plan-view TEM, no extended defects have been found, indicating a defect density below $10^5\,cm^{-2}$ for GOI realized with bulk Ge.

Light point defectivity was measured using a surface inspection tool from KLA-Tencor (Surfscan 6200) with an incident scanning laser on the wafer surface, with a defect size threshold at 0.25 μm. These light point defects (LPDs) were found to be mainly holes in the transferred film, whose origins are considered to be particles from the wafer bonding process and defects that exist in the donor wafer. On recent final GOI wafers an average of less than 10 defects cm^{-2} has been measured, which is approaching that of the Ge bulk wafers with about 3 defects cm^{-2}. It indicates that the donor wafer quality will play an important role on improving the LPD on GOI.

The GOI samples have also been characterized from an electrical point of view. The pseudo-MOSFET technique [50] enables to extract key electrical data directly from an SOI-type wafer with very little processing. In particular, the low field mobility is extracted from the drain current (I_D) versus gate voltage (V_G) curves [51]. Figure 1.14 shows $I_D(V_G)$ characteristics of a GOI wafer made from a bulk donor wafer. Typical results obtained are around $5 \times 10^{12}\,eV^{-1}cm^{-2}$ for D_{it} (interface state density) at the Ge/SiO_2 interface, around $350\,cm^2V^{-1}s^{-1}$ for hole mobility, and could be up to $300\,cm^2V^{-1}s^{-1}$ for electron mobility. The difficulty in electron mobility extraction suggests that passivation of the back-side interface is necessary to improve the carrier transport properties in GOI. Such a passivation is currently under investigation. This passivation will be the key point, essential for microelectronics applications, as seen below.

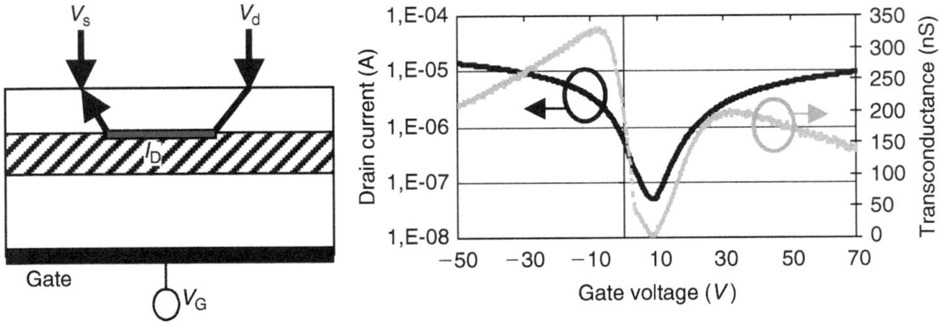

Figure 1.14. $I_D(V_G)$ characteristics of pseudo-MOSFETs on GOI wafers with the donor wafer coming from bulk material.

1.3.2.4. GOI MOSFETs

Submicron p- and n-MOSFETs with gate length down to 0.15 μm were demonstrated using a thin epitaxial Si layer as a Ge passivation layer on GOI made by Smart Cut™ technology from bulk Ge donor wafers [52]. The obtained mobility enhancement was comparable to that of bulk Ge CMOS. On GOI substrates obtained from epitaxial Ge donor wafers, the demonstration of fully depleted p-MOSFETs evidenced the high compatibility of GOI with a traditional low temperature full CMOS process [53]. Devices using nitridation for surface passivation [54] exhibit promising drain current ratios between ON and OFF states. As Ge passivation technology improves, GOI's advantages will play an important role in high-performance transistors technology.

1.3.2.5. GOI as III–V epitaxy template

Direct growth of GaAs on Ge (due to the small lattice mismatch) can be used for a variety of III–V applications (see Sections 1.1 and 1.2.3.1). The first demonstration of GaAs-based HBT devices on GOI demonstrated comparable device performances on GOI by Smart Cut™ technology, bulk GaAs and Ge substrates [2]. Since HBTs are very sensitive minority carrier devices, this demonstration confirmed the compatibility of GOI with high-quality III–V layer growth. The associated self-heating effect was smaller for GOI substrates than for GaAs or Ge substrates. This is due to the better heat dissipation by the Si substrate underneath ($1.5\,W\,cm^{-1}\,°C^{-1}$), even with the buried oxide layer in-between, compared to those on Ge and GaAs (0.6 and $0.46\,W\,cm^{-1}\,°C^{-1}$). Combining GaAs as a very high electron mobility n-channel with a Ge p-channel on the same GOI substrate would yield very high mobility structures. GOI allows not only III–V integration to mature Si devices but also can provide novel structures such as III–V and Ge on Si substrates.

1.4. General Conclusion

Although at the early transistor technology development the quality of Ge crystals was far better than for Si, the latter has been dominating the semiconductor market during the last

40 years. Ge materials improvements have been mainly focusing on other market segments such as e.g. detectors and solar cells. However, due to the potential revival of Ge for deep submicron CMOS applications, much effort has been devoted in recent years to fabricate high-quality 200 and 300 mm Ge wafers. This chapter gave a detailed overview of the present status and the different challenges to get there.

There is also a very strong interest in GOI wafers and different approaches have been critically discussed. Although further improvements are needed, the feasibility of obtaining high-quality GOI wafers enabling excellent device performance has been demonstrated. For future technologies there is also an interest for so-called hybrid approaches combining GOI with III–V epitaxy.

References

[1] A.J. Derluyn, K. Dessein, G. Flamand, Y. Mols, J. Poortmans, G. Borghs and I. Moerman, *J. Cryst. Growth* **247** (2003) 237.

[2] S.G. Thomas, E.S. Johnson, C. Tracy, P. Maniar, X. Li, B. Roof, Q. Hartmann and D.A. Ahmari, *IEEE Electron Dev. Lett.* **26** (2005) 438.

[3] C. Quaeyhaegens and M. D'Hondt, *Gorham Compound Semiconductor Outlook Conf.*, San Francisco, 2000.

[4] M.D Hondt, Z-O. Yu, B. Depreter, C. Sys, I. Moerman, P. Demeester and P. Mijlemans, *J. Cryst. Growth* **195** (1998) 655.

[5] O.I. Dosunmu, D.D. Cannon, M.K. Emsley, B. Ghyselen, J. Liu, L.C. Kimerling and M.S. Ünlü, *IEEE J. Sel. Topics Q. El.* **10** (2004) 694.

[6] J. Harris, Y. Kuo, Y. Lee, Y. Ge, S. Ren, J. Roth, G. Solomon, T. Kamins and D.A. Miller, *Proc. MRS Spring Meeting 2006* (in press).

[7] J. Scoyer, H. Guislain and H. Uwe Wolf, *Ulmann's Encyclopedia of Industrial Chemistry*, Wiley-VCH, Weinheim, 1987.

[8] G.K. Teal and J.B. Little, *Phys. Rev.* **78** (1950) 647.

[9] M. Azoulay and G. Gafni, *J. Cryst. Growth* **79** (1983) 326.

[10] D. Hoffman, T. Jung and G. Müller, *J. Cryst. Growth* **128** (1993) 213.

[11] O. Pätzold, K. Jenkner, S. Scholz and A. Cröll, *J. Cryst. Growth* **277** (2005) 37.

[12] L. Van Goethem, L.Ph. Van Maele and M. Van Sande, *Proc. SPIE* **683** (1986) 160.

[13] J. Völkl, *Handbook of Crystal Growth*, Vol. 2b, Elsevier Science, North-Holland, 1994, p. 821.

[14] M.P. Volz, M. Schweizer, B. Raghothamachar, M. Dudley, J. Szoke, S.D. Cobb and F.R. Szofran, *J. Cryst. Growth* **290** (2006) 446.

[15] O. Pätzold, B. Fischer and A. Cröll, *Cryst. Res. Technol.* **37** (2002) 1058.

[16] P. Dold, F.R. Szofran and K.W. Benz, *J. Cryst. Growth* **234** (2002) 91.

[17] D.T.J. Hurle and B. Cockayne, *Handbook of Crystal Growth,* Vol. 2b, Elsevier Science, North-Holland, 1994, p. 24.

[18] B. Depuydt, P.M. Boone, P. Union, P.F. Muys, D. Vyncke and C. Goessens, *Proc. European Symposium on Lasers and Optics in Manufacturing, SPIE Proc.* **3098** (1997) 559.

[19] P. Glasow and E.E. Haller, *IEEE Trans. Nucl. Sci.* **NS-23** (1) (1976) 92.

[20] R.N. Hall and T.J. Soltys, *IEEE Trans. Nucl. Sci.* **18** (1971) 160.

[21] E.E. Haller, G.S. Hubbard, W.L. Hasen and A. Seeger, *Inst. Phys. Conf. Ser.* **31** (1977) 309.

[22] http://www.ioffe.rssi.ru/SVA/NSM/Semicond/Ge/index.html

[23] I. Yonenaga and K. Sumino, *J. Appl. Phys.* **80** (1996) 3244.

[24] W.C. Dash, *J. Appl. Phys.* **29** (1958) 739.

[25] S. Hens, J. Vanhellemont, D. Poelman, P. Clauws, I. Romandic, A. Theuwis, F. Holsteyns and J. Van Steenbergen, *Appl. Phys. Lett.* **87** (2005) 061915-1.

[26] W. Geens, K. Dessein, N. Posthuma, G. Flamand, G. Raskin, H. Vercammen, B. De Meulemeester, C. Quaeyhaegens, P. Mijlemans, W. Köstler, G. Strobl, P. Sharps and S. Taylor, *Proc. of the 19th European PVSEC*, Paris, 2004, 3594.

[27] W. Geens, K. Dessein, W. Köstler, M. Meusel, S. Taylor, P. Mijlemans and G. Strobl, *Proc. Seventh European Space Power Conf.*, Stresa, May 2005.

[28] N.E. Posthuma, G. Flamand, W. Geens and J. Poortmans, *Sol. Energy Mater. Sol. Cells* **88** (2005) 37.

[29] G. Flamand, W. Geens, J. van der Heide, E. Simons and S. Taylor, *Proc. Seventh European Space Power Conf.*, Stresa, May 2005.

[30] B. Depuydt, *Encyclopedia of Materials: Science and Technology*, Elsevier Science Ltd., 2001 p. 3507.

[31] R. Lane, *Handbook of Semiconductor Silicon Technology*, edited by W.C. O'Hara, R.B. Herring and L.P. Hunt, Noyes Publisher, Park Ridge, New Jersey, 1990, Chapter 4.

[32] A.A. Chernov, *Modern Crystallography III: Crystal Growth (Springer Series in Solid-State Sciences)*, Vol. 36, 1984.

[33] R.K. Willardson and A.C. Beer, Indium phosphide: crystal growth and characterization, *Semicond. Semimet.* **31** (1990) Chapters 4 and 5.

[34] W. Kern, *RCA Rev.* **39** (1987) 278.

[35] A.F. Bogenschütz, *Ätzpraxis für Halbleiter*, Carl Hanser Verlag, München, 1967.

[36] R. Memming, *Semiconductor Electrochemistry*, Wiley-VCH, Weinheim, 2001.

[37] Y. Pleskov, *Semiconductor photoelectrochemistry*, 1st edition Springer, New York, London, 1986.

[38] K. Prabhakarana, T. Ogino, R. Hull, J.C. Bean and J. L Peticolas, *Surf. Sci. Lett.* **316** (1994) L1031.

[39] Y.L. Rivillon, F. Chabal, F. Amy and A. Kahn, *Appl. Phys. Lett.* **87** (2005) 253101.

[40] B. Onsia, T. Conard, S. De Gendt, M. Heyns, I. Hoflijk, P. Mertens, M. Meuris, G. Raskin, S. Sioncke, I. Teerlinck, A. Theuwis, J. Van Steenbergen and C. Vinckier, *Solid-State Phen.* **103–104** (2005) 27.

[41] R.R. Moskalyk, *Minerals Eng.* **17** (2004) 393.

[42] S. Nakaharai, T. Tezuka, N. Sugiyama, Y. Moriyama and S.-I. Takagi, *Appl. Phys. Lett.* **83** (2003) 3516.

[43] Y. Lui, K. Gopalakrishnan, P.B. Griffin, K. Ma, M.D. Deal and J.D. Plummer, *IEDM Techn. Digest* (2004) 1001.

[44] M. Bruel, *Electron. Lett.* **31** (1995) 1201.

[45] F. Letertre, C. Deguet, C. Richtarch, B. Faure, J.M. Hartmann, F. Chieu, A. Beaumont, J. Dechamp, C. Morales, F. Allibert, P. Perrau, S. Pocas, S. personnic, C. Lagahe-Blanchard, B. Ghyselen, Y.M. Levaillant, E. Jalaguier, N. Kernevez and C. Mazure, *Mat. Res. Soc. Proc.* **809** (2004) 153.

[46] C. Deguet *et al.*, *Electrochem. Soc. Proc.* **2005–2006** (2005) 78.

[47] C. Deguet, L. Sanchez, T. Akatsu, F. Alibert, J. Dechamo, F. Madeira, F. Mazen, A. Tauzin, V. Loup, C. Richtarch, D. Mercier, T. Signamarcheix, F. Letertre, B. Depuydt and N. Kernevez, *IEE Electronics Lett.* **42** (2006) 415.

[48] J.M. Hartmann, A. Abbadie, A.M. Papon, P. Holliger, G. Rolland, T. Billon, J.M. Fédéli, M. Rouvière, L. Vivien and S. Laval, *J. Appl. Phys.* **95** (2004) 5905.

[49] S. Cristoloveanu, D. Muntenaeau and S.T. Liu, *IEEE Trans. Electron Dev.* **47** (2000) 1018.

[50] G. Ghibaudo, *Electron. Lett.* **24** (1988) 543.

[51] B. De Jaeger, R. Bonzom, F. Leys, O. Richard, J. Van Steenbergen, G. Winderickx, E. Van Moorhem, G. Raskin, F. Letertre, T. Billon, M. Meuris and M. Heyns, *Microelectron. Eng.* **80** (2005) 26.

[52] L. Clavelier, C. Le Royer, C. Tabone, J.M. Hartmann, C. Deguet, V. Loup, C. Ducruet, C. Vizioz, M. Pala, T. Billon, F. Letertre, C. Arvet, Y. Campidelli, V. Cosnier and Y. Morand, *Proc. Silicon Nano Workshop*, Kyoto, 2005, p. 18.

[53] C. Le Royer, X. Garros, C. Tabone, L. Clavelier, Y. Morand, J.-M. Hartmann, Y. Campidelli, O. Kermarre, V. Loup, E. Martinez, O. Renault, B. Guigues, V. Cosnier and S. Deleonibus, *Proc. ESSDERC 2005*, edited by G. Ghibaudo and S. Cristiloveanu, 2005, p. 97.

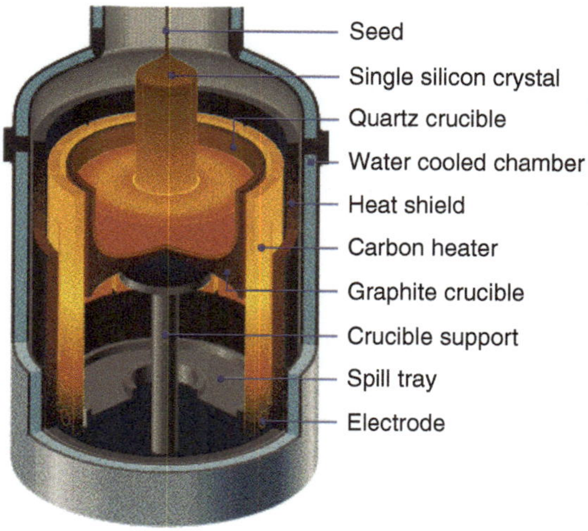

(a)

(b)

Plate 1. (a) Schematic principle of a Czochralski Si growth system. (b) View of the growing Ge crystal inside the furnace – one can discern the relatively bright (yellow) meniscus ring at the lower end of the crystal cylinder.

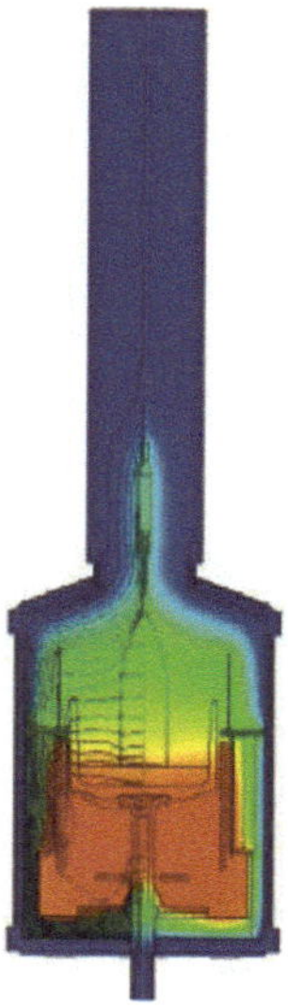

Plate 2. Temperature distribution in a Cz puller during the growth of a 200 mm dislocation-free germanium crystal.

CHAPTER 2

Grown-in Defects in Germanium

Jan Vanhellemont, Eddy Simoen, Igor Romandic and Antoon Theuwis

2.1. Introduction

Czochralski grown germanium crystals are used for various applications, ranging from detectors for nuclear radiation, requiring high-resistivity material, infrared optical components like mirrors and lenses, over germanium on insulator substrates that are being developed for advanced nano-electronics and request moderately doped substrates, to low resistivity substrates for GaAs epitaxy for application in advanced solar cells. Each of these applications imposes its own requirements with respect to crystal quality and more specifically with respect to grown-in defects.

In this chapter a review is given of the main grown-in lattice defect issues that can be encountered during germanium crystal growth and how they relate with the application one has in mind. In the course of the discussion it will become obvious that most of the available data on material properties of germanium are several decades old. A considerable research effort is needed in order to establish an updated knowledge base for germanium, which will surely be required to develop this material further for possible application in advanced nano-electronic devices [1].

2.2. Intrinsic Point Defects in Germanium

Experimental data on the properties of the intrinsic germanium point defects are scarce, especially for self-interstitials. For that reason *ab initio* simulation of intrinsic point defect

Germanium-Based Technologies: From Materials to Devices. C. Claeys and E. Simoen (Editors)

Table 2.1. Calculated and experimental formation (H^F) and migration (H^M) enthalpies in eV of intrinsic point defects in germanium [11].

H_{fV} (eV)	H_{mV} (eV)	H_{fI} (eV)	H_{mI} (eV)	References
***Molecular dynamics simulations with Stillinger and Weber potential* [11]**				
3.72	0.1	6.03	–	[12]
2.22	0.37	3.17	0.83	[14] set A
1.12	0.31	3.72	0.62	[14] set B
1.3	0.38	3.64	0.63	[14] set S
1.97	0.37	2.86	0.56	[13]
DFT calculations				
1.93		2.29		[4,8]
Experimental				
2.01	1.08			[16,17]
2	1			[18,19]
1.9	1.2			[20]
	0.2			[21]
1.9				[22]

properties is at the moment an important input to understand intrinsic point defect behavior, to design experiments for the study of the intrinsic point defect properties and for intrinsic point defect clustering simulations.

2.2.1. Simulation of intrinsic point defect properties

Already in 1982, Mashovets [2] published an extensive review on this topic of which the main results concerning formation and migration energies of self-interstitials and vacancies are represented in Table 2.1 together with more recent results.

Bernstein *et al.* [3] studied a number of properties of germanium using non-orthogonal tight-binding (TB) calculations. In case of point defects, these calculations include the vacancy, as well as three interstitial positions, the tetrahedral, hexagonal and <110> dumbbell. TB indicated that the lowest-energy interstitial was the tetrahedral configuration, in contrast with Density Functional Theory (DFT) [4] simulations which found that this was the case for the <110> dumbbell like for Si. Also the Stillinger–Weber [5] potential was used in order to study the self-interstitial structure [6] and from these investigations the <110> dumbbell configuration is the most stable one, like in silicon [7]. DFT [4,8] results show that the <110> dumbbell is significantly more stable in Ge than in Si, and that the formation energy for a germanium vacancy is smaller than the formation energy of a silicon vacancy. These two features explain in part why the vacancy is much more important for self-diffusion in germanium than in silicon [9,10] (see also Chapter 3). An overview of recent results is given in Table 2.1 [11]. The molecular dynamics simulations were performed for a 1000 atom cluster using the parameterizations of Ding and Andersen [12], Nordlund *et al.* [13] and the three parameter sets proposed by Yu *et al.* [14]. The values for the vacancy formation enthalpy obtained with the parameterization of Nordlund *et al.* and with one parameterization of Yu *et al.* (set A) are in good agreement with the results from DFT simulations and from experiments.

2.2.2. Experimental data on vacancy properties

Mainly three approaches have been used to directly study intrinsic point defect properties in germanium [15]: quenching from high temperatures followed by annealing at lower temperatures, plastic deformation at high temperatures and irradiation at low temperatures in order to create Frenkel pairs and individual point defects. In addition, the study of the diffusion and/or precipitation of fast-diffusing dopants, such as Cu and Zn, can provide indirect information on intrinsic point defects. Finally, also *ab initio* calculations can shed more light on the formation and migration energy and the electrical activity of the intrinsic point defects.

The quenching technique has been very successful in the assessment of vacancy properties in metals and has therefore initially also been used extensively for germanium. These preliminary results demonstrated that quenching of n-type Ge above 800°C resulted in p-type material, pointing to the creation of quenched-in acceptors. Annealing at 500°C recovered the original n-type, indicating the out-diffusion/annihilation of the quenched-in thermal acceptors. It was soon realized, however, that the early experiments suffered from copper contamination, whereby substitutional copper is a triple acceptor. Furthermore, it became also clear that quenching experiments cannot yield information on single vacancies as diffusion is so fast that vacancy clustering during quenching cannot be avoided.

Self-diffusion in Ge is the slowest-diffusion process and provides direct evidence of the existence of an equilibrium concentration of intrinsic point defects. Recent experiments of Giese *et al.* [16] showed that in contrast to silicon, the self-diffusion coefficient D_{Ge} in germanium ($= 13.6\,\mathrm{e}^{-3.09\,\mathrm{e}kT}\mathrm{cm}^2\mathrm{s}^{-1}$) is dominated by vacancies. They indeed presented strong experimental evidence that the vacancy-assisted Ge tracer diffusion coefficient $D_{\mathrm{Ge}}^{\mathrm{V}}$ is nearly equal to the self-diffusion coefficient D_{Ge}, or

$$D_{\mathrm{Ge}}^{\mathrm{V}} = 0.5\frac{C_{\mathrm{V}}^{\mathrm{eq}} D_{\mathrm{V}}}{C_{\mathrm{Ge}}} \approx D_{\mathrm{Ge}} = 0.5\frac{C_{\mathrm{V}}^{\mathrm{eq}} D_{\mathrm{V}}}{C_{\mathrm{Ge}}} + 0.73\frac{C_{\mathrm{I}}^{\mathrm{eq}} D_{\mathrm{I}}}{C_{\mathrm{Ge}}} \quad \text{and thus } C_{\mathrm{I}}^{\mathrm{eq}} D_{\mathrm{I}} \ll C_{\mathrm{V}}^{\mathrm{eq}} D_{\mathrm{V}} \tag{2.1}$$

This is quite different from the behavior in silicon. $C_{\mathrm{I}}^{\mathrm{eq}}$ and $C_{\mathrm{V}}^{\mathrm{eq}}$ are the thermal equilibrium self-interstitial and vacancy concentrations, respectively. C_{Ge} is the concentration of germanium lattice atoms.

Little information can be found on the thermal equilibrium vacancy concentration $C_{\mathrm{V}}^{\mathrm{eq}}$. On the basis of annealing experiments on thermal acceptors quenched into germanium, Mayburg [17] obtained a concentration of Frenkel defects (and thus vacancies) given by:

$$C_{\mathrm{V}}^{\mathrm{eq}} = 3 \times 10^{23}\,\mathrm{e}^{-2.01\,\mathrm{eV}/kT}\ (\text{in cm}^{-3}) \tag{2.2}$$

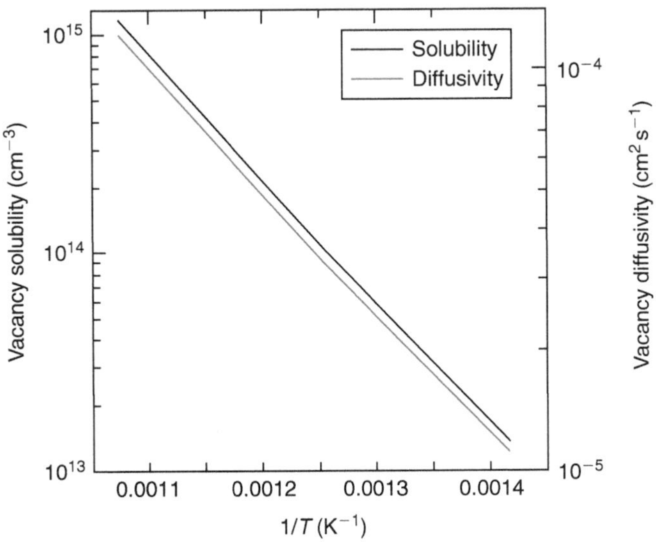

Figure 2.1. Solubility and diffusivity of vacancies according to Equations (2.2) and (2.3).

Equation (2.2) predicts an equilibrium vacancy concentration of $1.3 \times 10^{15}\,\text{cm}^{-3}$ at the melt temperature (937.4°C), in good agreement with the estimate of $2.9\text{–}3.9 \times 10^{15}\,\text{cm}^{-3}$, that Tweet [23] obtained based on Cu precipitation and diffusion behavior.

According to Giese *et al.* [16] one can write:

$$C_V^{eq} D_V = 118 \times 10^{22}\, e^{-3.09\,\text{eV}/kT} (\text{in cm}^{-1}\,\text{s}^{-1}) \tag{2.3}$$

Combining Equations (2.2) and (2.3) leads to

$$D_V = 3.09\, e^{-1.08\,\text{eV}/kT}\ (\text{in cm}^2\,\text{s}^{-1}) \tag{2.4}$$

Figure 2.1 shows the vacancy solubility and diffusivity data that have been used in this work, calculated with Equations (2.2) and (2.3).

2.2.3. Application of the Voronkov model to germanium

The present understanding of extended lattice defect formation by condensation of intrinsic point defects during the Czochralski pulling process of silicon crystals is based on the model of Voronkov [24–26]. The growing crystal can be considered as a solid-state reactor as schematically shown in Figure 2.2. In brief, the reasoning that can probably also be applied for germanium, is as follows. It is assumed that during the solidification process the thermal equilibrium values of both intrinsic point defects, i.e. vacancies and self-interstitials, exist at the melt/solid interface. These point defects are transported axially by

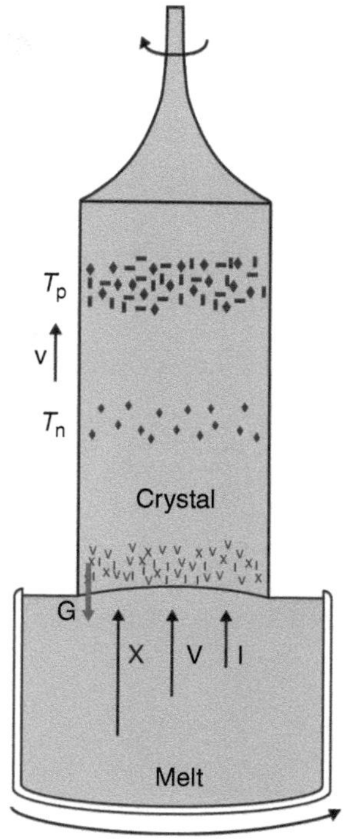

Figure 2.2. Schematic view of a growing germanium crystal as a solid-state reactor for point defects. At the melt temperature both intrinsic point defects are present with their thermal equilibrium concentrations. It is assumed that after recombination with self-interstitials (I) a supersaturation of vacancies (V) remains. This leads to the homogeneous nucleation of vacancy clusters at temperature T_n. At temperature T_p the supersaturation of extrinsic point defects (X) reaches also the critical value to have homogeneous nucleation of precipitates.

on the one hand thermal diffusion due to the thermal gradient (G) at the melt/solid interface and on the other hand by the crystal itself that is moving away from the melt with the pulling speed (v). The combination of these two transport mechanisms and the recombination of the intrinsic defects determine the dominant intrinsic point defect in the cooling crystal and thus also the type of extended defects formed by intrinsic point defect clustering. In cases where also a supersaturation of extrinsic point defects is present, such as H or O, clustering of these point defects during crystal cooling will also lead to the nucleation of precipitates that can grow by the absorption of extrinsic point defects during the further thermal history of the crystal. It is well accepted now, that silicon crystals pulled with a v over G ratio larger than a critical value are, e.g. vacancy-rich while below the critical value the crystal is interstitial-rich.

In a more quantitative way, the flux J of intrinsic point defects at the melt/solid interface can be written as [26]:

$$J_{\mathrm{I,V}} \approx -D_{\mathrm{I,V}} \frac{\mathrm{d}C_{\mathrm{I,V}}}{\mathrm{d}z} + \nu C_{\mathrm{I,V}} - \frac{D_{\mathrm{I,V}} \varepsilon_{\mathrm{I,V}}}{k(T_{\mathrm{m}})^2} G(0) C_{\mathrm{I,V}} \tag{2.5}$$

The first contribution is related to Fickian diffusion, the second to the movement of the crystal while the third one is related to uphill diffusion. $D_{I,V}$ and $C_{I,V}$ are the diffusivity and concentration of interstitials and vacancies, respectively. v is the pulling speed, $\varepsilon_{I,V}$ the drift energy, $G(0)$ the temperature gradient at the melt/solid interface and T_m the melt temperature. Assuming that the "mass action" law:

$$C_I(z)C_V(z) = C_I^{eq}(T)C_V^{eq}(T) \tag{2.6}$$

is also valid for germanium, the net flux J of vacancies is given by:

$$J(z) = J_V(z) - J_I(z) = D_V\frac{dC_V}{dz} + D_I\frac{dC_I}{dz} + \nu(C_V - C_I) - \frac{G(0)}{k(T_m)^2}(\varepsilon_V D_V C_V - \varepsilon_I D_I C_I) \tag{2.7}$$

$J > 0$ means that the crystal will be vacancy-rich while $J < 0$ is an indication of interstitial-rich material.

One can write:

$$\begin{aligned} C_{I,V} &\approx C_{I,V}^{i} e^{\frac{E_{fI,V}}{kT_m}} e^{-\frac{E_{fI,V}}{kT}} \\ \frac{dC_{I,V}}{dz} &\approx -C_{I,V}\frac{E_{fI,V}}{kT^2} \end{aligned} \tag{2.8}$$

$C_{I,V}^{i}$ is the point defect concentration at the melt/solid interface, while E_{fI} and E_{fv} are the formation energies of the interstitial and vacancy, respectively.

Vacancy-free material would be obtained if near the interface $J = 0$ leading to:

$$\xi_{crit} = \frac{\nu}{G(0)} \approx \frac{D_V^i C_V^i (E_{fV} - \varepsilon_V) - D_I^i C_I^i (E_{fI} - \varepsilon_I)}{(C_J^i - C_V^i)k(T_m)^2} \tag{2.9}$$

As for silicon one can assume that $\varepsilon_V,\varepsilon_I < E_{fI} \approx E_{fV}$. On the other hand, from the experiments of Giese *et al.* [16], it is known that $D_V C_V \gg D_I C_I$ so that a positive ξ_{crit} value can only be obtained if $C_I^i > C_V^i$. In that case D_I should be much smaller than D_V. $C_I^i > C_V^i$ would, however, mean that J is negative for $\xi > \xi_{crit}$ and thus the crystal would be interstitial-rich with the interstitial concentration increasing with increasing pulling speed. This is not in agreement with experimental observations which reveal only vacancy-related pits on the wafer surface. If on the other hand $C_I^i < C_V^i$, then the crystals are always vacancy-rich. For these reasons it is unlikely that vacancy-free or interstitial-rich germanium crystals can be prepared by only modifying the hot zone like is the case for silicon.

2.3. Extrinsic Point Defects

Extrinsic point defects can be divided in three classes: intentionally introduced dopants, neutral impurities having no direct impact on the electrical properties of the germanium crystal and unwanted electrically active (metallic) impurities. The discussion in the present chapter will be mainly on the neutral impurities C, N, H, O and Si that are introduced during crystal growth. The properties and behavior of oxygen and metal impurities in germanium are discussed more in detail in Chapters 4 and 5, respectively.

2.3.1. *Dopants*

At this moment the vast majority of commercial Cz germanium wafers is highly doped, serving as substrate for GaAs epitaxial growth, to be used in solar cells for space applications. For p-type substrates Ga is the standard dopant while for n-type substrates Sb and P (under development) are used. Typical dopant concentrations for solar cell application are $2 \times 10^{17}\,\text{cm}^{-3}$ (Sb, resistivity $20\,\text{m}\Omega\text{cm}$), $2 \times 10^{18}\,\text{cm}^{-3}$ (P, resistivity $4\,\text{m}\Omega\text{cm}$) and $2 \times 10^{19}\,\text{cm}^{-3}$ (Ga, resistivity $3\,\text{m}\Omega\text{cm}$) [27]. Intrinsic Cz germanium is used for detector applications and has a resistivity of about $50\,\Omega\text{cm}$ at room temperature due to the high concentration of thermally generated carriers (about $10^{13}\,\text{cm}^{-3}$ at 300 K). Dopant concentrations are typically in the range between 10^9 and $10^{10}\,\text{cm}^{-3}$.

2.3.2. *Neutral point defects*

Besides intentionally introduced shallow dopants and unintentionally introduced rapid diffusing metal contamination, there exists another class of impurities that do not produce direct electrical activity in germanium. These so-called "neutral" impurities can occur in high concentrations (up to $10^{18}\,\text{cm}^{-3}$ for oxygen) in as-grown material, depending on the crystal pulling conditions.

Most commercial Ge crystals are pulled from a melt in a graphite (acting as a source of carbon) or quartz crucible (acting as a source of silicon, oxygen and SiO_2 particles) [28,29]. Other impurities, like hydrogen, nitrogen or oxygen, may be introduced from the growth atmosphere. In fact, high-purity (HP) germanium (net doping concentration $N_{\text{dop}} < 10^{11}\,\text{cm}^{-3}$) can only be produced under a hydrogen atmosphere. One of the reasons is that a high amount of hydrogen can be incorporated (up to $10^{15}\,\text{cm}^{-3}$) [30], which passivates part of the electrically active impurities and contaminants, like, e.g. Cu. Oxygen can be introduced using an oxygen-containing atmosphere or can be dissolved in the melt from a quartz crucible. Nitrogen diffuses in from the gas phase.

In general, the impurities dissolved in the melt establish equilibrium with the ones in the crystal which is in contact with it. This thermodynamic equilibrium is described by the distribution or segregation coefficient k_{d}. As the crystal is pulled with a finite axial growth speed, one obtains in practice an effective distribution or segregation coefficient k_{deff}, given by [28]:

$$k_{\text{deff}} = \frac{C_{\text{S}}}{C_{\text{L}}} \tag{2.10}$$

with C_S and C_L the impurity concentration in the solid and in the melt, respectively. A value close to 1 as for Al implies that the impurity will be uniformly incorporated in the crystal, while a value >1 means that the impurity preferentially remains in the solid, so that there will be an enrichment at the seed end and a lower concentration near the tail of the crystal. For $k_{deff} < 1$, the opposite holds; examples are P (0.25) and O (0.11) [31]. Due to this segregation behavior, a HP-Ge crystal is generally p-type at the seed end (Al) and becomes n-type at the tail (P), with the occurrence of a p–n junction in between [28].

An important difference between Ge and Si is that the commonly used dopants in silicon, i.e. As, P, and B have a segregation coefficient that is close to unity. This allows obtaining a homogeneous distribution of dopants during crystal growth. Achieving homogeneous distributions of the Sb and Ga dopants in germanium is, however, not obvious in view of their small segregation coefficients causing a gradual increase of the dopant in the melt during the pulling of the crystal.

When present in supersaturation (i.e. a concentration larger than the equilibrium solubility at a specific temperature), these impurities will form clusters. A well-known example is the formation of the Oxygen Thermal Donors (OTDs), generated upon annealing in the range 300–500°C [32] or pairing with other foreign atoms also giving rise to electrically active complexes. In the case of hydrogen, both passivation and activation of impurities can occur [28].

2.3.3. *Carbon*

In the metallurgical sense, graphite is not miscible with Ge, explaining the use of graphite crucibles for crystal pulling [29]. Early investigations, extrapolating high-temperature data, pointed out that the solubility of C near the Ge melting point T_m should be at least $10^8\,cm^{-3}$ [33]. The presence of up to $10^{12}\,cm^{-3}$ A(H,C) shallow acceptors in HP-Ge [28,29] suggests that the total C concentration [C] should be higher.

A unique way to obtain near 100% detection efficiency for carbon is by growing an HP-Ge crystal in a quartz crucible coated by ^{14}C-spiked pyrolytic graphite. Close to all betas generated by the nuclear decay within the material are collected by the detector (autoradiography). Such study has revealed that [C] is $\sim 2 \times 10^{14}\,cm^{-3}$ [29]. Moreover, the presence of large carbon clusters has been demonstrated in this way. An effective segregation coefficient of 1.85 was found, indicating the preferential incorporation of carbon in the crystal with an increased carbon concentration towards the seed end.

2.3.4. *Hydrogen*

It was long thought that hydrogen is an inactive species in semiconductors, populating the T_d interstitial site. The study of hydrogen in high-resistivity germanium has, however, revealed that it can also introduce electrically active defects by forming pairs with substitutional neutral impurities like C and Si [28].

The solubility and diffusivity of hydrogen near the melting point has been measured by permeation experiments [34,35]. According to Frank and Thomas [35], the diffusion coefficient D_H and solubility C_H^{eq} of atomic hydrogen between 800°C and 910°C is:

$$D_H = 2.7 \times 10^{-3} e^{-0.38\,\mathrm{eV}/kT}\ \mathrm{cm^2\ s^{-1}} \tag{2.11}$$

$$C_H^{eq} = 1.6 \times 10^{24} e^{-2.3\,\mathrm{eV}/kT}\ \mathrm{cm^{-3}} \tag{2.12}$$

Extrapolation gives a solubility of about $4 \times 10^{14}\,\mathrm{cm^{-3}}$ at the melting point. The large D_H above 800°C is typical for interstitial diffusion and shows that trapping by other impurities is ineffective. At the same time, extrapolating Equation (2.12) to lower temperatures yields a solubility of only $10^5\,\mathrm{cm^{-3}}$ at 350°C. In practice, much higher concentrations have, however, been observed [30] which is explained by a trapping-related increase of the (equilibrium) hydrogen concentration. Also the experimental hydrogen diffusivity at 350°C is about 200 times lower than the one extracted from Equation (2.11) ($10^{-6}\,\mathrm{cm^2 s^{-1}}$), which can also be attributed to extensive trapping. The effective diffusion coefficient has been modeled by Hall [36], assuming a balance between atomic and molecular hydrogen whereby at low(er) temperatures, the equilibrium is towards H_2 formation with a dissociation energy of 2.06 eV.

As mentioned above, the solubility of hydrogen near T_m is about $4 \times 10^{14}\,\mathrm{cm^{-3}}$ [30,34,35]. Given an effective segregation coefficient of about 0.2, a higher H-concentration is measured near the tail ($\sim 1.5 \times 10^{15}\,\mathrm{cm^{-3}}$ in center tail) than near the seed end, as expected [30]. The solubility of hydrogen drops rapidly with temperature so that clustering of hydrogen is expected. It was observed that in dislocation-free high-resistivity Ge material, a homogeneous density of defects was present that could be revealed by etching and that was not observed in dislocated material [30]. It is known that dislocations are a good sink both for H and vacancies. A lower density of large pits is observed near the seed, while a higher density of smaller pits occurs near the tail end. Typical densities of these assumed hydrogen precipitates (or voids) are in the range $10^7\,\mathrm{cm^{-3}}$. Assuming that H-precipitation is complete, one arrives at a density of $10^8\,\mathrm{cm^{-3}}$ hydrogen atoms in each pit [30]. These etching defects can be dissolved by heating for 30 min at 900°C, indicating their high stability.

In addition to the pit defects, it has also been demonstrated that hydrogen gives rise to a so-called V_2H complex [28], with typical densities in the range $1\text{–}5 \times 10^{11}\,\mathrm{cm^{-3}}$ and a corresponding deep acceptor level at $E_V + 0.072$ eV (E_V = valence band maximum). It is believed that this acceptor defect could be responsible for the observed "coring" in large-diameter hydrogen-grown HP-Ge crystals [36]. Due to the segregation behavior of P, crystals are normally p-type at the top and n-type at the tail end. However, it was found that the p–n junction in the middle of the crystal was bent upward near the center of it, pointing to strong radial net doping profiles and giving rise to lower than expected carrier mobilities. These doping non-uniformities can be removed by thermal annealing at 357°C and a strong correlation with the presence of the V_2H deep acceptor has been established [36]. The origin of "coring" could thus be related to the distribution of some fast-diffusing quenched-in native lattice defects (probably vacancies in the form of V_2 and larger clusters). The concentration of quenched-in divacancies V_2 was estimated $\sim 10^{12}\,cm^{-3}$; they are stable up to 400°C. Interaction with hydrogen partly passivates these centers according to the reaction:

$$V_2 + H \rightarrow V_2H \tag{2.13}$$

These deep acceptors anneal at 200°C, following the reaction:

$$V_2H + H \rightarrow V_2H_2 \tag{2.14}$$

leading to their full passivation.

2.3.5. Oxygen

While oxygen is the most important neutral impurity in Cz-grown silicon, this is not the case for germanium. One of the reasons is that most commercial crystals are pulled in graphite crucibles. Another reason is the lower affinity of oxygen for germanium. The effective segregation coefficient is $\geqslant 0.11$ [31], indicating that oxygen has the tendency to stay in the liquid phase leading to an increased oxygen concentration near the tail of the crystal. High concentrations of oxygen (10^{16}–$10^{18}\,cm^{-3}$) can only be obtained in crystals grown under an oxygen-containing atmosphere [32].

As for silicon, the normal lattice site for oxygen is interstitial, whereby it is close to the bond centered (BC) position between two host atoms. Interstitial oxygen (O_i) is electrically inactive, but its vibrational modes can be detected in infrared absorption. Also like for silicon supersaturated oxygen tends to form clusters or GeO_2 precipitates. The small oxygen clusters formed in the temperature range 300–500°C have a donor nature, whereby a whole family of double donors is created by the subsequent attachment of an additional oxygen atom. Few results are available on oxygen precipitation in germanium and on the impact of oxygen on its mechanical strength [37–39].

The properties and behavior of oxygen in germanium are discussed in detail in Chapter 4.

2.3.6. Nitrogen

Commercial germanium crystals are often grown in a nitrogen ambient making this an important grown-in impurity. Although nitrogen belongs to the Group V elements, like P, As and Sb, it turns out to be a weak donor in both germanium [40] and silicon. This is partly related to its low solubility [41]. Moreover, the dominant nitrogen species is not substitutional N, but the N pair, which can be detected in infrared absorption spectra by its vibrational modes at 825.3 and $658.6\,cm^{-1}$ [41–43].

The nitrogen pair is not electrically active and becomes unstable at 600°C. The N pair defect consists of two <100> oriented Ge–N split interstitials, arranged in an anti-parallel

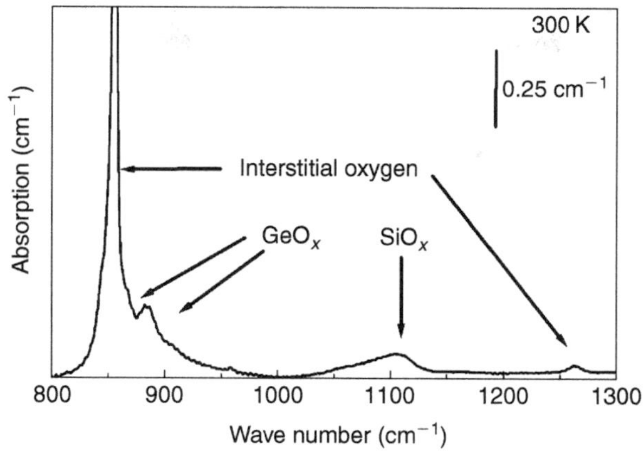

Figure 2.3. Typical absorption spectrum of oxygen precipitates and interstitial oxygen in germanium. The initial and final interstitial oxygen concentrations are $2.5 \times 10^{17}\,cm^{-3}$ and $1.66 \times 10^{17}\,cm^{-3}$, respectively. The heat treatment was 66 h, 610°C + 66 h, 560°C in Ar (after De Gryse *et al.* [37]).

configuration and with the four bonds forming a square lying on {011}. The corresponding vibrational modes are sharper than in silicon [41].

It should be remarked that in nitrogen-atmosphere- and graphite-crucible-grown HP-Ge crystals a number of N-related shallow acceptors has also been identified by photo-thermal ionization spectroscopy (PTIS) [28].

2.3.7. Silicon

Molten germanium in contact with quartzware will inevitably contain silicon, oxygen and SiO_x inclusions leading to typical absorption lines in the FTIR spectrum (Figure 2.3) as was already observed in the early work of Kaiser and Thurmond [44]. Crystal pulling under a hydrogen atmosphere reduces this tendency. As silicon shows a greater affinity with oxygen than Ge, additional SiO_x precipitates may form during high-temperature anneals.

The presence of SiO_x precipitates can be established from the 9.1 μm infrared absorption band. In addition, it is believed that a certain type of preferential etch pits – the so-called smooth pits (SPs) – is related to SiO_2 precipitates [45–47]. These SPs may have different diameters and are ~0.3 μm deep [47]. Annealing experiments did not have an effect on the density or shape of the etch defects, demonstrating that they are quite stable clusters. Moreover, radiation detectors fabricated from material showing a high density of SPs exhibit a poor spectral resolution due to hole trapping. Deep level transient spectroscopy (DLTS) has revealed an acceptor level at $E_V + 0.069\,eV$ associated with these SPs [48]. The trapping cross-section derived per SP is in the range $5 \times 10^{-9}\,cm^2$ or corresponding to a diameter of 1 μm. No evidence for electron trapping was observed [47]. Recent *ab initio* simulations of the stability of vacancy and oxygen clusters suggest an important difference between silicon and germanium: vacancy clusters are more stable in silicon, while oxygen clusters are more stable in germanium [49].

2.4. Dislocation Formation During Czochralski Growth

High-quality dislocation-free germanium single crystals are produced by the Czochralski pulling process as discussed in Chapter 1. In this paragraph the focus will be on possible sources of dislocation formation during the crystal growth process. Important factors hereby are the propagation of dislocations from the seed into the crystal and the development of thermal stress during crystal cooling. It is shown that in some cases a well-controlled density of dislocations is needed in order to obtain the desired material properties.

2.4.1. Thermal simulation

As discussed in Chapter 1, the Cz pulling process is well understood nowadays and can be simulated in great detail for germanium with respect to crystal geometry and temperature distributions [50–52]. This has been illustrated in Figure 1.6, showing the simulated temperature distribution in a crystal puller during the pulling process of a 200 mm crystal. The nucleation and growth of lattice defects during crystal pulling is much less clear as little research has been performed on this topic during the last decades.

2.4.2. Development of mechanical stresses

Due to the thermal gradients in the growing crystal, considerable mechanical stresses can develop and in extreme cases lead to the generation of dislocations. Although they have to be avoided for applications in electronic devices, they are allowed in crystals for optical applications albeit in well-defined conditions and they are even necessary in well-defined numbers and types for good quality detectors made from high-resistivity Ge. For that reason a good understanding of thermal stress distribution has become indispensable. One example is germanium for optical applications. Germanium is transparent for infrared light with wavelengths between 2 and 15 μm and in the ideal case it is also optically isotropic. Nevertheless, stresses in the material lead to local losses of symmetry, giving rise to non-isotropic effects like birefringence.

In order to produce germanium that is sufficiently optically uniform and isotropic as typically needed for applications in optics, the crystal birefringence must be minimized and the homogeneity of the refractive index maximized. Therefore the residual stresses in the as-grown germanium crystals must be minimized. These residual stresses are often associated to plastically deformed regions of the crystal resulting from localized stress fields during crystal growth. When the yield stress is exceeded, dislocations nucleate and can easily glide inwards from the rim of the crystal giving rise to the activation of various dislocation multiplication mechanisms. These dense dislocation networks, as for instance also associated with grain boundaries, will after cooling down lead to localized residual stress fields in the crystal [53].

Figure 2.4 shows the simulated temperature and stress distribution in a growing 200 mm crystal for optical applications. Although the temperature distribution looks quite uniform, the thermoelastic stress distribution shows a quite different picture with a high stress concentration

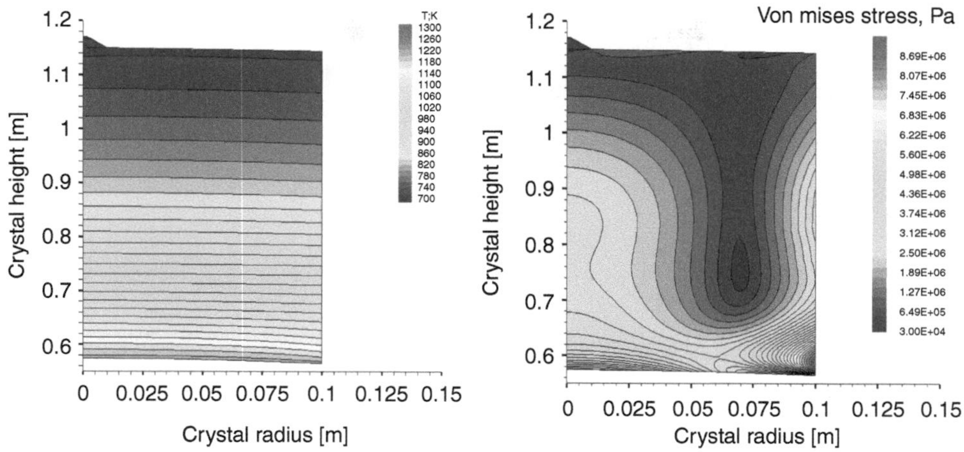

Figure 2.4. (a) Calculated temperature distribution in a growing 200 mm infrared germanium crystal. (b) The von Mises stress invariant in the same crystal [54].

at the crystal edge, close to the solid/melt interface. It is clear that this will be the favoured site for dislocation nucleation when a critical stress level is exceeded.

2.4.3. Mechanical properties of germanium

Germanium has a face-centered cubic (fcc) diamond lattice and as the other Group IV semiconductors is brittle at room temperature. It is only at higher temperatures that germanium becomes ductile and deforms plastically by the nucleation and multiplication of dislocations. As for silicon the preferred glide planes are the (111) planes which have the highest-packing density and the preferred slip direction is also <110>. The mechanical properties of germanium are less favorable for crystal pulling and for its use as semiconductor substrate compared to those of silicon as illustrated in Table 2.2, which is the same as Table 1.2 in Chapter 1. Especially the more than two times higher density of germanium in combination with a significantly lower mechanical strength (that decreases also more rapidly with increasing temperature) than silicon can lead to dislocation-related problems during crystal pulling and to warpage when using it as large diameter bulk substrate for device processing. The larger weight of the germanium crystal combined with the lower mechanical strength makes larger seed and neck diameters necessary for similar crystal lengths which increases the probability of dislocation propagation from the seed.

The dependence of first-order elastic constants on temperature is illustrated in Figure 2.5. For the temperature window which is relevant for crystal growth and device processing, the first-order elastic constants of germanium show a nearly linear dependence on temperature. The impact of n-type doping on C_{44} as is shown in Figure 2.6, is related to the electronic contribution to the strain energy function [55]. Since the elastic constants are the strain derivatives of the free energy of the crystal, it is assumed that heavy doping reduces the electronic contribution of the latter.

Table 2.2. Mechanical properties at 300 K of germanium versus silicon [56].

Material parameters	Germanium	Silicon
[100] Young modulus [10^{11} dyne cm^{-2}]	10.3	13.0
[100] Poisson ratio	0.26	0.28
Bulk modulus [10^{11} dyne cm^{-2}]	7.13	9.8
Shear modulus [10^{11} dyne cm^{-2}]	4.1	5.2
Hardness [Mohs]	6	7
Knoop hardness [kg mm^{-2}]	780	1150
Density [g cm^{-3}]	5.323	2.329
C_{11} [10^{11} dyne cm^{-2}]	12.60	16.60
C_{12} [10^{11} dyne cm^{-2}]	4.40	6.40
C_{44} [10^{11} dyne cm^{-2}]	6.77	7.96

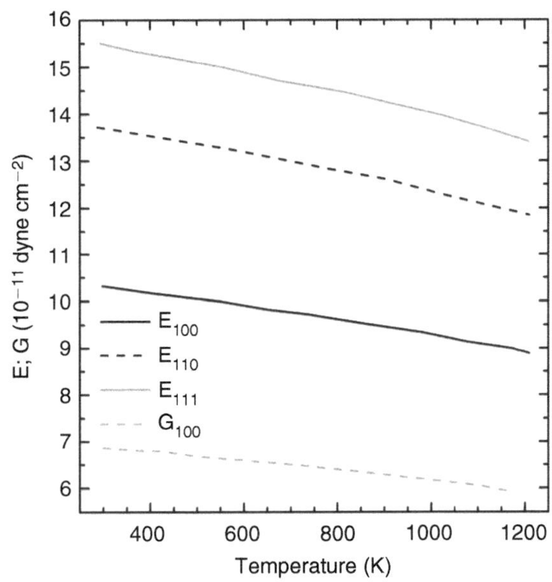

Figure 2.5. Temperature dependence of the Young modulus E (for <111>, <110> and <100> directions), as well as of the shear modulus G_{100} in <100> direction (after Burenkov *et al.* [57]).

2.4.4. Dislocation nucleation and multiplication during crystal pulling

The mechanical properties of dislocations in germanium and their relation with material strength and plastic flow have been studied in detail [58–60] and revealed much higher dislocation mobility in germanium. Recently, also *in-situ* transmission electron microscopy studies have been performed to study dislocation nucleation, propagation and interaction in germanium [61–62]. Optical microscopy and scanning tunneling microscopy have been used to study dislocations in plastically deformed germanium crystals revealing bands of steps on the crystal surface that are related with the emergence of groups of $a/2$ [110] type

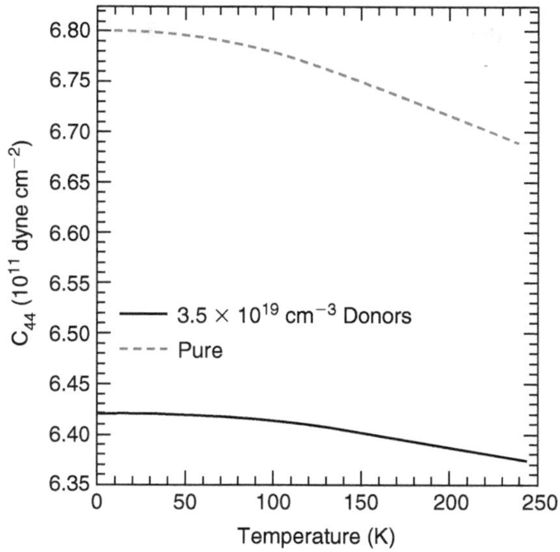

Figure 2.6. Temperature dependence of C_{44} in undoped and heavily doped germanium (after Bruner and Keyes [55]).

dislocations on the surface [63]. These bands show up in a crosshatched pattern of ridges and trenches in optical microscopy.

Nowadays, it is common practice to grow dislocation-free germanium crystals, except for the case of HP material, where a dislocation density in the range of several hundred cm^{-2} is usually required to achieve the specified detector performance. In the early days of crystal growth it was difficult to control the density and distribution of dislocations and it was only by the use of necking of the seed as proposed by Dash that one could avoid dislocation propagation from the seed crystal into the growing crystal. Besides the seed crystal, there are several other potential sources of grown-in dislocations, the dominant one being thermal stresses during growth and cooling of the crystal as discussed in Section 2.4.2. Even today, grown-in dislocations can be an issue especially in the case of alternative crystal-growth techniques, like detachment growth [64].

As mentioned before, an obvious cause of dislocation formation during crystal growth is the presence of thermal stresses caused by the radial temperature gradient across the solid–liquid interface [65–70] leading to an increase of thermoelastic stress at the rim of the crystal. Ideally, one should design a hot zone in order to have a perfect one-dimensional (axial) temperature gradient.

Provided that the cooling crystal cannot expand freely, a simple first-order estimate of the dislocation density n_{disl} related to the solidification process can be written as [65]:

$$n_{\mathrm{disl}} = \frac{\alpha_{\mathrm{e}}}{b}\frac{\partial T}{\partial r} \tag{2.15}$$

With $\partial T/\partial r$ the radial temperature gradient, α_e the linear expansion coefficient and b the length of the Burgers vector. In this model, the stress builds up during solidification due to the change in lattice parameter. As the diamond lattice is a quite open structure, germanium expands about 5% upon solidification [66]. Defects originate directly on solidification from the melt, i.e. at the interface between the liquid and solid. Regions of high n_{disl} are always associated with the highly curved parts of the liquid–solid interface, regardless whether it is concave or convex [66]. As the hot zone changes continuously during Cz crystal growth also the shape of the solid–liquid interface and thus the thermal gradients and stresses are continuously changing. Since these thermal gradients are the largest near the top and bottom end of the crystal and also at the perimeter for the cylindrical part, these will be the preferred sites for dislocation nucleation and multiplication [65]. Assuming that stress is concentrated in the direction of growth, one expects n_{disl} to increase from <110> over <111> to <100> crystal pulling directions [66]. As was mentioned before, dislocations can also propagate easily from the seed into the crystal [65,70]. Using a dislocation-free seed minimizes this problem and in combination with the necking of the seed at the start of crystal pulling this problem is now well under control. The pulling of heavy 200 and 300 mm diameter crystals requires however the use of larger seed and neck diameters in order to be able to sustain the larger weight of the crystal. This imposes more stringent demands on the necking procedure and seed quality.

Apart from the radial temperature gradient, there exists also an axial temperature gradient. During cooling the crystal will therefore shrink by an amount ε given by:

$$\varepsilon = \alpha_e \Delta T \tag{2.16}$$

and as the temperature in the central part is initially higher than in the outer part of the crystal, it will shrink more. As was shown in Figure 2.4 this differential shrinkage produces thermoelastic stresses during cooling of the ingot, which when they supersede the yield stress will cause plastic flow. These stresses will be compressive in the central part and tensile in the outer part of the crystal and are given by:

$$\tau = E\Delta\varepsilon \tag{2.17}$$

Wagner [71] reported that dislocations can also be produced by thermal shock after crystal growth when the crystal is detached from the melt, and not only by the cooling of the ingot itself.

Compositional fluctuations and condensation of a supersaturation of point defects has been shown as another cause of dislocation formation during crystal growth [72–75]. In the case of germanium–silicon alloy growth, arrays of dislocations were found, most likely arising at fluctuations in the solute concentration C. It was demonstrated that for an abrupt change ΔC, causing a change of the lattice parameter a, one expects a dislocation density n_{disl} given by [72]:

$$n_{disl} = \Delta C \frac{da_0}{dC}\left(\frac{\sqrt{2}}{a_0^2}\right) \tag{2.18}$$

For low-silicon concentration alloys, Equation (2.18) predicts that $\Delta C = 0.1\%$ leads to a dislocation density of $1200\,cm^{-1}$. This implies that at a discontinuity in the solute concentration of 0.1% one can expect an array of dislocations in germanium [72]. An important source of dislocation formation can also be particles or impurity clusters that are incorporated at the perimeter of the growing crystal. The localized stress fields associated with these particles add to the thermal stresses and make them preferential sites for dislocation formation.

2.4.5. Electrical impact of dislocations in germanium

Hornstra [76] has discussed in detail the dislocation types in fcc materials nearly half a century ago. Dislocations in the (111) primary glide plane tend to align along the <110> Peierls valleys thus minimizing the free energy.

Dislocation formation during Cz crystal growth is avoided by using Dash necking of the crystal seed and by suppressing excessive thermal stress during the crystal cooling. In high-resistivity germanium, however, a sufficient density of dislocations is needed in order to suppress the formation of hydrogen-related voids and V_2H centers that are detrimental for detector performance [27].

The electrical activity of dislocations in germanium has been studied extensively on plastically deformed material, as a relatively high-dislocation density is needed in order to have a measurable impact on the electrical properties. Plastic deformation of germanium has been studied extensively and yields predominantly $\frac{a}{2}\langle 011\rangle$ type dislocations as is typical for the diamond lattice structure. Plastic deformation is not required to study the electrical properties of dislocations in high-resistivity Ge where as-grown dislocated material can be used due to the higher sensitivity of the characterization tools related with the low background doping.

Although in general dislocations are considered to be harmful for devices, in high-resistivity germanium used for detector applications, their presence is vital to achieve the desired material properties. It is known that in order to obtain optimal detector resolution a grown-in dislocation density between 10^2 and $10^4\,cm^{-2}$ is needed. A typical example of a good quality detector crystal is shown in the optical micrograph of Figure 2.7 which was obtained after preferential etching to reveal the extended lattice defects [77]. Dislocation lineage, mosaic structures, as well as isolated dislocations are observed. A higher dislocation density will cause too much carrier trapping, leading to a loss of resolution; a too low dislocation density, on the other hand, gives rise to the creation of hydrogen-related complexes, lowering also the resolution of the detector.

Dominant dislocation types that can be introduced in a more or less controlled way during crystal pulling are 30°, 60° and 90° dislocations as listed in Table 2.3 [77]. In such material dislocations serve as sink for vacancies and thus the creation of harmful V_2H complexes can be avoided. Hydrogen is introduced in large quantities in this material, as the HP Cz crystals have to be pulled in hydrogen atmosphere.

The dislocations of Class 1 often lead to the formation of low angle grain boundaries that show up as regular patterns of etch pits after defect etching, so-called lineage or mosaic

Figure 2.7. Optical micrograph of a preferentially etched crystal slice revealing dislocation lineage and mosaic structures, as well as isolated dislocations (after Van Sande *et al.* [77]).

Table 2.3. Grown-in dislocations identified in [100] germanium crystals [77].

Direction	**[100]**	**[110]**	**[112]**	**[112]**
Type	90°	60°	30°	90°
Class	1	2	3	4

structures [78]. A typical example of a low angle grain boundary consisting of dislocations of Class 1 are shown in the high voltage transmission electron microscopy (HVEM) micrograph in Figure 2.8.

Class 2 dislocations are present throughout the crystal with densities on the order of a few thousand per cm^2. Class 3 and 4 dislocations are difficult to avoid and often reveal radial distributions. When the crystal is heavily dislocated, this class of dislocations is the dominant one [77].

The electrical activity of dislocations has also been studied extensively using the DLTS technique [77,79,80]. As in silicon, dislocations lead to broad DLTS bands of which the amplitude correlates well with dislocation-related etch pit densities. Trapping by dislocations is more effective in n- than in p-type high-resistivity germanium.

Due to this important impact of the number and types of dislocations on germanium detector performance, they are specified by the vendor as illustrated in Table 2.4. In moderately and heavily doped germanium, dislocation formation can easily be avoided during Cz crystal growth.

(a)

(b)

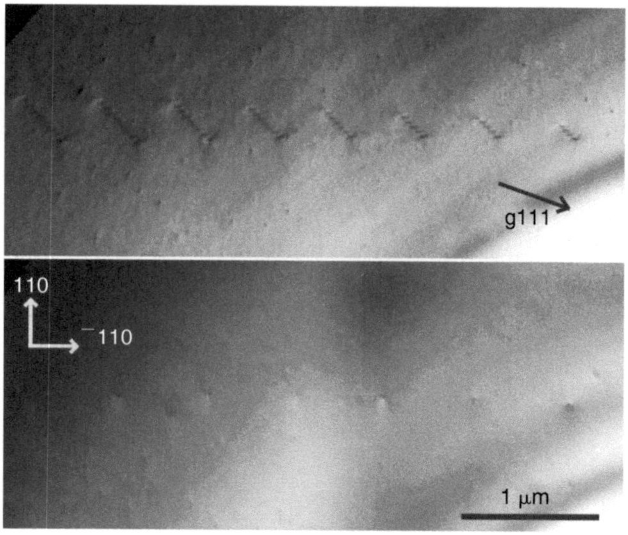

Figure 2.8. HVEM observations of typical lineage dislocations (Type 1 in Table 2.3) in an as-grown high-resistivity germanium crystal [1]. (a) A row of lineage dislocations in a sample tilted about 35° away from the [001] crystal pulling axis. (b) The same area viewed along the [001] axis. Dislocations are now seen end-on (= lie along the [001] axis) (courtesy Hugo Bender, IMEC).

Table 2.4. Typical specs for the crystallographic perfection of high-resistivity germanium for detector applications [27].

	p-type	n-type
Dislocation density (EPD, cm^{-2})	$\leqslant 10^4$	$\leqslant 5 \times 10^3$
Lineage (unit length = slice radius)	$\leqslant 3$	$\leqslant 2$
Mosaic structures (unit surface = 100 mm^2)	$\leqslant 5$	$\leqslant 2$
Saucers (cm^2)	$\leqslant 500$	$\leqslant 500$

2.5. Point Defect Clustering

2.5.1. Experimental observations of vacancy clustering

As discussed in the previous paragraphs, the vacancy is the dominant intrinsic point defect in germanium. Dislocation free Cz-grown germanium crystals can therefore contain vacancy clusters and even macroscopic voids especially when pulled in a hydrogen-rich atmosphere [81,82]. In high-resistivity-dislocation free germanium, V_2H traps can be formed during cooling of the crystal. These traps can lead to “coring”, a change of the resistivity in the central part of the crystal [82]. Without special measures, the excess of vacancies that is created during cooling of the crystal leads to the formation of larger agglomerates that are visible as pits on the wafer surface, comparable to the crystal-originated particles or so-called “COPs” observed on silicon wafers. Already in 1959, Tweet [83] reported the occurrence of surface pits with a size depending on the thermal history of the crystal. He observed a pit density reduction when pulling slower or by reducing the temperature gradient thus keeping the crystal longer at high temperatures or by post-heating the crystal in the puller. The observed pit density reduction was accompanied by an increase of size of the remaining pits. The observations were explained by a vacancy clustering mechanism. Three decades later, similar ideas were used to reduce the number of grown-in intrinsic point defect clusters in large diameter Cz silicon crystals [84].

A typical observation of imperfections on a polished germanium wafer obtained using a light scattering-based wafer surface inspection tool is given in Figure 2.9. The imperfections show

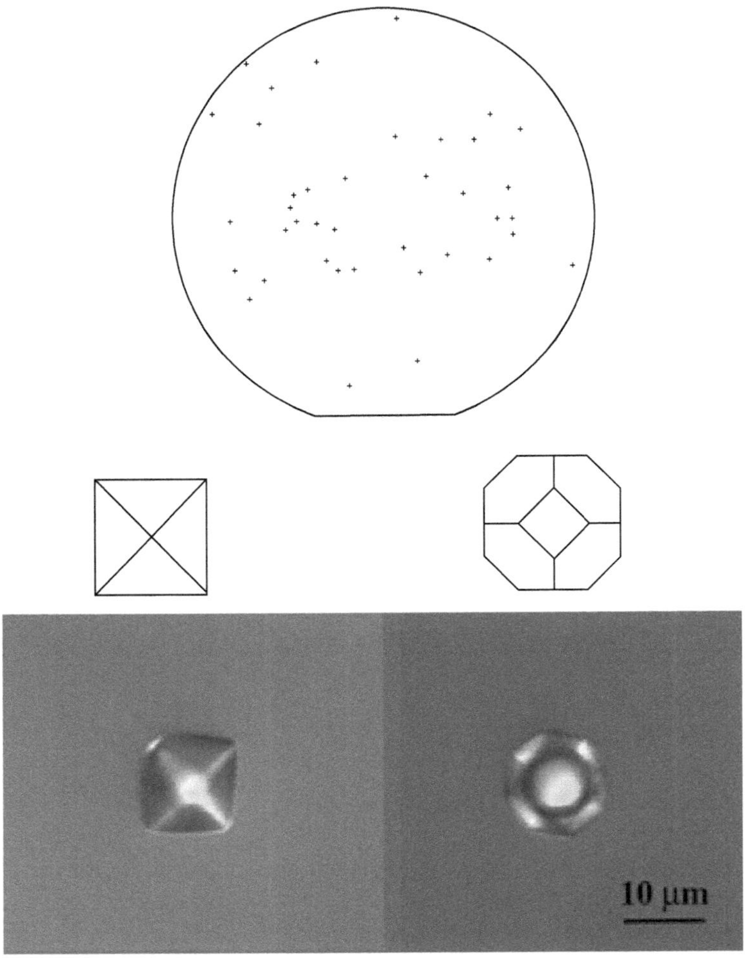

Figure 2.9. Typical observation with a surface inspection tool revealing the presence of large LPDs (between 10 and 15 μm LSE) on a polished germanium wafer. The optical micrographs below reveal that most of the large LPDs are crystallographic pits on the wafer surface [88]. The left image corresponds with an octahedral void while the right image is in agreement with a truncated octahedral void, both schematically shown viewed along the <001> axis.

up as so called light point defects (LPDs) of which the size is defined as LSE (latex sphere equivalent), based on the light scattering intensity due to latex spheres with calibrated size deposited on a polished wafer surface. A low density of LPDs is observed with a size of the order of 10 μm LSE and thus nearly two orders of magnitude larger than those typically observed on silicon wafers. Owing to their large size the surface defects can also easily be investigated using optical microscopy as is illustrated in Figure 2.9. Optical microscopy reveals that the defects are crystallographic surface pits that can be related to [001] projections of octahedral and truncated octahedral voids as illustrated.

2.5.2. Modeling and simulation of vacancy cluster formation

Let us assume that the germanium crystal in Figure 2.2 is vacancy-rich after the initial recombination of intrinsic point defects that occurs close to the melt/solid interface. During further cooling of the crystal (= with increasing distance from the melt/solid interface), the supersaturation of vacancies (V) and possibly also of extrinsic point defects (X) increases. At a certain moment a critical supersaturation of vacancies is reached at temperature T_n and homogeneous nucleation of clusters of vacancies occurs. Similarly a supersaturation of extrinsic point defects can be reached at a later stage, e.g. at temperature T_p and nucleation of precipitates occurs (e.g. GeO_2 in the case of interstitial oxygen, see further), possibly assisted by vacancies. During the further cooling process the nuclei of both defects continue to grow by point defect diffusion. At this moment only limited experimental data is available to further corroborate this qualitative model. A review on the impact of dopants and pulling conditions on grown-in intrinsic point defect clusters in silicon and on techniques to characterize and simulate these clusters in silicon has been published [85].

In silicon, the process of intrinsic point defect incorporation and agglomeration during crystal pulling can be modeled with great success, making effective hot zone design possible to tune point defect recombination and diffusion in order to avoid nucleation of agglomerates. In the case of silicon, simulation of the observed dependence of vacancy agglomerate size and density distributions over a wide range of pulling conditions and crystal diameters allowed also to determine accurate values of the diffusivity and solubility parameters of the intrinsic point defects [86,87].

A similar approach can also be used for germanium [88]. An initial result is given in Figure 2.10 showing the expected vacancy cluster size distribution for three different axial temperature gradients. The model developed by Sinno *et al.* [87] to simulate intrinsic point defect agglomeration in silicon has been used and adapted to the case of vacancy aggregation in germanium. Diffusion coefficients and thermal equilibrium concentrations for vacancies given by Equations (2.2) and (2.4) were used as input for the aggregation software. For the thermal simulations the CrysVUN simulator developed at the university of Erlangen [51,52] was used. The thus calculated axial temperature profile served as input for the vacancy cluster simulations. The void distributions were calculated for a steady-state simulation run. Using the simulated thermal history of the diameter germanium crystal, the simulations predict vacancy clusters with sizes between 1 and 6 μm.

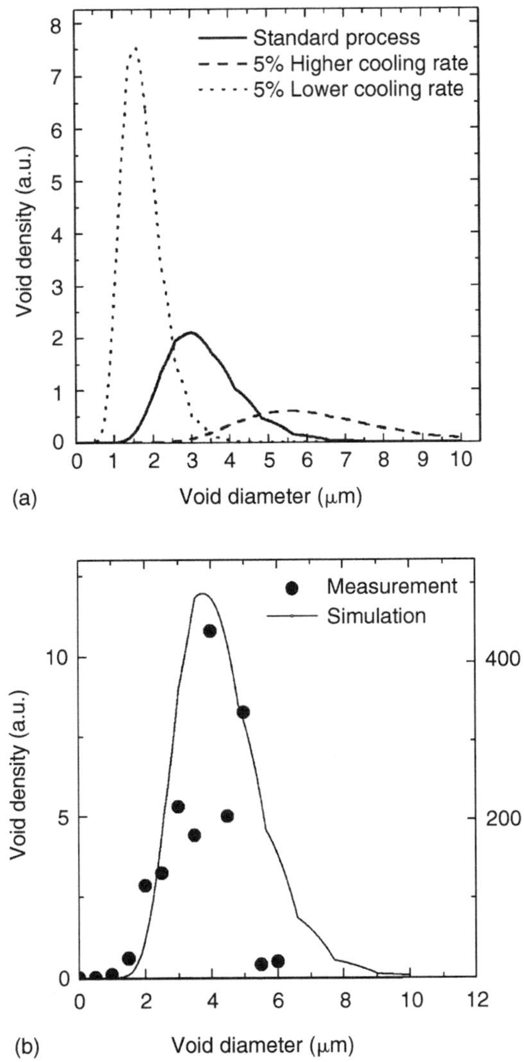

Figure 2.10. (a) Calculated vacancy cluster size distribution illustrating the large impact of the crystal cooling rate. (b) Simulation of the pit size distribution in an as-grown germanium crystal compared with the measured surface pit size distribution using optical microscopy (after Hens *et al.* [88]).

The diameter of each surface pit was determined by an optical review microscope (CRS of KLA-Tencor) to allow generating a pit-size-distribution. Prior to the review, the number and the position of these pits were measured with a light scattering inspection tool ($SP1^{DLS}$ of KLA-Tencor). The measured size distribution of these pits is represented in Figure 2.10 (b) with superimposed the simulated distribution. Figure 2.10 (a) shows simulations with three slightly different temperature gradients, illustrating the strong impact of the temperature gradient on the void size/density distribution.

2.6. Conclusions

In general, there is a good semi-quantitative understanding of grown-in defects in Ge. Dislocation control during Cz growth of germanium crystals is well established for up to 200 mm diameter crystals. Dislocation-free 300 mm crystal growth has been demonstrated.

During crystal growth both self- and dopant diffusion are mediated by vacancies, in contrast to the situation in silicon, where interstitial, vacancy and mixed behavior can be observed. The insight in the properties of the intrinsic point defects in germanium is less well developed especially for the self-interstitial. This is to a large extent due to a lack of experimental data. The study of the void defects (vacancy clusters) in germanium and their dependence on crystal pulling conditions and thermal treatments of wafers in combination with process simulation will allow to obtain more information on the thermodynamic parameters of the vacancy near the melt temperature. As for silicon, an important problem is that it is not clear whether one can extrapolate the high-temperature intrinsic point defect data to temperatures typically used for dopant activation and diffusion during device processing. Another phenomenon that will require substantial further work is the non-equilibrium dopant diffusion (enhanced diffusion). Some evidence points to vacancy-assisted diffusion enhancement at lower temperatures (300–500°C).

Acknowledgements

Part of this work was supported by the Institute for the Promotion of Innovation by Science and Technology in Flanders (IWT-Vlaanderen) and by the Science Foundation of Flanders (FWO-Vlaanderen).

References

[1] J. Vanhellemont, O. De Gryse, S. Hens, P. Vanmeerbeek, D. Poelman, P. Clauws, E. Simoen, C. Claeys, I. Romandic, A. Theuwis, G. Raskin, H. Vercammen and P. Mijlemans, *Defect Diffus. Forum* **230–232** (2004) 149.

[2] T.V. Mashovets, *Sov. Phys. Semicond.* **16** (1982) 1.

[3] N. Bernstein, M.J. Mehl and D.A. Papaconstantopoulos, *Phys. Rev. B* **66** (2002) 075212.

[4] A.J.R. da Silva, A. Janotti, A. Fazzio, R.J. Baierle and R. Motta, *Phys. Rev. B* **62** (2000) 9903.

[5] F.H. Stillinger and T.A. Weber, *Phys. Rev. B* **31** (1985) 5262.

[6] H.R. Schober, *Phys. Rev. B* **39** (1989) 13013.

[7] T. Sinno, Ph.D. Thesis, Massachusetts Institute of Technology, Cambridge, MA, 1998.

[8] A. Fazzio, A. Janotti, A.J.R. da Silva and R. Motta, *Phys. Rev. B* **61** (2000) R2401.

[9] W. Frank and N.A. Stolwijk, *Mater. Sci. Forum* **15–18** (1987) 369.

[10] H.D. Fuchs, W. Walukiewicz, E.E. Haller, W. Dondl, R. Schorer, G. Abstreiter, A.I. Rudnev, A.V. Tikhomirov and V.I. Ozhogin, *Phys. Rev. B* **51** (1995) 16 817.

[11] P. Śpiewak, K.J. Kurzydłowski, J. Vanhellemont, P. Clauws, P. Wabińnski, K. Młynarczyk, I. Romandic and A. Theuwis, *Mater. Sci. Semiconductor Processing* **9** (2006).

[12] K. Ding and H.C. Andersen, *Phys. Rev. B* **34** (1986) 6987.

[13] K. Nordlund, M. Ghaly, R.S. Averback, M. Caturla, T. Diaz de la Rubia and J. Tarus, *Phys. Rev. B* **57** (1978) 7556.

[14] W. Yu, Z.Q. Wang and D. Stroud, *Phys. Rev. B* **54** (1996) 13946.

[15] A. Seeger and K.P. Chik, *Phys. Stat. Sol.* **19** (1968) 455.

[16] A. Giese, N.A. Stolwijk and H. Bracht, *Appl. Phys. Lett.* **77** (2000) 642.

[17] S. Mayburg, *Phys. Rev.* **95** (1954) 38.

[18] R.A. Logan, *Phys. Rev.* **101** (1956) 1455.

[19] H. Letaw, W.M. Portnoy and L. Slifkin, *Phys. Rev.* **102** (1956) 636.

[20] A. Hiraki, *J. Phys. Soc. Jap.* **21** (1966) 34.

[21] J.A. Van Vechten, *Phys. Rev. B* **10** (1974) 1482.

[22] L.F. Konorova, *Soviet Phys. –Solid State* **10** (1969) 2233.

[23] A.G. Tweet, *Phys. Rev.* **106** (1957) 221.

[24] V.V. Voronkov, *J. Cryst. Growth* **59** (1982) 625.

[25] R. Falster, V.V. Voronkov and F. Quast, *Phys. Stat. Sol. (B)* **222** (2000) 219.

[26] V.V. Voronkov and R. Falster, *J. Appl. Phys.* **86** (1999) 5976.

[27] http://substrates.umicore.com

[28] E.E. Haller, W.L. Hansen and F.S. Goulding, *Adv. Phys.* **30** (1981) 93.

[29] E.E. Haller, W.L. Hansen, P. Luke, R. McMurray and B. Jarrett, *IEEE Trans. Nucl. Sci.* **29** (1982) 745.

[30] W.L. Hansen, E.E. Haller and P.N. Luke, *IEEE Trans. Nucl. Sci.* **29** (1982) 738.

[31] W.D. Edwards, *J. Electrochem. Soc.* **115** (1968) 753.

[32] P. Clauws, *Mater. Sci. Eng. B* **36** (1996) 213.

[33] R.I. Scace and G.A. Slack, *J. Chem. Phys.* **30** (1959) 1551.

[34] A. van Wieringen and N. Warmoltz, *Physica* **22** (1956) 849.

[35] R.C. Frank and J.E. Thomas, Jr., *J. Phys. Chem. Solid.* **16** (1960) 144.

[36] R.N. Hall, *IEEE Trans. Nucl. Sci.* **31** (1984) 320.

[37] O. De Gryse, P. Vanmeerbeek, J. Vanhellemont and P. Clauws, *Physica B* **376–377** (2006) 113.

[38] J. Vanhellemont, S. Hens, J. Lauwaert, O. De Gryse, P. Vanmeerbeek, D. Poelman, P. Śpiewak, I. Romandic, A. Theuwis and P. Clauws, *Solid State Phen.* **108–109** (2005) 683.

[39] P. Penning and G. De Wind, *Physica* **25** (1959) 765.

[40] A.B. Campbell, J.B. Mitchell, J. Shewchun, D.A. Thompson and J.A. Davies, *Can. J. Phys.* **53** (1975) 303.

[41] F. Berg Rasmussen, R. Jones and S. Öberg, *Phys. Rev. B* **50** (1994) 4378.

[42] H.J. Stein, *Appl. Phys. Lett.* **52** (1988) 153.

[43] F. Berg Rasmussen and B. Bech Nielsen, *Mat. Sci. Eng. B* **36** (1996) 241.

[44] W. Kaiser and C.D. Thurmond, *J. Appl. Phys.* **32** (1961) 115.

[45] J.R. Dale and J.C. Brice, *Solid-State Electron* **3** (1961) 105.

[46] R.N. Hall, *IEEE Trans. Nucl. Sci.* **19** (1972) 266.

[47] R.D. Westbrook, *Nucl. Instrum. Meth.* **108** (1973) 335.

[48] E.E. Haller, P.P. Li, G.S. Hubbard and W.L. Hansen, *IEEE Trans. Nucl. Sci.* **26** (1979) 265.

[49] K. Sueoka and J. Vanhellemont, *Mater. Sci. Semiconductor Processing* **9** (2006) 494.

[50] N. Van den Bogaert and F. Dupret, *J. Cryst. Growth* **171** (1997) 65, 77.

[51] G. Müller, *J. Cryst. Growth* **237–239** (2002) 1628.

[52] G. Müller and J. Friedrich, *J. Cryst. Growth* **266** (2004) 1.

[53] B. Depuydt, P. Boone, P. Union, P. Muys, D. Vyncke and C. Goessens, *SPIE* **3098** (1997) 559.

[54] K. Mlynarczyk, private communication.

[55] L.J. Bruner and R.W. Keyes, *Phys. Rev. Lett.* **7** (1961) 55.

[56] http://www.ioffe.rssi.ru/SVA/NSM/Semicond/Ge/mechanic.html and http://www.ioffe.rssi.ru/SVA/NSM/Semicond/Si/mechanic.html

[57] Yu. A. Burenko, S.P. Nikanorov and A.V. Stepanov, *Sov. Phys. – Solid State* **12** (1971) 1940.

[58] B.M. Mecs and A.S. Nowick, *Appl. Phys. Lett.* **8** (1966) 75.

[59] H. Kayano, K. Kamagaki and S. Koda, *J. Phys. Soc. Jpn.* **23** (1967) 1189.

[60] H.-J. Möller and J. Buchholz, *Phys. Stat. Sol. (A)* **20** (1973) 545.

[61] T. Kruml, D. Calliard, C. Dupas and J.-L. Martin, *J. Phys. Condens. Matter* **14** (2002), 12897.

[62] C. Dupas, PhD thesis 2853, École Polytechnique Fédérale de Lausanne (2003).

[63] K. Ohori and K. Sumino, *Phys. Stat. Sol. (A)* **14** (1972) 489.

[64] M. Schweizer, S.D. Cobb, M.P. Volz, J. Szoke and F.R. Szofran, *J. Cryst. Growth* **235** (2002) 161.

[65] E. Billig, *Proc. Royal. Soc. A* **235** (1956) 37.

[66] D.C. Bennett and B. Sawyer, *Bell Syst. Techn. J.* **35** (1956) 637.

[67] I.G. Cressell and J.A. Powell, *Prog. Semicond.* **2** (1957) 139.

[68] P. Penning, *Philips Res. Repts.* **13** (1958) 79.

[69] F.D. Rosi, *RCA Rev.* **19** (1958) 349.

[70] A.D. Belyaev, V.N. Vasilevskaya and E.G. Miselyuk, *Sov. Phys. – Solid State* **2** (1960) 208.

[71] R.S. Wagner, *J. Appl. Phys.* **29** (1958) 1679.

[72] A.J. Goss, K.E. Benson and W.G. Pfann, *Acta Metall.* **4** (1956) 332.

[73] V.G. Alekseeva and P.G. Eliseev, *Sov. Phys. – Solid State* **1** (1959) 1195.

[74] G.E. Brock and C.F. Aliotta, *IBM J.* **6** (1962) 372.

[75] J. Mitchell, *J. Appl. Phys.* **33** (1962) 406.

[76] J. Hornstra, *J. Phys. Chem. Sol.* **5** (1958) 129.

[77] M. Van Sande, L. Van Goethem, L. De Laet and H. Guislain, *Appl. Phys. A* **40** (1986) 257.

[78] F.L. Vogel, W.G. Pfann, H.E. Corey and K. Thomas, *Phys. Rev.* **90** (1953) 489.

[79] G.S. Hubbard and E.E. Haller, *J. Electron. Mater.* **9** (1980) 51.

[80] E. Simoen, P. Clauws and J. Vennik, *Solid State Commun.* **54** (1985) 1025.

[81] H. Widmer and G.R. Gunther-Mohr, *Helv. Phys. Acta* **34** (1961) 635.

[82] R.N. Ghoshtagore, *Phys. Rev. Lett.* **16** (1966) 890.

[83] A.G. Tweet, *J. Appl. Phys.* **30** (1959) 2002.

[84] R. Falster and V.V. Voronkov, *Mater. Sci. Eng. B* **73** (2000) 87.

[85] J. Vanhellemont, E. Dornberger, D. Gräf, J. Esfandyari, U. Lambert, R. Schmolke, W. von Ammon and P. Wagner, *Proc. Kazusa Akademia Park Forum on The Science and Technology of Silicon Materials*. Kazusa Akademia Park, Chiba, Japan, 1997 p. 173.

[86] E. Dornberger, D. Gräf, M. Suhren, U. Lambert, P. Wagner, F. Dupret and W. von Ammon, *J. Cryst. Growth* **180** (1997) 343.

[87] T. Sinno and R.A. Brown, *J. Electrochem. Soc.* **146** (1999) 2300.

[88] S. Hens, J. Vanhellemont, D. Poelman, P. Clauws, I. Romandic, F. Holsteyns and J. Van Steenbergen, *Appl. Phys. Lett.* **87** (2005) 061915.

CHAPTER 3

Diffusion and Solubility of Dopants in Germanium

Eddy Simoen and Cor Claeys

3.1. Introduction

The focus of this chapter is on the diffusion and equilibrium (maximum) solubility of dopants in germanium. Overviews on the early results, up to the mid-eighties, can be found elsewhere [1–3]. The relevant information will be recalled and an update of the recent literature will be made. The chapter is built up as follows: Section 3.2 gives a general introduction to diffusion in semiconductors. In Section 3.3, the properties of intrinsic point defects (vacancies V and interstitials I) in germanium will be discussed. Section 3.4 is devoted to the diffusion of Group IV dopants in germanium, with particular emphasis on the Ge self-diffusion. This is followed by a discussion on the solubility of the main dopants in germanium. In Section 3.6, the diffusion of the common Group V donor and Group III acceptor dopants in Ge will be described, with particular focus on ion implantation as a dopant source. Finally, some general conclusions are drawn in Section 3.7. Whenever relevant, a comparison will be made with the behavior of dopants in silicon, where for obvious reasons much more data is available. This may be helpful in understanding possible chemical trends in the diffusion coefficient or solubility of impurities in Group IV semiconductors.

3.2. Diffusion in Semiconductors

Diffusion is the process of migration of atoms through a solid. It is usually a thermally driven process, which can be described by an Arrhenius or activation type of law, if diffusion occurs

Germanium-Based Technologies: From Materials to Devices. C. Claeys and E. Simoen (Editors)

at thermodynamic equilibrium. It means that the diffusion time at a specific temperature T should be sufficiently long to enable establishing an equilibrium density of intrinsic point defects in the lattice. This stands in contrast to non-equilibrium diffusion, like the transient-enhanced diffusion (TED) phenomenon, which is supported by a non-equilibrium density of point defects, created by ion-implantation. As will be shown below, there is some evidence of anomalously fast dopant diffusion in Ge, e.g. radiation- or concentration-enhanced diffusion, but in this section, the theory will be outlined for the equilibrium thermal diffusion.

3.2.1. Diffusion mechanisms

The diffusion coefficient D (in $cm^2 s^{-1}$), which is the factor of proportionality relating the flow density (flux) of diffusing atoms with the concentration gradient of atoms, is usually given by the empirical law:

$$D = D_0 \exp(-H/kT) \tag{3.1}$$

In Equation (3.1), k is the Boltzmann constant and H the activation enthalpy of diffusion.[1] The pre-exponential factor D_0 can be considered as a jump frequency factor (in $cm^2 s^{-1}$).

There are different types of diffusion mechanisms [1–3], which depend on whether the diffusing atom resides in a lattice (substitutional) or interstitial position. A substitutional atom can migrate by exchange with a direct neighbor, while an interstitial can jump from one interstitial site to a nearest neighbor one. In this case, the activation enthalpy H is mainly composed of the migration enthalpy H_m. This enthalpy is required to enable the crossing of the barrier to migration and is illustrated schematically in Figure 3.1.

Certain metals in Ge diffuse through an interstitial mechanism (e.g. Cu, Li, Ni) with a low activation energy ~0.3 eV, so that they have very high diffusivities at low T. The direct exchange mechanism for substitutional atoms is generally very unlikely. However, most diffusion processes are mediated by intrinsic point defects: vacancies ("V"s) and interstitials ("I"s).

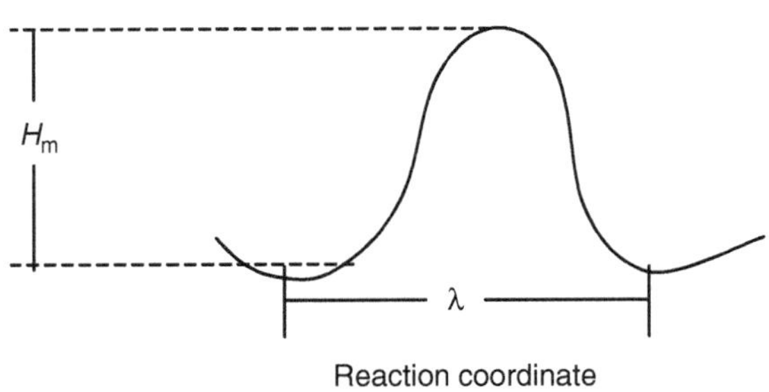

Figure 3.1. Standard free enthalpy versus system configuration as an atom jumps across a barrier to a new position. H_m is the barrier to migration.

[1] Often this is called the activation energy of the diffusion coefficient although it is thermodynamically more correct to term this the enthalpy.

Figure 3.2 illustrates the vacancy mechanisms for diffusion: a diffusing atom jumps into a neighboring vacant lattice site. It is generally believed that this mechanism is the dominant diffusion mechanism in germanium, for most common dopants and also for self-diffusion [1–3].

Another important mechanism, that has been proposed for the diffusion of Cu in Ge, which is the most common transition metal contaminant in as-grown and annealed germanium, is the dissociative or Frank–Turnbull mechanism [4], represented by:

$$X_i + V \Leftrightarrow X_s \tag{3.2}$$

Equation (3.2) means that a fast diffusing (mobile) interstitial atom of species X (X_i) recombines with a vacancy, resulting in a less mobile substitutional atom X_s. It is clear that vacancies play an important role in this diffusion mechanism. Conversely, studying the diffusion or precipitation of Cu in Ge may yield information about the properties of vacancies [1].

The diffusion of most dopants in silicon occurs through yet another mechanism, namely, the kick-out mechanism, according to the reaction:

$$I + X_s \Leftrightarrow X_i \tag{3.3}$$

In this case, the interstitials take over the role of the vacancies in Equation (3.2) to kick out an immobile substitutional impurity atom (X_s) from its lattice site into a highly (or more) mobile interstitial position. If a non-equilibrium density of interstitials is present, e.g. agglomerated at the end-of-range damage region after an ion implantation, mechanism (3.3) can lead to a strong enhancement of the dopant diffusivity. This is basically what happens during TED of boron in Si. Once the excess interstitials are consumed or have disappeared by surface recombination, the TED stops and normal (slow) equilibrium diffusion takes over.

3.2.2. *Self-diffusion*

In general, an atom can migrate through different reaction paths, using both interstitial and vacancy-related mechanisms. This is for example the case for self- and Group IV (Ge, Sn, . . .)

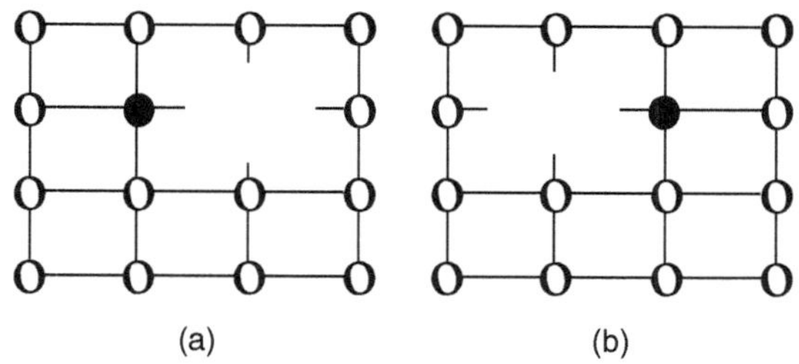

Figure 3.2. Vacancy mechanism for self-diffusion: (a) The tagged germanium atom (dark color) moves by jumping into the vacancy to its right. (b) After the jump, it has moved by one nearest-neighbor distance.

diffusion in silicon [5]. In the latter case, the diffusion coefficient is composed of at least two contributions, which can be written as [5]:

$$D = D_{0I}\exp(-H_I/kT) + D_{0V}\exp(-H_V/k\mathrm{T}) \tag{3.4}$$

The activation enthalpies in Equations (3.1) and (3.4) are composed of two terms: an enthalpy of formation (H_f) of the respective point defect and an enthalpy of migration (H_m). One can thus write [1,2,5]:

$$H_{I,V} = H_{fI,V} + H_{mI,V} \tag{3.5}$$

The fact that a single Arrhenius law is obtained empirically over a certain temperature interval indicates that one of the mechanisms (V or I) is dominant. In the case of self-diffusion in silicon, the fractional vacancy contribution is 0.4–0.5 in the temperature range 800–1100°C [6].

In germanium, on the other hand, it has been firmly established that self-diffusion is dominated by the monovacancy mechanism [1–3]. In that case, the diffusion coefficient or diffusivity is given by the product of the diffusion coefficient of the monovacancy D_V times its equilibrium concentration relative to the atomic density in Ge, C_V^{eq} [1]. In other words:

$$D_{Ge} = f_V D_V C_V^{eq}(T) \tag{3.6}$$

In Equation (3.6), f_V is the correlation factor, equal to ½ for the (mono)vacancy mechanism in the diamond (Ge) lattice. D_V is given by:

$$D_V = a^2\nu_V/8 \tag{3.7}$$

with a the lattice parameter (a = 5.65 Å) and ν_V is the jump frequency of an atom adjacent to a vacancy into that vacancy. The latter obeys [1]:

$$\nu_V = \nu_V^0\exp(S_{mV}/k)\exp(-H_{mV}/kT) \tag{3.8a}$$

with S_{mV} and H_{mV} the migration entropy and enthalpy of a single vacancy. The frequency factor ν_V^0 has been estimated as (M is the mass of the diffusing atoms):

$$\nu_V^0 = \sqrt{\frac{8H_{mV}}{3Ma^2}} \tag{3.8b}$$

The equilibrium vacancy concentration is given by [1]:

$$C_V^{eq}(T) = \exp(S_{fV}/k)\exp(-H_{fV}/kT) \tag{3.9}$$

In Equation (3.9), S_{fV} and H_{fV} are the entropy and enthalpy of formation of a single vacancy. Equation (3.9) implies that the equilibrium vacancy concentration is only determined by the diffusion temperature. Combining Equations (3.6)–(3.9) one arrives at [1]:

$$D_{Ge} = 1/16\, a^2 \nu^0_V \exp([S_{fV} + S_{mV}]/k)\exp(-[H_{fV} + H_{mV}]/kT) \quad (3.10)$$

Comparing Equation (3.10) with (3.1), one arrives at a physical meaning for the empirical parameters D_0 and H in the case of self-diffusion. One finds that [1]:

$$D_0 = a^2/16\, \nu^0_V \exp([S_{fV} + S_{mV}]/k) \quad (3.11a)$$

$$H = H_{fV} + H_{mV} \quad (3.11b)$$

However, the monovacancy in Ge may occur in different charge states. Assuming for the moment, that V can exist in the neutral (suffix 0), singly positive (donor, suffix +) and singly negative (acceptor, suffix −) state, the total equilibrium concentration of vacancies will be different from Equation (3.9) and becomes a function of the Fermi level in the material [1–3]. In other words, the self-diffusion coefficient is not only a function of T but becomes also dependent on the Fermi level (= doping density) for highly doped, extrinsic material. In the case that one has to consider different charge states of the vacancy, Equation (3.6) becomes [2]:

$$D_{Ge} = 1/2\left\{D^0_V C^{eq0}_V + D^-_V C^{eq-}_V + D^+_V C^{eq+}_V\right\} \quad (3.12)$$

In Equation (3.12), C^{eq0}_V is a function of T only, while C^{eq-}_V and C^{eq+}_V are also determined by Fermi–Dirac statistics and thus by the doping density. It can be readily shown that Equation (3.12) transforms into [2]:

$$D_{Ge} = D^0_{Ge} + D^-_{Ge}(n/n_i) + D^+_{Ge}(n_i/n) \quad (3.13)$$

with n the free electron concentration and n_i the intrinsic carrier concentration. In Equation (3.13), the following holds:

$$D^x_{Ge} = 1/2 D^x_V C^{eqx,i}_V \quad (x = 0, -, +) \quad (3.14)$$

with $C^{eqx,i}_V$ the equilibrium charged vacancy concentrations in intrinsic (undoped) material. For the neutral state: $C^{eq0}_V = C^{eq0,i}_V$, a function of temperature only.

In the case of multiple positive and negative charge states, formula (3.13) can be generalized to [2]:

$$D_{Ge} = D^0_{Ge} + \sum_{r=1} D^{r-}_{Ge}\left(\frac{n}{n_i}\right)^r + \sum_{s=1} D^{s+}_{Ge}\left(\frac{n_i}{n}\right)^s \quad (3.15)$$

From Equation (3.15), one can clearly derive that in extrinsic n-type material ($n \gg n_i$) the self-diffusion of germanium will be enhanced, due to the contribution of negatively charged vacancies. The same should in principle hold for p-Ge as well, if the vacancy donor state exists. As will be seen below, the experimental self-diffusion coefficient can be explained by considering neutral and negative charge states of V only. This means:

$$D_{\mathrm{Ge}} = D_{\mathrm{Ge}}^{0} + \sum_{r=1} D_{\mathrm{Ge}}^{r-} \left(\frac{n}{n_{\mathrm{i}}} \right)^{r} \tag{3.16}$$

Equation (3.16) indicates that the self-diffusion will be enhanced in heavily n-type doped Ge, compared with intrinsic (low doped) material, while it is expected to be lowest in heavily p-type doped Ge [1–3].

3.3. Intrinsic Point Defects in Germanium

It is clear from the previous section that intrinsic point defects are essential for the diffusion mechanisms in semiconductor materials. In order to model D_{Ge}, one needs a value for the equilibrium concentration and for the migration Gibbs free energy $G_{\mathrm{mV}} = H_{\mathrm{mV}} - TS_{\mathrm{mV}}$ of vacancies. Moreover, knowledge is required on the available charge states (which depends on the doping concentration of the material) and energy levels. This explains the large efforts, which have been spent in the early days in trying to determine these parameters (see also Chapter 2).

In the literature, three basic approaches have been followed to study intrinsic point defects in Ge [1]:

1. Quenching from high temperatures followed by annealing at lower temperatures.
2. Plastic deformation.
3. Irradiation at low temperatures, in order to create vacancy–interstitial pairs and simple point defects.

In addition, the study of the diffusion and/or precipitation of certain fast dopants (Cu in Ge) can provide indirect information on intrinsic points defects [1]. Finally, also theoretical calculations have shed more light on the electrical activity (charge states, energy levels) of the intrinsic point defects.

3.3.1. Quenching

The quenching technique has been very successful in the assessment of vacancy properties in metals. It has also been attempted in Ge [7–17]. For an overview of the early results (see e.g. Ref. [1]). The basic idea behind it is to heat up the material close to the melting point T_{m} for a sufficiently long time to establish the thermal equilibrium concentration of the intrinsic defects. Next, the sample is rapidly cooled to low temperature, where the mobility of the intrinsic point defects is low enough to freeze their concentration at the high-temperature value, provided negligible diffusion and precipitation can take place during the quench. If this is done for different temperatures, one can derive C_V and, hence according to Equation (3.9) S_{fV} and H_{fV}. If one, on the other hand, anneals the quenched samples at temperatures well below T_m, the quenched-in super-saturation of intrinsic defects will diffuse out, with activation enthalpy H_{mV} and entropy S_{mV}.

Initial results demonstrated that quenching of n-type Ge above 800°C resulted in p-type material, pointing to the creation of quenched-in acceptors. Annealing at 500°C recovered the original n-type, indicating the out-diffusion/annihilation of the quenched-in thermal acceptors. It was first thought that these acceptors were related to monovacancies. However, it was soon realized that the early experiments suffered from copper contamination effects [1], whereby one should take into account that substitutional copper (Cu_s) is a triple acceptor in Ge. Different methods were proposed to prevent copper contamination in order to study the "pure" thermal defects. In spite of these efforts, it is now clear that quenching does not yield information on monovacancies, for the following reason [1]: the diffusion coefficient of "V"s is so high that one cannot avoid their clustering during cooling from the quenching temperature. Therefore, properties originally ascribed to vacancies probably correspond to di- or multivacancy complexes.

Nevertheless, some information on the self-diffusion in Ge could be obtained indirectly, by studying the precipitation rate of copper [18]. Based on the dissociative mechanism of Equation (3.2), one can show that the effective diffusion coefficient for vacancies D_{eff} equals [18]:

$$D_{eff} = D_{Ge}/C_{Cu}^{eq} \tag{3.17}$$

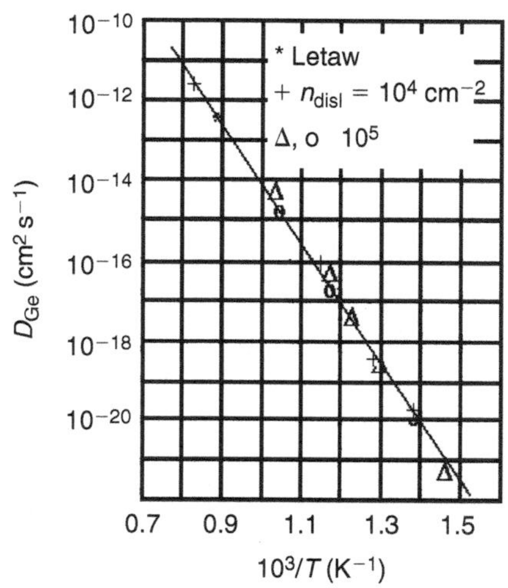

Figure 3.3. The coefficient of self-diffusion in germanium plotted versus the reciprocal of the absolute temperature. The data of Letaw *et al.* [19] have been obtained by diffusion of a radio-active germanium isotope. The other points were derived from the rate of precipitation of copper in germanium as measured by Tweet [20] (after Penning [18]).

with C_{Cu}^{eq} the equilibrium concentration of substitutional copper in Ge. On the other hand, D_{eff} can be derived from the decay constant τ of the vacancy concentration, which decays exponentially with annealing time, according to [18]:

$$1/\tau \pm b_0 n_{disl} D_{eff} \tag{3.18}$$

with b_0 a constant and n_{disl} the dislocation density in the material. Combining Equations (3.17) and (3.18) yields the self-diffusion coefficient of Figure 3.3 [18] which is described by:

$$D_{Ge} = 6.2\ \exp(-2.94\text{eV}/kT)\ \ (\text{cm}^2\text{s}^{-1}) \tag{3.19}$$

in good agreement with earlier reports [19]. The advantage of this technique is that it yields information on self-diffusion in Ge at temperatures well below T_m, which is otherwise much more difficult, given the slow nature of the process (low D_{Ge}) [1].

3.3.2. Irradiation

The basic idea is that by interaction between lattice atoms and an energetic particle, ion or photon, one can displace the former to create a vacancy, while the recoiled lattice atom comes at rest in an interstitial position. The combination of a radiation induced V and I is sometimes called a Frenkel pair. For that purpose, high-energy electron and γ irradiations are most suitable, since they lead to simple isolated displacements (no cluster damage). In addition, it turns out that the intrinsic point defects are highly mobile at room temperature and above, so that one should preferably perform the irradiations at low temperature (4.2 or 77 K) to immobilize the created damage. An overview of such experiments can be found in Chapter 3 of Ref. [21] or Chapter 7 of this book. It should be remarked that the lack of information about the electronic properties of the vacancy in Ge is to some extent related to the experimental difficulty in applying electron spin resonance (ESR) spectroscopy to irradiated germanium, in contrast to its successful application in silicon [1,21].

Here, attention will be drawn to some recent experimental results, which are claimed to produce for the first time clear information on the properties of vacancies and interstitials in germanium [22]. For that purpose, perturbed angular correlation spectroscopy (PACS) [23] is applied to low-temperature electron-irradiated Ge. The aim is to use a radioactive probe atom (^{111}In) in the immediate vicinity of the intrinsic point defects. Annealing following a low-temperature irradiation should enable the capture of a vacancy or interstitial by the probe atom, whereby it is assumed that the intrinsic point defects in Ge are immobile at $T_{irr} = 77$ K. Using this technique, it was observed that around 200 K, defects become trapped at the ^{111}In probes, which were stable up to ~400 K [22,24,25]. Two types of trapped defects have been identified, one occurring particular in n-type Ge and the other one present in both types.

From the temperature dependence of the ^{111}In atom fraction with a captured defect in its vicinity and the different behavior in n- or p-type Ge, the energy level scheme of Figure 3.4 has been deduced [22,23,25]. As shown in Figure 3.4, the vacancy in Ge has at least a single acceptor level at 0.2 eV above the valence band.

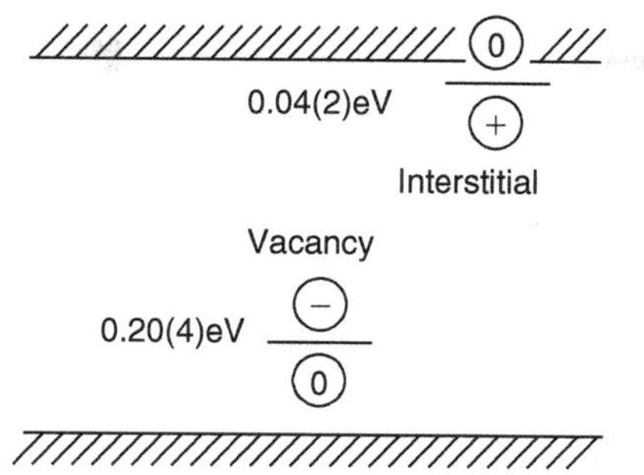

Figure 3.4. Electrical levels of vacancy and self-interstitial in Ge as deduced from PACS. For the self-interstitial, two options are possible: a donor or acceptor (after Hässlein *et al.* [22]). According to more recent interpretations, the acceptor nature for the self-interstitial seems to be more likely [24,25].

Some support for the level scheme in Figure 3.4 has been supplied by quantum mechanical calculations. Density functional theory (DFT) calculations [26] have pointed out that the vacancy in Ge is not a negative-U center, in contrast to the case of silicon. For the single acceptor level, Fazzio *et al.* find a value of 0.37 eV above the top of the valence band [26]. This is significantly higher than the $E_V + 0.2$ eV level in Figure 3.4, but close to the $E_V + 0.33$ eV level found in deep level transient spectroscopy (DLTS) and ascribed to the In–V pair [25]. The latter is believed to have a similar split vacancy structure as the Sn–V pair in silicon. The calculated formation energy for the singly negative vacancy in Ge was found to be 2.3 eV, which is of the same order as earlier calculations [27].

For the Ge self-interstitial, an acceptor level at $E_C - 0.12$ eV is found from DFT, while a donor level at $E_V + 0.11$–0.16 eV could explain the PACS results for p-type Ge [28].

3.4. Self- and Group IV Diffusion in Germanium and Silicon

Self-diffusion in Ge and Si is the slowest diffusion process in these materials. At the same time, it is a very important phenomenon as it provides direct evidence for the existence of an equilibrium concentration of intrinsic point defects. Therefore, its investigation should provide more insight in the dominant diffusion mechanism(s).

The fact that D_{Ge} is low implies that lengthy diffusion experiments at high temperatures, close to T_m, are generally required. Different experimental approaches have been followed in the past. One technique is to make use of radioactive tracer atoms and to determine the in-diffused profile for a gradually thinned sample by measuring the radioactivity in the removed layer. Another possibility is the use of different isotopes in combination with secondary ion mass spectrometry (SIMS). As outlined above, the study of the doping dependence of D_{Ge} should provide more insight in the dominant diffusion mechanism and the possible charge states involved. The same is expected when the self-diffusion is studied under hydrostatic pressure. The outcome of the different approaches will be briefly summarized here.

3.4.1. Radioactive tracer experiments

In Ge, a number of radioisotopes with a sufficiently long half-life are candidates for radiotracer diffusion studies. Most often, ^{71}Ge is produced from the stable ^{70}Ge isotope by neutron capture and γ-emission ((n,γ) reaction) [19,29–32]. It decays by electron capture to a stable ^{71}Ga atom, with a half-life of 11.2 days. Another candidate is ^{77}Ge, produced by a (n,γ) reaction with a ^{76}Ge atom [19,30]. The activity is 12 h, where ^{77}Ge decays to ^{77}As by electron (β) emission. The latter isotope is also radioactive, decaying in about 2 days by electron (β) emission to ^{77}Se, which undergoes a short-lived isomeric transition.

Generally, these radioisotopes are in-diffused from the surface for some time, where after a de-layering (sectioning) is performed and the radioactivity of the removed material measured, to determine the diffusion profile [19,29,30]. Since the breakthrough of SIMS, the latter technique is utilized to determine the profile of the diffused (radio)-isotopes [31–33]. The high(er) sensitivity of SIMS allows deriving more accurate profiles and diffusion coefficients at lower diffusion temperatures. Figure 3.5 summarizes the obtained D_{Ge} from radioactive tracer studies [19,29–32].

According to Werner *et al.* [32], the Arrhenius law of Figure 3.5 can be described by:

$$D_{Ge} = 13.6 \exp(-3.094\,\text{eV}/kT) \quad (\text{cm}^2\text{s}^{-1}) \tag{3.20}$$

This is in reasonable agreement with the result obtained from Cu precipitation studies and represented by Equation (3.19). Overall, one obtains a $D_0 \sim 10\,\text{cm}^2\text{s}^{-1}$ and an activation enthalpy of ~ 3 eV. The most recent ^{71}Ge radiotracer studies for self-diffusion yielded an even larger D_0 of $81\,\text{cm}^2\text{s}^{-1}$ for a $H = 3.14$ eV [34].

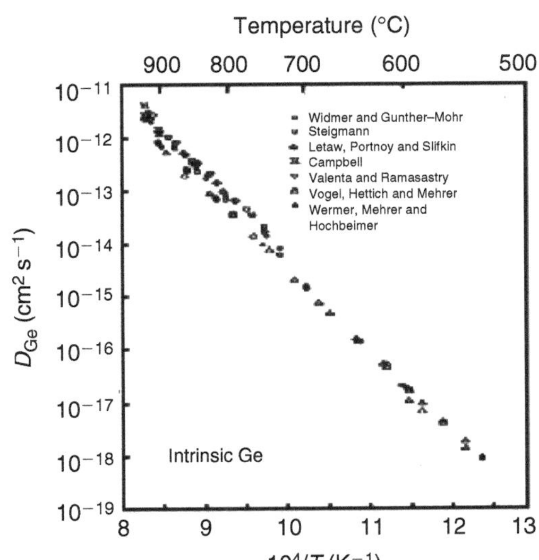

Figure 3.5. Arrhenius plot of self-diffusion in intrinsic Ge at ambient pressure (after Werner *et al.* [32]).

From D_0, one arrives at a rather high value for the entropy term $S_{mV} + S_{fV} \sim 10\,k$, which is significantly larger than for the vacancy diffusion mechanism in face centered cubic (fcc) metals (S ~ 2–4 k) [1]. This has led Seeger and Chik to propose that self-diffusion in Ge is through an extended vacancy mechanism, explaining the much slower self-diffusion and a possible temperature dependence of the activation enthalpy [1]. Later theoretical studies have criticized this view strongly [35] and ruled out diffusion through extended vacancies. It was demonstrated that the lattice distortion around the Ge vacancy could explain the rather high entropy value. To date, the interpretation of the large entropy for self-diffusion in Ge is still a matter of debate [34] but there is a general consensus that the vacancy mechanism is the self-diffusion vehicle in germanium. The same goes for dopant impurity diffusion, as will be seen below.

3.4.2. Isotope effects and Group IV (Si;Sn) diffusion in Ge

Besides radioisotopes, one can also consider studying the diffusion behavior of stable isotopes with a different mass [33], relying on the mass separation capabilities of SIMS. The isotope effect on the diffusion coefficient is given by [1,33]:

$$\frac{\left[\dfrac{D_{\mathrm{I}}}{D_{\mathrm{II}}}\right] - 1}{\left[\dfrac{M_{\mathrm{II}}}{M_{\mathrm{I}}}\right]^{1/2} - 1} = f_{\mathrm{V}}\Delta K \tag{3.21}$$

with D_{I} and D_{II} the diffusion coefficient corresponding to the species with mass M_{I} and M_{II}, respectively. In Equation (3.21), f_{V} is the correlation factor and ΔK is the fraction of the kinetic energy which is associated with the atomic motion in the jump direction. In the work by Campbell [30], $f_{\mathrm{V}}\Delta K$ values in the range between 0.26 and 0.3 were found, which translates into a ratio of $D^{70}_{\mathrm{Ge}}/D^{74}_{\mathrm{Ge}}$ between 1.007 and 1.008. This small difference is below the detection limit of the SIMS technique employed [33]. As shown in Figure 3.6, a good agreement with previous work was found, based on the study of self-diffusion in isotopically engineered Ge heterostructures [33]. The corresponding fit yields:

$$D_{\mathrm{Ge}} = 12\,\exp(-3.0(5)\mathrm{eV}/kT)\ \ (\mathrm{cm^2 s^{-1}}) \tag{3.22}$$

It is clear that the isotope mass effect is rather small. A way to enhance this is by studying the diffusion properties of other Group IV impurities (Si, Sn, Pb) in Ge [36–38]. Since these impurities are electrically neutral, one expects at first sight that the same mechanism rules the diffusion behavior. The main difference should come from the difference in atom size (elastic interaction).

Figure 3.7 and Table 3.1 summarize the diffusion coefficients of Group IV elements in germanium [37]. The following trends have been found: the activation enthalpy in Table 3.1 follows the order $H_{\mathrm{Si}} < H_{\mathrm{Ge}} < H_{\mathrm{Sn}}$, while the same applies for the frequency factor: $D_{0\mathrm{Si}} < D_{0\mathrm{Ge}} < D_{0\mathrm{Sn}}$. It is concluded that the same vacancy diffusion mechanism operates for the Group IV elements

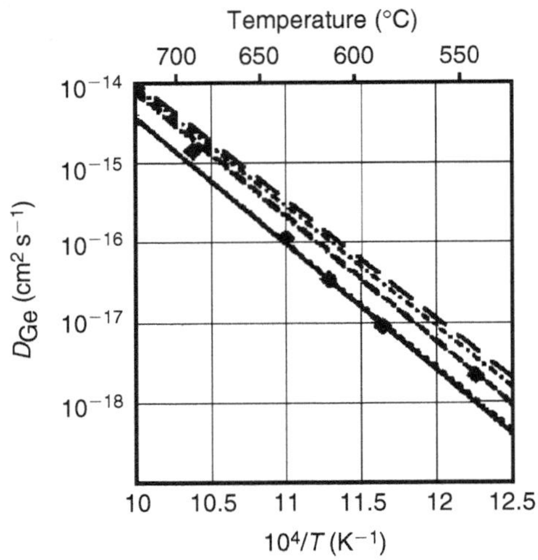

Figure 3.6. Arrhenius plot of the Ge self-diffusion coefficient as a function of temperature. The symbols correspond with experimental work, while the other lines refer to literature data (after Fuchs *et al.* [33]).

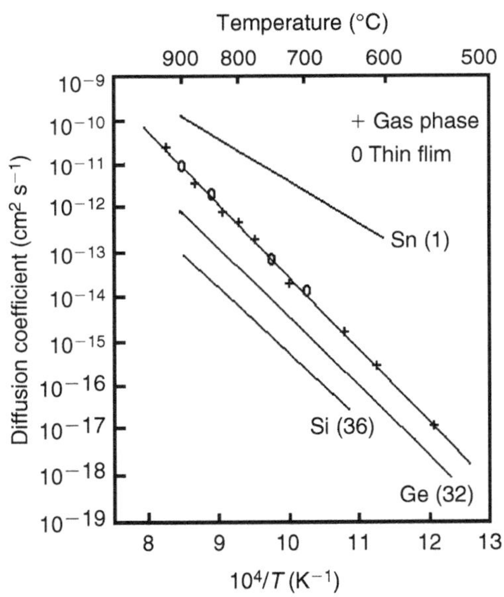

Figure 3.7. The diffusion coefficients of Sn in Ge as a function of temperature for the gas-phase samples (+) and thin-film samples (0), and a comparison with the literature data (after Friesel *et al.* [37]).

in Ge, whereby the diffusion coefficient for Si is even smaller than for Ge. In other words, the higher the atom size, the higher D is. At the same time, it was found that $D_{Sn}{}^{p} < D_{Sn}{}^{i} < D_{Sn}{}^{n}$ [37], which has also been observed for self-diffusion in highly n- and p-doped or intrinsic (i) Ge (see Figure 3.7). This again supports the vacancy mechanism for (self-)diffusion in Ge.

Table 3.1. Activation enthalpies, H, and pre-exponential factors, D_0, for diffusion in germanium.

Element	H (eV)	D_0 (cm²/s)	Technique
Al	3.45	1.3×10^3	SIMS
Ga	3.35	140	SIMS
In	3.67	1.8×10^4	SIMS
Ge	3.09	13.6	Radio tracer
Si	3.19	43	Radio tracer
Si	2.90	0.24	Resonance method
Sn	3.05	70	Radio tracer
Sn	3.26	840	SIMS

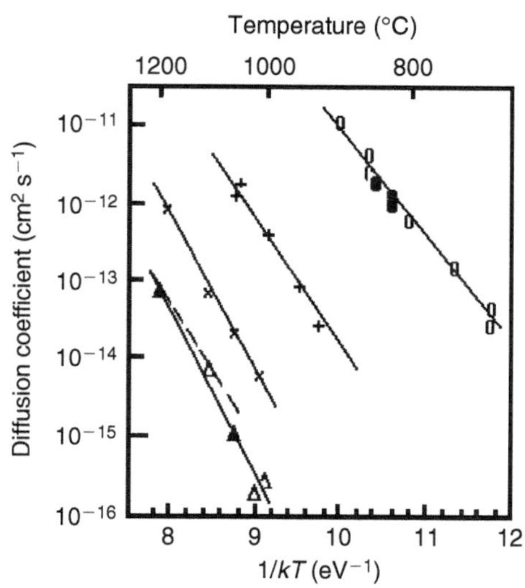

Figure 3.8. Diffusion coefficient of Sn as a function of temperature: (▲,△) is for Si, (x) is for $Si_{0.79}Ge_{0.21}$, (+) is for $Si_{0.47}Ge_{0.53}$, and (o) is for Ge. The dashed and dotted lines are literature values for Sn in Si and (●) is literature value for Sn in Ge [2] (after Kringhøj and Elliman [38]).

Table 3.2. Activation enthalpy and pre-exponential factor for the Si diffusion in various $Si_{1-x}Ge_x$ "alloys". The uncertainties given are statistical fitting errors only (after Kringhøj and Elliman [38]).

	Si	$Si_{0.79}Ge_{0.21}$	$Si_{0.47}Ge_{0.53}$	Ge
Activation enthalpy (eV)	4.91 ± 0.09	4.61 ± 0.08	3.88 ± 0.09	3.05 ± 0.08
Pre-exponential factor ($cm^2 s^{-1}$)	$5 \pm 4 \times 10^3$	$8 \pm 6 \times 10^3$	$8 \pm 7 \times 10^2$	$1.5 \pm 1.3 \times 10^2$

Another interesting trend is that the diffusion coefficient of Sn in Si, $Si_{1-x}Ge_x$ or Ge increases when moving to a higher Ge content, i.e. a larger lattice parameter [38]. This is shown in Figure 3.8 and Table 3.2 [38] and clearly demonstrates the impact of elastic interactions between the host and foreign diffusing atoms.

3.4.3. Doping and pressure effects

Equation (3.15) gives the general expression for the self-diffusion coefficient through neutral and charged vacancies. It means that the study of the doping dependence of D_{Ge} should yield information on the relevant charge states of the vacancy. Investigation of heavily n- and p-doped Ge has revealed that $D^{p}_{Ge} < D^{i}_{Ge}, D^{n}_{Ge}$ [29,32] similar as for the case of Sn diffusion in Ge [37]. This is illustrated by Figure 3.9 [32], where $n/n_i > 1$ means *n*-Ge and vice versa for *p*-Ge. Such behavior can be understood in the frame of the vacancy diffusion mechanism, assuming that neutral and single negative charged vacancies take part in the migration process. In that case, Equation (3.16) can be simplified to [1–3,29,32]:

$$D^{dop}_{Ge} = D^{0}_{Ge} + (n/n_i)D^{-}_{Ge} \tag{3.23a}$$

The relative contribution of charged defects to the extrinsic (doped) diffusivity then becomes [32]:

$$D^{dop}_{Ge}/D^{i}_{Ge} = D^{0}_{Ge}/D^{i}_{Ge} + (n/n_i)D^{-}_{Ge}/D^{i}_{Ge} \tag{3.23b}$$

which is represented in Figure 3.9.

Valuable information about the microscopic diffusion mechanism can also be gained by studying D as a function of the hydrostatic pressure P. From such investigations, one may hope to derive the activation volume of self-diffusion, according to [32,39]:

$$v_{Ge} = kT\left[-\frac{\partial \ln D_{Ge}}{\partial P} + \frac{\partial v_0}{\partial P}\right]_T \tag{3.24}$$

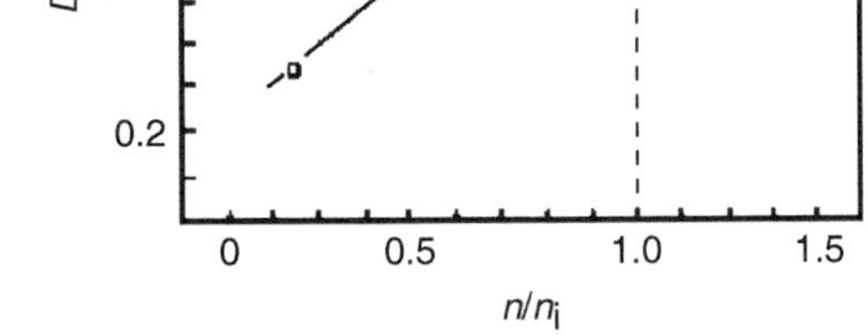

Figure 3.9. Doping dependence of the diffusion coefficient of ^{71}Ge in Ge at $T = 973$ K (after Werner *et al.* [32]).

For a diffusion mechanism mediated by thermally generated defects, the activation volume becomes [32]:

$$\nu_{\mathrm{Ge}} = \nu_{\mathrm{fV}} + \nu_{\mathrm{mV}} \tag{3.25}$$

In other words, it contains a term associated with the formation and the migration of the point defect. ν_{fV} should be considered as the volume change in the system upon formation of one defect in its standard state and ν_{mV} is the corresponding change when the defect reaches the saddle point in its migration path [39]. The second term in Equation (3.24) has been shown to be small, so that ν_{Ge} is approximately equal to the pressure dependence of the diffusivity at constant T. The resulting ν_{Ge} as a function of doping concentration is represented in Figure 3.10 [32] and turns out to be significantly smaller than for noble or fcc metals, which could point to a spread out or a relaxation of the Ge vacancy [1]. In addition, the doping dependence indicates that in Equation (3.24) both charge states of the vacancy have to be included, so that in practice a weighted average of the individual activation volumes ν_{Ge}^{0} and ν_{Ge}^{-} is obtained [32].

3.4.4. Diffusion of Ge in Si

As noted in Section 3.4.2 strain can also have an impact on the diffusivity of foreign atoms in a crystalline matrix. The results of Figures 3.7 and 3.8 suggest that a higher D is obtained for a larger Group IV atom in the germanium lattice, while, if one considers the same impurity (Sn, Figure 3.8), D increases by going to the material with the highest lattice parameter. Similar studies have also been performed for the diffusivity of Ge in $\mathrm{Si}_{1-x}\mathrm{Ge}_x$, with $0 \leqslant x \leqslant 1$ [5,34,40–43]. The same picture as for Sn diffusion emerges: the diffusivity of Ge increases with an increase in Ge content [5,34]. This seems to be also associated with the fractional vacancy contribution: in silicon, this increases from 0.4–0.5 for Si (800–1100°C), to 0.6–0.7 for Ge (1050°C) and to 1 for Sn. The corresponding effective activation enthalpy is found to reduce for increasing Ge content [41], as demonstrated in Figure 3.11a. The pre-exponential factor (entropy) behavior shows a turn/around at the 70/30 alloy region [41]. It could be interpreted as a change over from a vacancy-only diffusion mechanism to a mixed vacancy–interstitial mediated Ge migration.

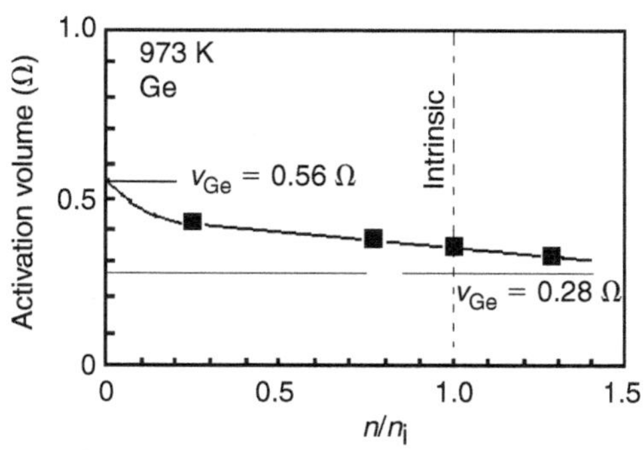

Figure 3.10. Activation volume of self-diffusion in Ge at 973 K as a function of n/n_i. Ω is the atomic volume in Ge (after Werner *et al.* [32]).

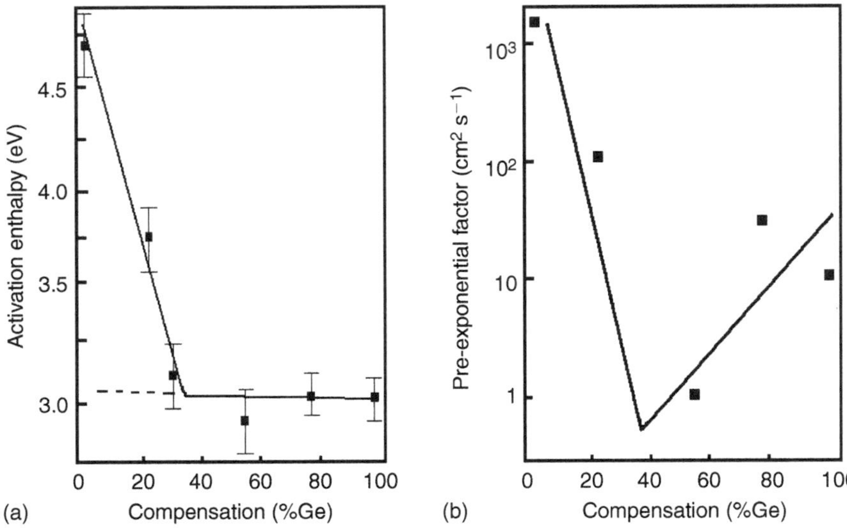

Figure 3.11. Compositional dependence of the activation enthalpy for ^{71}Ge diffusion in SiGe alloys (a) and compositional dependence of the pre-exponential factor for ^{71}Ge diffusion in SiGe alloys (b) (after McVay and DuCharme [41]).

The origin of the compositional dependence of the activation energy (enthalpy) has been ascribed to an elastic strain term, defined by [41]:

$$H_{\mathrm{SiGeimp}} = H_{\mathrm{Si}} + H_{\mathrm{s}} \tag{3.26}$$

with H_{Si} the activation enthalpy for self-diffusion in Si and H_{s} the difference of the formation enthalpy of the defect associated with diffusion in the SiGe alloy compared with silicon. At the same time, SiGe layers epitaxially deposited on a silicon (or germanium) substrate can be under a compressive strain, a tensile strain or relaxed, depending among others on the thickness of the layers. It has been suggested that in the strained case, the activation enthalpy can be written as [5,44]:

$$H_{\mathrm{SiGe}} = H(x) + H_{\mathrm{s}}'s \tag{3.27}$$

with s the strain in the layers. $H(x)$ is the separate effect of the Ge content, while $H_{\mathrm{s}}' = 1.6(4)\,\mathrm{eV}/\%$ strain for compressive and $1.1(5)\,\mathrm{eV}/\%$ strain for tensile strain. As a result, the Ge diffusion coefficient increases for compressive (Figure 3.12 [44]) and reduces for tensile strained SiGe, compared with a relaxed layer. This corresponds to a reduction (increase) of H_{SiGe} [5]. Meanwhile, it has been demonstrated that strain has a strong impact on the diffusivity of most common dopants in SiGe layers, which diffuse predominantly through the interstitialcy mechanism (B, P, As, . . .) [44]. On the other hand, vacancy dominated diffusers (Sb) show little dependence on strain [5]. This is related to the sensitivity of the activation enthalpy on s, where little dependence is found for purely vacancy diffusing species [5].

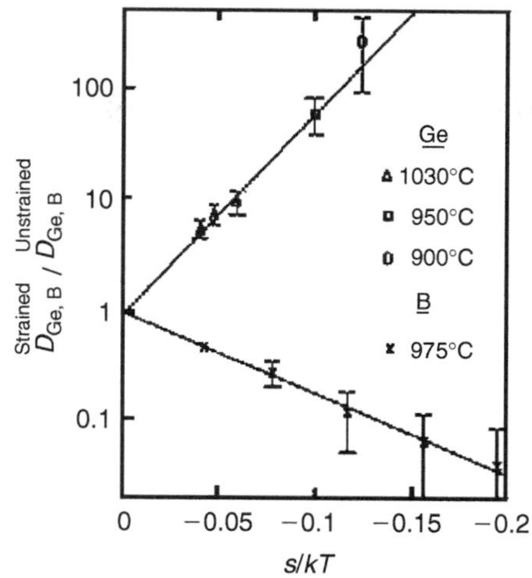

Figure 3.12. Diffusion of Ge and B in compressively strained Si(Ge), as a function of s/kT. The positive slope for Ge reflects a decrease in activation enthalpy of Ge diffusion under compression. The negative slope for B reflects an increase in the activation enthalpy of B diffusion (after Cowern *et al.* [44]).

3.5. Solubility of Impurities in Germanium

The solubility of an impurity in a solid phase is defined as the maximum concentration of that impurity in equilibrium with another phase, usually the melt (liquid), where the impurity is dissolved.[2] This thermodynamic equilibrium between the crystal and its melt is described by the distribution or segregation coefficient k_d, which is a function of temperature. As a crystal is pulled with a finite axial growth speed, one obtains in practice an effective distribution or segregation coefficient k_{deff}, given by [45]:

$$k_{deff} = C_S/C_L \tag{3.28}$$

with C_S and C_L the fractional impurity concentration in the solid and in the melt, respectively, i.e. the concentration per cm^3 divided by the atomic density of the solid or liquid. A value for k_{deff} equal to 1 implies that the impurity will be uniformly incorporated in the crystal (example: Al in Ge), while a value >1 means that it preferentially diffuses in the solid, so that there will be an enrichment at the seed end and a lower concentration near the tail of the crystal. For $k_{deff} < 1$, the opposite holds; an example is *P* (0.25) in Ge. This segregation behaviour gives rise to the fact that an undoped crystal is generally p-type at the seed end (Al) and becomes n-type at the tail (*P*), with the occurrence of a p–n junction in between [45].

In the early days of germanium crystal growth development, a substantial amount of data has been gathered concerning the distribution (segregation) coefficient and solid solubility

---

[2] If the second phase is another solid (i.e. ion-implanted layer or a germanide) or the gas phase, another equilibrium concentration or solubility may be obtained.

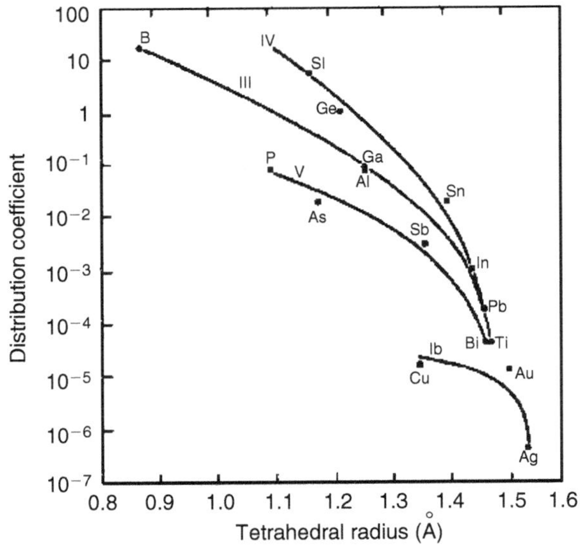

Figure 3.13. Distribution coefficients of impurities at the melting point of germanium as a function of the tetrahedral radii (after Trumbore [48]).

of various technologically relevant impurities (see, e.g., Refs. [46–48]). It was soon realized that there exists a relationship between the covalent (or ionic) radius r_{cov} of a solute atom and its segregation coefficient [46]: the larger r_{cov}, the lower k_d at the melting point of Ge (T_m). This is shown in Figure 3.13 for most of the relevant impurities in germanium [48]. One can also derive from that figure that the distribution coefficient aligns according to the different Groups in the periodic Table. There is a fairly rapid decrease of k_d with r_{cov} between 1.35 and 1.5 Å (=15% size difference between solvent and solute). A correlation has been derived between the heat of sublimation and k_d in Ge for the Group III, IV, and V elements [48]. In addition, solutes with a melting point higher than Ge and not forming an eutectic tend to have $k_d > 1$ [49].

Exceptions to the established trends are the Group Ib elements (Cu, Ag, Au), which have a small k_d for the same tetravalent radius (compare, e.g., Cu and Sb). This can be explained by considering the fraction of Cu, Au, and Ag occurring in interstitial sites, for which the ionic radii are of importance. The rule here is that an interstitial ion will take up its smallest charge state, whereby a positive ion has a smaller radius than a negative one and is, therefore, the preferred configuration.

The solubility (S_0) of most elements in germanium exhibits a retrograde behavior, meaning that S_0 increases when lowering the temperature from T_m and becomes maximum at some intermediate temperature, before dropping again. A simple explanation for this behavior has been developed by Hall [50] and results of his calculations for Ge are shown in Figure 3.14.

Besides the retrograde nature of S_0, another trend can be explained by the same model, namely, that elements with a smaller k_d tend to reach their maximum solubility closer to the

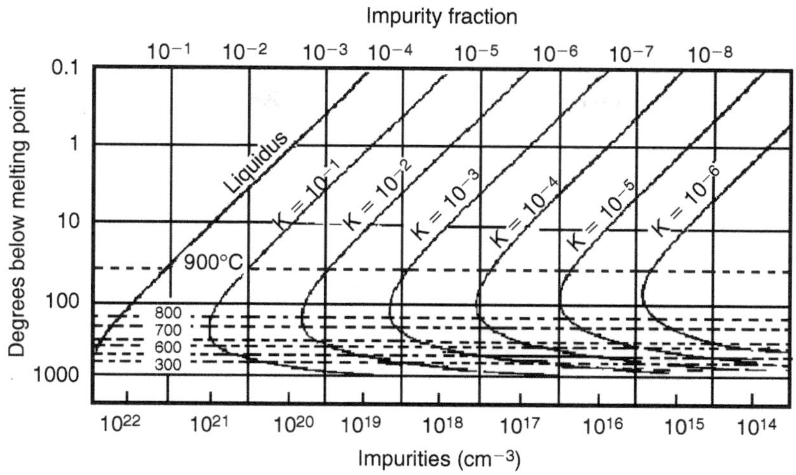

Figure 3.14. Calculated solid solubility curves for germanium (after Hall [50]).

melting point of Ge. The theory is valid for small distribution coefficients ($k_d < 0.1$) and assumes that k_d varies with temperature according to an activation law, i.e., $\log k_d = B - E/T$, whereby the energy E is proportional to the energy required to transfer a solute atom from the liquid phase to an appropriate position in the lattice. It can be easily demonstrated that k_d at any temperature T is given by [50]:

$$k_d = (k_{dm})^{T_m/T} \tag{3.29}$$

with k_{dm} the equilibrium distribution coefficient at T_m. From Equation (3.29) also follows that the lower k_{dm}, the lower the solubility of the element is and the higher the temperature where the impurity goes into the retrograde regime.

It has been remarked by Fischler [51] that there exists an empirical correlation between the distribution coefficient and the maximum solid solubility of impurities (S_{0max}) in germanium and silicon. This is shown in Figure 3.15 and can be represented by a fit:

$$S_{0max} = 4.4 \times 10^{21} k_d \ (\text{in cm}^{-3}) \tag{3.30}$$

for germanium. According to this rule, the values for Fe in Ge look anomalous.

The maximum equilibrium solubility of the common dopanting impurities in Ge are summarized in Table 3.3. Also shown is the expected sheet resistance for a diffusion (junction) depth x_j of around 100 nm [52], indicated in column 4 of the Table. This sheet resistance is an important technological parameter, which can be used to determine the suitability of dopanting elements for shallow junction formation in a germanium metal-oxide semi conductor field effect transistor (MOSFET). Based on these data, Ga and P stand out as the most promising

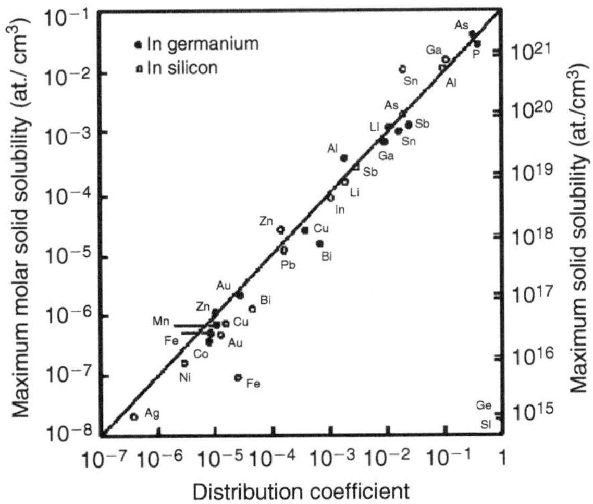

Figure 3.15. Maximum molar solid solubility and maximum solid solubility versus distribution coefficient at the melting point for impurities in Ge and Si (after Fischler [51]).

Table 3.3. Maximum equilibrium solubility of common dopants in Ge and corresponding calculated sheet resistance for a junction depth indicated in column 4.

Doping element	Maximum equilibrium solid solubility (at./cm^3)	Expected *rs* (Ohm/sq)	x_j (nm)
B	5.5×10^{18}	315	140
Ga	4.9×10^{20}	34	115
P	2.0×10^{20}	42	92
As	8.1×10^{19}	79	120
Sb	1.2×10^{19}	163	100

candidates. As will be seen in Chapter 10 other criteria like ion-implantation related defect removal, amorphization and dopant activation also play an important role [52].

Finally, it should be remarked that the solubility also depends on the doping concentration due to the Fermi level effect and the possible occurrence of ion pairing reactions.

3.6. Diffusion of Group III and V Dopants in Germanium

Self-diffusion (and Si diffusion) is the slowest process in germanium. With respect to impurities, a distinction is made between slow diffusers, which are up to 100 times faster than self-diffusion and fast diffusers, which follow generally an interstitial mechanism (Cu, Li, Ni, . . .) [1]. For doping purposes, Group III (p-Ge) and Group V (n-Ge) are the most

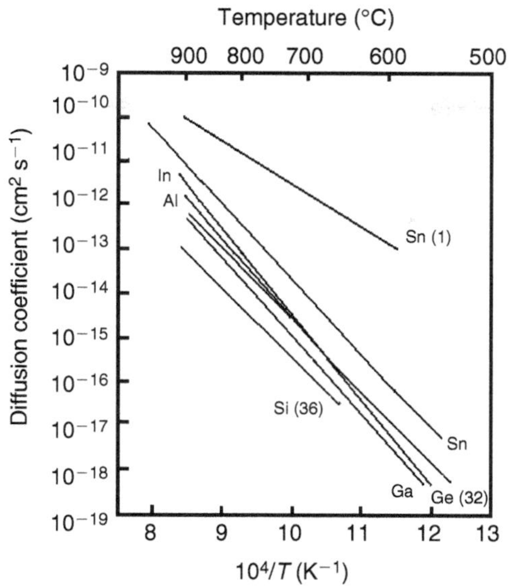

Figure 3.16. Diffusivities of impurities in germanium and germanium self-diffusion (after Friesel *et al.* [37]).

important elements. As can be seen from Figure 3.16 [37], Group III impurities are among the slow diffusers, while Group V dopants are typically 100 times faster, having an activation enthalpy $< H_{Ge}$ and a pre-factor $< D_{0Ge}$ [1].

From the similarity of the diffusivity of Group III elements and Ge, one can derive a vacancy mediated diffusion mechanism. The same has been derived for Group V dopants, whereby in this case a doping dependence is expected based on Equation (3.23). The reduction of H can be explained by considering the Coulomb attraction between a positive Group V ion and a negatively charged vacancy. Nevertheless, also in the case of Group III diffusion, enhancement effects have been observed, which will be described in the following.

3.6.1. Group III acceptor diffusion

Here, the diffusion coefficients of the main Group III impurities (B, Al, Ga, In) will be summarized, with emphasis on the recent literature. Early results have been reviewed in Refs. [3] and [53].

3.6.1.1. Boron

The activation and diffusion behavior of ion-implanted B in Ge has been studied in Refs. [54–60]. Boron has an equilibrium solubility of $\sim 5 \times 10^{18}\,\mathrm{cm}^{-3}$ at 850°C, although much higher concentrations can be activated if a rapid thermal anneal (RTA) step is utilized: in Ref. [59] up to $1.5 \times 10^{20}\,\mathrm{cm}^{-3}$ B atoms can be activated (Spreading Resistance Probe, SRP measurement) after a 650°C/10 s RTA and 20 keV BF_2 ion implantation to a dose of $4 \times 10^{15}\,cm^{-2}$. Recent studies confirm the low diffusivity of B, both under furnace anneal (FA) [57,58] or RTA [59]. Table 3.4 gives some recent data on the equilibrium diffusion coefficient of B at 850°C [58].

Quite often, a long diffusion tail has been observed in the B profile [57–59], which cannot be explained by channeling alone and points to some enhanced diffusion mechanism. One such mechanism, observed in the past for B in Ge [60], is the so-called radiation-enhanced diffusion. It occurs when an ion-implantation is performed at elevated temperatures. The enhancement factor can be several orders of magnitude, compared with a room temperature implantation followed by an anneal at the same T. The enhanced diffusivity has been found to be independent of temperature in the range 200–500°C [60]. The enhancement can be explained by assuming a higher than equilibrium concentration of vacancies under the high-temperature ion-implantation conditions [60].

A general observation regarding ion-implantation in n-type Ge is that as-implanted one often finds a p–n junction. This is related to the highly damaged near surface region, which is doped by ion-implantation-induced vacancies and vacancy complexes, expected to have an acceptor nature.

3.6.1.2. Aluminum

Ion-implantation of Al in Ge has been studied on a few occasions [56,61–64]. Aluminum should have a high solid solubility $\sim 4 \times 10^{20}\,cm^{-3}$ and is, therefore, suitable as a p-type dopant. Aluminum profiles also suffer severely from tailing [56,61,62], although no radiation-enhanced diffusion has been noted in this case [61,62]. Another problem with Al is its strong out-diffusion behavior, probably explaining the difficult activation [61,62]. Especially reaction with oxygen to form Al_2O_3 should be avoided [63]. Quite disperse values for H and D_0 have been reported [63]. Later data suggests the following value [64]:

$$D_{Al} = (1 \pm 0.5) \times 10^3 \exp(-3.45 \pm 0.04\,\mathrm{eV}/kT)\ (\mathrm{cm^2 s^{-1}}) \quad (3.31)$$

Table 3.4. Diffusion coefficient (D) of B in Ge at 850°C. The D values at 850°C have been calculated from prefactor (D_0) and activation enthalpy (H_a) values given by Stolwijk [3] (after Uppal *et al.* [58]).

D_0 (cm²s⁻¹)	H (eV)	D (cm²s⁻¹)	Temperature range (°C)	Wafer orientation	Experimental technique
6×10^8	4.5	3.88×10^{-12}	700–900	–	p–n junction
9.5×10^6	4.5	6.14×10^{-14}	750–900	(111)	Incremental sheet resistance
–	–	1.8×10^{-16}	850	(111)	SIMS profiling
–	–	1.3×10^{-16}	850	(100)	SIMS profiling

3.6.1.3. Indium and gallium

The diffusion of In and Ga has been studied in Refs. [65–68]. D_{Ga} is slightly smaller than D_{Al}, while the opposite holds for D_{In} [68]. The maximum solid solubility of In ($\sim 4 \times 10^{18}$ cm^{-3}) is lower than for Ga, Al or B.

3.6.2. Group V donor diffusion

3.6.2.1. Phosphorus

Quite some studies have been devoted to the diffusion of phosphorus in Ge [56,59,69–72]. Ion implanted P also suffers from tailing [56,59,69], which is a drawback if shallow n–p junctions are to be fabricated. At the same time, a large fraction of the implanted dose may be lost during subsequent annealing by out-diffusion [69]. A SiO_2 cap layer has been utilized to prevent this to happen [69]. Evidence has also been provided for radiation-induced enhanced diffusion, due to the presence of an excess of ion-implantation-induced vacancies [71]. The non-equilibrium diffusion coefficient at 300°C implantation was $D_P \sim 6 \times 10^{-14}$ cm^{2}s^{-1} compared with $D_P^{eq} = 4 \times 10^{-22}$ cm^{2}s^{-1}. At low concentrations, the diffusivity can be represented by [72]:

$$D_P = (0.09 \pm 0.03)\exp(-2.1 \pm 0.02\,\mathrm{eV/kT})\ (\mathrm{cm^2 s^{-1}}) \qquad (3.32)$$

However, a clear doping dependence has been found at large concentrations [70,72], pointing to the Coulomb attraction between V^- and P^+, so that at high concentration, part of the mass transport occurs through the diffusion of $E\,(= P - V)$ centers [72]. Consequently, D_P becomes higher at high phosphorus concentrations, as illustrated in Figure 3.17. At the same time, the obtained doping profiles cannot be described by the standard complementary error function (erfc) but rather by a box-like profile, which lends itself to a Boltzmann–Matano analysis [70]. In other words, the profiles can be normalized according

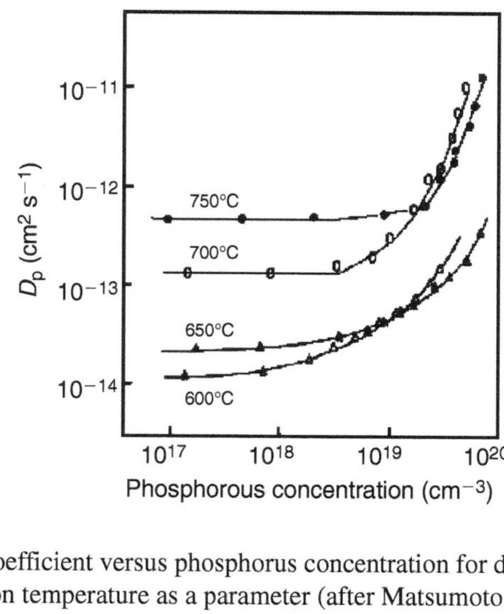

Figure 3.17. Diffusion coefficient versus phosphorus concentration for diffusions at source temperature 430°C, with diffusion temperature as a parameter (after Matsumoto and Niimi [70]).

to $x/t^{1/2}$, with x the depth and t the diffusion time. Recently, it has been demonstrated that during RTA at 650°C, P diffuses through a double minus charge state of the vacancy by deriving the best fit of Equation (3.16) to the experimental profiles [59].

Anomalous enhanced P diffusion has been observed in Ge, when exposed to 15 keV proton irradiation [73]. The diffusivity was increased by 3.5 decades above its thermal equilibrium value at 580°C, which was assumed to be related to the vacancy creation by the light ion implantation [73]. Below 500°C, no radiation enhanced P-diffusion was observed. Apparently, a minimum temperature is required for the created vacancies to leave their point of origin and not to form cluster defects [73]. It was proposed to employ this method for the engineering of the diffusion (junction) profile.

3.6.2.2. Arsenic

Arsenic has a high solubility $\sim 10^{20}\,\text{cm}^{-3}$ [56] and seems, therefore, a suitable candidate for n doping [74–82]. It has been pointed out that optimal electrical performance (activation + damage removal) can be obtained by a dual process, such as 430°C FA/650°C RTA [78]. At 650°C (FA), it was found that $D_P = 4 \times 10^{-14}\,\text{cm}^2\text{s}^{-1}$, $D_{As} = 9 \times 10^{-14}\,\text{cm}^2\text{s}^{-1}$ and $D_{Sb} = 1.8 \times 10^{-13}\,\text{cm}^2\text{s}^{-1}$ [69]. Of these three elements, only As showed up 100% activation at 500°C furnace anneal [69].

The diffusion profiles of As can be better described by assuming contributions from V^0 and V^{2-} [59,81]. The corresponding diffusivities are shown in Figure 3.18 [81]. It led Vainonen-Ahlgren *et al.* to propose that the vacancy in Ge is a negative-U center like in silicon [81]. This picture is, however, not supported by theory [26].

Studies of the activation volume of As diffusion in Ge under hydrostatic pressure has cast some doubt on the simple vacancy diffusion mechanism [39,79,80].

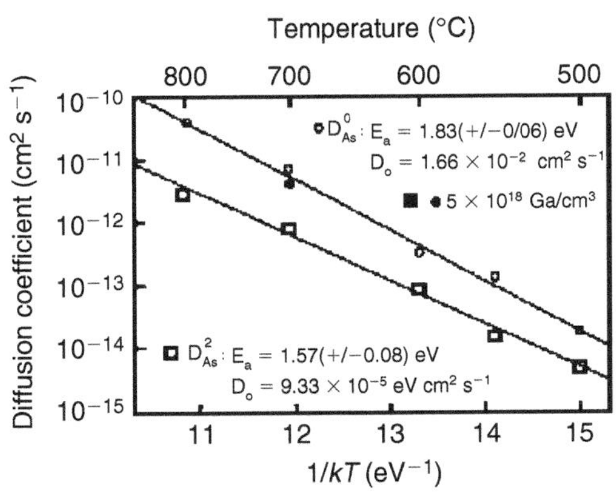

Figure 3.18. Arrhenius plots for diffusion of As atoms via neutral and double negatively charged Ge vacancies. The open and closed symbols correspond to the Ga doping levels 2×10^{17} and $5 \times 10^{18}\,\text{cm}^{-3}$, respectively. The fitted activation energies and pre-exponential factors are also given (after Vainonen-Ahlgren *et al.* [81]).

Also radiation enhanced diffusion of As in Ge was observed at 700°C using a 2 MeV electron beam [83]. The enhancement amounted to a factor of 3.83, in agreement with the expected rate of vacancy creation.

3.6.2.3. Antimony

Antimony (Sb) has a smaller solubility than As ($\sim 10^{19}\,cm^{-3}$) [56], which is a drawback. It also has a higher diffusivity than P or As [69] and a tendency to out-diffuse. In contrast with P and As diffusion, the profiles after 650 or 675°C RTA can be well described by assuming neutral and single negative charged vacancies [59]. The resulting model parameters are shown in Table 3.5 [59].

3.6.3. Electric field effects on dopant diffusion in Ge

The study of the co-diffusion of high concentrations of n- and p-type dopants has shown that there is an effect of the built-in electric field on the resulting profile [84–91]. This could be relevant for deep submicron complementary metal-oxide semi-conductor (CMOS) on germanium wafers, when junctions are created in a highly doped well (substrate) region. Usually, a retardation of the faster diffusing Group V dopants is found in a highly p-type doped Ge substrate. This is the case for As diffusion in $5 \times 10^{20}\,cm^{-3}$ Al-doped Ge [84], for Sb and Al [86], Sb and Ga [85,87] or In [87,88]. The retardation of the Sb-diffusion in Ge by a $10^{20}\,cm^{-3}$ Ga background doping is illustrated in Figure 3.19 [85], showing that the effect becomes more pronounced at lower diffusion temperatures, while at higher temperatures,

Table 3.5. Extracted diffusion coefficients and the associated model for various n-type dopants in Ge (after Chui *et al.* [59]).

Dopants	**Model**	***H*** **(eV)**	D_0 **($cm^2 s^{-1}$)**
P	$(n/n_i)^2$	2.07	4.38×10^{-2}
As	$(n/n_i)^2$	3.32	1.45×10^{6}
Sb	n/n_i	2.28	11.89

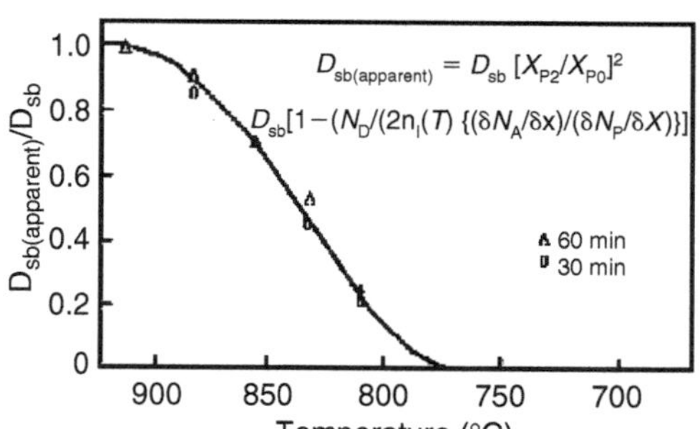

Figure 3.19. Dependence of the retarded diffusion of antimony on time and temperature for $N_A = 10^{20}\,cm^{-3}$ and $N_D = 10^{19}\,cm^{-3}$ (after Fa and Zuleeg [85]).

where the intrinsic carrier concentration becomes higher than the acceptor concentration in the substrate, the normal diffusion coefficient is obtained.

This retarded diffusion could be explained by several mechanisms [84]: the formation of a second phase, the creation of donor–acceptor pairs by Coulombic attraction or the enhancement of the solubility in a highly doped region. However, the main driving force in most cases is believed to be the impact of the built-in electric field on the diffusion, which becomes non-Fickian [85,90]. An additional drift term has to be included in the diffusion equations, given by $-\mu_D N_D F$, with μ_D the mobility of the donors atoms and the average force acting on a positively charged donor atom [85]. The resulting apparent diffusion coefficient then becomes:

$$D_{\text{apparent}} = D\left[1 - \frac{N_D}{2n_i(T)} \frac{\delta N_A}{\delta x} / \frac{\delta N_D}{\delta x}\right] \tag{3.33}$$

which according to the second term in Equation (3.33) depends on the ratio of the acceptor (A) and donor (D) concentration gradients. It is clear from Equation (3.33) that $D_{\text{apparent}} = D$ for $n_i \gg N_D$, while for $N_D = n_i(T)$, D_{apparent} becomes $D/2$ [85].

The theory also predicts that when the concentration gradients are in opposite directions, a field-enhanced diffusion of the donor impurities results [85,87], which has been observed for the case of Sb diffusion in an In-doped substrate [87]. In principle, one could expect similar effects on p-type dopants, although this was found in practice negligible for the slow diffusers.

It has also been observed that the flow of a DC current during In and Sb diffusion at high temperatures can cause anomalously fast transport of a fraction of the diffusing species [91]. This anomalously fast diffusion disappears below 450°C for Sb and is found down to 300°C for In [92].

3.6.4. Summary

Doping by ion implantation of most of the standard Group III and V elements in Ge is reasonably well studied, so that it can be used as a production technique for p–n junction fabrication. There is a good understanding of the problems related to out-diffusion, defect removal and dopant activation by furnace annealing. Only recently, results on the application of RTA became available, which is important for today's technology developments (see Chapter 10). At the moment knowledge is being collected about dopant activation by techniques like laser annealing, solid-phase epitaxial regrowth (SPER) and spike annealing on Ge.

Based on the available literature data, one can come to some recommendations for an optimal choice of n- and p-type dopant. B (BF_2) seems by far to be the best choice for p^+ ion implantation, while for n-type, P and As are in the balance, with a slight preference for As, given its full activation potential and high solubility. Of course, the picture may change completely for non-equilibrium annealing techniques, like RTA, spike or laser annealing. It has for example been demonstrated that good quality n^+–p junctions can be obtained after a 100 keV $4 \times 10^{15}\,cm^{-2}$ P ion implantation, followed by a 500°C 5 min anneal [82].

3.7. General Conclusion

In general, there is a good understanding of the basic thermodynamic equilibrium diffusion mechanism of impurities in Ge. Both self- and dopant diffusion are mediated by vacancies, in contrast to the situation in silicon, where interstitial, vacancy and mixed behavior can be observed. The insight in the properties of intrinsic point defects is less well developed, due to experimental difficulties. To some extent, the same applies for silicon, where, however, the vacancy and interstitial behavior is well understood near the melting temperature ($T >$ 1100°C). This is based on the study of void defects formed during cooling of the grown crystal (vacancy clusters), which allowed a careful modeling of the equilibrium concentrations near T_m. Currently, a similar study is undertaken for the grown-in defects during germanium crystal growth (see Chapter 2). The main problem remains that it is not clear whether one can extrapolate the high-temperature data to temperatures typically used in dopant activation and diffusion. Another issue which requires further work is the non-equilibrium dopant diffusion (enhanced diffusion) that can occur during short time anneals. Some evidence points to vacancy-assisted diffusion enhancement at lower temperatures (300–500°C), which should be well controlled when fabricating the shallow junctions necessary for deep submicron Ge FETs.

References

[1] A. Seeger and K.P. Chik, *Phys. Stat. Sol.* **29** (1968) 455.

[2] D. Shaw, *Phys. Stat. Sol.* **72** (1975) 11.

[3] N. Stolwijk, *"Germanium", Landolt-Börnstein New Series* III/22b (1988) 439.

[4] F.C. Frank and D. Turnbull, *Phys. Rev.* **104** (1956) 617.

[5] N.R. Zangenberg, J. Lundsgaard Hansen, J. Fage-Pedersen and A. Nylandsted Larsen, *Phys. Rev. Lett.* **87** (2001) 125901-1.

[6] A. Ural, P.B. Griffin and J.D. Plummer, *Phys. Rev. Lett.* **83** (1999) 3454.

[7] C.S. Fuller, H.C. Theuerer and W. van Roosbroeck, *Phys. Rev.* **85** (1952) 678.

[8] C. Goldberg, *Phys. Rev.* **88** (1952) 920.

[9] L. Esaki, *Phys. Rev.* **89** (1953) 1026.

[10] G. Finn, *Phys. Rev.* **91** (1953) 754.

[11] R.A. Logan, *Phys. Rev.* **91** (1953) 757.

[12] S. Mayburg and L. Rotondi, *Phys. Rev.* **91** (1953) 1015.

[13] S. Mayburg, *Phys. Rev.* **95** (1954) 38.

[14] R.L. Hopkins and E.N. Clarke, *Phys. Rev.* **100** (1955) 1786.

[15] R.A. Logan, *Phys. Rev.* **101** (1956) 1455.

[16] A. Hiraki, *J. Phys. Soc. Jpn.* **21** (1966) 34.

[17] F. Hashimoto and Y. Kamiura, *J. Phys. Soc. Jpn.* **33** (1972) 271.

[18] P. Penning, *Phys. Rev.* **110** (1958) 586.

[19] H. Letaw Jr., W.M. Portnoy and L. Slifkin, *Phys. Rev.* **102** (1956) 636.

[20] A.G. Tweet, *Phys. Rev.* **106** (1957) 221.

[21] C. Claeys and E. Simoen, '*Radiation effects in advanced semiconductor materials and devices*', Springer Series in Materials Science, Springer, Berlin (2002).

[22] H. Hässlein, R. Sielemann and Ch. Zistl, *Phys. Rev. Lett.* **80** (1998) 2626.

[23] R. Sielemann, *Nucl. Instrum. Method. Phys. Res. B* **146** (1998) 329.

[24] R. Sielemann, H. Hässlein and Ch. Zistl, *Physica B* **302–303** (2001) 101

[25] R. Sielemann, H. Hässlein, Ch. Zistl, M. Müller, L. Stadler and V.V. Emtsev, *Physica B* **308–310** (2001) 529.

[26] A. Fazzio, A. Janotti, A.J.R. da Silva and R. Mota, *Phys. Rev. B* **61** (2000) R2401.

[27] C.J. Hwang and L.A.K. Watt, *Phys. Rev.* **171** (1968) 958.

[28] A.J.R. da Silva, A. Janotti, A. Fazzio, R.J. Baierle and R. Mota, *Phys. Rev. B* **62** (2000) 9903.

[29] M.W. Valenta and C. Ramasastry, *Phys. Rev.* **106** (1957) 73.

[30] D.R. Campbell, *Phys. Rev. B* **12** (1975) 2318.

[31] G. Vogel, G. Hettich and H. Mehrer, *J. Phys. C: Solid State Phys.* **16** (1983) 6197.

[32] M. Werner, H. Mehrer and H.D. Hochheimer, *Phys. Rev. B* **32** (1985) 3930.

[33] H.D. Fuchs, W. Walukiewicz, E.E. Haller, W. Dondl, R. Schorer, G. Abstreiter, A.I. Rudnev, A.V. Tikhomirov and V.I. Ozhogin, *Phys. Rev. B* **51** (1995) 16817.

[34] A. Strohm, T. Voss, W. Frank, P. Laitinen and J. Räisänen, *Z. Metallkde* **93** (2002) 737.

[35] J.C. Bourgoin and M. Lannoo, *Radiat. Eff.* **46** (1980) 157.

[36] J. Räisänen, J. Hirvonen and A. Anttila, *Solid-State Electron* **24** (1981) 333.

[37] M. Friesel, U. Södervall and W. Gust, *J. Appl. Phys.* **78** (1995) 5351.

[38] P. KringhØj and R.G. Elliman, *Appl. Phys. Lett.* **65** (1994) 324.

[39] S. Mitha, S.D. Theiss, M.J. Aziz, D. Schiferl and D.B. Poker, *Mat. Res. Soc. Symp. Proc.* **325** (1994) 189.

[40] G.L. McVay and A.R. DuCharme, *J. Appl. Phys.* **44** (1973) 1409.

[41] G.L. McVay and A.R. DuCharme, *Phys. Rev. B* **9** (1974) 627.

[42] P. Dorner, W. Gust, B. Predel, U. Roll, A. Lodding and H. Odelius, *Phil. Mag. A* **49** (1984) 557.

[43] P. Fahey, S.S. Iyer and G.J. Scilla, *Appl. Phys. Lett.* **54** (1989) 843.

[44] N.E.B. Cowern, P.C. Zalm, P. van der Sluis, D.J. Gravesteijn and W.B. de Boer, *Phys. Rev. Lett.* **72** (1994) 2585.

[45] E.E. Haller, W.L. Hansen and F.S. Goulding, *Adv. in Physics* **30** (1981) 93.

[46] J.A. Burton, E.D. Kolb, W.P. Slichter and J.D. Struthers, *J. Chem. Phys.* **21** (1953) 1991.

[47] C.D. Thurmond and M. Kowalchik, *Bell Syst. Techn. J.* **39** (1960) 169.

[48] F.A. Trumbore, *Bell Syst. Techn. J.* **39** (1960) 205.

[49] W.C. Dunlap, Jr., *Progr. Semicond.* **2** (1957) 165.

[50] R.N. Hall, *J. Phys. Chem. Solids* **3** (1957) 63.

[51] S. Fischler, *J. Appl. Phys.* **33** (1962) 1615.

[52] A. Satta, E. Simoen, T. Janssens, T. Clarysse, B. De Jaeger, A. Benedetti, I. Hoflijk, B. Brijs, M. Meuris and W. Vandervorst, *J. Electrochem. Soc.* **153** (2006) G229.

[53] L. Sharma, *Defect Diff. Forum* **70–71** (1990) 1.

[54] G.D. Alton and L.O. Love, *Can. J. Phys.* **46** (1968) 695.

[55] V.M. Gusev, M.I. Guseva, E.S. Ionova, A.N. Mansurova and C.V. Starinin, *Phys. Stat. Sol.* (a) **21** (1974) 413.

[56] K. Schmid, H. Kranz, H. Ryssel, W. Müller and J. Dathe, *Phys. Stat. Sol.* (a) **23** (1974) 523.

[57] S. Uppal, A.F.W. Willoughby, J.M. Bonar, A.G.R. Evans, N.E.B. Cowern, R. Morris and M.G. Dowsett, *Physica B* **308–310** (2001) 525.

[58] S. Uppal, A.F.W. Willoughby, J.M. Bonar, A.G.R. Evans, N.E.B. Cowern, R. Morris and M.G. Dowsett, *J. Appl. Phys.* **90** (2001) 4293.

[59] C.O. Chui, K. Gopalakrishnan, P.B. Griffin, J.D. Plummer and K.C. Saraswat, *Appl. Phys. Lett.* **83** (2003) 3275.

[60] M.I. Guseva and A.N. Mansurova, *Radiat. Eff.* **20** (1973) 207.

[61] T. Itoh and I. Ohdomari, *Jpn. J. Appl. Phys.* **10** (1971) 1002.

[62] I. Ohdomari and T. Itoh, *Jpn. J. Appl. Phys.* **11** (1972) 1709.

[63] J. Räisänen, *Solid-State Electron* **25** (1982) 49.

[64] P. Dorner, W. Gust, A. Lodding, H. Odelius, B. Predel and U. Roll, *Acta. Metall.* **30** (1982) 941.

[65] K.C. Jones and P.R.C. Stevens, *Electron. Lett.* **5** (1969) 499.

[66] N.G.E. Johansson, D. Sigurd and K. Björkqvist, *Rad. Eff.* **6** (1970) 257.

[67] P. Dorner, W. Gust, A. Lodding, H. Odelius, B. Predel and U. Roll, *Z. Metallkde* **73** (1982) 325.

[68] U. Södervall, H. Odelius, A. Lodding, U. Roll, B. Predel, W. Gust and P. Dorner, *Phil. Mag.* **A54** (1986) 539.

[69] A. Axmann, M. Schulz and C.R. Fritzsche, *Appl. Phys.* **12** (1977) 173.

[70] S. Matsumoto and T. Niimi, *J. Electrochem. Soc.* **125** (1978) 1307.

[71] G.G. Devyatykh, V.K. Vasil'ev, V.A. Gavva, A.V. Gusev, Yu.A. Danilov, E.S. Zharkov and G.A. Maksimov, *Inorganic Mat.* **32** (1996) 1258.

[72] U. Södervall and M. Friesel, *Defect Diff. Forum* **143–147** (1997) 1053.

[73] R.L. Minear, D.G. Nelson and J.F. Gibbons, *J. Appl. Phys.* **43** (1972) 3468.

[74] W. Albers, *Solid-State Electron.* **2** (1961) 85.

[75] R. Wölfle and H. Dorendorf, *Solid-State Electron.* **5** (1962) 98.

[76] A.R.H. Niedermeyer, *Phys. Stat. Sol.* **6** (1964) 741.

[77] N. Isawa, *Jpn. J. Appl. Phys.* **7** (1968) 81.

[78] S.V. Hattangady, G.G. Fountain, E.H. Nicollian and R.J. Markunas, *J. Appl. Phys.* **63** (1988) 68.

[79] S. Mitha, M.J. Aziz, D. Schiferl and D.B. Poker, *Appl. Phys. Lett.* **69** (1996) 922.

[80] S. Mitha, M.J. Aziz, D. Schiferl and D.B. Poker, *Defect Diff. Forum* **143–147** (1997) 1041.

[81] E. Vainonen-Ahlgren, T. Ahlgren, J. Likonen, S. Lehto, J. Keinonen, W. Li and J. Haapamaa, *Appl. Phys. Lett.* **77** (2000) 690.

[82] H. Shang, K.-L. Lee, P. Kozlowski, C. D'Emic, I. Babich, E. Sikorski, M. Ieong, H.-S.P. Wong, K. Guarini and W. Haensch, *IEEE Electron Device Lett.* **25** (2004) 135.

[83] N.L. Peterson and R.E. Ogilvie, *Trans. Metall. Soc. AIME* **215** (1959) 873.

[84] J.O. McCaldin, *J. Appl. Phys.* **31** (1960) 89.

[85] C. Fa and R. Zuleeg, *Solid-State Electron.* **3** (1961) 18.

[86] I.P. Akimchenko and L.S. Milevskii, *Sov. Phys. Solid State* **2** (1961) 1891.

[87] B.I. Boltaks and T.D. Dzhafarov, *Sov. Phys. Solid State* **5** (1964) 2061.

[88] B.I. Boltaks, V.P. Grabchak and T.D. Dzhafarov, *Sov. Phys. Solid State* **6** (1965) 2542.

[89] P.V. Pavlov and V.A. Uskov, *Sov. Phys. Solid State* **8** (1967) 2377.

[90] K. Lehovec and A. Slobodskoy, *Solid-State Electron.* **3** (1961) 45.

[91] F.M. Smits, *Proc. IRE* **46** (1958) 1049.

[92] B.P. Konstantinov and L.A. Badenko, *Sov. Phys. Solid State* **2** (1961) 2400.

CHAPTER 4

Oxygen in Germanium

Paul Clauws

4.1. Introduction

In comparison with a typical Cz-grown silicon crystal, where oxygen is a dominant impurity, a germanium crystal normally contains only a moderate concentration of oxygen. There are several reasons for this difference, e.g. the lower melting point of germanium which results in less interaction with the crucible, the lower affinity for oxygen in comparison with silicon, the formation of volatile germanium oxide and the low segregation coefficient of oxygen in germanium. Taking into account that oxygen in the normal interstitial position (O_i) is also electrically inactive, it is obvious that oxygen in germanium is in many cases not considered as a problematic impurity.

In order to obtain germanium with high oxygen concentrations, i.e. 10^{16}–10^{18} at cm^{-3}, it is necessary to dope the crystal deliberately, for instance by mixing oxygen gas or water vapour into the growth atmosphere. In similar material, defects consisting of agglomerated oxygen are formed beside O_i during crystal growth or thermal treatments. Small clusters which form in the 300–500°C temperature range and contain a limited numbers of oxygen atoms, display donor activity and are called thermal donors (TDs). During prolonged treatments at 550–650°C, GeO_x precipitates are formed instead. Other oxygen containing defects, some of which are electrically active, are formed when oxygen-doped germanium is subjected to energetic irradiation. The practical possibilities of germanium with high oxygen content have not yet been sufficiently explored; it has, e.g. still to be established whether oxygen precipitates have the capability to getter metal contaminants as is the case in silicon.

Germanium-Based Technologies: From Materials to Devices. C. Claeys and E. Simoen (Editors)

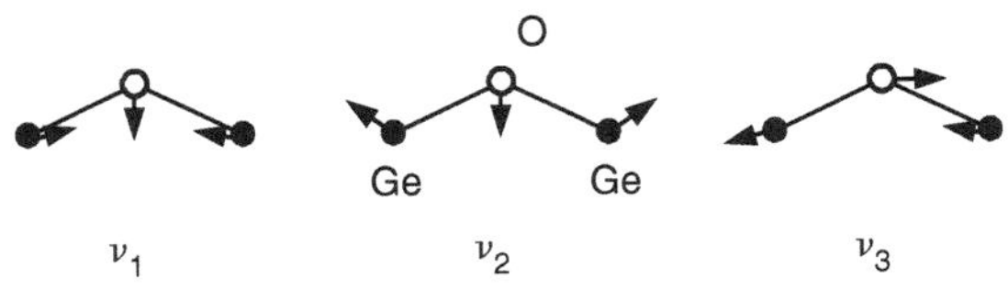

Figure 4.1. Vibrational modes (VLMs) of Ge—O—Ge quasi-molecule.

At the other extreme, oxygen concentrations of 10^{13} at cm^{-3} or lower may be realised by growing the crystal in a reducing atmosphere (hydrogen) and avoiding direct contact between the melt and silica components. Similar concentrations are obtained in high-purity germanium crystals for gamma detector fabrication. In this case electrically active complexes of oxygen with other impurities, such as Li, H or Si, may become important as shallow dopants or trapping centres.

In recent years the knowledge about oxygen-related defects in germanium has rapidly increased and the discrepancy with silicon diminished. Most often the defects are very similar in the two semiconductors and the properties and models in the case of silicon have been very helpful to the study of the corresponding defects in germanium. In this chapter a summary will be given of the properties of oxygen defects in germanium, with emphasis on results obtained during the last two decades.

4.2. Interstitial Oxygen

It is generally agreed that oxygen interstitials occupy bond-centred positions, with the oxygen displaced from the original Ge—Ge bond axis. In the simplest model the defect may be regarded as a non-linear symmetric Ge_2O quasi-molecule, with vibration modes as shown in Figure 4.1 [1]. The mode labels as given in the figure will be used throughout this chapter, even if the true form of the modes may be somewhat different depending on the exact site symmetry of the defect and on the elastic force field.

4.2.1. Measurement of oxygen concentration

The most convenient way to detect interstitial oxygen in germanium is by the infrared (IR) absorption spectrum, as shown in Figure 4.2. The main feature present is the band at 856 cm^{-1} which is due to the ν_3 antisymmetric stretching vibration of O_i (Figure 4.1) and is equivalent to the 1107 cm^{-1} band in silicon [2]. Early observations of the band in germanium have been reported in Refs. [3–6]. Also seen in Figure 4.2 is a weak band at 1264 cm^{-1}, which shows a constant amplitude ratio with respect to the main band [6] and has been attributed to the $\nu_1 + \nu_3$ sum mode of O_i [7].

The oxygen concentration $[O_i]$ (in at cm^{-3}) may be calculated from the measured amplitude of the 856 cm^{-1} absorption band α_{max} (in cm^{-1}) using the formula:

$$[O_i] = f_c \alpha_{max} \tag{4.1}$$

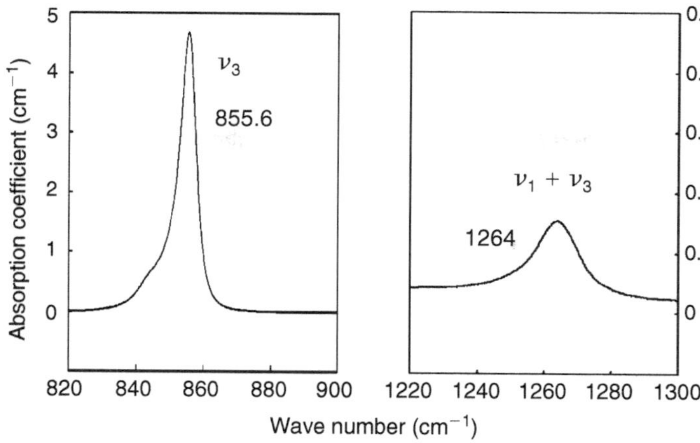

Figure 4.2. Infrared absorption bands due to interstitial oxygen in germanium.

provided that a reliable value for the calibration factor f_c is available. The calibration factor (applicable to the spectrum at room temperature) has been determined in several ways with different values as a result [4,5,8–10]. Mainly two "old" values have so far been in use: 5×10^{16} cm^{-2} derived from vacuum fusion analysis [5] on one hand and a theoretical value of 1.25×10^{17} cm^{-2} [4] confirmed by vacuum fusion analysis [8] on the other. Recently a new value of 0.95×10^{17} cm^{-2} based on secondary ion mass spectrometry (SIMS) experiments has been published as well [10]. The "old" value of 5×10^{16} cm^{-2} has, however, been most widely used, e.g. in articles dealing with the solubility or the agglomeration of oxygen in germanium [5,6,11–15]. Obviously, some of the conclusions in the latter articles would need to be adapted if the "new" value of Ref. [10] is accepted. The reason for the differences in calibration factor is not very clear: experimental errors as well as neglect of contributions from surface oxides or precipitated oxide phases seem possible. In any case the "new" calibration factor appears to be in better agreement with the calculated number of oxygen atoms in TDs [10] and with the simulation of GeO_x precipitate absorption [16] (see Section 4.4).

When the sample is cooled to near liquid helium temperature (LHeT) the ν_3 band shifts to 862.5 cm^{-1} and becomes about three times narrower and higher. This sharpening is less important than in the case of O_i in silicon, so that the effect of cooling on the detection of small oxygen concentrations is smaller in germanium. Using a cooled sample of sufficient thickness and if the shallow dopant concentration is not too high, the detection limit of O_i in germanium may be estimated at a few 10^{15} at cm^{-3}.

For O_i concentrations below the IR detection limit the lithium precipitation technique may be applied [17–20]. The solubility of lithium in germanium is increased by oxygen through the formation of Li–O donor complexes which is expected to proceed until exhaustion of all available O_i [17]. If the sample is saturated with lithium at elevated temperature and equilibrated at a sufficiently low temperature such that the concentration of Li–O donors exceeds that of free interstitial Li donors, then the original O_i concentration may be retrieved through an electrical measurement of the free electron concentration. The method requires, however,

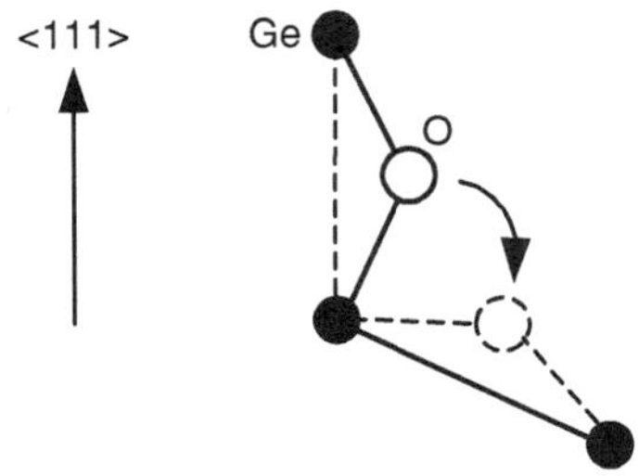

Figure 4.3. Diffusion jump of interstitial oxygen in germanium.

that the shallow dopant concentration is sufficiently low, so that it is especially suitable to high-resistivity germanium. For O_i concentrations lower than $10^{13}\,cm^{-3}$, the Li–O concentration becomes lower than the intrinsic carrier concentration and the free Li solubility at room temperature, so that the equilibration and measurement have to be done well below room temperature [19]. Another complication with high-purity germanium is that precautions have to be taken in order to avoid contamination with impurities like Cu during the Li saturation treatment [20].

Other methods have been applied occasionally, often within the context of the determination of the IR calibration factor. Beside SIMS [10] and vacuum fusion analysis [5,8] mentioned above, also helium-3 and alpha particle activation analysis have been used to determine doped or residual oxygen in germanium [9,21]. All those methods have in common that all kinds of oxygen species present are measured instead of O_i only.

4.2.2. Diffusion and solubility

Data concerning the diffusion coefficient of interstitial oxygen in germanium have been provided by two kinds of experiments, measuring the relaxation time for reorientation of interstitial oxygen between neighbouring sites. The idea is that a similar reorientation corresponds with the most probable diffusion jump, requiring that only one Ge—O bond is broken as illustrated in Figure 4.3. In internal friction experiments of oxygen containing germanium, a loss peak occurring at 770°C for 100 kHz vibrations has been observed, which was attributed to interstitial oxygen [22]. A value for the relaxation time at lower temperature has been obtained from measurement of the stress-induced dichroism of the ν_3 IR absorption band [2]; the latter experiments also support the model of the bond-centred oxygen configuration. The diffusion coefficient may be calculated from the relaxation time τ as:

$$D = \frac{a^2}{12\tau} \tag{4.2}$$

with a the Ge–Ge distance [2,23]. The experimental D values at 770°C and 280°C are plotted in Figure 4.4. The straight line connecting the data points corresponds with a temperature dependence given by [2]:

$$D_O = 0.4\exp(-2.076\,\mathrm{eV}/kT)\ \ (\mathrm{cm^2\,s^{-1}}) \tag{4.3}$$

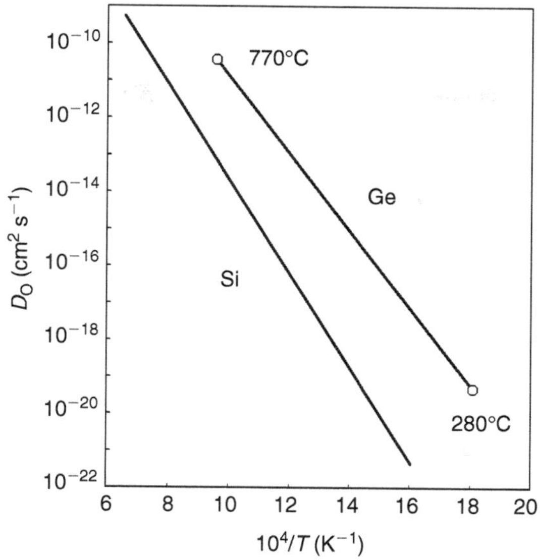

Figure 4.4. Temperature dependence of the diffusion coefficient of interstitial oxygen in germanium according to Equation (4.3), compared with the dependence for silicon according to Newman [26]. The data points at 770°C and 280°C indicate results obtained from friction experiments [22] and stress-induced dichroism [2], respectively.

A very similar dependence has been calculated from the internal friction experiments only, making use of scaled silicon data [23].

The activation energy for the diffusion jump may be identified with the energy barrier represented by the saddle point through which the configuration has to move between the two stable positions shown in Figure 4.3. *Ab initio* calculations on a Ge supercell containing oxygen resulted in a value of 1.70 eV for the diffusion barrier, assuming the Y-lid configuration with threefold coordinated oxygen in a C_{2v} symmetry as the saddle point [24]. The importance of the threefold coordination in the oxygen diffusion process has been confirmed by another theoretical modelling using semiempirical and *ab initio* methods; in this case a dependence in excellent agreement with equation (4.3) was found [25]. Although the diffusion mechanism is considered the same in silicon and germanium, due to the different activation energy the diffusion coefficient for oxygen in germanium is 2–3 orders of magnitude higher than in silicon in the same temperature range, as may be seen in Figure 4.4. As a consequence the agglomeration and clustering of oxygen will occur on a faster time scale in germanium.

The solubility of oxygen in germanium has been studied using prolonged annealing of oxygen-rich specimens at temperatures between 650°C and 810°C [5]. The treatment results in oxide precipitate formation with a corresponding loss of dissolved oxygen (see further). The amount of oxygen remaining in solid solution after each anneal was measured from the amplitude of the 856 cm^{-1} absorption band, resulting in the exponential temperature dependence shown in the diagram of Figure 4.5. The maximum solubility obtained by extrapolating the plot to the melting point would then amount to 2.2×10^{18} cm^{-3}. It should be noticed that

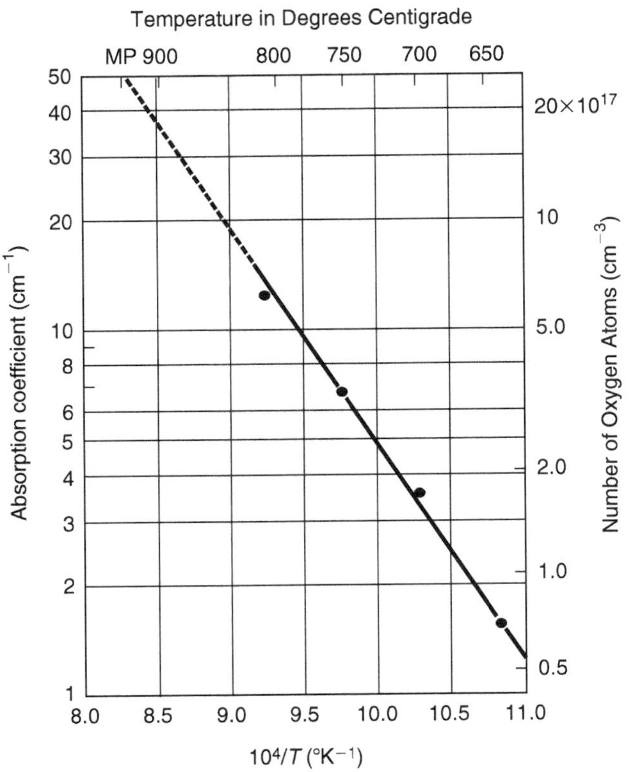

Figure 4.5. Solubility of oxygen in germanium (reproduced with permission from Kaiser and Thurmond [5], Copyright 1961, American Institute of Physics).

the diagram has been drawn assuming a calibration factor $f_c = 5 \times 10^{16}\,cm^{-2}$ which may underestimate the solubility by about a factor of two. The slope of the diagram corresponds with an energy value of 1.2 eV which is identified with the "heat of solution" of oxygen in germanium [5].

The oxygen incorporation during crystal growth of germanium depends primarily on the amount of oxygen in the growth atmosphere. In Figure 4.6 the relation between the oxygen concentration in a Czochralski germanium crystal as measured by the lithium precipitation method and the water content of the furnace gas is shown [19–20]. The measured data are in good agreement with the theoretical dependence given by the straight line, which was calculated using the solubility at the melting point derived from Figure 4.5 assuming $f_c = 1.25 \times 10^{17}\,cm^{-2}$.

4.2.3. Structure of the vibration spectrum and defect model

When measured near LHeT the ν_3 band at about $862\,cm^{-1}$ can be resolved into a fine structure with 25–28 lines [7,27–30], as shown in Figure 4.7. Natural germanium is a mixture of

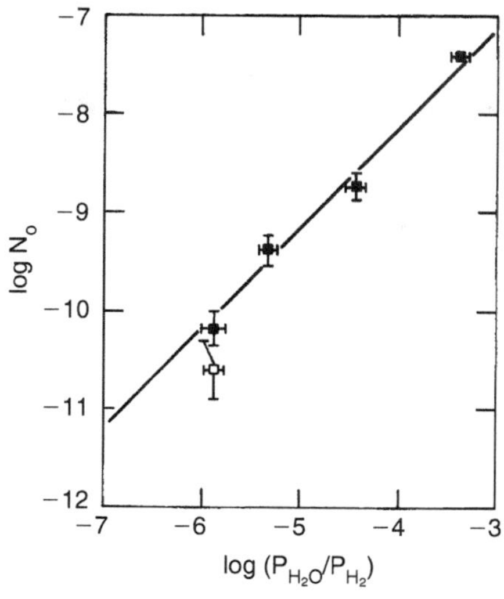

Figure 4.6. Atomic fraction of oxygen in germanium depending on the water content of the furnace gas during growth (reproduced with permission from L. Darken [20], Copyright 1982, ECS-The Electrochemical Society).

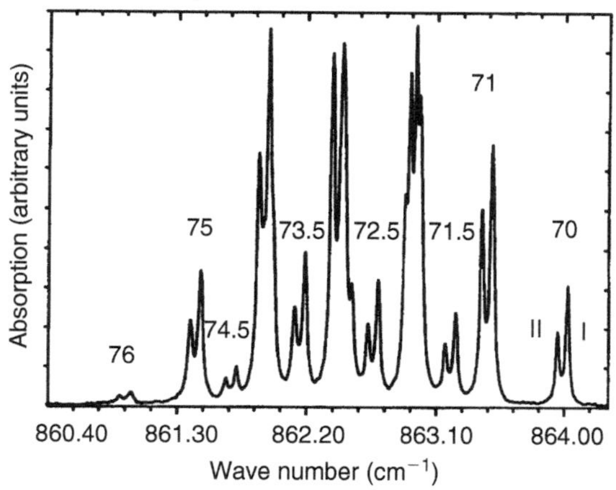

Figure 4.7. Absorption band of the ν_3 mode of oxygen in germanium measured at 1.6 K using high resolution (reproduced with permission from Artacho *et al.* [7], Copyright 1997, American Physical Society).

the stable isotopes ^{70}Ge, ^{72}Ge, ^{73}Ge, ^{74}Ge and ^{76}Ge. One origin of the fine structure is therefore the mode shift due to the 15 possible $^{M(i)}Ge—O—{}^{M(j)}Ge$ combinations, M(i) and M(j) indicating the mass of the Ge neighbours bonded to the oxygen. The combinations give rise to 11 distinct positions belonging to a different average value of the Ge masses, as indicated by the numbers in Figure 4.7. The second origin of fine structure is the coupling of the ν_3 mode with a nearly free rotation of the oxygen atom around the Ge—Ge axis. The latter movement is depicted in Figure 4.8. A dramatic simplification of the spectrum occurs in

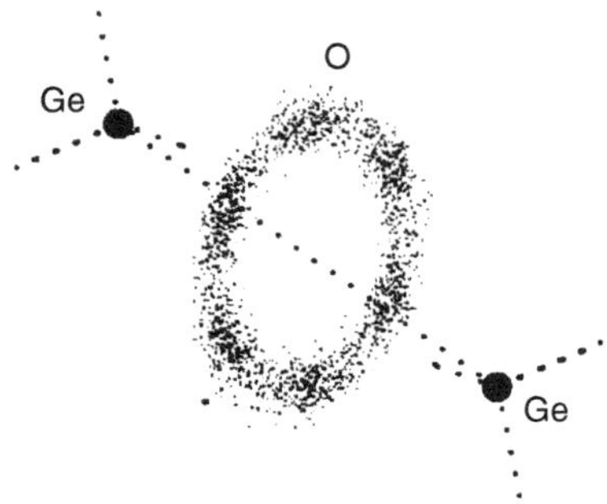

Figure 4.8. Quantum delocalisation (nearly free rotation) of interstitial oxygen in germanium in the plane perpendicular to the Ge—Ge axis (reproduced with permission from Artacho *et al.* [7], Copyright 1997, American Physical Society).

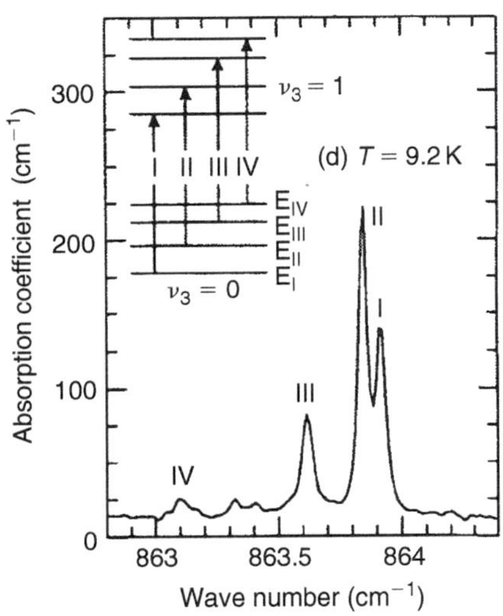

Figure 4.9. Fine structure due to vibration–rotation coupling in the ν_3 band of qmi ^{70}Ge (reproduced with permission from Mayur *et al.* [28], Copyright 1994, American Physical Society).

quasi-monoisotopic germanium (qmi Ge) samples, leaving only the effect of the vibration–rotation coupling [7,28,29,31,32]. It turns out that each isotopic combination gives rise to up to four components numbered I to IV, as indicated in Figure 4.9.

Low-energy excitations in the 0.18–5.8 meV range corresponding with the rotational transitions only, have been observed in phonon spectroscopy of oxygen-doped germanium using superconducting tunnelling junctions [33–35]. The transitions which occur without strict selection rules, allowed to deduce the first rotational levels in the vibrational ground state.

The IR and phonon spectra have been explained using the model of a non-rigid rotor, subjected to a weak hindering potential between equivalent positions along the rotation path of Figure 4.8 [7]. The calculated radial potential for the oxygen motion is shown in Figure 4.10.

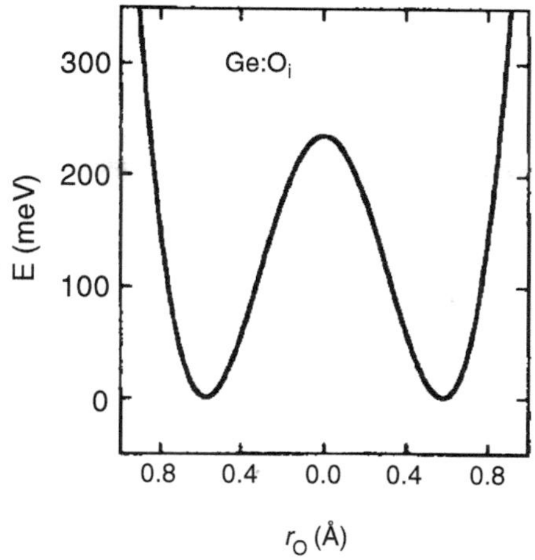

Figure 4.10. Radial potential in the non-rigid rotor model; r_O is the distance of the oxygen atom to the Ge—Ge axis (see Figure 4.8) (reproduced with permission from Artacho *et al.* [7], Copyright 1997, American Physical Society).

The barrier on the Ge—Ge axis amounts to 235 meV and the oxygen is located at a distance of 0.058 nm from this axis, with a Ge—O—Ge angle of 140°. The puckered structure is in contrast to that of interstitial oxygen in silicon, where the barrier is very small and an effective on-axis position with a Si—O—Si angle of 180° is found [7,36]. In the non-rigid rotor model, energy levels numbered $l = 0,1,2\ldots$ and with twofold degeneracy characterised by $m = -l,+l$ are predicted at:

$$E_0^1 = Bl^2 - Dl^4 \tag{4.4}$$

where B is the rotational constant and D the centrifugal distortion coefficient [7]. The hindering potential as a function of the azimuthal angle around the Ge—Ge axis gives rise to a splitting of the levels $l = \pm3,\pm6,\pm9\ldots$ The theoretical level spectrum according to this model can be very well fitted to the experimental data obtained from phonon spectroscopy, as shown in Figure 4.11. In this figure three possible hindering potentials with minima at different symmetry positions along the rotation path and compatible with the data are also shown. In this scheme the IR components I–IV correspond with transitions $|0,l\rangle \rightarrow |1,l\rangle$ with $l = 0,\pm1,\pm2,-3$, respectively, between the ground and first excited state of the ν_3 vibration and with conservation of the rotational quantum number l. The transitions are indicated in the inset of Figure 4.9. In addition to the usual ν_3 band, other very weak sideband series labelled H_1–H_4 and L_1–L_3 have been found that have been attributed to $|0,l\rangle \rightarrow |1,l'\rangle$ transitions with $l \neq l'$ [30]. The excitations observed in phonon spectroscopy are due to $|0,l\rangle \rightarrow |0,l'\rangle$ transitions.

Theoretical calculations of the IR absorption spectrum using a cluster lattice approach [7] predict a strong ν_3 mode at 877 cm^{-1} corresponding with the main absorption peak at 862.5 cm^{-1}. In the same calculation weak ν_1 and ν_2 modes occur at 416 and 230 cm^{-1}, respectively, which so far could not be observed experimentally. Nevertheless, according to the assignment of the 1270 cm^{-1} band (LHeT) to a $\nu_1 + \nu_3$ combination, ν_1 is expected near

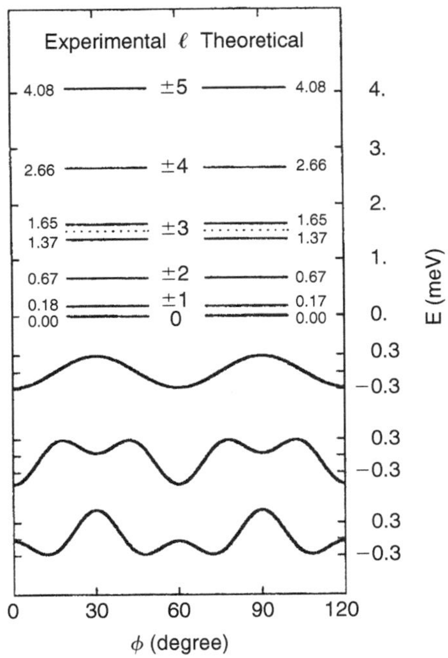

Figure 4.11. Theoretical level spectrum calculated in the hindered rotor model compared with experimental data obtained from phonon spectroscopy (reproduced with permission from Artacho *et al.* [7], Copyright 1997, American Physical Society).

407 cm^{-1} in good agreement with the calculated value. The latter assignments are substantiated by the observed isotopic shifts of the 1270 cm^{-1} band between qmi Ge samples of different Ge mass [7]. The application of density-functional calculations using a Ge supercell has yielded similar results concerning the ν_3 and ν_1 modes as in Refs. [7,24].

In germanium enriched with ^{18}O a ν_3 band shifted to 818 cm^{-1} (LHeT value) has been observed [37]. This band is also present as a small trace in normal oxygen-doped germanium, in agreement with the natural abundance of ^{18}O [28,30,38] and a similar band from ^{17}O with even smaller abundance has been observed at 839 cm^{-1} [30]. In both the bands from ^{18}O and ^{17}O similar fine structure has been resolved as in the 862.5 cm^{-1} band. The ν_3 mode positions corresponding with the different Ge and O isotopes represent an extended set of data, which have been fitted in a quasi-free molecule model, without coupling with the rotation [30]. Assuming a non-linear symmetric Ge_2O molecule in a central force model, the ν_3 frequency ratio for an isotopic molecule with respect to a standard molecule may be expressed as [1]:

$$\frac{\nu_3^{(i)}}{\nu_3} = \frac{\left[\dfrac{M_{Ge}M_O(M_O^{(i)} + 2M_{Ge}^{(i)}\sin^2\alpha)}{M_{Ge}^{(i)}M_O^{(i)}(2M_{Ge}\sin^2\alpha)}\right]^{1/2}}{M_O} \tag{4.5}$$

where the superscript (i) designates quantities of the isotopic molecule and 2α is the Ge—O—Ge apex angle. In a similar model, the binding of the quasi-molecule with the Ge

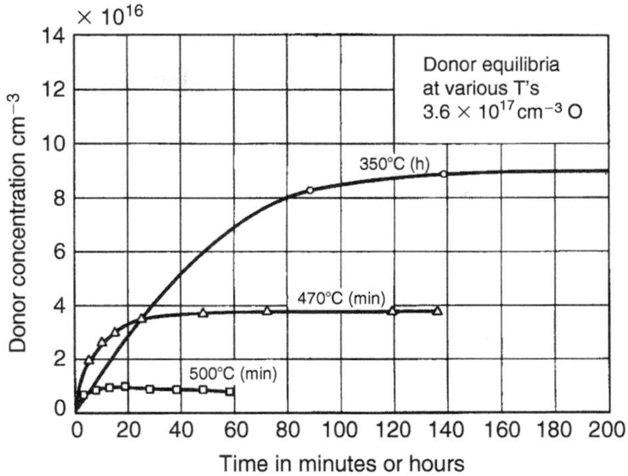

Figure 4.12. Dependence of TD concentration on temperature of anneal of oxygen-doped Ge; the donor concentration has been taken equal to the measured electron concentration (reproduced with permission from Fuller *et al.* [11], Copyright 1961, Elsevier).

lattice may be taken into account by adding an interaction mass m′ to the Ge masses in the formula [39]. The best fit using the most extended data set yields m′ = 11.65 atomic mass units (amu) and $2\alpha = 140°$ [27,30]. Assuming a Ge—O bond length of 0.174 nm, the O atom is found a distance of 0.06 nm away from the Ge—Ge axis. These values are in perfect agreement with the calculated result of Ref. 7 and shown in Figure 4.10.

In ^{18}O enriched germanium, an isotopic shift of the $\nu_1 + \nu_3$ combination band from 1270 to 1216 cm^{-1} is found, implying a shift of about 9 cm^{-1} of the ν_1 mode between ^{16}O and ^{18}O [30]. This small shift is indicative of a small amplitude of the oxygen atom in the mode.

4.3. TDs and the Oxygen Dimer

Oxygen-doped germanium becomes highly supersaturated at temperatures well below the melting point (see Figure 4.5). At room temperature the O_i atoms are immobile and no change occurs. When the material is treated at more elevated temperatures however, the increasing diffusivity of the O_i atoms leads to agglomeration and precipitation of the oxygen. In this section the defects formed during the first stages of oxygen agglomeration are described, i.e. O_2 dimers and TDs consisting of somewhat larger oxygen clusters. Germanium oxide precipitates which form above 550°C will be treated separately.

As a result of extended electrical measurements made around 1960, it was concluded that thermal annealing of oxygen-doped germanium in the range of 300–500°C results in the formation of donors at the expense of the concentration of O_i atoms [4–6,11,12,40]. In the temperature range mentioned, the conversion of O_i into these TDs is reversible and occurs much faster than in silicon, in agreement with the higher oxygen diffusivity. The equilibrium donor concentration decreases with increasing temperature, as shown in Figure 4.12. At 350°C an

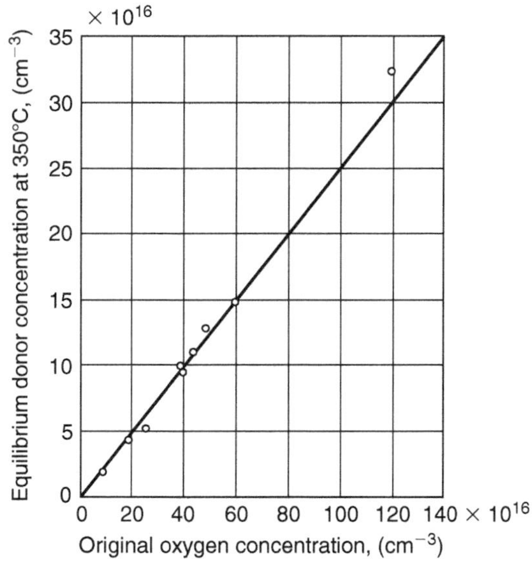

Figure 4.13. Equilibrium TD concentration at 350°C versus oxygen concentration in germanium; the donor concentration has been taken equal to the measured electron concentration (reproduced with permission from Fuller *et al.* [11], Copyright 1961, Elsevier).

almost complete conversion of O_i into TDs may be obtained [11]. In Figure 4.13 the equilibrium donor concentration at this temperature is plotted against the original O_i concentration measured by IR absorption. The dependence is almost linear with slope ¼, which agrees with the early model of TDs as consisting of a cluster of four oxygen atoms, similar to the Kaiser–Frisch–Reiss model for TDs in silicon [41]. Other support for this model was provided by the apparent fourth order of reaction at temperatures above 400°C [12]. The basic O_4 donor was identified with a level at $E_C - 0.017\,\mathrm{eV}$ observed in Hall effect, deeper levels were attributed to species with two or three oxygens [13].

It should be mentioned that in these early studies the donor concentration was taken identical with the measured free electron concentration, assuming single donors. Later investigations have shown, however, that the TDs instead consist of a sequence of double donors (DD) TDn with $n = 1, 2, 3$, etc. similar to the TDs in silicon. Another source of possible misinterpretation is that in Figure 4.13 and in the articles mentioned, the O_i concentration was determined assuming a calibration factor $f_c = 5 \times 10^{16}\,\mathrm{cm}^{-2}$. With our present knowledge it is obvious that the model for the TDs that resulted from the early studies is incorrect; nevertheless the data collected at that time remain a valuable source of information.

4.3.1. Electronic states of TDs

As-grown germanium crystals with $[O_i]$ of the order $10^{17}\,\mathrm{cm}^{-3}$ and with a sufficiently low shallow dopant concentration, invariably are n-type with free electron concentrations up to $n \approx 10^{16}\,\mathrm{cm}^{-3}$, indicating the presence of TDs formed during the cooling of the crystal [14]. The initial TD concentration can be reduced to a low value by the application of a dispersion

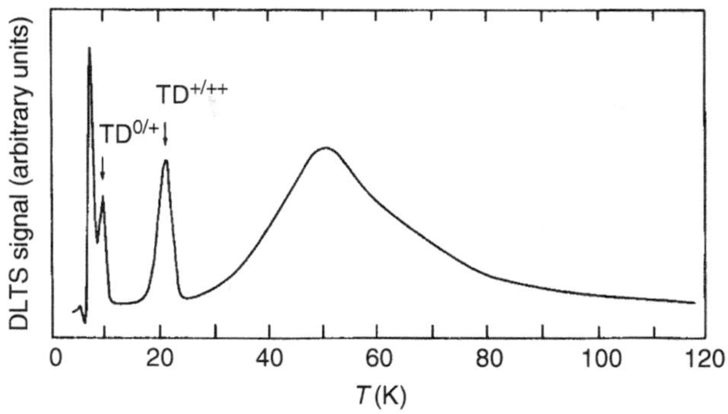

Figure 4.14. DLTS of TD centres in germanium (reproduced with permission from Clauws *et al.* [43], Copyright 1982, Elsevier).

treatment near 900°C followed by a quench to room temperature, after which the total O_i concentration can be measured and the formation of TDs studied.

For a long time the only data on the energy levels of TDs in germanium were provided by Hall effect experiments. Donor levels were reported at 0.017, 0.037 and 0.2 eV below the conduction band [13,42]. More details and an overall picture of the electronic states of the TD family were obtained in the eighties and early nineties, following spectroscopic studies including deep level transient spectroscopy (DLTS), far-IR absorption and photothermal ionisation spectroscopy (PTIS) and electron paramagnetic resonance (EPR); the details may be found in Refs. [43–52].

The first detection of TDs in germanium using DLTS is represented in Figure 4.14. The peaks labelled are due to the 0/+ and +/++ levels of DD with positions in fair agreement with the levels found by Hall effect at $E_C - 0.017$ and $E_C - 0.037$ eV [42]. Peak $TD^{0/+}$ is only clearly observed in moderately oxygen-doped samples with small donor concentration, not suffering from Poole–Frenkel shifts. The far-IR spectrum corresponding with this peak was recorded using PTIS and revealed three series due to effective-mass like shallow donors with slightly different binding energies of 17.2–18.1 meV [14,43]. Definite proof that TDs in germanium are DD was obtained using samples with higher O_i concentration, where the donors are accessible with far-IR absorption [44,45]. An example is shown in Figure 4.15. At low measurement temperature, the spectrum of neutral shallow donors with binding energies of 16–17.3 meV is observed. When the temperature is raised, a gradual transformation is seen into donor series located at higher wavenumber and with four times higher ($2p_{\pm}$–$2p_0$) splitting compared to a neutral donor. This is exactly the behaviour to be expected when a helium-like DD is thermally ionised, in the present case corresponding with the transformation of the TD^0 into the TD^+ charge state.

The donor series in Figure 4.15 correspond with "early" TD*n* species with low *n*, formed during a short thermal treatment at 350°C. Longer treatments result in stronger absorption corresponding with higher donor concentrations in agreement with Figure 4.12 and with donor ground states becoming shallower on the average. The level diagram of TDs constructed from the far-IR spectra is shown in Figure 4.16. Not all the series correspond with different donor

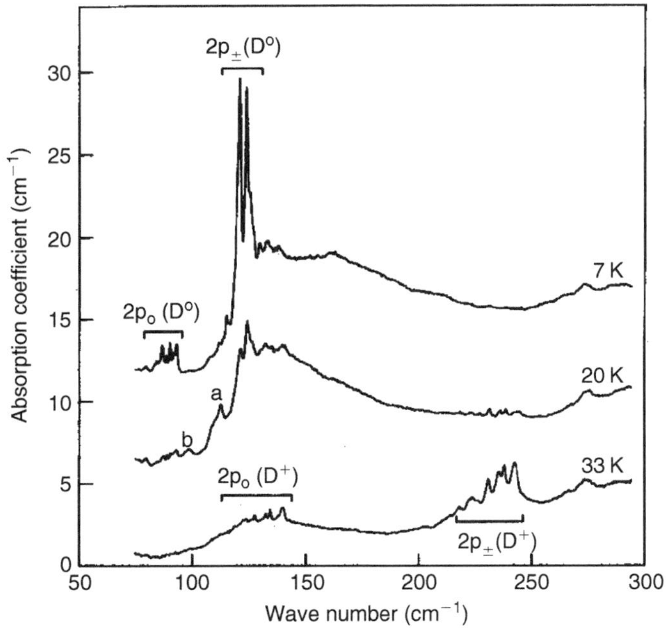

Figure 4.15. Far-IR absorption spectrum of early TDs in germanium (reproduced with permission from Clauws and Vennik [45], Copyright 1984, American Physical Society).

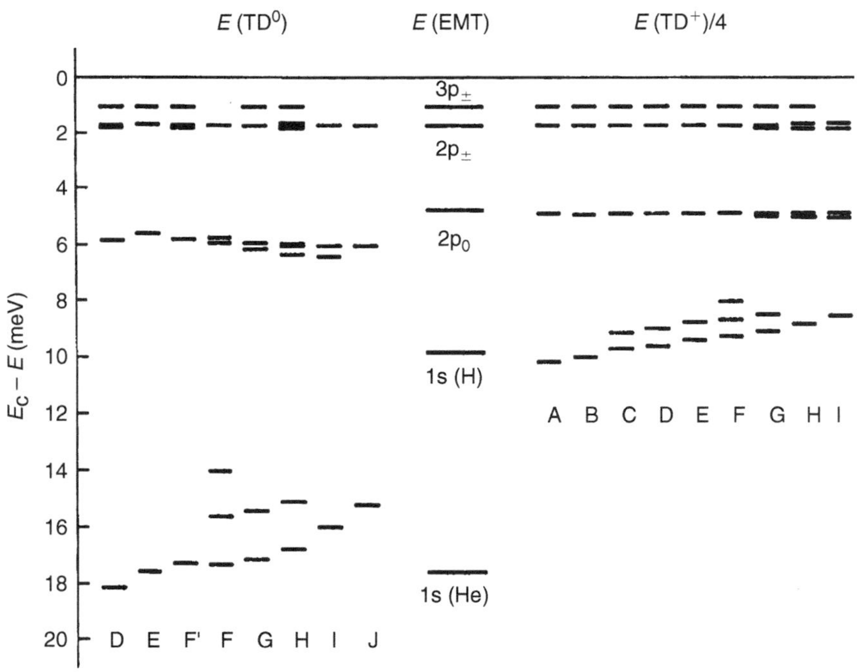

Figure 4.16. TD level spectrum according to the far-IR data of Clauws *et al.* [46,51].

species however: there are, for example, strong indications that the neutral donor series D, E and F′all three belong to the earliest TD (TD1). The diagram demonstrates that TDs in germanium are to be regarded as a family of almost ideal helium-like shallow donors in agreement with effective-mass theory (EMT), with a sequence of ground states becoming gradually more shallow as the donor formation proceeds from early to later formed TD species. The central cell correction for the ground state is small, in contrast with, e.g. column VI DD which have deep ground states. A very similar level system has been found for the TD family in silicon [14,53].

The EPR spectrum of as-grown oxygen-doped germanium is very complicated due to the superposition of spectra belonging to many different TD species [48]. The spectra become simpler in samples submitted to short anneals in order to produce a limited number of different species. Surprisingly, the spectra are observed at the lowest temperatures when the TDs are in the neutral charge state and occupied with two electrons. The spectrum identified with TD2 has been the best documented. The resonances in this spectrum show a doublet splitting and angular dependence that could only be fitted assuming a centre with spin $S = 1$ [48,49,52]. The g-factors are in very good agreement with the ground state wave function of an EMT donor with local C_{2v} symmetry [48]. In total four different spectra could be separated which were assigned to TD2–TD5 and display orthorhombic C_{2v} or <111> axial symmetry (see further).

The EPR spectrum of the TD^+ charge state has been investigated using partially compensated samples. In this case one simple spectrum was observed that could be fitted using $S = ½$ in agreement with a donor with one electron and with axial <111> symmetry [48].

It was observed already long ago that in oxygen-rich germanium slow relaxations of the electric conductivity take place following rapid cooling or illumination [12,54,55]. Hall effect measurements have demonstrated that this phenomenon is due to bistable TD species transformable between two configurations. The first configuration corresponds with the shallow DD states described above, the second is a negative-U configuration with a deep neutral state (X) [56–59]. The bistable behaviour has been a very helpful tool to study the TD system in germanium and to link TD species observed by different experimental methods. Switching between configurations may then be made using light, changes of diode bias or cooling rates and temperature sweeps. Details concerning the transformation mechanisms and activation energies in relation to the configuration coordinate diagram, obtained from DLTS and far-IR spectroscopy, may be found in Refs. [47,50,51].

Figure 4.17 shows the occurrence of bistability in the far-IR spectrum of early TDs, in three samples with a different degree of oxygen aggregation. For each sample, the lower spectrum (1) was taken after slowly cooling in the dark, the upper spectrum (2) after cooling under illumination or quenching to low temperature. For simplicity only the 1 s to $2p_0$ excitation lines are shown. In Figure 4.17(a) the lines (D, E, F′) display 100% bistability, while the other lines remain practically unchanged. In Figure 4.17(b) and (c) the F-lines show full bistability while the G-lines are about 50% bistable in Figure 4.17(c). The behaviour of the spectra has been explained in relation to the Fermi level position in the different samples during cooling [50]. Comparing this information and the time constants of the transformation with similar data on TD species identified by Hall effect analysis [57], the bistable spectra could be linked

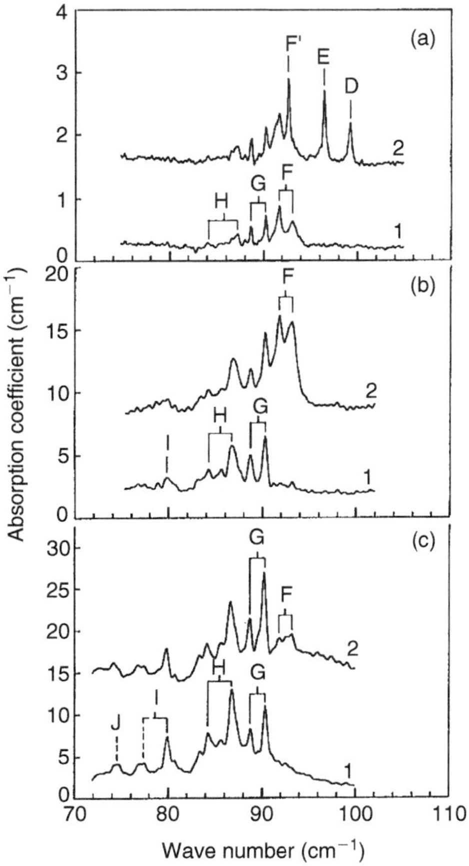

Figure 4.17. Bistability in the far-IR spectrum of TD^0 in germanium samples with different TD content (reproduced with permission from Clauws *et al.* [51], Copyright 1991, American Physical Society).

to TD1–TD3 as indicated in Table 4.1. EPR measurements using the same kind of samples as in Figure 4.17 allowed to make a similar correlation of far-IR and EPR spectra [49,51]. The EPR spectrum of the same sample as in Figure 4.17(c) is shown in Figure 4.18. The spectrum contains four kinds of resonances with doublet splitting in agreement with a $S = 1$ Hamiltonian. The resonances occur at different positions thanks to different symmetries and doublet splittings. The doublets indicated TD2 and TD3 display the same bistability as the far-IR F and G lines, respectively, in Figure 4.17(c). The final result of the comparison is also included in Table 4.1. It turns out that the EPR spectrum of TD2 and TD5 is in agreement with orthorhombic symmetry (C_{2v}) of the centres, while the spectrum of TD3 and TD4 indicates axial <111> symmetry (C_{3v}).

Despite many similarities, the electronic properties of TDs in germanium and silicon also show some differences. One difference is the possibility to observe EPR spectra from neutral TDs with $S = 1$ in germanium, which is not the case in silicon. The occurrence of a state with

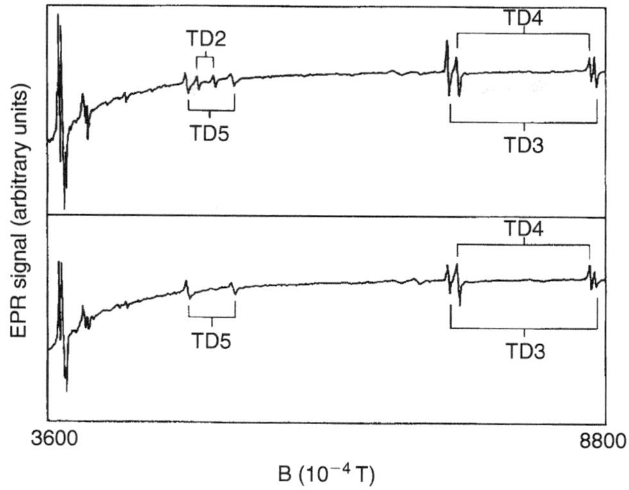

Figure 4.18. EPR spectra corresponding with the far-IR spectrum in Figure 4.17(c), displaying bistability of TD2 and TD3; the lower spectrum was taken after slowly cooling in the dark, the upper spectrum after quenching to 70 K (reproduced with permission from Clauws *et al.* [51], Copyright 1991, American Physical Society).

Table 4.1. Assignment of spectra of the TD^0 state to TD*n* species, resulting from comparison of bistable behaviour observed in electrical, far-IR and EPR measurements [50,51,57]; see Figures 4.17 and 4.18 for identification of the spectra.

TD species	Far-IR	EPR (symmetry)
TD1	D, E, F′	not observed
TD2	F	C_{2v}
TD3	G	<111> axial
TD4	H	<111> axial
TD5	I	C_{2v}
TD6	J	not observed

parallel spins has been attributed to the helium-like nature combined with small level splitting in the ground state of TDs in germanium. The important deviation of the $2p_0$ level of TD^0 from the EMT value, visible in the level diagram of Figure 4.16, is also thought to be a consequence of electron correlation [14]. Another peculiarity of TDs in germanium concerns the compensation by shallow acceptors [60].

4.3.2. Vibrational spectrum of TDs

The vibrational absorption due to TDs has been studied in detail only since the late nineties. Before, weak absorption at room temperature near 609 and 783 cm^{-1} had been reported, with

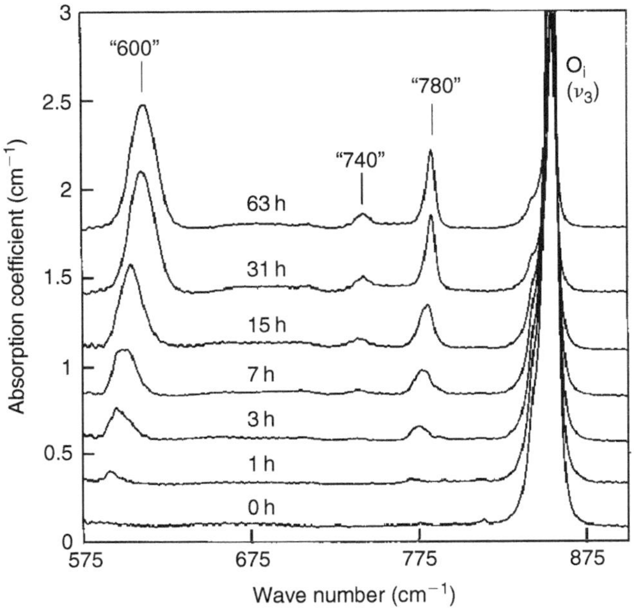

Figure 4.19. LVM spectrum (room temperature) of TDs in oxygen-doped germanium at different times of annealing at 350°C (after Clauws and Vanmeerbeek [15]).

amplitude and exact position depending on the stage of TD formation [6,14]. Figure 4.19 shows the evolution of this absorption in the course of isochronal annealing at 350°C [15]. Two broad bands labelled "600" and "780" develop at the expense of the O_i band at 856 cm^{-1} and shift to higher wavenumber as the annealing proceeds. A third and weaker band labelled "740" displays similar behaviour. The growth of the bands is accompanied by an increase of the electron concentration indicating the formation of TDs. Spectra taken at low temperature and with sufficient resolution reveal that the "600" and "780" bands are in fact composed of different lines which form sequentially in the course of the annealing, in agreement with the assignment of the lines to local vibrational modes (LVMs) of individual TD*n* species [15,61,62]. The evolution of the components in the "600" band is shown in Figure 4.20. A similar behaviour, however with fewer resolved components is displayed by the "780" band. The slightly different wavenumber of LVMs belonging to different TD*n* species, results in an average position of the lines moving to higher wavenumber as the TD formation proceeds and explains the shifts in the spectra at room temperature where the lines are unresolved.

The spectra in Figure 4.20 have been obtained after cooling the sample under illumination, so the observed LVMs correspond with the DD configuration. After cooling in the dark some of the lines disappear or become weaker and new lines appear instead in a nearby wavenumber range; for obvious reasons the new absorption is attributed to the X configuration [15,61,62]. The appearance of the bistable behaviour is shown in Figure 4.21. Again bistability has been used to assign spectra to individual TD*n* species, by comparison with electrical data [61,62] or using the far-IR spectrum as a reference [15]. It follows from the LVM spectrum

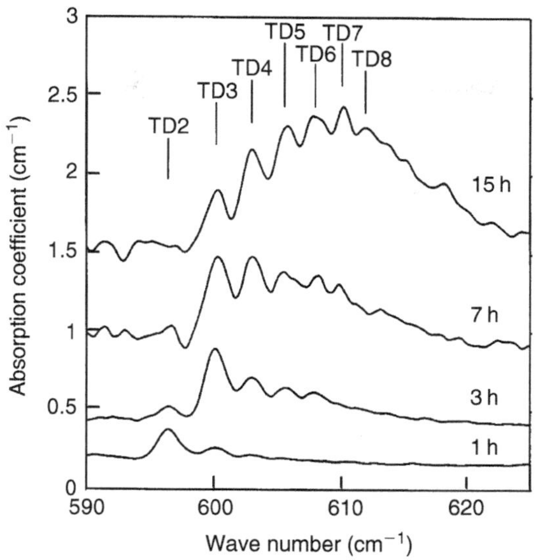

Figure 4.20. Resolved LVM components of TD*n* species in the "600" band measured at 80 K; same sample as in Figure 4.19 (after Clauws and Vanmeerbeek [15]).

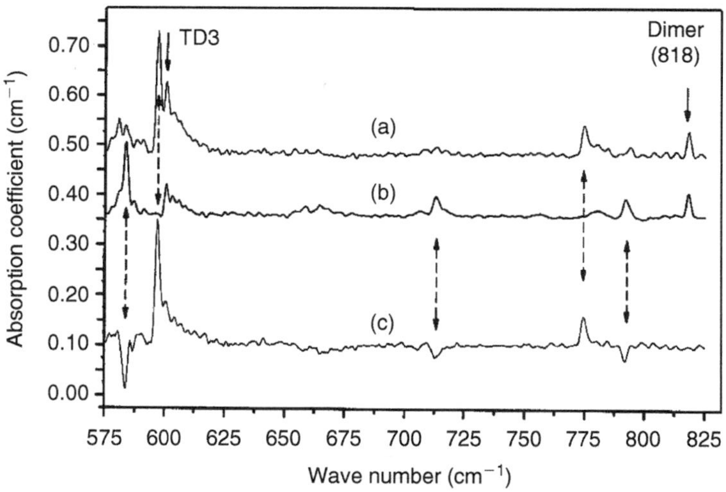

Figure 4.21. Bistability of TD2 in the infrared spectrum of the sample subjected to 1 h annealing at 350°C (see Figure 4.19); measurement at 5 K; (a): sample cooled under band gap illumination; (b): sample cooled in the dark; (c): difference of spectrum (a) and (b); the arrows indicate the LVM lines of TD2 (after Vanmeerbeek [63]).

that in germanium the four earliest TD species TD1 to TD4 are bistable. It should be noted that the LVM absorption of TDs has been the first opportunity for direct experimental observation of the low energy X configuration. The LVM positions assigned to TD*n* species are summarised in Table 4.2; no distinct modes could be separated in the "740" band.

Table 4.2. Experimental LVM positions (in cm^{-1}) of TD*n* species in the DD and X configuration (LHeT) compiled from References [38,61,63].

	TD1	TD2	TD3	TD4	TD5	TD6	TD7	TD8	TD9
DD	593.0	597.0	600.6	603.5	606.1	608.6	610.6	612.5	614.5
	769	774.5	777.6	780.4	782.7	785.1	789.4		
X	581.0	583.6	587.4	590.6					
		712.9	706.0	706.0					
	794.1	791.9	790.5	790.5					

In germanium samples doped with ^{18}O the LVM bands are shifted to lower wavenumbers, confirming that the modes indeed belong to defects incorporating oxygen atoms [61,64]. The shifted line positions may be found in Ref. [61].

It was found using electrical measurements that the TD formation in germanium occurs faster when the samples have been previously irradiated with γ-rays [65,66]. This enhancement has later been confirmed and studied in more detail using LVM spectroscopy of germanium irradiated with MeV electrons [38,67,68]. In similar samples the LVM lines of early TDs already develop at annealing temperatures as low as 200°C, much lower than in unirradiated samples. This item is further addressed in Section 4.3.3 dealing with oxygen dimers.

The LVM spectrum may be used to discuss structural models of the TDs. The fact that distinct modes of DD species are resolved up to TD9 offers an advantage in comparison with the far-IR electronic spectrum, where the spectra suffer from line broadening when the TD concentration is high. The average number of oxygen atoms incorporated in the TD*n* species appearing at a particular annealing stage may be calculated as $\Delta\,[O_i]/N_{TD}$, the interstitial oxygen concentration that has disappeared divided by the TD concentration. The result for the isothermal annealing experiment corresponding with Figure 4.19 is shown in the upper plot of Figure 4.22. In this plot Δ [Oi] has been calculated from the ν_3 band using $f_c = 5 \times 10^{16}\,cm^{-2}$ and N_{TD} has been taken equal to half the free electron concentration measured at room temperature. The average of eight oxygen atoms after long annealing times is in agreement with Figure 4.13 where the TDs were still assumed to be single donors. Comparison of the plot with the average TD*n* number estimated from the "600" band and assuming that at least one oxygen is added between successive TD*n* species, suggests however that the average number of oxygen atoms in the plot may be underestimated and that the calibration factor used for $\Delta\,[O_i]$ may be too low. Another correction that would apply to the early annealing stages may be related to the concentration of bistable TD species; this may explain the smaller average number of oxygens per TD derived in Ref. [10].

The lower plots in Figure 4.22 represent the average number of oxygen atoms per TD that are active in the "600" and "780" modes. This number has been estimated from the integrated absorption of the corresponding bands at room temperature (Figure 4.19), in the assumption that the same conversion factor may be applied as for the integrated absorption of the ν_3 band of O_i (the latter conversion factor is proportional to the calibration factor f_c). Even if an error may be expected to occur on the absolute number of oxygens calculated in this way, the trends that

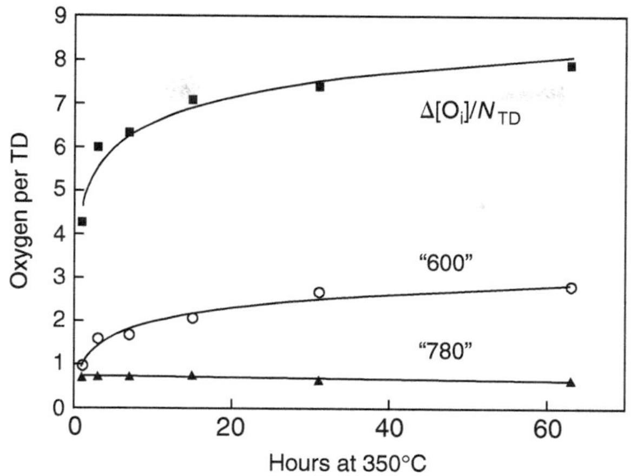

Figure 4.22. Evolution of the number of oxygen atoms in the TDs during isothermal annealing, calculated from the disappeared O_i and the absorption of the "600" and "700" bands (after Clauws and Vanmeerbeek [15]).

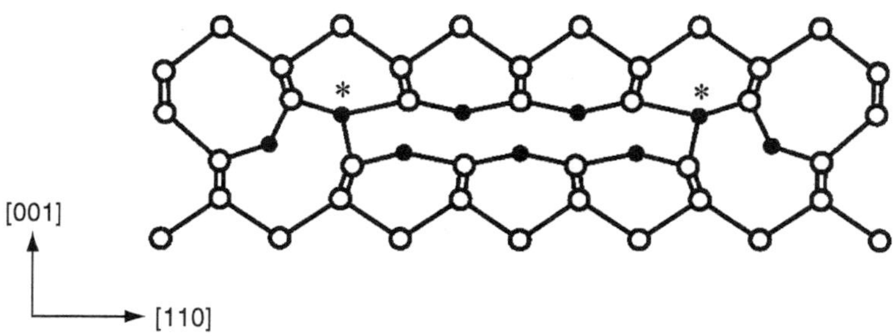

Figure 4.23. TD model in silicon and germanium; open circles: Si/Ge; black circles: O (reproduced with permission from Jones *et al.* [69], Copyright 2001, Elsevier); an asterisk has been added to indicate the over-coordinated oxygen atoms.

follow from a similar analysis are remarkable. The results for the "600" band follow the same trend as the total number of oxygen atoms per TD, suggesting that this mode is due to the agglomerating oxygen. The number of oxygen atoms that oscillate in the "780" mode however remains essentially constant, independent of the average TD*n* species. Taking account of the uncertainty on the exact conversion factor to be used for these modes and a possible underestimation of the calibration factor f_c [10], the number of oxygens per TD vibrating in the "780" mode should be one or two. This suggests that the latter mode is due to the part of the TD that does not change with agglomeration.

Interestingly, the latter conclusions are in agreement with the predictions of a recent model for TDs in silicon and germanium [69–71]. In this model the TDs are composed of a double chain along <110> of clustering O atoms linking second Si or Ge neighbours, terminated by two O atoms with twofold coordination similar to the case of O_i (Figure 4.23). At each side of the chain an over-coordinated O atom is located at the interface between chain and end atoms, being the source of the donor electrons. The calculations predict a lower frequency mode (identified with the "600" band in germanium) due to the vibrations of the O atoms in the chain and a higher frequency mode (identified with the "780" band) due to the two O atoms at the ends. The increasing trend for the "600" mode in Figure 4.22 would then reflect the increasing length of the chain with increasing TDn number, while the constant dependence for the "700" mode is in agreement with vibration of the two end atoms. The occurrence of "mixed modes" in ^{16}O and ^{18}O co-doped samples are interpreted as an indication of interacting O atoms in the chain [70].

In principle other TD models consisting of a constant "core" surrounded by clustering oxygen may also be in agreement with the experimental LVM data. The experiments suggest however that in this case the core should contain one or two O atoms with normal twofold coordination giving rise to the "780" mode.

4.3.3. Vibrational spectrum of the oxygen dimer

The oxygen dimer consisting of a pair of oxygen atoms is considered to be the first stage of oxygen agglomeration starting from dispersed O_i atoms. The rate of loss of O_i atoms from solution is then expected to be proportional to $[O_i]^2$. In silicon the initial rates of oxygen loss and of TD formation are found to be proportional to $[O_i]^n$ with n about 2 in the temperature range below 400°C, which is a strong indication of the involvement of dimers [26]. Similar results have been reported in an early study of the kinetics of TD formation in germanium: the initial rate of TD formation at 350–385°C shows a power dependence with n between 1 and 3 [12], indicating that probably also in this case the dimer formation represents the rate limiting step of oxygen agglomeration at low annealing temperatures.

There is general agreement now that the oxygen dimer in silicon and germanium consists of two O_i atoms in neighbouring sites, noted here as O_{2i}. *Ab initio* calculations predict that in germanium the *staggered* dimer consisting of two Ge—O—Ge entities with a Ge atom in common as shown in Figure 4.24, represents the most stable configuration, with a binding energy of 0.6 eV [24]. Another configuration called the *skewed* dimer would be less stable with a binding energy of 0.1 eV; in this configuration the Ge—O—Ge entities are situated in a different (110) plane and are separated by a Ge—Ge bond [24]. Both dimer configurations are electrically inactive. A third configuration called the double Y-lid was reported

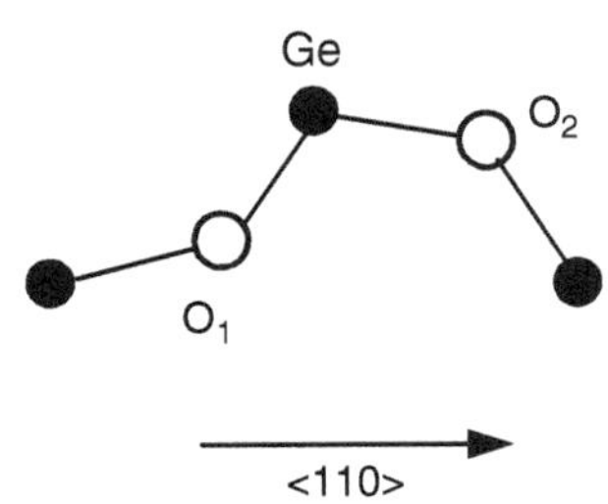

Figure 4.24. Model of the "staggered" oxygen dimer in germanium.

to display shallow donor activity in germanium [72], however another study showed that this defect should be unstable [24].

As a consequence of the electrical inactivity of the dimer, the defect is only directly accessible by its LVM spectrum. The IR vibration bands belonging to the dimer, however, are rather weak and they have only recently been identified in silicon [73,74] and in germanium [38,75–77]. The wavenumbers of the LVM bands that so far have been assigned to oxygen dimers in germanium are listed in Table 4.3, where they are compared with calculated values and with experimental data for silicon.

The modes at 780 and 818 cm^{-1} always occur together in oxygen-doped germanium samples after a dispersion quench from 900°C to room temperature [15]. In similar samples also small concentrations of TD1 and traces of TD2–TD3 are found in the far-IR electronic spectrum, indicating that the bands may be due to an early agglomeration stage of oxygen. The same bands have also been found after thermal anneals at 500°C followed by a quench, in agreement with the expected stability of the oxygen dimer at elevated temperatures [76,77]. In the latter experiments, a subsequent anneal at 300°C resulted in the disappearance of the bands and the appearance of LVM bands due to TD1 and TD2 instead. In samples doped with ^{18}O, shifted counterparts of the two bands have been observed at 741 and 776 cm^{-1}, confirming that the modes are due to vibrating oxygen [76,77]. Interestingly a "mixed" mode at 757 cm^{-1} has been reported to occur in samples doped with both ^{16}O and ^{18}O, indicating that possibly two coupled oxygen atoms are involved in the 780 cm^{-1} mode, in agreement with the assignment to the oxygen dimer. As may be seen in Table 4.3, the positions of the 780 and 818 cm^{-1} bands are in fair agreement with the wavenumber of modes calculated for the staggered dimer. The agreement with calculated isotopic shifts is also excellent [24]; according to the latter the mixed mode at 757 cm^{-1} would be due to the $^{18}O_1$–$^{16}O_2$ combination, with O_1, O_2 defined in Figure 4.24. Further evidence in favour of the assignment to the oxygen dimer lies in the scaling of the peak positions in germanium to those of known dimer peaks in silicon, also listed in Table 4.3; for both peaks a scaling factor of 1.30 is found, close to the value for the ν_3 band of O_i. With this observation it is possible to assign the LVM bands to specific modes of O_{2i}, by comparison with calculated results for silicon [78]. In agreement with the latter, the 818 cm^{-1} mode may be assigned to the stretching vibration with amplitude mainly on the O_1 atom, while the 780 cm^{-1} mode would also be a stretching vibration but with amplitude mainly on O_2.

Table 4.3. Experimental positions (in cm^{-1}, low T) and assignment of infrared LVM bands attributed to oxygen dimers in germanium [38,75–77], compared with calculated values [24] and with corresponding modes in silicon [73,74].

	Ge (experimental)	Ge (calculated)	Si (experimental)
Skewed	857	843–849	1105
Staggered	817.9	784	1059.8
	780.3	749	1012.4
	766.0	–	–
	525.5	517	690.1

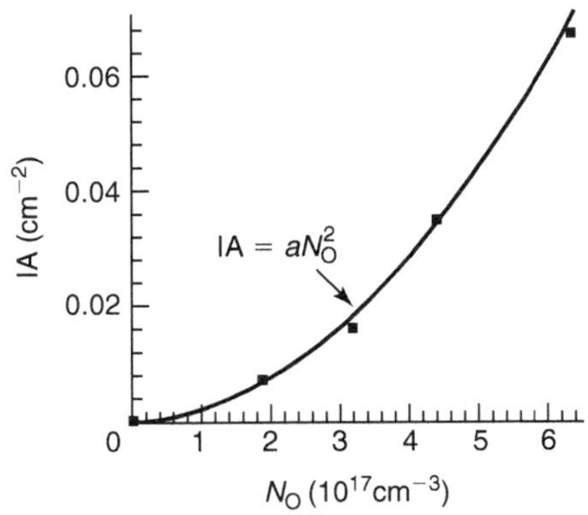

Figure 4.25. Integrated absorption of dimer band in germanium as a function of oxygen concentration, in samples equilibrated at 500°C (reproduced with permission from Litvinov *et al.* [76], Copyright 2001, Springer Science and Business Media).

In the assumption of equilibrium in the reaction $O_i + O_i \leftrightarrow O_{2i}$ and in the absence of competing reactions, the density of dimer centres is expected to vary as [76,77]:

$$[O_{2i}] = [O_i]^2 \ \exp\left(\frac{E_b}{kT}\right) \tag{4.6}$$

where E_b represents the binding energy of the dimer. In Figure 4.25 the evolution of the absorption of one of the dimer modes as a function of the oxygen concentration is shown for samples equilibrated at 500°C. The agreement with the dependence predicted by Equation (4.6) is obvious. From these experiments the binding energy has been estimated as $E_b = 0.35$–0.4 eV [76,77] to be compared with the theoretical value of 0.6 eV [24].

The dimer bands at 780 and 818 cm^{-1} appear with higher amplitude in samples irradiated with fast electrons [38,75–77]. In Figure 4.26 an overview is given of the LVM absorption spectrum of oxygen-doped germanium that has been submitted to 2 MeV electron irradiation followed by isochronal heat treatments. The resolution has been improved in comparison with earlier studies [37,67,79–81] so that overlapping modes are separated and as many as 62 oxygen-related modes could be observed [38].

The dimer absorption at 780 and 818 cm^{-1} reaches a maximum at an annealing temperature of 220–240°C and disappears between 280°C and 320°C. At maximum occurrence, the dimer absorption is one order of magnitude higher than in the as-quenched sample, offering the opportunity to look for weaker dimer modes in the spectrum of the irradiated sample. Two bands at 766 and 525 cm^{-1} have been found to display exactly the same annealing behaviour as the 780 and 818 cm^{-1} bands and are considered likely additional candidates for assignment to the dimer in germanium. As may be seen in Table 4.3, the 525 cm^{-1} band is again in good agreement with the calculated result for the staggered dimer and scales also with a similar band in silicon. Similarly, to the assignment of the 690 cm^{-1} mode in silicon [78], the 525 cm^{-1}

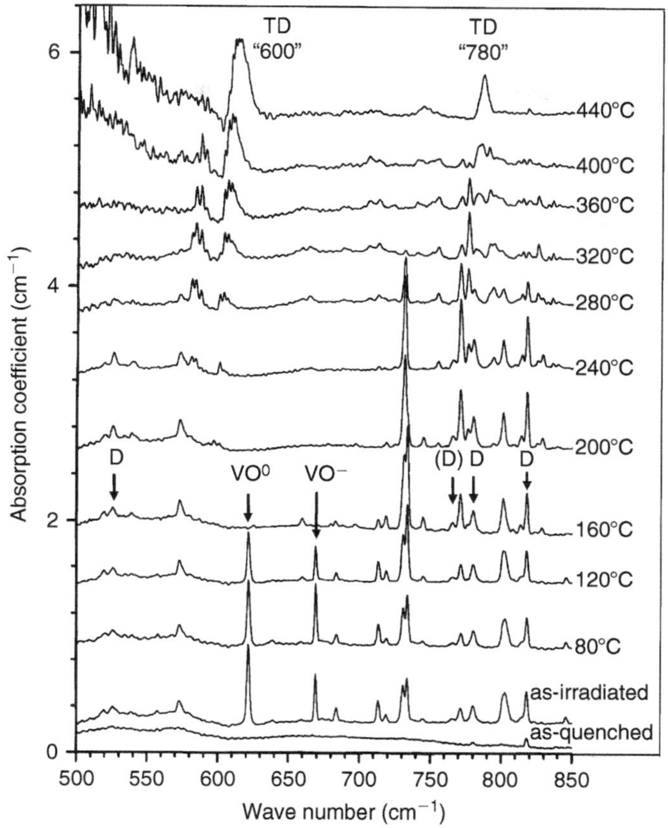

Figure 4.26. LVM absorption spectrum of oxygen-doped germanium irradiated with 2 MeV electrons and subjected to 20 min. isochronal annealing; measurement at 5 K; D: peaks assigned to oxygen dimer (after Vanmeerbeek and Clauws [38]).

mode may then be considered to correspond to an angle-deformation of the Ge—O_1—Ge and Ge—O_2—Ge groups. No counterparts have, however, been found for the 766 cm^{-1} band, so that assignment to the same defect as for the other modes seems less obvious in this case. A mode predicted at 466 cm^{-1} [24] would occur in a less favourable spectral range and has not been observed. Finally, a very weak mode at 857 cm^{-1} with a different behaviour with respect to irradiation and annealing and which scales with a similar mode in silicon (see Table 4.3) has been assigned to the stretching vibration of the skewed dimer [76,77].

As may also be seen in Figure 4.26, the first traces of LVM bands of early TDs already appear after the annealing at 200°C and the "600" and "780" bands develop much faster than in unirradiated samples. The evolution of the dimer absorption and of the absorption due to early TD*n* species during post-irradiation annealing is compared in Figure 4.27. In this diagram a sequential formation of the dimer and of TD1–TD3 appears, supporting the assumption of the dimer being a precursor of the TDs.

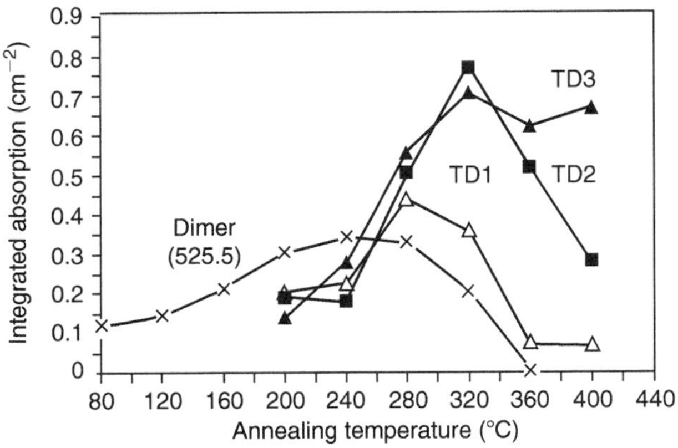

Figure 4.27. Sequential appearance of oxygen dimer and early TD bands during isochronal annealing (see Figure 4.26) (after Vanmeerbeek *et al.* [38,75]).

Similar results concerning the oxygen dimer and the TDs in irradiated germanium have been obtained in Refs. [38,75] and Refs. [76,77] on the other, however, with differences in the detailed behaviour. This is probably due to the differences in the oxygen concentration and in the sample temperature during irradiation.

4.4. Infrared Absorption of Oxygen Precipitates

When Czochralski-grown silicon is treated at elevated temperatures, the supersaturation with oxygen gives rise to the formation of oxide precipitates which may be observed using transmission electron microscopy [82,83]. Another way to address these extended defects is by the IR absorption in the 1100–1200 cm^{-1} range. Polyhedral and platelet precipitates are found to correspond with absorption maxima around 1100 and 1225 cm^{-1}, respectively, and the absorption could be fitted assuming a substoechiometric phase SiO_x [84].

In an early article concerning the solubility of oxygen in germanium, it was reported that in oxygen-doped germanium thermally treated above 600°C, a broad absorption band centred at 11.5 μm could be observed which was attributed to a precipitated GeO_2 phase [5]. This absorption, which is broad and weak, has recently been re-investigated in more detail [16]. Figure 4.28 shows the absorption before and after a prolonged thermal treatment at 560°C. In the spectrum taken before the anneal, LVM absorption due to interstitial oxygen and minor peaks due to oxygen dimers appear. A weak band which remains stable during the thermal treatment is also observed around 1100 cm^{-1}; it is attributed to SiO_x inclusions introduced during crystal growth. After 240 h of anneal, the original O_i concentration has dropped to 20% of the original value and new absorption has developed instead, consisting mainly of a broad band extending between 800 and 1100 cm^{-1}. A secondary band is also seen around 675 cm^{-1}. The integrated absorption of the broad band grows linearly with the concentration of interstitial oxygen lost, in agreement with the formation of a precipitated germanium oxide phase.

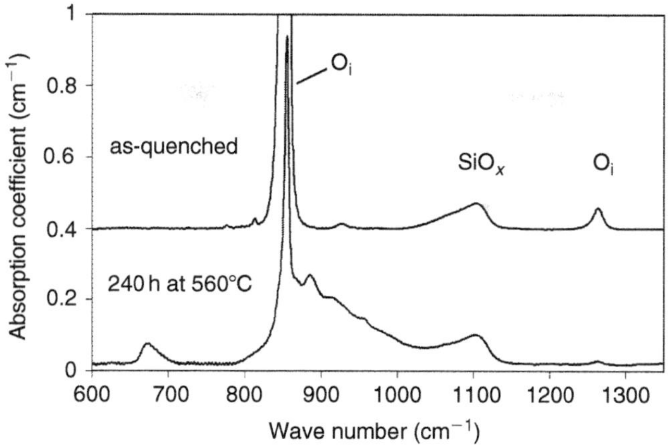

Figure 4.28. Absorption spectrum of oxygen-doped germanium after dispersion quench (upper spectrum) and subsequent precipitation treatment (lower spectrum) (reproduced with permission from De Gryse *et al.* [16], Copyright 2006, Elsevier).

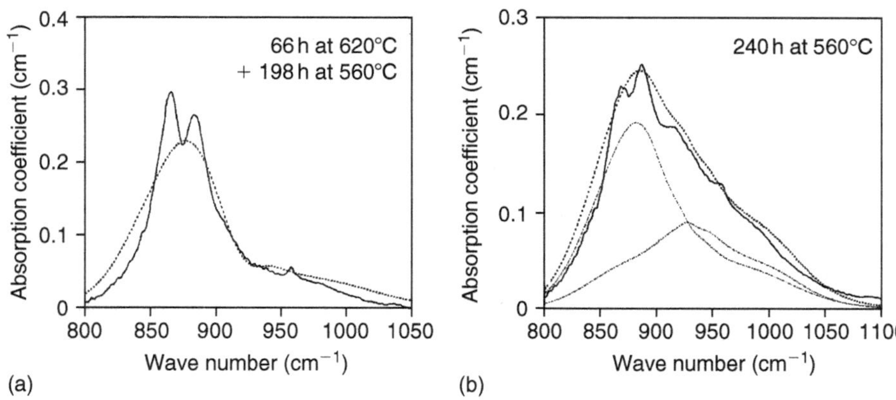

Figure 4.29. Fitting of oxygen precipitate spectra in germanium assuming only spherical (a) or a mixture of spherical and platelet precipitates (b) (reproduced with permission from De Gryse *et al.* [16], Copyright 2006, Elsevier).

The net absorption of the treated sample after the remaining O_i absorption and the SiO_x band have been removed is shown in Figure 4.29b. In Figure 4.29a the result is shown of another sample that has been pre-treated at 620°C before the anneal at 560°C. In the latter case the main band is narrower than in the former spectrum, with a clear double-peaked maximum. In fact the main difference seems to be that in the spectrum of Figure 4.29b an additional broad absorption is present between 900 and 1000 cm^{-1}.

The spectra have been analysed using the same simulation and fitting procedure as was used before to determine the morphology and stoechiometry of SiO_x precipitates in silicon [84]. The precipitates were assumed to be composed of a mixture of Ge and GeO_2 and the measured concentration $\Delta[O_i]$ of the interstitial oxygen lost during the anneal was used as an input parameter. No acceptable fit to the spectra in Figure 4.29a and b could be obtained assuming a calibration factor $f_c = 5 \times 10^{16}\,cm^{-2}$. It was, however, possible to obtain a fair overall fit to the data when the calibration factor was assumed twice as large, which is close to the factor obtained in Ref. [10]. The results of the fitting procedure are shown by the dashed curves in Figure 4.29a and b. A reasonable fit to the spectrum of the pre-treated sample could be obtained assuming spherical precipitates only, while the spectrum of the other sample was fitted assuming both spherical and platelet precipitates. The results seem to indicate that also in germanium oxide precipitates of different morphology may be formed, depending on the annealing scheme. More experiments including TEM observations are, however, needed to confirm the results and to determine the phase of the precipitates. It also remains to be investigated whether oxide precipitates in germanium may offer similar opportunities as oxide precipitates in silicon, e.g. with respect to intrinsic gettering of metal impurities.

4.5. The Vacancy-Oxygen Defect

Apart from the oxygen dimer, the vacancy-oxygen defect (VO) or A-centre is the only oxygen-related irradiation defect that has been experimentally identified in germanium. In an early EPR study, the spectrum of a centre with orthorhombic symmetry was attributed to the single negative charge state VO^- [85], similar to the case of the A-centre in silicon [86]. A LVM at $620\,cm^{-1}$ in electron irradiated germanium was also attributed to VO [37]. More recently, it was demonstrated using DLTS that the A-centre in germanium may occur in three charge states (double negative, negative and neutral). The corresponding energy levels located at $E_C - 0.21\,eV$ and $E_V + 0.27\,eV$ were assigned to the acceptor states $VO^{2-/-}$ and $VO^{-/0}$, respectively [87]. More details on the electronic properties of VO may be found in Chapter 7.

In the IR spectrum of irradiated germanium, three LVM bands occurring at 621.6, 669.1 and $716.2\,cm^{-1}$ have respectively, been assigned to the neutral, negative and double negative charge state of VO [38,75,87–90]. The former two bands are indicated as V^0 and V^- in Figure 4.26. The bands are observed in p-type germanium (doped or as a consequence of the irradiation) and anneal out together around 135°C. The counterparts in silicon, occurring at 835.8 and $885.3\,cm^{-1}$ [91], scale with a factor close to 1.3 as in the case of the bands from O_i and the oxygen dimer. In ^{18}O-doped germanium the modes are shifted to 589.6 and $635.4\,cm^{-1}$ [89]. The absence of mixed modes in the case of doping with ^{16}O and ^{18}O is in agreement with a defect containing a single oxygen atom. The band at $716.2\,cm^{-1}$ assigned to VO^{2-} is only observed in n-type samples, i.e. when the Fermi level is in the upper part of the band gap. In ^{18}O-doped samples the latter band shifts to $680.4\,cm^{-1}$ [89].

When the $669\,cm^{-1}$ (^{16}O) and $635\,cm^{-1}$ (^{18}O) modes from VO^- are measured near LHeT and at high spectral resolution, the bands are resolved into 10 components, as illustrated in Figure 4.30 for one of the modes [90]. The relative strength of the components is in perfect agreement with the abundance of the Ge isotopes in natural germanium, if each component is assigned to a different average mass of two Ge atoms (in fact 11 components are expected, however the component with an average mass of 76 is predicted too weak to be observed). Taking account of all observations, it is straightforward to assign the VO^- LVM bands to a

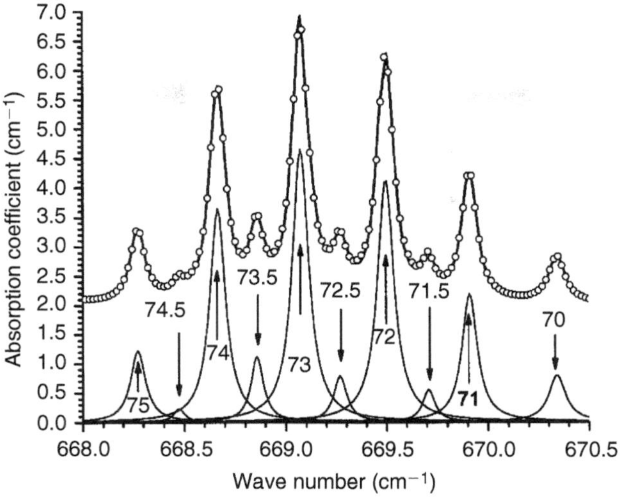

Figure 4.30. Fine structure due to Ge isotopes in the LVM of VO^- in ^{16}O-doped germanium. Upper trace, full line: experimental spectrum. Lower traces: fitted Gaussian components. The sum of the fitted components is represented by dots in the upper trace (after Vanmeerbeek *et al.* [90]) (reproduced with permission from Vanmeerbeek *et al.* [90], Copyright 2004, American Physical Society).

defect containing a Ge—O—Ge quasi-molecule with two equivalent Ge atoms. The situation is in fact very similar to that of the ν_3 mode of O_i, except for the rotational fine structure which is absent in the case of VO. The fine structure is in agreement with the model of the A-centre following from the EPR measurements, i.e. with the O atom forming bonds with two of the Ge-neighbours of the vacancy, while the other two Ge-neighbours form a reconstructed bond [85,86]. It has been possible to observe the same fine structure in the LVM band of VO^{2-} at 716 cm^{-1}, indicating that the same model applies as for VO^-. Remarkably however, no such fine structure occurs for the 621/589 cm^{-1} modes from VO^0 [90].

The components in the 669 and 635 cm^{-1} bands from VO^- form an extended data set of Ge and O isotopic shifts, which may be fitted using Formula (4.5). This results in a fitted Ge—O—Ge apex angle $2\alpha = 102°$, corresponding with the O atom displaced a distance of 0.033 nm from the vacancy. In a similar way a value of $2\alpha = 107°$ is found for VO^{2-}, corresponding with a slightly higher V—O distance [90]. These results seem to indicate that the charge from the one or two extra electrons which are situated in the reconstructed anti-bonding orbitals, push the oxygen away from the vacancy site. If this interpretation is correct, it may be expected that in the case of VO^0 the O atom would be closest to the vacancy site (i.e. closer to fourfold coordination) which may explain the absence of fine structure in the LVM band.

The shift of the VO modes in germanium depending on the charge state is in agreement with similar shifts in the case of silicon. Theoretical calculations of the LVMs of VO in germanium originally failed to predict a substantial shift between the neutral and single negative charge state [24]. Recent progress has, however been made in this respect making use of a Ge cluster model [92].

It may be remarked that fine structure indicative of a symmetric Ge—O—Ge quasi-molecule with $2\alpha = 123°$, has also been found in a mode at $731.4\,cm^{-1}$, due to a defect different from VO. Regarding the higher annealing temperature, this mode may be due to a V_nO defect, although VO_2 may not be excluded [90].

4.6. Conclusions

The important gap in the knowledge concerning oxygen-related defects in germanium compared to silicon has been diminished for a great deal in the last two decades. In general the similarity between those defects in silicon and germanium is amazing, despite some differences in details which seem related to differences in band structure parameters, lattice constant or chemical properties. The knowledge of the properties of interstitial oxygen, oxygen dimers, TDs and the A-centre may be considered as very satisfactory. This is much less the case, however for oxygen precipitates for which only first results have been obtained and for the oxygen-related irradiation defects giving rise to yet unidentified LVM bands, as may be judged from Figure 4.26. Several of the latter are probably due to higher order complexes of the type V_nO_m.

References

[1] G. Herzberg, *Molecular Spectra and Molecular Structure II. Infrared and Raman Spectra of Polyatomic Molecules*, Krieger, Malabar, FL, 1991.

[2] J.W. Corbett, R.S. McDonald and G.D. Watkins, *J. Phys. Chem. Solids* **25** (1964) 873.

[3] W. Kaiser, P.H. Keck and C. Lange, *Phys. Rev.* **101** (1956) 1264.

[4] J. Bloem, C. Haas and P. Penning, *J. Phys. Chem. Solids* **12** (1959) 22.

[5] W. Kaiser and C.D. Thurmond, *J. Appl. Phys.* **32** (1961) 115.

[6] W. Kaiser, *J. Phys. Chem. Solids* **23** (1962) 255.

[7] E. Artacho, F. Ynduráin, B. Pajot, R. Ramírez, C.P. Herrero, L.I. Khirunenko, K.M. Itoh and E.E. Haller, *Phys. Rev. B* **59** (1997) 3820.

[8] E.J. Millett, L.S. Wood and G. Bew, *Brit. J. Appl. Phys.* **16** (1965) 1593.

[9] G.I. Aleksandrova, L.A. Goncharov, M.A. Il'in, E.P. Rashevskaya, P.A. Leonov and A.M. Khorvat, *Ind. Lab.* **42** (1976) 1398.

[10] V.V. Litvinov, B.G. Svensson, L.I. Murin, J.L. Lindström and V.P. Markevich, *Solid State Phen.* **108–109** (2005) 735.

[11] C.S. Fuller, W. Kaiser and C.D. Thurmond, *J. Phys. Chem. Solids* **17** (1961) 301.

[12] C.S. Fuller, *J. Phys. Chem. Solids* **19** (1961) 18.

[13] C.S. Fuller and F.H. Doleiden, *J. Phys. Chem. Solids* **19** (1961) 251.

[14] P. Clauws, *Mater. Sci. Eng. B* **36** (1996) 213.

[15] P. Clauws and P. Vanmeerbeek, *Physica B* **273–274** (1999) 557.

[16] O. De Gryse, P. Vanmeerbeek, J. Vanhellemont and P. Clauws, *Physica B* **376–377** (2006) 113.

[17] R.J. Fox, *IEEE Trans. Nucl. Sci.* **13** (1966) 367.

[18] L.A. Goncharov and N.G. Chavleishvili, *Sov. Phys. Semicond.* **6** (1973) 1888.

[19] L.S. Darken Jr., *J. Electrochem. Soc.* **126** (1979) 827.

[20] L.S. Darken Jr., *J. Electrochem. Soc.* **129** (1982) 226.

[21] C. Vandecasteele and J. Hoste, *Anal. Chim. Acta* **78** (1975) 121.

[22] P.D. Southgate, *Phys. Rev.* **110** (1958) 855.

[23] C. Haas, *J. Phys. Chem. Solids* **15** (1960) 108.

[24] J. Coutinho, R. Jones, P.R. Briddon and S. Öberg, *Phys. Rev. B* **62** (2000) 10824.

[25] V. Gusakov, *J. Phys.: Condens. Matter* **17** (2005) S2285.

[26] R.C. Newman, *J. Phys.: Condens. Matter* **12** (2000) R335.

[27] B. Pajot and P. Clauws, *Proc. ICPS-18*, World Scientific, Singapore (1987) 911.

[28] A.J. Mayur, M.D. Sciacca, M.K. Udo, A.K. Ramdas, K. Itoh, J. Wolk and E.E. Haller, *Phys. Rev. B* **49** (1994) 16293.

[29] B. Pajot, E. Artacho, L.I. Khirunenko, K. Itoh and E.E. Haller, *Mater. Sci. Forum* **258–263** (1997) 41.

[30] B. Pajot, P. Clauws, J.L. Lindström and E. Artacho, *Phys. Rev. B* **62** (2000) 10165.

[31] L.I. Khirunenko, V.I. Shakotsov, V.K. Shinkarenko and F.M. Vorobkalo, *Sov. Phys. Semicond.* **24** (1990) 663.

[32] L.I. Khirunenko, V.I. Shakotsov and V.K. Shinkarenko, *Mater. Sci. Forum* **83–87** (1992) 425.

[33] M. Gienger, M. Glaser and K. Lassmann, *Solid State Com.* **86** (1993) 285.

[34] N. Aichele, U. Gommel, K. Lassmann, F. Maier, F. Zeller, E.E. Haller, K.M. Itoh, L.I. Khirunenko, V. Shakovtsov, B. Pajot, E. Fogarassy and H. Müssig, *Mater. Sci. Forum* **258–263** (1997) 47.

[35] K. Lassmann, C. Linsenmaier, F. Maier, F. Zeller, E.E. Haller, K.M. Itoh, L.I. Khirunenko, B. Pajot and H. Müssig, *Physica B* **263–264** (1999) 384.

[36] H. Yamada-Kaneta, C. Kaneta and T. Ogawa, *Phys. Rev. B* **42** (1990) 9650.

[37] R.E. Whan, *Phys. Rev.* **140** (1965) A690.

[38] P. Vanmeerbeek and P. Clauws, *Phys. Rev. B* **64** (2001) 245201.

[39] B. Pajot and B. Cales, *Mater. Res. Soc. Symp. Proc.* **59** (1986) 39.

[40] G. Elliott, *Nature* **180** (1957) 1350.

[41] W. Kaiser, H.L. Frisch and H. Reiss, *Phys. Rev.* **112** (1958) 1546.

[42] L.A. Goncharov, P.A. Leonov and A.M. Khorvat, *Sov. Phys. Semicond.* **13** (1979) 1336.

[43] P. Clauws, J. Broeckx, E. Simoen and J. Vennik, *Solid State Com.* **44** (1982) 1011.

[44] P. Clauws, E. Simoen and J. Vennik, *Proc. ICDS-13*, Coronado, 1984, Metallurgical Society AIME (1985) 911.

[45] P. Clauws and J. Vennik, *Phys. Rev. B* **30** (1984) 4837.

[46] P. Clauws and J. Vennik, *Mater. Sci. Forum* **10–12** (1986) 941.

[47] P. Clauws and J. Vennik, *Mater. Sci. Forum* **38–41** (1989) 483.

[48] F. Callens, P. Clauws, P. Matthys, E. Boesman and J. Vennik, *Phys. Rev. B* **39** (1989) 11175.

[49] H.H.T.P. Bekman, T. Gregorkiewicz, I.F.A. Hidayat, C.A.J. Ammerlaan and P. Clauws, *Phys. Rev. B* **42** (1990) 9802.

[50] P. Clauws and J. Vennik, *Mater. Sci. Forum* **65–66** (1990) 339.

[51] P. Clauws, F. Callens, F. Maes, J. Vennik and E. Boesman, *Phys. Rev. B* **44** (1991) 3665.

[52] D. Israel, F. Callens, P. Clauws, and P. Matthys, *Solid State Com.* **82** (1992) 215.

[53] P. Wagner and J. Hage, *Appl. Phys. A* **49** (1989) 123.

[54] E. Adachi, *J. Phys. Chem. Solids* **28** (1967) 1821.

[55] V.E. Gusakov, V.I. Urenev and V.I. Yavid, *Sov. Phys. Semicond.* **10** (1976) 580.

[56] V.V. Litvinov and V.I. Urenev, *Sov. Phys. Semicond.* **17** (1983) 689.

[57] V.V. Litvinov, G.V. Pal'chick and V.I. Urenev, *Sov. Phys. Semicond.* **19** (1985) 841.

[58] V.V. Litvinov, G.V. Pal'chick and V.I. Urenev, *Sov. Phys. Semicond.* **21** (1987) 1281.

[59] V.V. Litvinov, G.V. Pal'chick and V.I. Urenev, *Phys. Status Solidi A* **108** (1988) 311.

[60] D.M. Watson, M.T. Guptill, J.E. Huffman, T.N. Krabach and S.N. Raines, *Phys. Rev. B* **49** (1994) 16361.

[61] V.P. Markevich, L.I. Murin, V.V. Litvonov, A.A. Kletchko and J.L. Lindström, *Physica B* **273–274** (1999) 570.

[62] V.V. Litvinov, A.A. Klechko, V.P. Markevich, L.I. Murin and J.L. Lindström, *Solid State Phen.* **69–70** (1999) 303.

[63] P. Vanmeerbeek, PhD thesis, Ghent University (2004).

[64] L.I. Murin, J.L. Lindström, V.P. Markevich, T. Hallberg, V.V. Litvinov, J. Coutinho, R. Jones, P.R. Briddon and S. Öberg, *Physica B* **308–310** (2001) 290.

[65] V.V. Litvinov and V.I. Urenev, *Sov. Phys. Semicond.* **18** (1984) 445.

[66] V.V. Litvinov, G.V. Pal'chick and V.I. Urenev, *Phys. Status Solidi A* **115** (1989) K9

[67] A.A. Klechko, V.V. Litvinov, V.P. Markevich and L.I. Murin, *Semiconductors* **33** (1999) 1163.

[68] V.V. Litvinov, L.I. Murin, V.P. Markevich, J.L. Lindström and A.A. Klechko, *Physica B* **302–303** (2001) 201.

[69] R. Jones, J. Coutinho, S. Öberg and P.R. Briddon, *Physica B* **308–310** (2001) 8.

[70] L.I. Murin, J.L. Lindström, V.P. Markevich, T. Hallberg, V.V. Litvinov, J. Coutinho, R. Jones, P.R. Briddon and S. Öberg, *Physica B* **308–310** (2001) 290.

[71] J. Coutinho, R. Jones, L.I. Murin, V.P. Markevich, J.L. Lindström, S. Öberg and P.R. Briddon, *Phys. Rev. Lett.* **87** (2001) 235501.

[72] D.J. Chadi, *Phys. Rev. Lett.* **77** (1996) 861.

[73] L.I. Murin, T. Hallberg, V.P. Markevich and J.L. Lindström, *Phys. Rev. Lett.* **80** (1998) 93.

[74] S. Öberg, C.P. Ewels, R. Jones, T. Hallberg, J.L. Lindström, L.I. Murin and P.R. Briddon, *Phys. Rev. Lett.* **81** (1998) 2930.

[75] P. Vanmeerbeek, P. Clauws and W. Mondelaers, *Physica B* **308–310** (2001) 517.

[76] V.V. Litvinov, L.I. Murin, J.L. Lindström, V.P. Markevich and A.A. Klechko, *Semiconductors* **35** (2001) 864.

[77] V.V. Litvinov, L.I. Murin, J.L. Lindström, V.P. Markevich and A.A. Klechko, *Solid State Phen.* **82–84** (2002) 105.

[78] M. Pesola, J. von Boehm and R.M. Nieminen, *Phys. Rev. Lett.* **82** (1999) 4022.

[79] R.E. Whan and H.J. Stein, *Appl. Phys. Lett.* **3** (1963) 187.

[80] J.F. Becker and J.C. Corelli, *J. Appl. Phys.* **36** (1965) 3606.

[81] R.E. Whan, *Appl. Phys. Lett.* **6** (1965) 221.

[82] H. Bender, *Phys. Status Solidi A* **86** (1984) 245.

[83] K. Sueoka, N. Ikeda, T. Yamamoto and S. Kobayashi, *J. Appl. Phys.* **74** (1993) 5437.

[84] O. De Gryse, P. Clauws, J. Van Landuyt, O. Lebedev, C. Claeys, E. Simoen and J. Vanhellemont, *J. Appl. Phys.* **91** (2002) 2493.

[85] J.A. Baldwin Jr., *J. Appl. Phys.* **36** (1965) 793.

[86] G.D. Watkins and J.W. Corbett, *Phys. Rev.* **121** (1961) 1001.

[87] V.P. Markevich, I.D. Hawkins, A.R. Peaker, V.V. Litvinov, L.I. Murin, L. Dobaczewski and J.L. Lindström, *Appl. Phys. Lett.* **81** (2002) 1821.

[88] V.P. Markevich, V.V. Litvinov, L. Dobaczewski, J.L. Lindström, L.I. Murin and A.R. Peaker, *Phys. Stat. Sol (C)* **0** (2003) 702.

[89] V.V. Litvinov, L.I. Murin, J.L. Lindström, V.P. Markevich and A.N. Petukh, *Semiconductors* **36** (2002) 621.

[90] P. Vanmeerbeek, P. Clauws, H. Vrielinck, B. Pajot, L. Van Hoorebeke and A. Nylandsted Larsen, *Phys. Rev. B* **70** (2004) 035203.

[91] J.L. Lindström, L.I. Murin, V.P. Markevich, T. Hallberg, and B. Svensson, *Physica B* **273–274** (1999) 291.

[92] A. Carvalho, R. Jones, J. Coutinho, M. Shaw, V.J.B Torres, P.R. Briddon and S. Oberg, *Mat. Sci In Semicond. Process.* **9** (2006) 489.

CHAPTER 5

Metals in Germanium

Eddy Simoen and Cor Claeys

5.1. Introduction

While Group III and V elements are widely employed in dopant engineering, the use of Group I and II metals is mostly avoided, as they generally introduce deep levels in the band gap of silicon or germanium. These deep levels can act as leakage current (*generation*) centers or carrier lifetime killing (*recombination*) centers. In other cases, charge can be trapped for some time (*traps*). The exact electrical behavior of such deep-level center will largely depend on its energy level (E_T), the concentration (N_T) and the capture cross section for electrons (σ_n) and holes (σ_p). For most applications, a low leakage current and/or a high lifetime are required, implying that low concentrations of metals in the bulk of the wafer should be preserved in the starting material and during processing. Moreover, for some common metals, like e.g. interstitial iron (Fe_i) in silicon, only trace amounts ($\sim 10^{11}$–$10^{13}\,cm^{-3}$) suffice to control the electrical properties of the substrate. The fact that the diffusivity (D) of metals in Group IV semiconductors is high indicates that they can be easily introduced from the surface during a heat treatment step. In addition, their solubility (S_0) is usually quite small, so that precipitates are normally formed upon subsequent cooling from high temperatures.

All this strongly suggests that it is worthwhile to avoid the introduction of metals by applying optimized surface cleans and, in the case of inadvertent contamination, the implementation of a gettering procedure, where the unwanted elements are removed from the device areas and trapped in a predetermined defective zone, sufficiently far from the electrically active layer. However, in current deep submicron complementary-metal-oxide-semiconductor (CMOS), metals are utilized in various processing steps, from gate dielectric

Germanium-Based Technologies: From Materials to Devices. C. Claeys and E. Simoen (Editors)

deposition, over silicidation (or germanidation), to metallization. Certain metals may also be introduced during the crystal growth or the wafering process (Cu in-diffusion during polishing, for example). This emphasizes that the study and understanding of the behavior of metals in silicon or germanium is still of key importance.

In this chapter, a summary will be given concerning the properties of transition metals (TMs) in germanium which are the fastest diffusing metals and, hence, potential contaminants, as they are often encountered in process equipment. Data is given on the distribution coefficient, the solubility and the diffusion coefficient versus temperature. A second class of parameters that will be addressed relates to the electrical properties, describing the energy levels and corresponding capture cross sections, the impact on the carrier lifetime or trapping, etc.

As a first element, Cu will be discussed, since this is by far the most notorious and best studied metal contaminant in germanium. Subsequently, attention will be given to Ag and Au, belonging to the same Group Ib as copper. This should reveal possible chemical trends. Moreover, Au doping has been used in the past to modify the optical properties of germanium photoconductors. Next, the properties of Ni are described, which behaves quite analogous to copper in Ge. The other TMs are the subject of a following section. Finally, chemical trends in the electrical and optical properties of metals in Ge will be discussed, followed by some general conclusions.

5.2. Copper in Germanium

It was soon realized that Cu is the dominant contaminant in germanium for temperature treatments above 500°C (see, e.g., Chapter 3 and also Refs [1,2]). The reason for this fact is the high diffusivity, which at high temperatures can amount to a few times $10^{-5}\,cm^2 s^{-1}$, a value typical for diffusion via interstitial sites in solids (interstitial diffusion) and not so far below the diffusion coefficient of solutes in liquids. Based on the assumption of copper as an interstitial impurity in the germanium lattice, it was expected that it behaves like a donor, similar as Li. Surprisingly, material doped with copper in the melt turned out to show p-type behavior, pointing to an acceptor nature of Cu and, therefore, a substitutional lattice site. This apparent contradiction puzzled many researchers for some time and, finally, became resolved by Frank and Turnbull [3], who proposed the so-called dissociative diffusion mechanism for copper (and many other metals) in germanium. In the following section, the different properties of Cu will be detailed.

5.2.1. Distribution coefficient k_d

Figure 5.1 represents the phase diagram of Cu in Ge [4]. A distribution coefficient $k_d = C_S/C_L \sim 1.5 \times 10^{-5}$ has been obtained [4,5] demonstrating that copper has a tendency to stay in the liquid phase and will enrich toward the tail end of the growing crystal. Nevertheless, the k_d of Cu is anomalously low compared with Sb for example [5], an element with a similar tetrahedral covalent radius. This turns out to be connected to its transport via interstitial sites and its rather large radius, which limits the copper solubility in the

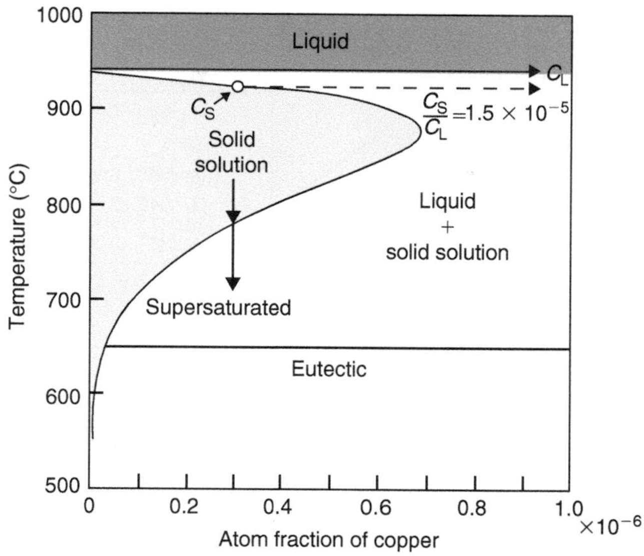

Figure 5.1. Phase diagram for Cu in Ge (after Burton [4]).

lattice. At the same time, an interstitial impurity will take up its lowest possible charge state in order to lower its ionic radius.

From Figure 5.1, one can also derive a maximum solubility around 875°C, so that Cu exhibits a retrograde solubility in Ge as many other elements. In addition, one should note that the Cu–Ge system has an eutectic point at 650°C.

5.2.2. Configurations of atomic Cu in Ge

The substitutional character of Cu in germanium was derived from the observation that after quenching from high temperatures of Cu-diffused (or even undiffused) n-type material, p-type conduction was systematically found in resistivity or Hall effect measurements (see, e.g., Refs [6–9] for some of the earlier work). Likewise, crystals grown from a melt doped with Cu inevitably yielded p-type material at room temperature. This suggests that Cu shows acceptor behavior, pointing to a predominant substitutional incorporation in the lattice. As Cu_s can accept three electrons to complete the covalent bonding, one expects triple acceptor behavior. This was finally confirmed by Woodbury and Tyler [10], who found three acceptor levels by Hall measurements versus temperature study: at $E_V + 0.04\,eV$ ($Cu_s^{0/-}$), $E_V + 0.32\,eV$ ($Cu_s^{-/2-}$) and at $E_C - 0.26\,eV$ ($Cu_s^{2-/3-}$), with E_V the top of the valence band and E_C the bottom of the conduction band. This conclusion was further validated by a comparison between resistivity and radiotracer measurements showing an acceptable agreement between the two techniques [11].

Proof for the existence of interstitial Cu (Cu_i) in Ge came from electric-field drift experiments (1–10 V cm^{-1}) at high temperatures (800–850°C), while passing a large heating current

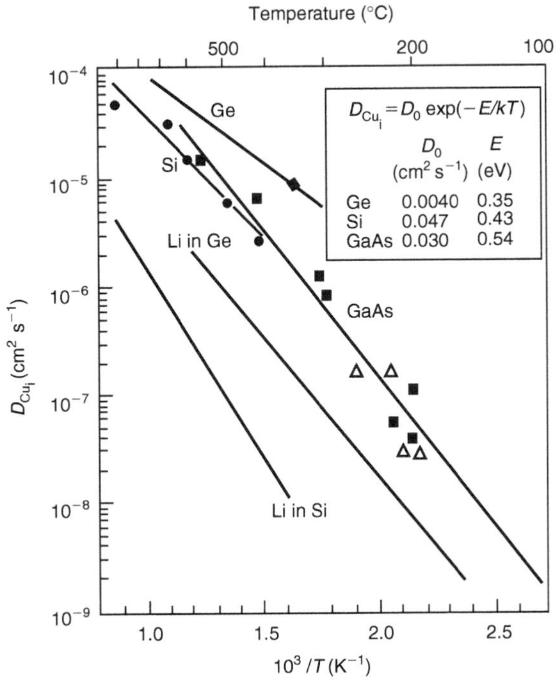

Figure 5.2. Interstitial Cu diffusion measured in extrinsic p-type Ge (◆), Si (●), GaAs (■). ■ Out-diffusion in GaAs. Δ Drift in GaAs. ● Cu diffusion in intrinsic Si. Diffusion of Li in Ge and Si is included for comparison (after Hall and Racette [14]).

through the Ge sample [12]. It was observed that Cu drifted toward the negative electrode and, hence, moved as a positive ion (donor) in the electric field at 800°C. The mobility of the ions was found to obey Einstein's relation [12], given by:

$$D = \mu kT/q = (kT/q)(x_d/Ft_A) \tag{5.1}$$

with μ the ion mobility, k the Boltzmann's constant, T the absolute temperature and q the elementary charge, assuming Cu_i is single positive charged (smallest interstitial radius). Further, we have that F is the electric field and x_d the average distance over which the ions drift during the (effective) annealing time t_A [13]. It was concluded that Cu moves as a singly charged positive ion toward the negative electrode with a diffusion constant $D_{Cu_i} = 2.2 \times 10^{-5} cm^2 s^{-1}$, in line with part of the data in Figure 5.2 [12,13]. Tracer experiments in extrinsic p-type Ge, where Cu_i is the dominant species, pointed out that the activation energy for Cu_i diffusion amounts to ~0.33 eV [14], so that D_{Cu_i} is only weakly temperature dependent, as shown in Figure 5.2.

The high value of the interstitial copper diffusivity (even higher than for Li) in Ge has been theoretically explained by Weiser [15], who considered the contribution from electrostatic interaction between the positive ion and the host lattice to the related activation energy. The polarization energy and Coulomb repulsive energy have opposite effects, so that for some optimum sized ion, a nearly zero activation energy can be obtained. Copper is close to this optimum [15]. It was, furthermore, pointed out that the equilibrium interstitial site of Cu_i is probably the hexagonal (H) and not the tetragonal (T) one, so that the diffusion path goes from H → T → H, with T as the saddle point (energy maximum). A theoretical value for the activation energy of 0.22 eV was derived, in reasonable agreement with the measured value of Figure 5.2.

Further evidence for the existence of Cu_i came from low-temperature (77 K) electron irradiation experiments [16], where the carrier removal rate in n-type Ge was studied as a function of temperature in Cu-doped and undoped material. Removal of one Cu_s by interaction with a germanium interstitial I through the kick-out mechanism:

$$Cu_s + I \rightarrow Cu_i \tag{5.2}$$

should add four electrons to the conduction band (remove three holes + add one electron for the expected Cu_i donor). The irradiations were performed at 77 K to prevent Cu_i from migrating, whereby Cu_i diffuses out at $T > 300°C$. Annealing studies at different temperatures demonstrated that up to 260 K, a recovery of Cu_s took place by interaction with radiation-induced Ge vacancies, through the reaction:

$$Cu_i + V \rightarrow Cu_s \tag{5.3}$$

However, only 14 % of the originally kicked-out Cu_s returned to a substitutional position, while the other 86% persists in some other form, either as Cu_i or in Cu-related complexes, which were shown up to 300 K. It was speculated that the pairing of Cu_s^{3-} and Cu_i^{+} was highly probable due to the Coulombic attraction [16].[1]

5.2.3. *The dissociative copper diffusion mechanism*

The early diffusion data showed a large spread in Cu diffusivities even at a single temperature (for an overview of the early results see Ref. [18]). It was soon discovered that one of the key factors determining the diffusivity was the structural perfection (or defectiveness) of the crystals: germanium with a high dislocation density (n_{disl}) exhibits a high D_{Cu} ($D > 4 \times 10^{-5} cm^2 s^{-1}$ for material with small-angle grain boundaries), while low defect density material shows a $D \sim 2 \times 10^{-8} cm^2 s^{-1}$ [19]. At the same time, it was demonstrated that there was no preferential diffusion along the dislocation lines (so-called pipe diffusion) [20]. The main role of the dislocations is to act as a source of vacancies, for assisting the Cu transport. van der Maesen and Brenkman were the first to realize the importance of the equilibrium between Cu_s and Cu_i, whereby the fast transport could be ascribed to the interstitial species and the solubility was mainly substitutional [21]. Finally, a coherent model

[1] Note that the kick-out reaction was first demonstrated for shallow acceptors (B, Al) in silicon after 4.2 K electron irradiation and subsequently also in Ge for shallow dopants [17].

was developed by Frank and Turnbull [3], who acknowledged the important role of vacancies in the so-called dissociation reaction:

$$Cu_s \leftrightarrow Cu_i + V \tag{5.4}$$

In accordance with reaction (5.4), the fractional concentrations of substitutional and interstitial copper (C_{Cu_i}, C_{Cu_s}) and of vacancies (C_V) will be related by the law of mass action:

$$C_{Cu_s} = KC_{Cu_i}C_V \tag{5.5}$$

where K is a reaction constant, depending only on the temperature.

In the frame of this model, the effective diffusion coefficient in dislocated material, where a sufficient supply of vacancies exists to maintain C_V at its equilibrium value, becomes equal to the interstitial diffusivity (i.e. $D_{Cu_i} = 4 \times 10^{-3}\,cm^2 s^{-1}$) times the fraction of the time a copper atoms spends in the interstitial state of high diffusivity, i.e. $C^{eq}_{Cu_i}/(C^{eq}_{Cu_i} + C^{eq}_{Cu_s})$ with $C^{eq}_{Cu_i}$ and $C^{eq}_{Cu_s}$, the equilibrium fractional concentration (atomic fraction) of copper in the interstitial and substitutional state, respectively. This yields:

$$D_{Cu} = D_{Cu_i}\frac{C^{eq}_{Cu_i}}{C^{eq}_{Cu_i} + C^{eq}_{Cu_s}} = D_{Cu}\frac{1}{1+\alpha_i} \tag{5.6}$$

Identifying the observed diffusivities in the relatively imperfect crystals with D_{Cu}, one arrives at $\alpha_i = C^{eq}_{Cu_s}/C^{eq}_{Cu_i} \approx 100$, which is in the first instance confirmed by the data of Figure 5.3 [18]. The corresponding absolute solubilities can be written as [18]:

$$S^{eq}_{Cu_s} = C^{eq}_{Cu_s} n_a = 4.8 \times 10^{23} \exp(-1.58\,eV/kT)\ (cm^{-3}) \tag{5.7a}$$

and:

$$S^{eq}_{Cu_i} = C^{eq}_{Cu_i} n_a = 8.5 \times 10^{21} \exp(-1.46\,eV/kT)\ (cm^{-3}) \tag{5.7b}$$

In Equations (5.7), n_a is the atomic density (4.413×10^{22} at cm^{-3} for Ge). Equation (5.7b) is in fair agreement with results derived from internal friction measurements [22] or the tracer data on highly extrinsic p-Ge by Hall and Racette [14] (see Figure 5.2).

On the other hand, with an initial fractional concentration of vacancies equal to its equilibrium value C^{eq}_V and no immediate supply of vacancies by, e.g. dislocations (highly perfect material), one will have for the dissociative mechanism of in-diffusing copper that [3]:

$$D_{Cu} = D_V\frac{C_V}{C_V + C_{Cu_s}} \tag{5.8}$$

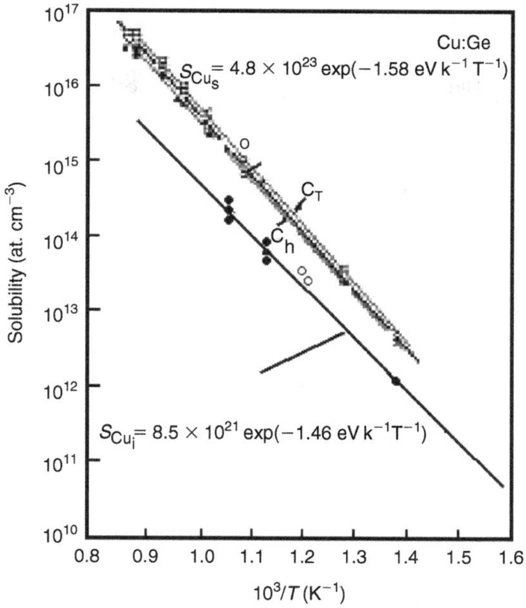

Figure 5.3. Temperature dependence of the interstitial and substitutional solid solubility of copper in germanium (after Huntley [18]).

which is valid for diffusion in near perfect crystals, with only the surface as a source of vacancies. The migration rate in low-defective material is thus equal to that of the slower moving species (vacancies compared with Cu_i) times the proportion of the time that the vacancies are free (from being occupied by a copper atom). Equation (5.8) can be further transformed into [3]:

$$D_{Cu} = \frac{D_V}{1 + KC_{Cu_i}} = \frac{D_{Ge}}{C_V^{eq} + \alpha_i C_{Cu_i}} \tag{5.9}$$

In Equation (5.9), D_{Ge} is the self-diffusion coefficient of Ge, indicating that the substitutional transport is occurring at a similar slow rate. This diffusivity is not a constant but diminishes with interstitial copper concentration, i.e. it changes with the time of the in-diffusion anneal (t_A) and is expected to increase with increasing t_A. If C_{Cu_i} is low, due to a high generation rate of vacancies by defects (see Equation (5.7)), D_{Cu} becomes equal to D_V, about $6 \times 10^{-5} cm^2 s^{-1}$. When the high interstitial diffusivity maintains C_{Cu_i} close to its equilibrium value $C_{Cu_i}^{eq}$, one has essentially $D_{Cu} \approx D_{Ge}/D_{Cu_s}^{eq} \sim 6 \times 10^{-8}$ cm^2 s^{-1}. Combining Equations (5.5) and (5.6) one can readily see that the temperature dependence of the copper diffusivity in defective material will be less than the one for D_{Cu_i}, since [18]:

$$\frac{C_{Cu_i}^{eq}}{C_{Cu_s}^{eq}} = 1.77 \times 10^{-2} \exp\left(\frac{0.12 \text{ eV}}{kT}\right) \tag{5.10}$$

As according to Figure 5.3 the activation enthalpy of D_{Cu_i} is 0.33 eV, it follows from Equation (5.10) that D_{Cu} will be relatively temperature independent in highly imperfect material (i.e. 0.33–0.12 eV). One important requirement for the validity of the above analysis is that $C_{Cu_s} + C_{Cu_i} > C_V$ [3], or, in other words, $C_{Cu_s} > C_V$. In general, it follows from the dissociative mechanism that the effective diffusivity of copper in Ge will largely depend on the crystal perfection and may change locally and with time.

5.2.4. Impact of doping density on Cu diffusion and solubility

In their classical paper, Hall and Racette pointed out another important factor [14], namely the impact of the Fermi level (doping density) on the solubility ratio between Cu_i and Cu_s. When Cu_i is a single donor, the enthalpy of solution includes the energy gained when the electron drops to the Fermi level. In extrinsic p-type Ge (heavily doped), the Fermi energy is in the lower half of the band gap and the electron can gain an energy ΔE in reaching its equilibrium energy E_F. Consequently, the fractional solubility of Cu_i^+ will increase by the Boltzmann factor $\exp(\Delta E/kT)$ and is given by [14]:

$$C_{Cu_i} = C_{Cu_i}^{eq}(p/n_i) \tag{5.11}$$

with p the free hole concentration and n_i the intrinsic carrier concentration. Equation (5.11) holds for a non-degenerate doping density; for degenerate doping an additional factor has to be included [14]. The corresponding solubility of Cu_i is represented in Figure 5.4 and compared with the one in intrinsic (lowly doped) Ge. A strong enhancement is found for an acceptor doping density of $10^{20}\,cm^{-3}$, which, as shown by Williams [23], can lead to a pronounced redistribution of Cu in the neighborhood of a p^+–p junction in Ge.

Likewise, in extrinsic n-type Ge, the equilibrium solubility will be shifted toward Cu_s, according to the rule [14]:

$$C_{Cu_s} = C_{Cu_s}^{eq}(n/n_i)^r \tag{5.12}$$

with r the multiplicity of the acceptor level: three in germanium and two in silicon. Ion pairing between Cu_s and shallow donors or acceptors may further enhance its solubility, as theoretically suggested by Shockley and Moll [24]. The diffusivity, on the other hand, will be reduced both for Cu_s and Cu_i by the ion pairing mechanism [14]. Evidence for such an effect was obtained by Potemkin and Potapov [25], who observed in $10^{17}\,cm^{-3}$ Sb-doped n-Ge a retardation of the Cu diffusion at 550°C, that disappeared above 600°C. It was proposed that Sb–Cu pairing caused the reduction of the effective diffusivity. Above solubility, co-precipitation of a Cu_2Sb phase can also take place [25]. A similar enhancement was obtained for even lower Sb concentrations ($\sim 1 \times 10^{14} - 5 \times 10^{16}\,cm^{-3}$), where the formation of Cu_2Sb was ruled out [26]. It was rather assumed that $Sb^+ - Cu^-$ pairs are the origin of the solubility enhancement.

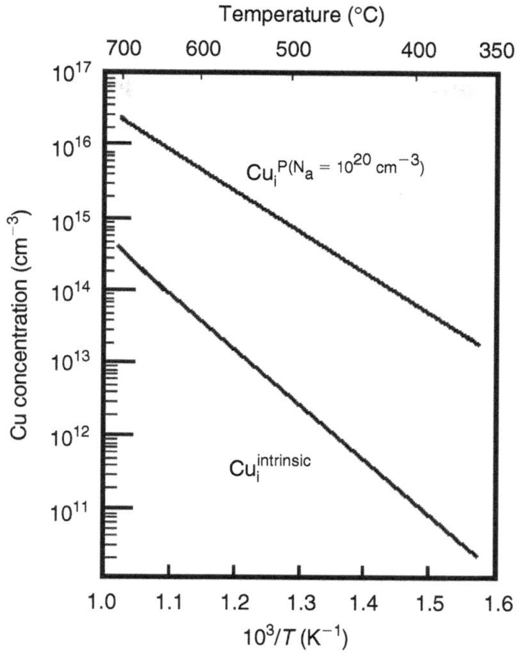

Figure 5.4. Solubility of interstitial copper in germanium. The solubility is much higher in Ge containing a p-type impurity at a concentration of $10^{20}\,\mathrm{cm}^{-3}$. The solubility of copper in both pure and doped material is a strong function of temperature (after Williams [23]).

In the case that there is ample supply of vacancies, so that local equilibrium exists between Cu_i and Cu_s, the effective copper diffusivity in intrinsic Ge is given by [3]:

$$D_{Cu} = D_{Cu_i}/(1+\alpha_e) \tag{5.13}$$

In extrinsic material, α_e is given by:

$$\alpha_e = \alpha_i(n/n_i)^{r+1} \tag{5.14}$$

This leads to the anticipation that in extrinsic n-type material such that $\alpha_e \gg 1$, the effective diffusivity will drop with the inverse fourth power of the donor concentration. In extrinsic p-Ge on the other hand, we have that $\alpha_e \ll 1$ and $D_{Cu} = D_{Cu_i}$, so that in such case, the diffusivity will be solely determined by the interstitial one. This is only valid in the absence of ion pairing. The resulting D_{Cu_i} was found to be independent of the acceptor concentration in heavily doped p-Ge [14]. One should also take into account that the interstitial diffusivity is electric field dependent, since the positive ions can drift according to Einstein's law (see above).

The experimental data for extrinsic n-type Ge are shown in Figure 5.5, where an enhancement of the copper solubility with increasing donor concentration is clearly demonstrated [14]. The behavior of D_{Cu} is rather surprising in the sense that after the expected strong

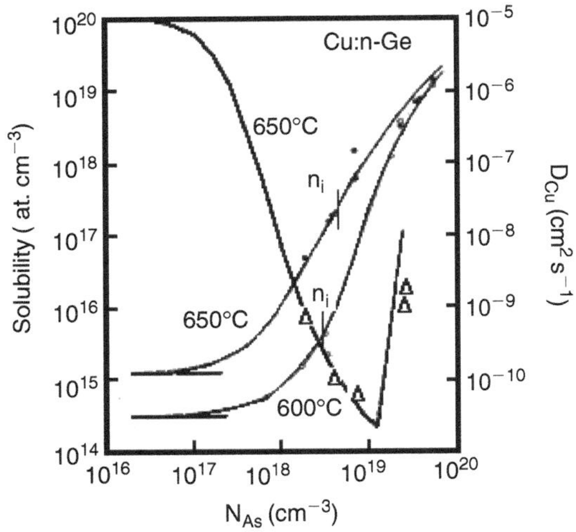

Figure 5.5. Solubility and diffusion of Cu in n-type Ge. The two high-concentration points (triangles) are the lower limits, since the samples were uniformly saturated (after Hall and Racette [14]).

decrease with n (Equations (5.13) and (5.14)), a sharp increase follows at $\sim 10^{19}\,\text{cm}^{-3}$ donor concentration. A possible explanation for this effect could be the enhanced vacancy concentration related to the high donor concentration and/or the precipitation of shallow dopants, injecting vacancies in the crystal [14].

Although not explicitly mentioned by the authors, Figure 5.5 shows the potential of a heavily n-doped region for the segregation gettering of copper from a lowly doped Ge region. The reduced diffusivity at the same time keeps the in-diffused Cu in the highly doped region. A further increase of the gettering potential can occur through ion pairing, stabilizing the copper in solution. This mechanism is the counterpart of the well-known Fe segregation gettering in heavily p^+-doped silicon, where Fe–B pairing plays a crucial role. As mentioned before, a redistribution of Cu will take place at a p^+–p junction [23], although the stability of the gettering (binding energy) will be smaller than in n-type material, as Cu_i^+ remains highly mobile.

5.2.5. Dissociative versus kick-out mechanism for copper diffusion in germanium

A further test of the kick-out mechanism has more recently been undertaken by Stolwijk *et al.* [27,28]. Again, it was demonstrated that the diffusion of Cu in Ge can be well described by the dissociative mechanism (Figure 5.6), while kick-out does not at all produce a good fit to the measured diffusion profiles. Figure 5.6 clearly illustrates the different diffusion behavior of copper in dislocation-free or dislocation-rich germanium. It was shown that in high dislocation density material (4–8 $\times$ $10^3\,\text{cm}^{-2}$), the internal equilibrium vacancy concentration is maintained by generation through dislocations. The effective copper diffusion is then interstitial, with a diffusion coefficient given by [28]:

$$D_{\text{eff}}(\text{Cu}_i) = 7.8 \times 10^{-5} \exp(-0.084/kT)\ \text{cm}^2\text{s}^{-1} \tag{5.15}$$

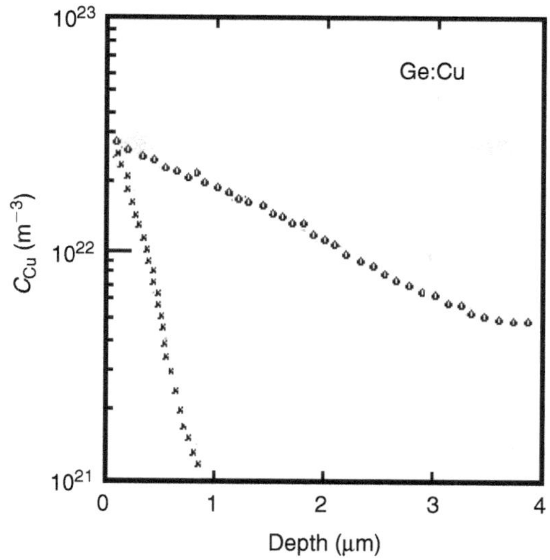

Figure 5.6. Comparison of Cu penetration profiles in (almost) dislocation-free Ge (x, 1126 K, 900 s and in a crystal with a virtually high dislocation density (o, 1124 K, 780 s) (after Bracht *et al.* [28]).

exhibiting a weak temperature activation, in line with previous estimates [18]. For low dislocation densities, the copper diffusion is described by Equation (5.8) and is ruled by the slowest component, namely, the transport of vacancies. The corresponding effective diffusivity is represented in Figure 5.7a and obeys [28]:

$$D_{\text{eff}}(\text{Cu}) = 5.5 \exp(-1.55\,\text{eV}/kT)\ \text{cm}^2\text{s}^{-1} \tag{5.16}$$

The corresponding solubility is represented in Figure 5.7b and can be approximated by:

$$S^{\text{eq}}_{\text{Cu}_\text{s}} = 3.44 \times 10^{23} \exp(-1.55\,\text{eV}/kT)\ (\text{cm}^{-3}) \tag{5.17}$$

in good agreement with earlier assessments and with Equation (5.7a) [18]. In addition, it is found that the equilibrium substitutional copper solubility is not very sensitive to the dislocation density n_{disl} (see boundary concentration at $x = 0$ in Figure 5.6) [28].

The diffusion profiles can be understood by considering that the near surface diffusion is controlled by vacancy diffusion, while deep in the interior of the wafer, dislocations dominate the copper migration, resulting in a higher diffusivity, determined by Cu_i. In the intermediate part of the profiles, a concentration plateau $(C_{\text{Cu}_\text{s}})_{\text{plateau}}$ is predicted by the dissociative mechanism, which can be described by [28]:

$$\frac{(C_{\text{Cu}_\text{s}})_{\text{plateau}}}{C^{\text{eq}}_{\text{Cu}_\text{s}}} = \gamma_\text{d} n_\text{d} (D_{\text{eff}})_\text{V} t \tag{5.18}$$

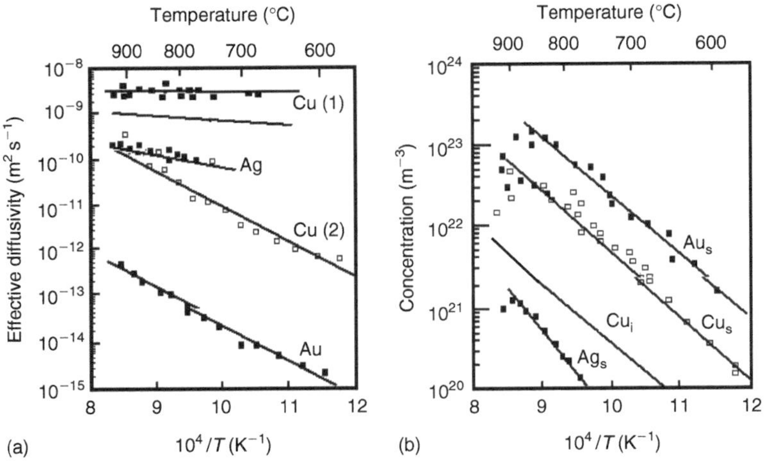

Figure 5.7. (a) Effective diffusivities of Cu, Ag and Au in Ge. Cu(1) in virtually highly dislocated crystals; Cu(2) in virtually dislocation-free Ge. (b) Equilibrium solubilities of Cu, Ag and Au in Ge (after Bracht *et al.* [28]).

with $\gamma_d \sim 1$. From Equation (5.18) one derives that in the intermediate region the vacancy flow from the dislocations determines the incorporation rate of Cu_s.

5.2.6. Precipitation of copper in germanium

Due to the combination of the retrograde solubility of copper (see e.g. Figure 5.7b), its interstitial solubility becoming higher than the substitutional one below 500°C and the high interstitial diffusivity, one can expect Cu precipitation when cooling down a contaminated or doped sample from high annealing temperatures. Early studies pointed out that the rate of precipitation was a function of the cooling rate, the supersaturation rate and the structural perfection of the material [29,30]. Tweet has studied more systematically the impact of the different factors [31–33] and came to the following conclusions. In highly dislocated (plastically deformed) Ge, the precipitation of Cu, with annealing time, monitored by the decrease of the assumed copper single acceptors ($N_A(t)$) through resistance measurements, follows in first approximation an exponential decay, represented by [31]:

$$N_A(t) = N_A(0)\exp(-t/\tau_p) \tag{5.19}$$

with $N_A(0)$ the initial acceptor (Cu_s) concentration. The precipitation time constant (τ_p) depends strongly on the dislocation density n_{disl} for temperatures above ~500°C, as illustrated by Figure 5.8 [31]. For low precipitation temperatures, on the other hand, all curves come together and are represented by the theoretical boundary g^{-1}. The high-temperature curves are thermally activated and can be represented by an Arrhenius law with a constant activation enthalpy of ~1.3 eV. This is close to the activation energy reported, e.g., in Figures 5.5 and 5.7 and demonstrates that in this regime the precipitation is diffusion limited and governed by the dissociative diffusion coefficient of copper, given by Equation (5.8).

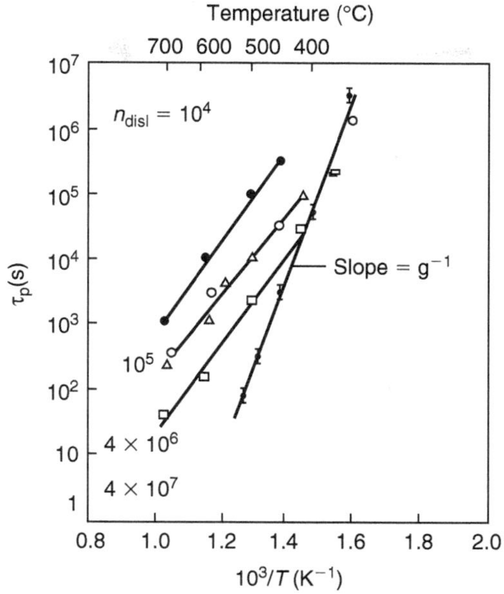

Figure 5.8. Plot of τ_p, the time constant for Cu precipitation versus $1/T$. Curves are labeled with dislocation densities. Smaller densities were measured from etch pit counts. Larger densities were estimated crudely from amount of deformation (after Tweet [31]).

The precipitation is thus rate limited by the transport of Cu to the dislocation sinks and τ_p can be interpreted as follows [31]:

$$D\tau_p = L^2 = 1/n_{disl} \tag{5.20}$$

The low-temperature behavior has been explained by assuming that the rate limiting step is now the dissociation of Cu_s into Cu_i and V. This is possible if the activation enthalpy $E_{diss} > E_{diff}$ [31]. The boundary line labeled g^{-1} in Figure 5.8 thus corresponds to the reciprocal of the vacancy–interstitial copper generation rate g. As pointed out later by Penning [34], the precipitation time constant in this region can be used to estimate the self- and vacancy diffusion constant in Ge (see Chapter 3). The fact that the break in the Cu-precipitation time constant shifts to lower temperature for a higher n_{disl} demonstrates the impact of the dislocations on the generation of vacancies. At temperatures above the g^{-1} line, the dissociation rate is rapid enough, so that the vacancy and interstitial densities are above the thermal equilibrium values. In the g^{-1} regime, the vacancy–interstitial dissociation rate is small enough and the D_V large enough, so that vacancy equilibrium with the lattice can be maintained by unhindered vacancy diffusion to the dislocations [31].

Tweet also noticed that at high supersaturation (Σ_{Cu}), the Cu-precipitation rate was strongly dependent on Σ_{Cu}, provided the dislocation density was smaller than $10^4\,cm^{-2}$ [32]. In such material, precipitation is supposed to occur at nucleation sites other than dislocations. The possibilities could be either homogeneous nucleation (although this appeared to be rather unlikely), nucleation at oxygen or other neutral impurities and, finally, defects (e.g. vacancy clusters) created during the solution anneal (in-diffusion of Cu). Evidence in support of the latter hypothesis was provided by the observation that the precipitation in quenched samples occurred at a different rate than in slowly cooled material [32].

5.2.7. Energy levels and capture cross sections of substitutional copper

The energy of the three acceptor levels associated with Cu_s was originally studied by Hall effect versus temperature [10,35,36], leading to thermal activation enthalpies of 0.04, 0.32 and -0.26 eV. More recently, deep level transient spectroscopy (DLTS), using the thermal emission rate of majority of carriers from the three Cu_s levels enabled to obtained the activation energy from the slope of an Arrhenius plot [37–41]. The hole capture cross section can in principle be obtained from the intercept of the Arrhenius plot, although this merely leads to an order of magnitude estimate. More accurate values for σ_p have been derived by measuring the trap filling rate through the DLTS peak amplitude as a function of the filling pulse duration [40,41]. This resulted in the values of Table 5.1. Note that the parameters for the third acceptor level corresponding with $Cu_s^{2-/3-}$ have been obtained in n-type Ge [41], while the other two charge states were studied in p-Ge [40].

Comparing the activation energies of Table 5.1 with the Hall effect data, it is noted that the former ones are generally slightly lower. As pointed out by Kotina *et al.* [42], this could be due to the difference in both techniques: DLTS measurements are usually performed at a non-negligible electrical field F, while a weaker field is normally applied during Hall measurements. Second, the DLTS activation energy may also depend on the activation energy (or in general, the temperature dependence) of the relevant capture cross section.

In the case of the single and double Cu_s acceptor level in p-type Ge, the presence of an electric field will lower the activation energy, according to the Poole–Frenkel (PF) effect, resulting in a reduction of E_A by [44]:

$$\Delta E = q\,(ZqF/\pi\varepsilon_{Ge})^{1/2} \tag{5.21}$$

with Zq the charge of the unoccupied impurity atom (q for the singly ionized 0.04 eV level and $2q$ for the doubly ionized 0.32 eV acceptor), while ε_{Ge} is the permittivity of Ge ($=16\varepsilon_0$). The measured PF activation energy lowering is represented in Figure 5.9 [44] and

Table 5.1. Deep-level parameters of the three acceptor levels of copper as derived from DLTS [40,41].

Charge state	Energy level	Capture cross section (cm^2)		Temperature (K)
		Holes	Electrons	
$Cu_s^{0/-}$	$E_V + 0.037$ eV	2.4×10^{-12}	–	22
$Cu_s^{-/2-}$	$E_V + 0.322$ eV	2.0×10^{-13}	–	145
$Cu^{2-/3-}$	$E_C - 0.259$ eV	4.0×10^{-16}	$5.3 \times 10^{-19*}$	183

*Determined by Paramonova and Rzhanov [43], extrapolating from 300 K assuming a T^{-3} law.

is close to the theoretical expectations ($\Delta E = 1.9 \times 10^{-4}\,qF^{1/2}$ for the singly ionized level). However, the data of Table 5.1 have been obtained in high-purity Ge, corresponding with a low $F \sim$ a few $10^3\,V\,cm^{-1}$ and, hence, resulting in a small energy lowering (<0.01 eV). Moreover, electron capture in the upper Cu level occurs by a repulsive Cu_s^{2-} center, where in principle, no PF lowering should happen. This strongly suggests that the temperature dependence of the respective cross sections should be responsible for the differences in the experimental energy levels between DLTS and Hall effect [42].

Dividing the emission rate by the capture cross section at the same temperature allows one to calculate the change in free energy G of the deep level, which is related to the enthalpy (H) by the following relationship [45]:

$$G = H - TS \tag{5.22}$$

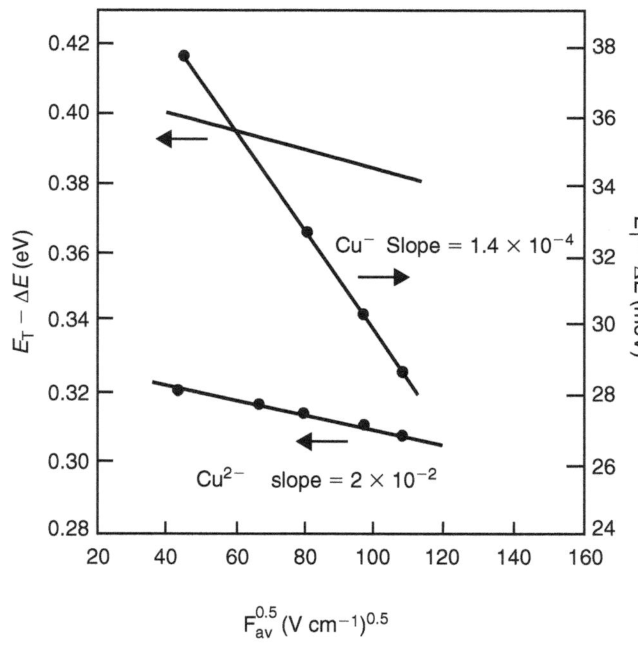

Figure 5.9. Observed Coulombic barrier lowering for Cu^-, Cu^{2-}, and the A impurities[2] as a function of the average electric field in the junction F_{av}. The slope of the line is equal to the PF constant for germanium, for the different charge states on the negative ion remaining after hole emission (after Pearton *et al.* [44]).

[2] Based on the agreement between the experimental and theoretical PF shift for a triply charged impurity, the authors identified this level tentatively with the third acceptor level of copper in p-Ge [44]. It has meanwhile been shown that this assignment is wrong [41]: the third acceptor level of Cu_s can only be seen in n-Ge as an electron trap or in p-Ge as a minority carrier trap to the left of the $E_V + 0.32\,eV$ level.

with S the entropy change upon capture of a free carrier in the center. The resulting values are summarized in Table 5.2 for Cu_s [45]. The experimental activation energy derived from an Arrhenius plot (E_A) and the one found for the capture cross section $E_{A\sigma}$ (if applicable, see below), are related through [45]:

$$E_A = H + E_{A\sigma} \tag{5.23}$$

Finally, at very low temperatures (~1 K), Cu_s can behave as an overcharged acceptor in the Cu_s^+ charge state, thus carrying a fourth hole [46,47]. The binding energy was found to be 2.0 meV. This is too low to consider this as a donor level for Cu_s. This charge state could in principle open perspectives for long wavelength far infrared (IR) detection.

Besides the concentration and the energy levels, the cross sections for hole (σ_p) and electron (σ_n) capture are important parameters to characterize the electrical activity of a recombination center. The magnitude of $\sigma_{n,p}$ depends largely on the charge state of the empty trap: $>10^{-14}\,cm^2$ when the potential is attractive; 10^{-15}–$10^{-14}\,cm^2$, neutral or $<10^{-16}\,cm^2$ repulsive. Moreover, $\sigma_{n,p}$ is generally temperature and electric field dependent. The nature of this dependence is dictated by the dominant recombination mechanism, i.e. capture by an attractive Coulombic potential, following a cascade recombination [48,49], tunneling through a repulsive barrier, etc. The electric field dependence will mainly occur in the hot carrier regime, for fields $F \sim 10^3\,V\,cm^{-1}$ and higher. Particularly for p-type Ge, inelastic collisions are important in strong electric fields [50], leading to a clear F-dependence of σ_p.

For an extensive summary of the temperature dependence of the different capture cross sections, the reader is referred to Table A1. It is clear from this data set that the hole capture cross section is generally much larger than the corresponding electron cross section, due to the negative (or neutral) charge state of Cu_s. As a consequence, the temperature dependence shows the opposite trend for σ_n and σ_p: while σ_n usually increases for increasing T, σ_p reduces normally with T. Globally speaking, one expects a thermally activated σ_n, whereby electrons can overcome the repulsive barrier of the charge $-Zq$ more easily at higher thermal energies. Electric-field-assisted tunneling may also aid in electron recombination at Cu_s centers [51–53].

The temperature dependence of σ_p increases for lower T, as illustrated in Figure 5.10 for the two lower Cu_s acceptor levels, combining data collected from the literature [54]. Generally, the temperature and charge state dependence of the capture cross section for an

Table 5.2. Summary of the enthalpy and entropy changes H and S for the copper related levels. The entropy S is given in units Boltzmann constant k [45].

Charge state	***H* (meV)**	***S* (units *k*)**
$Cu_s^{0/-}$	41.3	−0.8
$Cu_s^{-/2-}$	333	3.2

attractive trap is described within the Lax cascade theory [48,49], leading for Ge to a simplified analytical expression [54–56]:

$$\sigma_{\text{cascade}} = \frac{1.1 \times 10^{-9}\, Z^3}{T(T + 2.00)^2} \tag{5.24}$$

which is valid for $T < 30\,\text{K}$ ($Z = 1$) and $T < 120\,\text{K}$ ($Z = 2$). The basic idea behind the cascade theory is that a carrier is first loosely bound in an excited state of the Coulomb attractive trap, next descends the staircase of hydrogen-like excited states by one-phonon emissions and, finally, decays to the ground state with a transition that may be single-phonon, multi-phonon or radiative [48,49].

However, as shown in Figure 5.11a for singly ionized and Figure 5.11b for doubly ionized acceptors, the cascade theory does not at all fit the experimental temperature dependence [56]. Moreover, also the Z-dependence is not correctly reproduced, as evidenced from Figure 5.11, where $\sigma_p(Cu^-)$ can be higher than $\sigma_p(Cu^{2-})$ for the higher temperatures assessable by the measurements. Equation (5.24) predicts just the opposite behavior, namely an increase by a factor 8. The same discrepancy has been noted for other double acceptors, like Zn and CuH_x [54]. Moreover, the cross sections were found to be independent of the doping concentration and of any background IR irradiation. In brief, it is found that for most deep acceptors in p-type Ge, a larger hole capture cross section is obtained than predicted by the generally accepted theory, which may form a lower bound at lower temperatures.

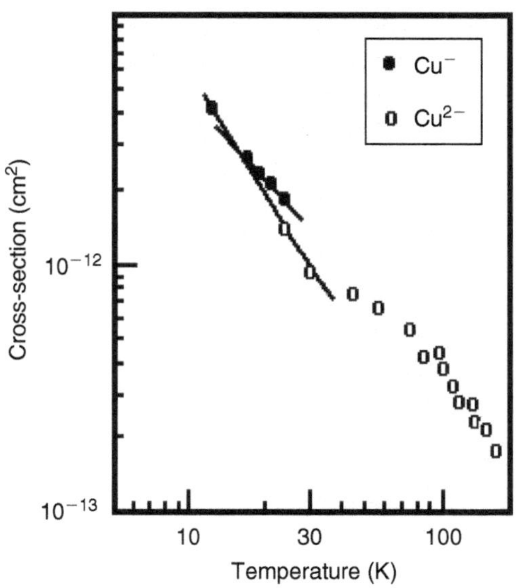

Figure 5.10. Hole capture cross sections versus temperature for the single and double ionized copper center measured in the same sample (after Darken *et al.* [54]).

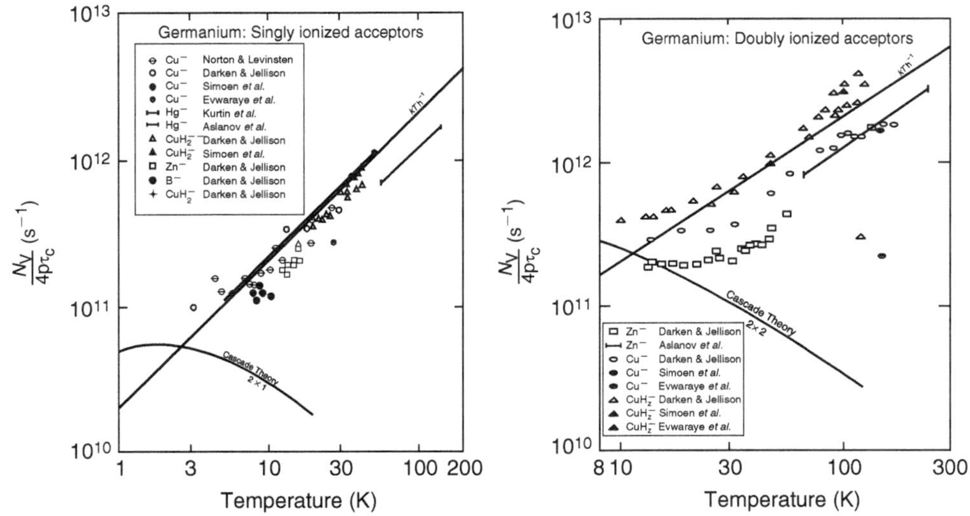

Figure 5.11. (a) Hole capture at single ionized acceptors in germanium and (b) hole capture at double ionized acceptors in germanium (after Darken [56]).

In order to explain these anomalously large hole capture cross sections of acceptors in p-Ge, the following empirical relationship was established [54–56]:

$$N_V/4p\tau_c \approx kT/h \tag{5.25}$$

with N_V is the effective density of states at the valence band edge, τ_c the capture time constant for a hole at the ionized acceptor, p the free hole density and h Planck's constant. As can be seen in Figures 5.11a and b, Equation (5.25) provides a much better fit to the data than the curve derived from the cascade theory. The left-hand side of Equation (5.25) has the dimensions of a frequency and can be considered as an attempt-to-escape frequency in the thermal emission rate of holes from an occupied acceptor, following from detailed balance [55,56]. The right-hand side is none other than the frequency of a thermal phonon. This implies that the rate limiting step in the emission and, conversely, in the capture, is a transition from or to the ground state. The model of Equation (5.25) predicts a $\sigma_p \sim 1/T$ rather than $\sim 1/T^3$, becoming thus relatively higher with increasing temperature.

For some deep acceptors, it was shown that a better fit is obtained by implementing some additional correction factors in Equation (5.25), so that the general behavior can be described by [54–56]:

$$\frac{N_V}{4p\tau_c} = \frac{kT}{h}\left\{\eta \exp\left(\frac{-E_a}{kT}\right)\right\} \tag{5.26}$$

The parameter η can be interpreted as a generalized degeneracy factor 4η, while the enthalpy of the activated complex is E_a instead of 0 [54].

The fundamental question remains to be answered what the exact physical capture mechanism is. It is in this context important to remark that the optical-phonon energy in Ge is 37 meV, which is smaller than the ionization energy of Cu_s^0 for example [54]. The explanation of this phenomenon awaits further theoretical modeling. It should be noted, however, that a modified cascade theory for deep attractive centers, including optical-phonon-assisted capture at higher T predicts a $\sigma \sim 1/T$ in better agreement with the observations [57–59]. As far as we know, this trend has not yet been studied/reported for deep attractive traps in silicon or in n-type Ge. Candidate centers for a closer investigation could be, e.g. the oxygen thermal donors.

It should finally be remarked that in high electric fields, where carrier heating occurs, the recombination rate of holes with deep acceptor levels reduces; in other words, the carrier lifetime for holes becomes larger [50]. The recombination coefficient for Cu_s^{3-} is fairly well represented by $c_p(F)/c_p(0) \sim (kT_hF)^{-3/2}$, a relation typical for the optical capture mechanism at high hole temperatures T_h. The corresponding electron capture coefficient ($c_n{}^{2-}$), on the other hand, strongly increases with field and the field dependence becomes stronger at lower T [53]. The data could be well described by the model for electron tunneling through a Coulomb repulsive barrier, taking into account the approximation of an electron temperature and an isotropic effective mass [53].

5.2.8. *Energy level for interstitial copper and Cu_s–Cu_i pairs*

While the literature on the electrical activity of substitutional copper is extensive, far less is known about the properties of interstitial copper, for the obvious reason that its solubility at room temperature is negligible. Only by quenching an annealed sample from high temperatures, one may hope to freeze a certain concentration of Cu_i. Figure 5.12 reports the DLT-spectrum of a p-type Ge sample heated for several hours at 500°C and quenched to room temperature [60]. Besides the well-known Cu_s levels at 0.04 (H1) and 0.33 eV (H4), two more unknown peaks have been revealed, H2 (0.09) and H3 (0.23), with about the same concentration N_T. Annealing at 280°C removes both the H2 and H3 traps (Figure 5.12), while the H1 and H4 levels grow accordingly [60]. H2 is shown to exhibit a PF shift, indicating its acceptor character (attractive hole trap), while H3 does not move with variation of the electric field, from which its deep donor nature is derived. This is furthermore supported by the small hole capture cross section of the level. These observations have more recently been confirmed in another study, assigning a hole capture cross section of $1 \times 10^{-14}\,cm^2$ to these so-called "slow" states of copper [61]. They were typically observed in the low dislocation density regions of high-purity germanium crystals ($n_{disl} < 10^3\,cm^{-2}$), while after heat treatment at 450°C for 30 h only Cu_s ($E_V + 0.33$ eV) was observed by DLTS [61].

The annealing behavior of the H2 and H3 levels in Figure 5.12 has been explained by the following picture [60]:

$$H2 \leftrightarrow Cu_s + H3 \tag{5.27a}$$

and:

$$H3 \rightarrow \text{sinks} \tag{5.27b}$$

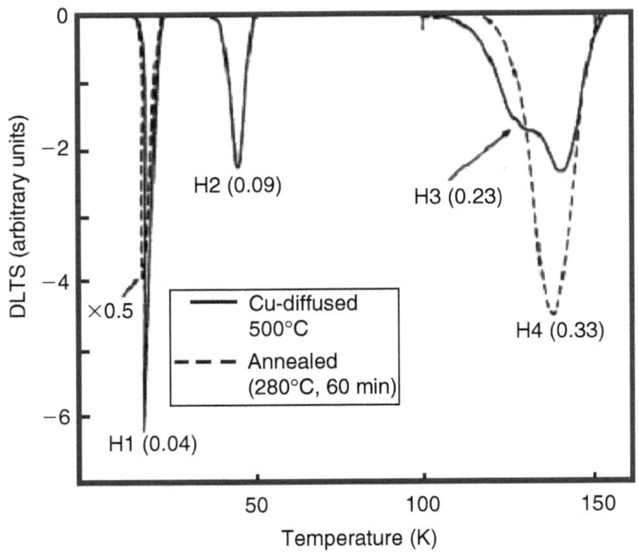

Figure 5.12. DLT-spectra for a sample that was copper diffused at 500°C and then annealed at 280°C for 60 min. Both reverse and pulse bias voltages are 1 V for the solid curve and 3 V for the dashed curve (after Kamiura and Hashimoto [60]).

whereby it is assumed that H3 is a mobile species. The rate constant for the second reaction (the out-annealing of H3 (Equation (5.27b)) was shown to depend on the sample thickness, demonstrating that the surface acts as a sink for H3. The activation enthalpy corresponding with the rate constant for reaction (5.27b) amounts to 0.35 eV and should be the migration energy of H3. This is very close to the migration energy of Cu_i reported in Figure 5.2, identifying H3 with interstitial copper. H2 is then assigned to a Cu_s–Cu_i pair (the $(Cu_i^+Cu_s^{2-})^-$/$(Cu_i^+Cu_s^{2-})^0$ charge state). This assignment is consistent with the observed acceptor nature (PF shift) and with the binding energy of the complex of 0.85 eV, which corresponds in the first approximation with the binding energy of two attractive point charges [60].

The schematic picture arising from this is represented in Figure 5.13. Although there appears to be a strong analogy with the case of FeB pairing in silicon, the difference for Cu_i–Cu_s pairing is that the dissociation of the pair and the precipitation of Cu_i are taking place in the same temperature region, which is due to the difference in the migration energy of the mobile partners and in the binding energy [60].

Cu_i–Cu_s pair formation has also been studied by the internal friction method [62]. Under the application of an external elastic stress, these pairs reorient, giving rise to a lattice relaxation. It was concluded that 90% of the pairs consists of the nearest neighbor configuration, while 10% is second nearest and higher neighbor positions. A binding energy for the pairs of 0.34 eV was obtained, which is significantly smaller than the value derived

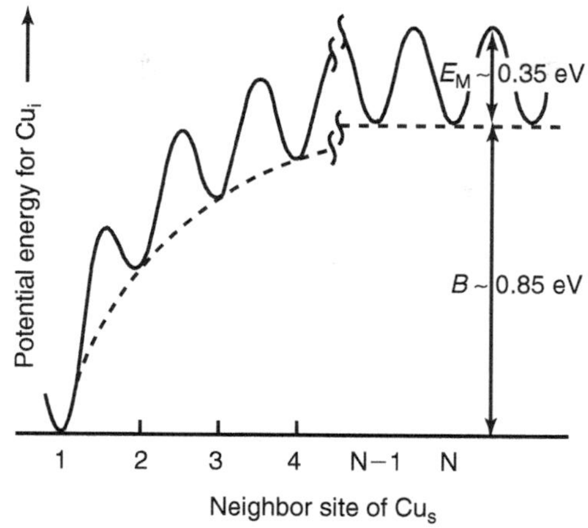

Figure 5.13. Schematical diagram of potential energy for an interstitial copper atom, Cu_i, around a substitutional copper atom, Cu_s. E_M is the migration energy of Cu_i and B is the binding energy between Cu_i and Cu_s (after Kamiura and Hashimoto [60]).

from DLTS studies [60]. The equilibrium constant corresponding with the reaction $Cu_i^+ + Cu_s^{-/2-} \rightarrow Cu_p = Cu_i^+Cu_s^{-/2-}$ was found to obey [62]:

$$K_p = \frac{C_p}{C_{Cu_i}C_{Cu_s}} = 0.08\exp(0.34/kT) \tag{5.28}$$

5.2.9. Impact of copper on carrier lifetime in germanium

Early studies on the room-temperature recombination lifetime in germanium doped in the melt with copper were shown to be well described by a recombination energy level at $E_T - E_V = 0.25 - 0.30\,\text{eV}$ (doubly ionized copper) [63].[3] This follows from the doping concentration dependence illustrated in Figure 5.14, where one can also observe the pronounced impact of Cu doping in the melt on the lifetime. A hole capture cross section of $10^{-16}\,\text{cm}^2$ was obtained, while the corresponding σ_n was one decade smaller [63]. Another study pointed out that Auger (or impact) recombination in Ge does not play a role up to a doping density of $10^{17}\,\text{cm}^{-3}$ [64].

The recombination lifetime dominated by the third acceptor level (Cu_s^{3-}) has also been studied as a function of temperature in n-type Ge and partly compensated Ge:Cu [52,53]. In contrast to normal extrinsic n-type Ge, the recombination lifetime in high resistivity Cu-doped material shows an exponential increase for lower temperatures. This was explained by the thermally activated behavior of the capture cross section for holes and electrons [65,66]. It was believed that the upper copper level transforms from a recombination center at room temperature to a

[3]This assignment is questionable: due to the more or less symmetrical position of the third copper level at $E_C - 0.26\,\text{eV}$, this can also determine the recombination lifetime, certainly in n-type Ge.

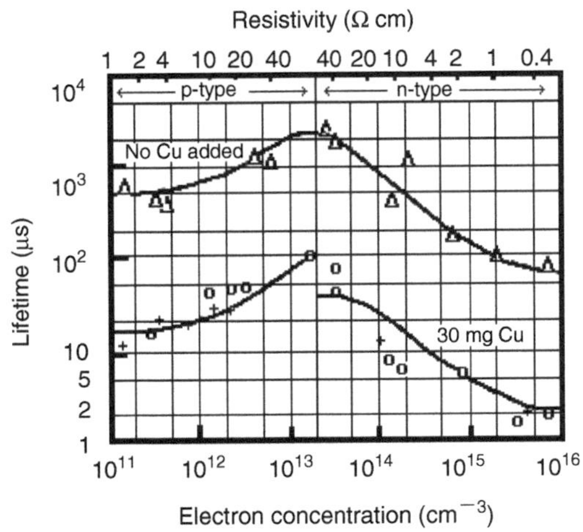

Figure 5.14. Lifetime at 297 K versus electron concentration for Ge crystals, measured at the fraction solidified $x = 0.50$: triangles, only Sb or In added to the melt; circles: 30 mg Cu plus Sb or In added to the melt; crosses, Cu introduced by diffusion at 590°C (after Burton *et al.* [63]).

hole (minority carrier) trap[4] at low temperatures, where the electron lifetime drastically enhances. Further confirmation was obtained by another study [67], where it was shown that σ_p/σ_n at 300 K is larger than 1000 and rapidly increases upon cooling. This explains the transformation, since at low temperatures, the rate for electron capture through Cu_s^{2-} becomes very small. It means that upon capture of a hole by Cu_s^{3-}, the probability for thermal re-emission to the valence band becomes much higher, compared with the recombination with an electron, so that the upper copper level behaves as a hole trap in n-type Ge.

Another parameter to consider is the injection level dependence of the lifetime, where high injection corresponds with an injected excess carrier concentration much larger than the doping concentration of the material. In Cu-doped n-Ge, it was found that both the high- (τ_∞) and low-injection level lifetime (τ_0) were nearly independent on T [68]. The same was observed for τ_∞ in p-Ge, while τ_0 reduced for lower T [68]. This stands in contrast to previous studies, where an increase of τ was observed [65,66]. At the same time, it was concluded that carrier recombination was governed by trapping of holes in the upper Cu_s level (Cu_s^{3-} at $E_C - 0.26$ eV), followed by electron trapping in the middle level (Cu_s^- at $E_V + 0.32$ eV) [69].

The different temperature dependencies of the lifetime in Ge:Cu, reported in the works cited above, were resolved by Schulz [70], who pointed out the role of compensating impurities (donors), on the one hand, and the multiple levels (charge states) of Cu_s on the other. In partly compensated p-type material, the charge state of Cu_s remains constant: all centers are in the singly negative state. This gives rise to an increase of τ with lower T, as found previously [65,66]. In non-compensated p-Ge, the charge state of copper will change upon

[4] In case of trapping, the density and lifetimes of free excess carriers become unequal; in this case: $n > p$.

cooling. Part of the Cu_s centers will become neutral, which may open an additional recombination channel through the capture of electrons by Cu_s^0. This gives then rise to a weakly reducing lifetime for lower T [70]. From the above it is clear that one should use the Shockley–Read–Hall (SRH) statistics for multi-level impurities in order to describe accurately the temperature dependence of the lifetime of copper-doped Ge [71]. Moreover, activation energies derived from τ versus $1/T$ should be considered with great care. They can reflect a charge state transition and not necessarily a thermally activated capture coefficient (cross section).

5.3. Ag, Au and Pt in Germanium

Ag and Au belong to the same Group Ib elements as Cu and are, therefore, expected to behave in a similar manner as copper. In fact, in the early days of germanium research, besides copper, Au was one of the most studied metal impurities, because of its optical properties and potential application in IR detectors [72–74]. Another reason for its past popularity is that it is one of the few amphoteric elements in Ge, i.e. substitutional gold behaves both as a (deep) donor and as a (triple) acceptor [75]. This unique behavior offered the possibility for some interesting studies regarding the physics of defects in semiconductors [76]. In fact, more recently evidence was reported for the amphoteric behavior of Ag_s in germanium, whereby the donor level is at 0.035 eV from the valence band, derived from DLTS [77].

Ag and Pt have drawn less attention, partly because of the low distribution coefficient and solubility. This renders research more difficult as the preparation of samples with a suitable impurity concentration is not easy. It should bear no surprises that most of the results summarized in the following paragraphs deal with Au_s in Ge.

5.3.1. Distribution coefficient, solubility and diffusivity

The first studies of Au doping in Ge indicated a $k_d \sim 1.5 \times 10^{-5}$ and a maximum solubility in the range $10^{15}\,cm^{-3}$ [78]. In the phase diagram of the Au–Ge system an eutectic point exists at 356°C [79]. More detailed studies indicated that $k_{dm} = 2.1 \times 10^{-5}$, while the temperature dependence of k_d was found to obey [80]:

$$kd = 8.2 \times 10^4 \exp(-(2.3 \pm 0.6)/kT)\ (T > 700°C) \tag{5.29a}$$

$$kd = 1.17 \times 10^{-5} \exp(-(0.38\ \pm .08)/kT)\ (T < 700°C) \tag{5.29b}$$

These data were found to be in agreement with Hall's simple theory [81], assuming a negative temperature coefficient for the heat of solution of Au in Ge. At the same time, a higher maximum solubility in the range $4.1 \times 10^{16}\,cm^{-3}$ was found, which was even higher than Syed's data (~$2.8 \times 10^{16}\,cm^{-3}$ [82]).

For the case of Ag in Ge, a maximum solid solubility of $10^{15}\,cm^3$ at 875°C was reported [83], while for Pt, $k_{dm} = 10^{-6}$ and a maximum S ~$2 \times 10^{14}\,cm^{-3}$ was determined from crystals doped in the melt [84].

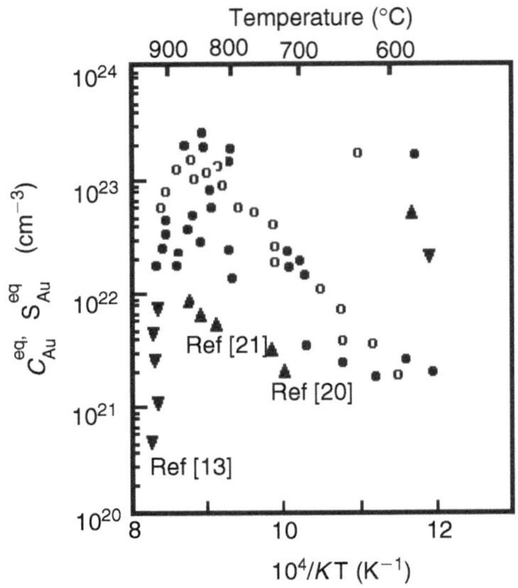

Figure 5.15. Arrhenius diagrams of the Au concentration in Ge: (•) C^*_{Au} values from inductive coupled plasma and radiotracer measurements and (o) C^*_s values from spreading resistance measurements (after Almazouzi *et al.* [84]).

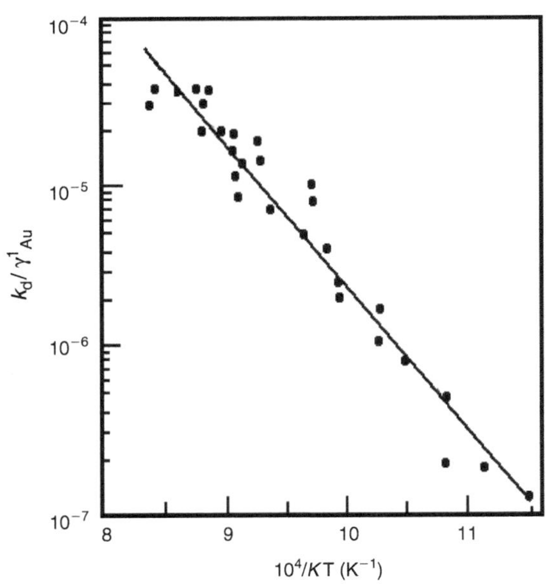

Figure 5.16. An Arrhenius plot of the distribution coefficient (k_d) reduced by the activity coefficient of Au ($\gamma^1{}_{Au}$) of the Ge(Au) liquidus (after Almazouzi *et al.* [84]).

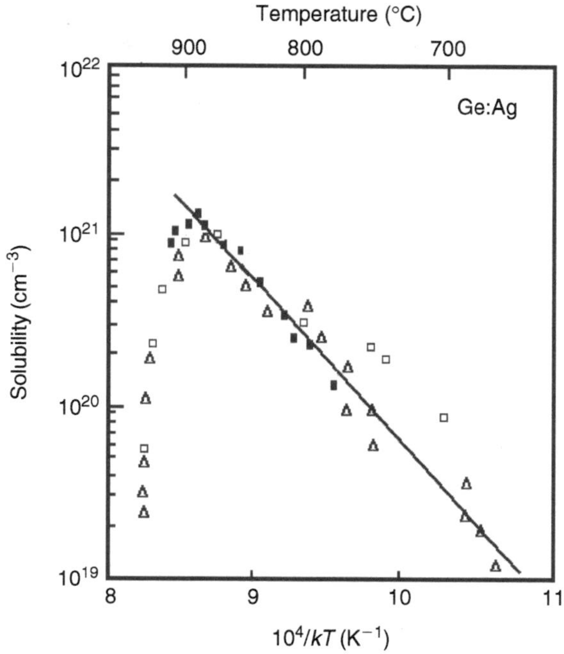

Figure 5.17. Solubility data of Ag in Ge. ●, Ref. [83]; Δ , Hall effect data [86]; □, ^{110}Ag tracer data [83]; solid line: (Equation (5.29b) (after Bracht *et al.* [28]).

More recent data for the case of Au [28,84] and Ag [28] indicate that the respective solubility, below the maximum solubility temperature can be represented by:

$$S(\text{Au}) = 12 \times 10^{-6} \exp(-1.46\,\text{eV}/kT) \text{ (in cm}^{-3}\text{)} \tag{5.30a}$$

$$S(\text{Ag}) = 2.8 \times 10^{-6} \exp(-1.85\,\text{eV}/kT) \text{ (in cm}^{-3}\text{)} \tag{5.30b}$$

This leads to a maximum solubility for Au_s of $1.5 \times 10^{17}\,\text{cm}^{-3}$ at 1131 K [28], as represented in Figure 5.15, derived from ^{195}Au radiotracer and spreading resistance analysis [85]. The distribution coefficient reduced by the activity coefficient of Au in the liquid phase ($\gamma^{1}{}_{\text{Au}}$) is given in Figure 5.16 [84], from which a distribution coefficient at T_m of $k_{dm} = 4.3 \times 10^{-5}$ can be derived. It was also found that the relative partial molar enthalpy of gold in germanium is close to that of copper in germanium [84].

The solubility of Ag in Ge yields a maximum near 1160 K, as given in Figure 5.17 [28], close to $10^{15}\,\text{cm}^{-3}$. As was already shown in Figure 5.7b, the highest maximum solid solubility is found for Au > Cu > Ag.

The effective diffusivity of Au in germanium is given by [28]:

$$D_{\text{eff}}(\text{Au}) = 1.05 \times 10^{-2} \exp\left[\frac{-1.52 \pm 0.04\,\text{eV}}{kT}\right] (\text{cm}^2\,\text{s}^{-1}) \tag{5.31a}$$

while for Ag one finds [28]:

$$D_{\text{eff}}(\text{Ag}) = 1.62 \times 10^{-4} \exp\left[\frac{-0.45 \pm 0.09\ \text{eV}}{kT}\right](\text{cm}^2\ \text{s}^{-1}) \tag{5.31b}$$

Equation (5.31a) has been represented in Figure 5.7b. Comparing with the data for copper, one finds the chemical trend that the activation energy is smallest for Cu< Ag< Au. It has been shown that the diffusion of the three Group Ib metals can be completely understood in the frame of the dissociative model [28,84]. However, for the case of Au and Ag, the concentration of interstitial atoms is low enough so that the condition:

$$C^{\text{eq}}_{\text{Au}_i,\text{Ag}_i} D_{\text{Au}_i,\text{Ag}_i} + C^{\text{eq}}_{\text{Au}_i-\text{V},\text{Ag}_i-\text{V}} D_{\text{Au}_i-\text{V},\text{Ag}_i-\text{V}} << C^{\text{eq}}_{\text{V}} D_{\text{V}} \tag{5.32}$$

with $C^{\text{eq}}_{\text{Au}_i,\text{Ag}_i}$, $C^{\text{eq}}_{\text{Au}_i-\text{V}}$, $\text{Ag}_{i-\text{V}}$ the fractional equilibrium concentration of Au or Ag interstitials and the equilibrium concentration of Au_i–V or Ag_i–V pairs, respectively. $D_{\text{Au}_i,\text{Ag}_i}$ and $D_{\text{Au}_i-\text{V},\text{Ag}_i-\text{V}}$ are the corresponding diffusion coefficients. Under the conditions of Equation (5.32), the Au (Ag) diffusion is always dominated by the interstitials and independent of the vacancy concentration or the dislocation density n_{disl} of the material. In other words, the effective diffusion coefficient is given by [84]:

$$D_{\text{eff}} = \frac{C^{\text{eq}}_{\text{x}_i} D_{\text{x}_i} + C^{\text{eq}}_{\text{x}_i-\text{V}} D_{\text{x}_i-\text{V}}}{C^{\text{eq}}_{\text{x}_s}} \tag{5.33}$$

irrespective of n_{disl}. $C^{\text{eq}}_{\text{x}_s}$ is the equilibrium concentration of substitutional x-atoms. The x_i–V pair-diffusion mechanism has been proposed by Millea for Au in Ge [86]. It is in fact a special case of the dissociative mechanism [84], which can be represented by the reaction:

$$\text{Au}_i\text{V} + \text{V} \leftrightarrow \text{Au}_s + \text{V} \tag{5.34}$$

and by an effective diffusion coefficient [85]:

$$D_{\text{eff}} = \frac{C^{\text{eq}}_{\text{Au}_i-\text{V}} D_{\text{Au}_i-\text{V}}}{C^{\text{eq}}_{\text{Au}_s}} \tag{5.35}$$

However, based on the experimental data obtained for Au diffusion, no conclusive evidence for the contribution of pair diffusion to the transport of Au atoms in Ge was obtained, so that the pair term can be neglected in Equation (5.34) [28].

As shown by Bracht *et al.* [28], the $C_i{}^*D_i$ product exhibits the chemical trend represented by Figure 5.18 and by the equations:

$$C^*_i D_i(\text{Cu}) = 6.1 \times 10^{-4} \exp\left[\frac{-1.64\ \text{eV}}{kT}\right] \text{cm}^2\text{s}^{-1} \tag{5.36a}$$

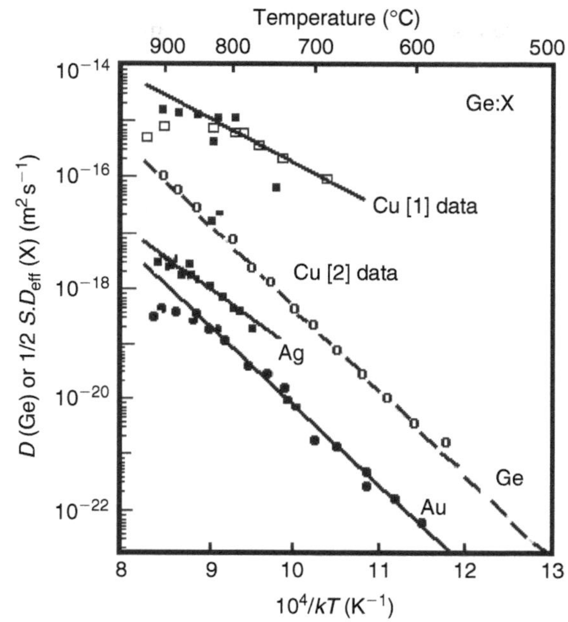

Figure 5.18. Half products of solubility and effective diffusivity of Cu, Ag and Au in Ge with various dislocation densities compared to the germanium self-diffusivity (dashed line) (after Bracht *et al.* [28]).

$$C_i^* D_i(\mathrm{Ag}) = 4.5 \times 10^{-4} \exp\left[\frac{-2.30\,\mathrm{eV}}{kT}\right] \mathrm{cm^2\,s^{-1}} \quad (5.36b)$$

$$C_i^* D_i(\mathrm{Au}) = 1.3 \times 10^{-1} \exp\left[\frac{-2.98\,\mathrm{eV}}{kT}\right] \mathrm{cm^2\,s^{-1}} \quad (5.36c)$$

This trend can be explained by assuming that the interstitial metal ions carry a positive charge that corresponds to the smallest possible closed-shell structure. The corresponding ionic radii are: 0.96 Å (Cu_i^+), 1.26 Å (Ag_i^+) and 1.37 Å (Au_i^+). From this, one can derive that the large spacing in the $C_i^* D_i$ data between Cu and Ag in Figure 5.18 originates from the rather large difference in size between Cu_i^+ and Ag_i^+, while a smaller difference exists between Ag_i^+ and Au_i^+ [28].

At high temperatures (800–900°C), a "slow" diffusion of Ag in Ge has also been observed, using radiotracer experiments. It was found to correspond with a solubility of $4 \times 10^{18}\,\mathrm{cm^{-3}}$, which is about 10^4 times larger than the solubility for the "fast" (normal) diffusion [87]. Little interaction was observed between the two diffusion "streams".

It should finally be remarked that both Ag [83, 88] and Au [84] exhibit a high diffusivity along germanium surfaces, so that back-side contamination may occur if the metals have only been deposited on the front side of the wafer.

5.3.2. *Energy levels and capture cross sections*

The acceptor nature and energy levels of Au_s have first been established by Hall effect and resistivity measurements [76,89]. Shortly after, a deep donor level at $E_V + 0.05\,eV$ was also ascribed to Au_s [75]. The complete level scheme for substitutional gold in Ge is represented schematically in Figure 5.19 [75]. One can see there that Ag_s is also a triple acceptor, with energy levels at $E_V + 0.13\,eV$ (A1 = $Ag_s^{0/-}$), $E_C - 0.29\,eV$ (A2 = $Ag_s^{-/2-}$) and at $E_C - 0.09\,eV$ (A3 = $Ag_s^{2-/3-}$). The chemical trend is obvious: for a larger tetravalent radius, the corresponding level is shifted further away from the valence band and closer to the conduction band. It indicates the lower binding energy for the first (second, third) bound electron when increasing the size of the substitutional atom. For the case of Pt_s, Hall measurements have identified only two acceptor levels at $E_V + 0.04$ eV and $E_C - 0.2\,eV$ [36,90], corresponding with the singly and triply negative charge state. No intermediate level has been reported so far. It has been suggested that there exists a strong Jahn–Teller distortion, explaining the missing middle level [90].

More recently, DLTS has been applied for the study of Au [42,91,92] and Ag [42,77,91] in metal-diffused Ge. The spectra corresponding to p-Ge and n-Ge, with gold in-diffused at 700°C for 6 h are represented in Figures 5.20 and 5.21, respectively [92]. The activation energies, derived from the slope of an Arrhenius plot and the corresponding pre-exponential factor for the emission rate (K_T) are summarized in Table 5.3 and compared with the other available DLTS data. The energy levels of Table 5.3 are in reasonable agreement with the Hall values of Figure 5.19. Slight differences can be ascribed to the temperature dependence of the capture cross sections for holes (two lower levels in p-Ge) or electrons (two upper

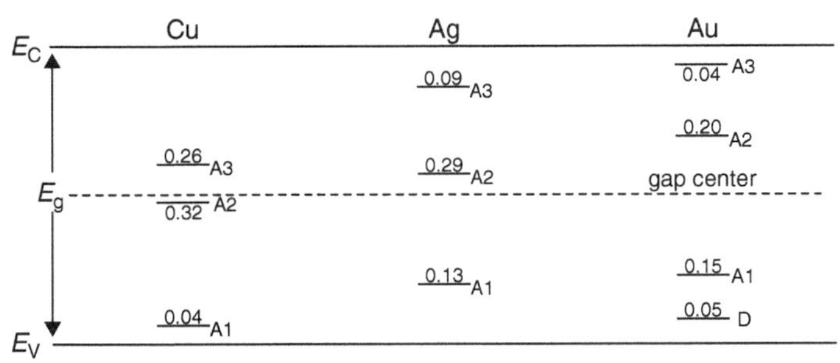

Figure 5.19. Energy level scheme for Cu_s, Ag_s and Au_s in Ge, showing the triple acceptor behavior of the Group Ib metals. Au_s is an amphoteric impurity with a shallow donor level at $E_V + 0.05\,eV$. More recently, it was demonstrated that a donor level found in DLTS at $E_V + 0.035\,eV$ could be associated with Ag_s [77]. The corresponding level for Cu is then resonant with the valence band and can thus not be detected electrically. Note, however, that an overcharged Cu_s^+ acceptor has been detected by far-infrared absorption measurements [46,47], although the binding energy is too small (~2 meV) to consider this as a stable donor state.

levels for n-Ge) [42]. In the case of the first acceptor level at $E_V + 0.150$ eV, the lower activation energy derived from DLTS (Table 5.3) results most likely from a PF lowering of the emission barrier in the electric field of the Schottky diode employed [92]. The other three peaks should not be sensitive to this effect, since they offer a neutral or repulsive potential for the majority carriers emitted from the neutral single acceptor (p-Ge) or the double or triple negatively charged center (n-Ge).

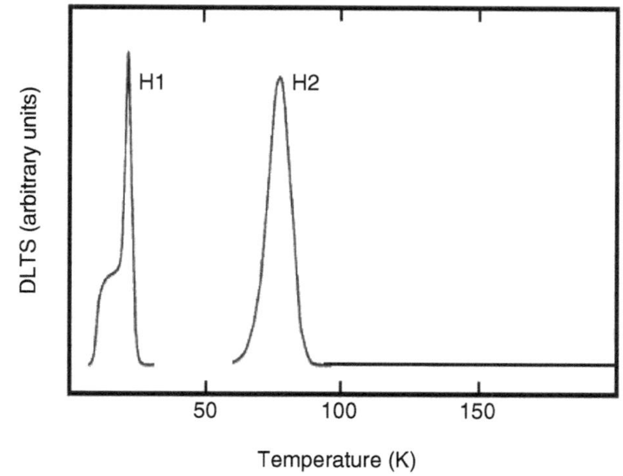

Figure 5.20. DLTS of boron-doped p-type germanium (1–3 × 10^{14} cm^{-3}) following gold diffusion (700°C, 6 h). The In Schottky barrier was applied after removal of a 40 μm germanium layer. $N_{H1} = 3.9 \times 10^{14}$ cm^{-3} and $N_{H2} = 3.4 \times 10^{14}$ cm^{-3} (after Simoen *et al.* [92]).

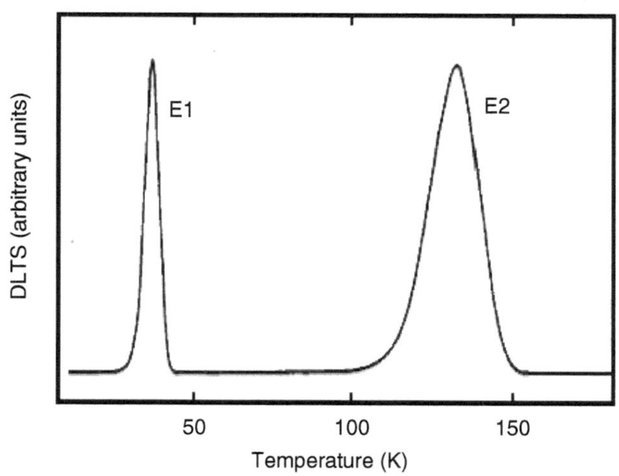

Figure 5.21. DLTS of antimony-doped n-type germanium (1–3 × 10^{14} cm^{-3}) following gold diffusion (700°C, 6 h). The Au Schottky barrier was applied after removal of a 55 μm germanium layer. $N_{E1} = 5 \times 10^{13}$ cm^{-3}; $N_{E2} = 5.2 \times 10^{13}$ cm^{-3} (after Simoen *et al.* [92]).

Table 5.3. Energy levels and assignments of the DLTS bands of gold-diffused germanium [92], compared with DLTS data from literature.

Level assignment	Energy level	K_T ($s^{-1}K^{-2}$)	Pearton [91]	Kotina [42]
$Au_s^{(2-/3-)}$	$E_C - 0.056\,eV$ (A)	2.7×10^6		
$Au_s^{(-/2-)}$	$E_C - 0.215\,eV$ (A)	1.2×10^6	$E_C - 0.22\,eV$ (D)	$E_C - 0.20\,eV$
$Au_s^{(0/-)}$	$E_V + 0.135\,eV$ (A)	1.8×10^7	$E_V + 0.14\,eV$ (A)	$E_V + 0.15\,eV$
$Au_s^{(+/0)}$	$E_V + 0.044\,eV$ (D)	3.9×10^9		

The DLTS level corresponds to the signature (E_T, K_T) derived from Arrhenius diagrams.
A: acceptor level; D: donor level.

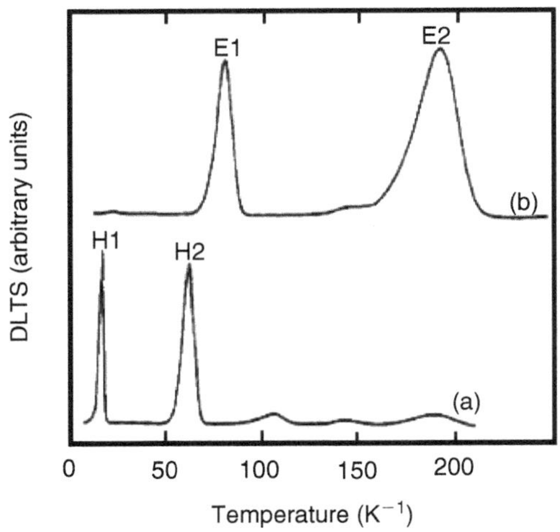

Figure 5.22. DLT-spectrum of silver-diffused (700°C, 5 h) p- and n-type germanium. (a) p-type sample ($8.3 \times 10^{13}\,cm^{-3}$ boron doped) and (b) n-type sample ($1.5 \times 10^{14}\,cm^{-3}$ antimony doped). Minor peaks in the spectrum of the p-type sample are $Ni_s^{0/-}$ (106 K) and $Cu_s^{-/2-}$ (145 K). $N_{H1} = 6.2 \times 10^{12}\,cm^{-3}$; $N_{H2} = 6.1 \times 10^{12}\,cm^{-3}$; $N_{E1} = 9.0 \times 10^{12}\,cm^{-3}$ and $N_{E2} = 8.9 \times 10^{12}\,cm^{-3}$ (after Huylebroeck *et al.* [77]).

The DLT-spectra for silver-diffused n- and p-Ge (700°C for 5 or 24 h) are represented in Figure 5.22 [77]. The very similar concentration corresponding to peaks H1 and H2 in Figure 5.22 and the fact that silver doping gives rise to a compensation of the original p-type doping at low temperatures has led to the conclusion that H1 could be a deep donor level associated with Ag_s [77]. If this is true, both substitutional silver and gold act as amphoteric impurities in Ge, whereby the chemical trend is still respected, i.e. the up-shift of the corresponding levels for increasing atom size. Table 5.4 gives the activation energies derived from an Arrhenius plot for the Ag_s levels from Refs [42,77,91].

DLTS of Pt-doped n-type germanium revealed a donor level at $E_C - 0.17\,eV$ ($\sigma_n = 10^{-17}\,cm^2$) and a second donor at $E_C - 0.21\,eV$ ($\sigma_n = 3 \times 10^{-19}\,cm^2$) [91].[5] It should finally be

[5]The donor assignment in Ref. [91] can be questioned, based on the corresponding small electron capture cross sections. There was probably confusion between the meaning of a deep donor and an electron trap in n-type Ge.

Table 5.4. Energy levels and assignments of the DLTS bands of silver-diffused germanium, compared with DLTS data from the literature.

Level assignment	Energy level	K_T ($s^{-1}K^{-2}$)	ΔH (eV)	Kotina [42]	Pearton [91]
$Ag_s^{(2-/3-)}$	$E_C - 0.144$ eV (A)	2.9×10^7	0.113		
$Ag_s^{(-/2-)}$	$E_C - 0.302$ eV (A)	5.7×10^5	0.261	$E_C - 0.288$ eV	$E_C - 0.270$ eV
$Ag_s^{(0/-)}$	$E_V + 0.116$ eV (A)	8.5×10^7		$E_V + 0.10$ eV	$E_V + 0.10$ eV
$Ag^{(+/0)}$	$E_V + 0.035$ eV (D)	9.2×10^8			

The DLTS level corresponds to the signature (E_T,K_T) derived from an Arrhenius diagram.
A: acceptor level; D: donor level.
ΔH is the enthalpy for ionization after correction for the activation energy of the capture cross section.
After Huylebroeck *et al.* [77].

mentioned that also palladium has been investigated by DLTS [91], yielding energy levels at $E_C - 0.23$ eV ($\sigma_n = 3 \times 10^{-17}$ cm^2); $E_C - 0.32$ eV ($\sigma_n = 4 \times 10^{-15}$ cm^2); $E_C - 0.42$ eV ($\sigma_n = 10^{-14}$ cm^2) and $E_V + 0.19$ eV ($\sigma_p = 8 \times 10^{-20}$ cm^2) [91].

An overview of the capture cross-section data corresponding with the four levels of Au_s is given in Appendix Table A2. Whenever available, the temperature dependence of the respective σ_n and σ_p is mentioned. By comparing the capture cross sections of several impurities in a single attractive charge state, Rupprecht concluded that there is not only a Coulombic factor determining σ but also a size effect [93]. The larger the atom, the larger σ, which goes hand in hand with a smaller maximum solubility and a lower distribution coefficient k_{dm} [93]. It can be derived from Appendix Table A2 that to our knowledge no reliable data are available for σ_p^{3-} and for σ_n^{+}. There is quite some scatter in the magnitude of the capture cross sections, which can be partly explained by considering the measurement temperature. As will be seen below, the electron capture cross section for the acceptor levels is generally a strong function of the electric field, which plays a role for the characterization of deep levels in the depletion region of a junction during DLTS. Moreover, different temperature dependences are reported for the same capture cross section, which could be related to the different measurement techniques used. In general, the hole capture cross sections are expected to show a T^{-n} dependence, with $n \sim 3$–4, while the electron cross sections for the three acceptor levels tend to follow an exponential increase with *T*, exhibiting a small Coulombic barrier in the range of 10 meV or a behavior typical for tunneling through a barrier [94].

As discussed already for the case of copper-doped Ge, electron heating in a high electric field can increase the probability for tunneling through a repulsive barrier, enhancing the capture rate. It was first theoretically predicted by Ridley and co-workers [95,96] that hot electron capture by repulsive centers could lead to a so-called negative differential resistance (NDR). NDR can occur if the mobility or the carrier density reduces with increasing electric field, leading to a reduction of current with voltage. It was first postulated and later demonstrated experimentally that NDR exists in gold [97,98] and copper-doped [99] n-type germanium. In order to separate the high-field reduction of mobility and carrier density, one normally compares pulsed *I*–*V* measurements, only affected by a change in

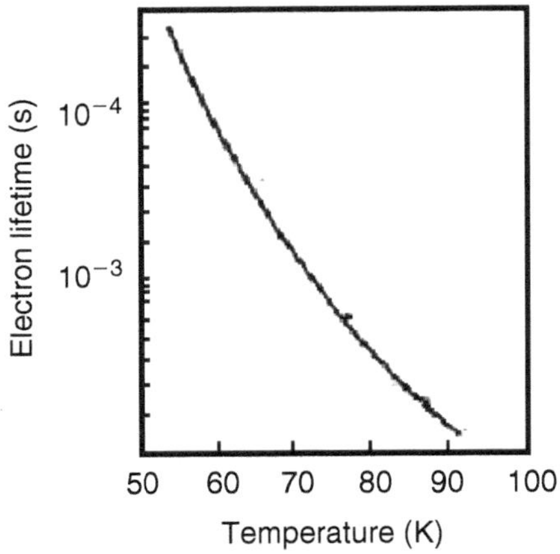

Figure 5.23. Temperature dependence of the electron lifetime in n-type Ge:Sb:Ag (after Aksyanov *et al.* [103]).

mobility that can follow rapid changes in F, with static measurements, allowing also the carrier density to reach steady state [99].

5.3.3. Impact on carrier lifetime

Gold or silver doping can have a strong impact on the lifetime and trapping characteristics of Ge. For example, in n-type Ge the same $E_C - 0.2$ eV Au-related level may cause carrier trapping and recombination [100]. Particularly at low T ($\sim$100 K) severe hole trapping by Au_s^{2-} centers can occur [101]. A continuous reduction of τ_{rec} with increasing T was observed in n-Ge, which became steeper closer to room temperature. In the same work [101], it was demonstrated that a high dislocation density may impact on the recombination properties of Au doping: hole trapping becomes more pronounced and a stronger temperature dependence of τ_{rec} has been observed [101]. This could point to some interaction (decoration) of dislocations with Au, to form a Cottrell atmosphere. In gold-doped p-Ge, the recombination lifetime is governed by the $E_V + 0.15$ eV and $E_C - 0.2$ eV acceptor states [102]. This gives rise to two plateaus in the dependence of τ_{rec} on T. At room temperature, electron capture occurs through Au_s^-, while at low T, through Au_s^0. The former process is more efficient, so that the lifetime increases upon cooling for gold-doped p-Ge, from a lower plateau at 300 K to a higher one below 200 K. The intermediate transition region is determined by the Fermi level shift with T, which changes the occupancy of Au_s from the singly negative to the neutral state. No trapping occurs in that case. This means that for p-Ge $\tau_n = \tau_p$.

The effect of silver doping on the lifetime and trapping in Ge was studied in more detail in Refs [103–105,106]. At 300 K, the $E_V + 0.24$ eV and $E_C - 0.25$ eV Ag_s acceptors play a dominant role in the lifetime of n-Ge [106]. A continuous decrease of the electron lifetime has been found in n-Ge:Sb:Ag as shown in Figure 5.23 [103]. As shown by Figure 5.24, two plateaus are observed in the PCD lifetime versus temperature of p-type Ge doped with Ag [105].

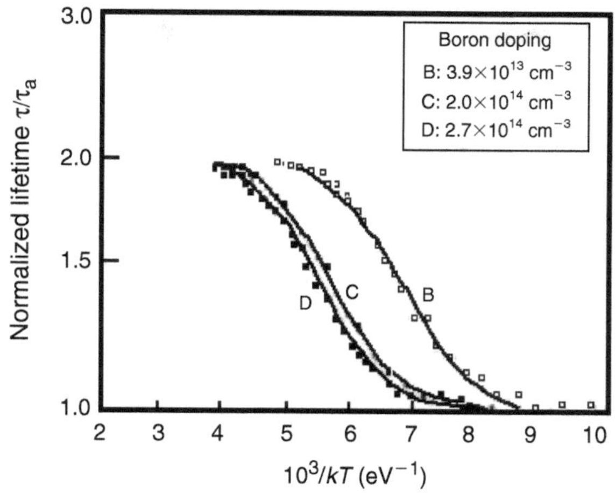

Figure 5.24. Experimental lifetime and line fits with $c_{n1}/c_{n2} = 2$, $E_T - E_V = 0.14\,eV$ and for the boron doping concentration: (B) $3.9 \times 10^{13} cm^{-3}$; (C) $2.0 \times 10^{14} cm^{-3}$ and (D) $2.7 \times 10^{14} cm^{-3}$ (after Schoenmaekers and Henck [105]).

While the magnitude of the plateaus is only determined by the silver concentration, the transition temperature is sensitive to the background p-type doping density (Fermi level). The room temperature lifetime is dominated by the second acceptor level ($E_C - 0.29\,eV$) while the low-temperature plateau is governed by the $E_V + 0.13\,eV$ state [105]. Here, opposite behavior compared with Au in p-Ge is found: a reduction of τ_{rec} with lower T. The ratio of two between the low- and high-temperature plateaus in Figure 5.24 comes from the ratio of the electron capture rates or – in a first approximation – from the ratio of the electron capture cross sections σ_n^0/σ_n^- (Appendix Table A3). It also means that electron capture by the neutral charge state of silver is more efficient than by the single negative charge state [105].

5.4. Nickel in Germanium

As the neighboring TM element of copper in the periodic table, nickel behaves very similar in germanium and has often been studied in conjunction with it [21,107,108]. Nickel has a slightly lower distribution coefficient than Cu – values of 1.8×10^{-6} from resistivity measurements [109] or 5×10^{-6} from ^{63}Ni radiotracer measurements [4] can be found in the literature – and, in agreement with the simple theory by Hall [81], a lower maximum solid solubility. The diffusivity is close to the one of copper, while it behaves electrically as a double acceptor [4]. In the following sections, the properties of Ni will be discussed to some detail.

5.4.1. Solubility and diffusivity of Ni in Ge

The solid solubility of Ni in the temperature range 700–900°C and derived from resistivity measurements is represented in Figure 5.25 [109]. A maximum retrograde solubility is

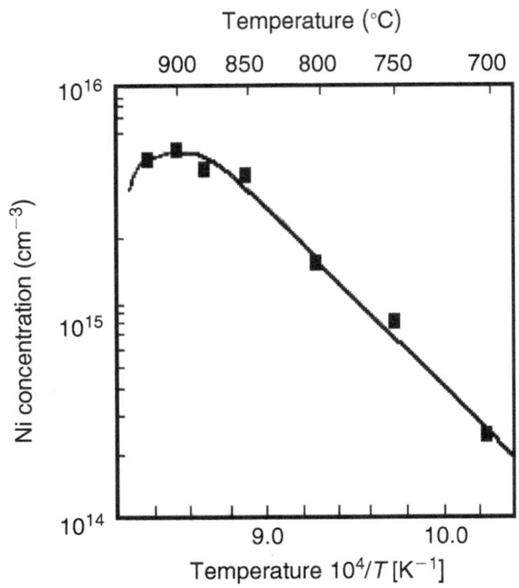

Figure 5.25. Plot of the concentration of nickel in Ge versus $1/T$ (after van der Maesen and Brenkman [109]).

observed near 900°C ($\sim 8 \times 10^{15}$ cm^{-3}) [85,109,110], while and eutectic point occurs at 775°C. At a given temperature, the solubility of nickel amounts to about 10% of the value for copper [107]. Regarding the precipitation of nickel, it was concluded that the precipitation rate is faster for Ni at 500°C compared with Cu [107]. However, the precipitation rate slows down in the presence of copper doping below supersaturation. At the same time, there will be an enhanced Ni precipitation in the presence of precipitating copper, at 500°C, through the release of vacancies, facilitating the diffusion of nickel in the material [21].

Nickel is also a fast diffuser in germanium, whereby the interstitial ion (positively charged) controls the transport, while the substitutional fraction determines the solubility and acceptor nature of the impurity. Figure 5.26 compares the diffusion coefficient of Ni in Ge determined at 1073 K with the one for some other elements [111]. The diffusion behavior of nickel can be understood in the frame of the dissociative mechanism, whereby an interstitial nickel (Ni_i) and a vacancy-limited (Ni_s) diffusivity can be defined. Additionally, it was observed that the structural perfection of the Ge material has a strong impact on the diffusivity of nickel, as dislocations are a known source of vacancies, necessary for the conversion of Ni_i into Ni_s [111]. It was also found that $N_{Ni}^{eq} = 3.3 \times 10^{15}$ cm^{-3} at 800°C (1073 K). Using Ni (or Cu) diffusion profiles, it has been possible to extract some basic information regarding the self-diffusion in Ge and the equilibrium concentration of vacancies [112].

5.4.2. Energy levels and capture cross sections of Ni in Ge

Based on resistivity and Hall measurements and the observation that n-Ge converted to p-type when nickel doped in the melt, it was concluded that substitutional nickel behaves

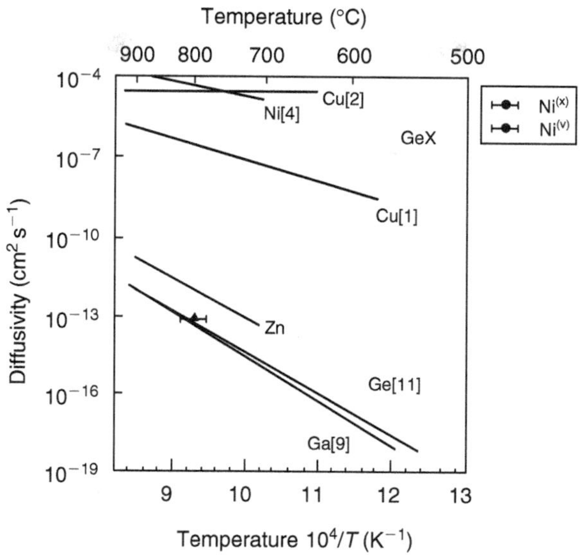

Figure 5.26. Diffusivities of several elements of the fourth row in the Ge matrix (after Giese *et al.* [111]).

as a double acceptor, with the single acceptor at $E_V + 0.22$ eV and the double acceptor at $E_C - 0.30$ eV [4,21,85,113]. Comparing Ni, Zn and Cu, three neighboring elements in the periodic table behaving as multiple deep acceptors, quite different energy position and spacing of the energy levels are found. As will be further discussed in Section 5.6, chemical trends within elements of the same group can be clearly observed, while this is not so much the case within one row of the periodic system.

In the eighties and nineties, some DLTS studies of nickel-doped Ge have been performed [42,45,91,114]. The resulting activation energies are summarized in Table 5.5. A reasonable agreement between the different studies can be noted. In addition, some other deep levels, typical for Ni-doped Ge have been observed. Two levels are ascribed to Ni–H, similar as for Cu–H [45]. It was also observed that Ni and Li show a tendency to pair [91]. The passivation of nickel-related deep levels by deuterium has been studied by Pearton *et al.* [115]. A 3 h plasma treatment at 300°C was shown to reduce significantly the concentration of Ni in p-Ge. However, as can be derived from Table 5.5, some Ni–H centers may be formed as well. Complete passivation is expected to occur for two hydrogen atoms giving up their electron to the nickel double acceptor.

Similar as for the case of copper in a Ge n^+–p junction, nickel impurities may be removed by drifting in a large electric field at room temperature, toward the n^+ – contacts [116]. A mobility of the $E_V + 0.14$ eV level[6] of $(2.0 \pm 1.1) \times 10^{-13}$ cm²/Vs^{-1} at 25°C was derived, corresponding to a diffusion coefficient of $(5.1 \pm 2.7) \times 10^{-15}$ cm^{2}s^{-1}.

[6] This value has not been included in Table 5.5, because it is far off the expected $E_V + 0.22 - 0.23$ eV range. Therefore, the results in these papers [91,115,116] have to be considered with great care.

Table 5.5. K_T, E_T and enthalpy ΔH derived from Arrhenius plots in DLTS for Ni-related deep levels in Ge.

Level	E_T (eV)	K_T ($K^{-2}s^{-1}$)	ΔH (eV)	References
$Ni_s^{0/-}$	0.217	5.8×10^8	–	[114]
	0.223	1.5×10^9	–	[42]
	0.215	1.5×10^8	0.207	[45]
Ni related	0.414	2.2×10^8	–	[114]
Ni–Li*	0.40	–	–	[91]
Ni–Li	0.21	–	–	[91]
Ni–H	0.154	8.85×10^7	0.159	[45]
Ni–H	0.416	2.69×10^8	0.299	[115]
$Ni_s^{-/2-*}$	0.299	7.0×10^6	–	[114]
	0.303	7.0×10^7	–	[42]
	0.18	–	–	[91]

* In n-Ge, energies with respect to the conduction band. The other levels have been observed in p-Ge and are with respect to the valence band.

It should finally be remarked that upon nickel precipitation in the 400–500°C temperature range, shallow acceptors have frequently been observed in the early stages of annealing [107,108,110]. Their ionization energies are approximately 50 and 20 meV and their concentration is equal to the original Ni concentration. They have also been observed after a quench of Ge:Ni from 800°C [107], but their concentration increases upon subsequent annealing at 500°C. It was speculated at that time that these shallow acceptors could be related to vacancies, released by precipitating nickel atoms.

The capture cross sections for the two acceptor levels have been rather intensively studied. A summary of the literature data is given in Appendix Table A4.

5.4.3. Impact on carrier lifetime

Early studies demonstrated that Ni_s is an efficient lifetime killer in Ge [63,70,117–121]. Figure 5.27 shows the lifetime at 300 K for nickel-doped and undoped material [63]; addition of Ni causes the lifetime to drop by a factor of 40. Already at concentrations of $10^{12}\,cm^{-3}$, substitutional Ni lowers τ_{rec}. As seen in Figure 5.27, the data versus doping concentration can be accurately described by the SRH theory for carrier recombination. The $E_V + 0.23$ eV level was shown to be the predominant recombination level [63].

When studied as a function of temperature, two plateaus are usually observed for n-type Ge:Ni, whereby t_{rec} reduces for lower T in the case of Figure 5.28 [117]. The transition (or slope) between the two plateaus can be described by the passage of the Fermi level across the 0.30 eV second acceptor level. The high-temperature lifetime plateau is determined by the two coupled acceptor levels. This is different from the τ behavior with two independent

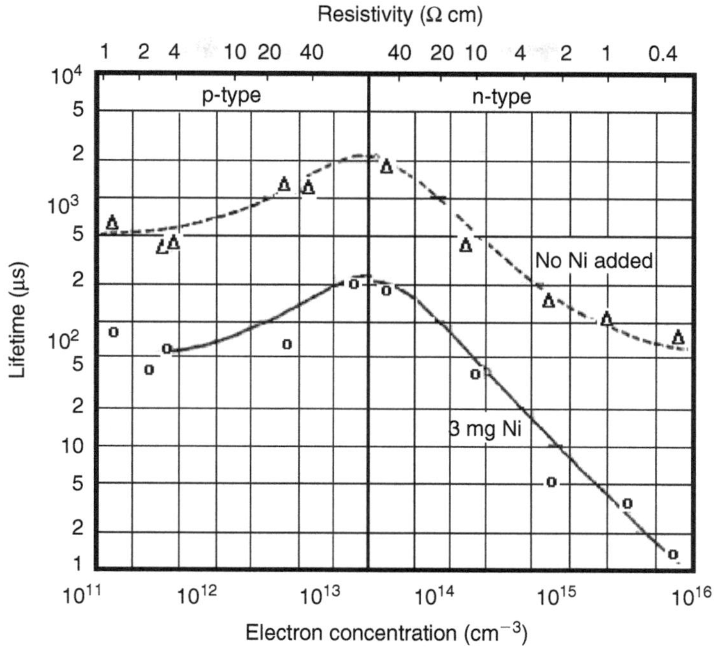

Figure 5.27. Lifetime versus electron concentration for Ge crystals, measured at 300 K and fraction solidified $x = 0.5$. Δ: only Sb or In added to the melt; o, 3.0 mg Ni plus Sb or In added to the melt. The curves were calculated according to the SRH theory (after Burton *et al.* [63]).

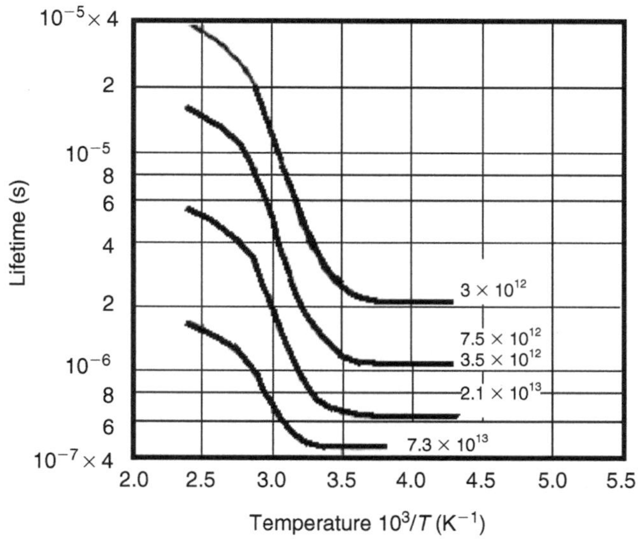

Figure 5.28. Lifetime in n-type germanium containing nickel impurities, computed for a net-doping concentration of $3.0 \times 10^{14}\,\mathrm{cm}^{-3}$, $E_C - E_{T2} = 0.31\,\mathrm{eV}$, $c_{n2} = 8.3 \times 10^{-9}\,\mathrm{cm^3 s^{-1}}$; $c_{p2} = 2.0 \times 10^{-7}\,\mathrm{cm^3 s^{-1}}$ (after Wertheim [117]).

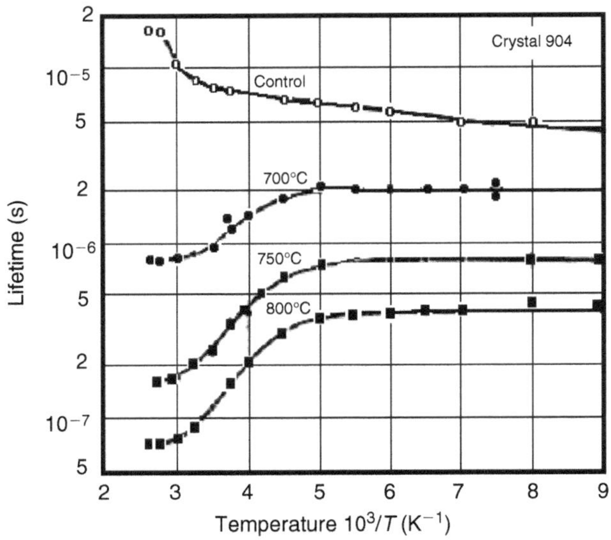

Figure 5.29. Bombardment-conductivity decay time in p-type germanium containing nickel introduced by diffusion at three different temperatures, corresponding with an increasing Ni concentration (after Wertheim [117]).

levels. At 350 K, the magnitude of the high-temperature plateau is inversely proportional to the Ni concentration, in line with SRH theory.

Typical lifetime behavior versus $1/T$ for p-type Ge:Ni is shown in Figure 5.29 [117]. In this case, the lifetime increases for lower T. At low T, the recombination is limited by electron capture in the neutral state of the $E_V + 0.23$ eV acceptor level [122]. Upon increasing the temperature, a second recombination channel is opened by the production of Ni_s^-, thereby populating the upper acceptor level. However, as noted by Schulz [70], the early lifetime results on p-type Ge:Ni were rather controversial in the sense that compensated material showed an increase of τ_{rec} upon cooling, while in non-compensated p-Ge, just the opposite was found. The explanation he forwarded was that in compensated material, the low-temperature recombination will be determined by a fixed amount of Ni_s^-, while in non-compensated germanium, the neutral fraction continuously increases for lower T.[7] As it is expected that $\sigma_n^0 > \sigma_n^-$, this will speed up the recombination at lower T [70]. Moreover, it was pointed out that the dislocation density n_{disl} plays a strong role in the recombination lifetime versus temperature behavior in p-type nickel-doped Ge [122]. For higher n_{disl}, the temperature dependence of τ_{rec} becomes stronger, whereby τ_{rec} reduces significantly with n_{disl} at low T. At room temperature on the other hand, dislocations were found to be relatively inefficient recombination centers [122]. The conclusion was that for a reliable determination of the electron capture rate by neutral Ni, a low dislocation density is required. Moreover, at high "n_{disl}"s,

[7] Schulz performed most of his experiments on low dislocation density material [70].

the global effect of Ni and dislocations is not just the sum of the two contributions. It was found that an interaction occurs [122], similar as for the case of gold doping, yielding a stronger lifetime reduction.

It should finally be remarked that Ni_s is one of the few metallic impurities in Ge where successful electron spin resonance (ESR) measurements have been performed [123,124]. Substitutional nickel is paramagnetic in the single negative charge state, whereby the ^{61}Ni isotope (abundance 1.25%) leads to hyperfine structure in the spectra, according to a spin 3/2. It has been found that Ni_s is a Jahn–Teller distorted center, whereby the atom moves out of the substitutional site into any cubic direction [123].

5.5. TMs in Germanium

In the following section, the known properties of the so-called TMs Ti, V, Cr, Mn, Fe and Co in germanium will be described. Most of them are characterized by a rather low solubility and high diffusivity. The former explains why the available information on some of the TMs is rather scarce.

5.5.1. Iron

Fe is a double acceptor with levels at $E_V + 0.34\,eV$ and $E_C - 0.27\,eV$ [125]. The distribution coefficient is quoted to be $<10^{-6}$ [113]. When the atomic concentration reaches 0.3% Fe in the melt, lineage (= a small angle grain boundary) occurs in the grown crystal, associated with Fe precipitation. This was confirmed by radiotracer measurements [125]. The maximum solubility of Fe appears to be $\leqslant 10^{13}\,cm^{-3}$ [125] at ~800°C [126]. A few studies on the Fe diffusivity in Ge have been reported [126,127]. Values for D in the range $10^{-6} - 10^{-7}\,cm^2$ at 800°C have been observed, about one decade smaller than for Cu and Ni. The corresponding activation energy was ~1 eV [126] and it was believed that trap-assisted diffusion played a strong role in the two-stream diffusion process [126].

In spite of the low S_0, Fe has been shown to reduce the minority carrier lifetime [128]. In n-type Ge, the lifetime of Ge:Fe at room temperature is determined by the $E_C - 0.27\,eV$ level [128]. In p-type iron-doped Ge, the carrier lifetime could be described by the single charged Fe_s acceptor level at $E_V + 0.34\,eV$ and a corresponding electron capture cross section of $\sim 10^{-15}\,cm^2$ [128]. The capture cross section at 300 K was found to be $\sim 10^{-15}\,cm^2$ for either holes or electrons [129]. A more detailed study [130] revealed the following values:

$$\sigma_p^- = 3 \times 10^{-15}\,cm^2\ (300\,K)$$
$$\sigma_p^{2-} = 10^{-14}\,cm^2\ (300\,K)$$
$$\sigma_n^0 = 10^{-15}\,cm^2\ (300\,K)$$
$$\sigma_n^- \sim \exp(-\Delta E/kT)\ \text{with}\ \Delta E \sim 0.05\,eV\ \text{and}\ \sigma_n^- \sim 10^{-16}\,cm^2\ (300\,K)$$

It should be remarked that one of the first observations of NDR in semiconductor devices was in a p–n junction made in Fe-doped Ge [131]. At 77 K, the forward I–V shows an unusual breakdown behavior, for a voltage which is much larger than the built-in potential of ~0.7 V. Such a behavior could later be explained in terms of the impact ionization of occupied Fe acceptors.

5.5.2. Cobalt

Cobalt introduces two acceptor levels, at $E_V + 0.25\,\text{eV}$ and $E_C - 0.31\,\text{eV}$ and has a $k_d \sim 10^{-6}$ [113,132]. Moreover, a deep donor at $E_V + 0.09\,\text{eV}$ has been ascribed to Co_s [85], although it is not generally accepted. It has been reported that lineage develops in the crystal for the same atomic concentration as for Fe in the Ge melt (~0.3%) [132]. A solubility limit of $2 \times 10^{15}\,\text{cm}^{-3}$ can be found in the literature [133]. The diffusivity of Co has been studied using the ^{58}Co tracer [126]. Very similar results as for Fe have been obtained: $D \sim 10^{-6} - 10^{-7}\,\text{cm}^2\text{s}^{-1}$ at 800°C and a trap-assisted double-stream diffusion mechanism. Likewise, an activation energy of ~1 eV was found.

A few studies were devoted to the recombination properties of Co in Ge [133,134]. It was observed that Co gives rise to efficient recombination [134], with the following values for the capture cross sections:

$\sigma^0{}_n = 10^{-15}\,\text{cm}^2$
$\sigma_n^- = 10^{-19}\,\text{cm}^2$ at 80 K; = $(0.05\text{–}2) \times 10^{-16}\,\text{cm}^2$ at 145 K; $<10^{-15}\,\text{cm}^2$ at 300 K
$\sigma_p^{2-} \sim T^{-n}$ ($n = 3.5 - 4.5$) ($T < 200$ K) and $= 22 \times 10^{-16}\,\text{cm}^2$ at 145 K; for $T > 200$ K, the temperature dependence weakens, probably due to the dissipation of the excess energy by optical phonons [134].

It was, furthermore, observed that the Co levels were responsible for both carrier trapping (n-type Ge) and recombination, through the $E_C - 0.3\,\text{eV}$ level [134]. The trapping ratio increases for lower T. Moreover, it was found that σ_p^0 is practically field independent at 51 K and equal to $8 \times 10^{-14}\,\text{cm}^2$ [133]. The degeneracy ratio for the upper Co_s level was estimated to be ~10 [134].

In DLTS, an electron trap at $E_C - 0.18\,\text{eV}$ has been reported in Co-doped n-type Ge [135]. This level was hard to passivate by hydrogen: it shows a passivation rate 20 times slower than for Cu or Ni in Ge.

5.5.3. Manganese

As shown in Figure 5.30, Mn is a double acceptor in Ge, with levels at $E_V + 0.16\,\text{eV}$ and $E_C - 0.37\,\text{eV}$ (mid-gap) [136]. An effective distribution coefficient of $\sim 10^{-6}$ was obtained and a maximum solubility of $10^{16}\,\text{cm}^{-3}$. It has been reported that there exists shallow Mn-related acceptor levels as well [85]. Mn doping reduces the recombination lifetime

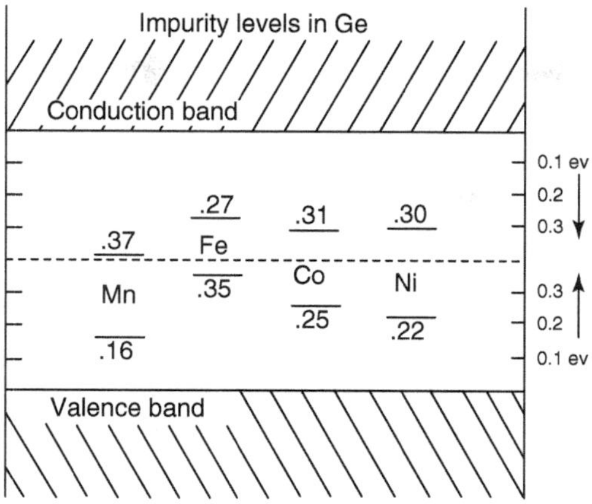

Figure 5.30. Impurity levels introduced into germanium by various elements of the fourth row of the periodic chart. In addition to these elements, it is to be noted that vanadium is electrically inactive or insoluble in germanium (after Woodbury and Tyler [136]).

according to the inverse of the concentration [136]. The capture cross section of the Mn levels has been studied by several groups [100,137,138]. The following results have been obtained:

$$\text{For the } E_C - 0.37\,\text{eV level: } \sigma_n^- = 4\times10^{-17}\,\text{cm}^2\ (300\,\text{K; weakly } T \text{ dependent})$$

$$\sigma_p^{2-} = 10^{-16}\,\text{cm}^2 \text{ at } 300\,\text{K}$$

$$\text{For the } E_V + 0.16\,\text{eV level: } \sigma_n^0 \sim 10^{-16}\,\text{cm}^2\ (300\,\text{K});\ = 2\times10^{-16}\,\text{cm}^2\ (90\,\text{K})$$

$$\sigma_p^- = 2\times10^{-12}\,\text{cm}^2\ (100\,\text{K})$$

The electron cross section for the neutral charge state can be explained in terms of the cascade theory, although the polarization forces create little excited states [136]. In p-type Ge, no marked trapping was observed [137], while in n-Ge, the upper Mn level is responsible for both hole trapping and recombination [138]. It was also found that the degeneracy ratio of the lower level of Mn equals $g^-/g^0 = 3$ [137].

5.5.4. *Other TMs*

5.5.4.1. Chromium

Chromium is characterized by a small distribution coefficient $<10^{-6}$ [73]. It seems to introduce shallow levels in p-type Ge, between 0.12 eV and the valence band [85]. Levels at $E_V + 0.07$ and 0.12 eV have been reported [73]. DLTS studies of Cr-diffused n-type Ge reveal an electron trap at $E_C - 0.31$ eV, which can be partially passivated by hydrogen [135].

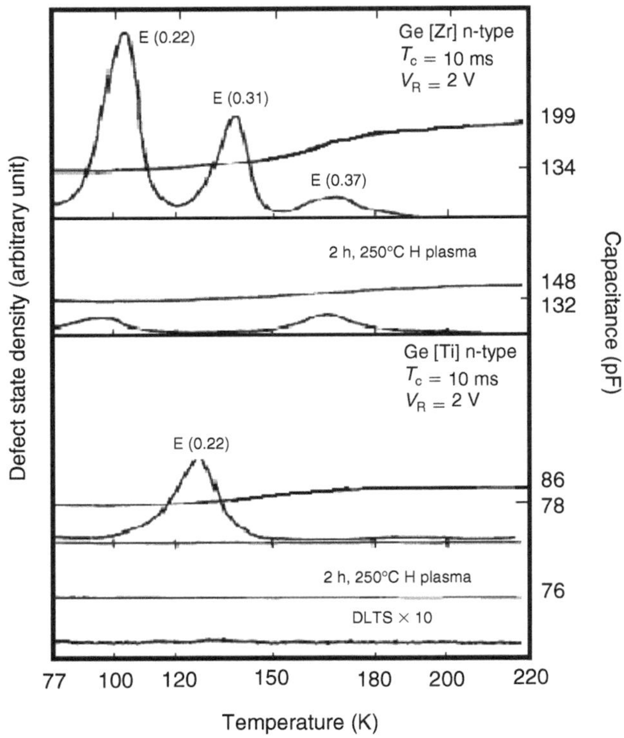

Figure 5.31. DLT-spectra and TSCAP scans recorded for (a) Zr-diffused n-type Ge and (b) Ti-diffused n-type Ge. The E(0.37) Zr-related centers are unaffected by the atomic hydrogen, whilst the other states show varying susceptibilities to passivation (after Pearton and Tavendale [135]).

5.5.4.2. Zirconium

Diffusion resulted in the introduction of three electron traps in n-Ge, at $E_C - 0.22$ eV, $E_C - 0.31$ eV and $E_C - 0.37$ eV [135]. The latter comes in after storage at room temperature for several months. Weak neutralization by hydrogen plasma was found for these levels [135].

5.5.4.3. Titanium and vanadium

Early attempts to study the electrical activity of Ti and V in Ge were unsuccessful [85,136]. Pearton and Tavendale show the existence of an electron trap at $E_C - 0.22$ eV of n-type Ti-doped Ge, which according to Figure 5.31 can be fully passivated by hydrogen [135].

5.6. Chemical Trends in the Properties of Metals in Ge

5.6.1 Electrical properties

As mentioned before, the electrical activity of a substitutional impurity can be understood based on its electronic structure [4,76,85,139]. Similar as for the Group III and V elements,

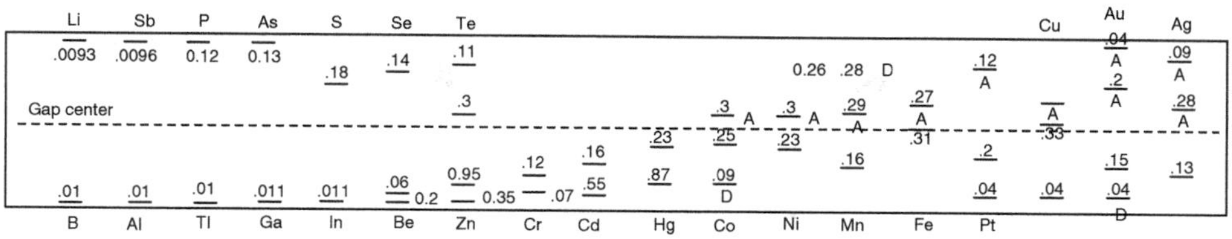

Figure 5.32. Impurity energy levels for various elements in Ge. The levels below the gap center are measured from the top of the valence band and are acceptor levels unless indicated by D for a donor. The levels above the gap center are measured from the bottom of the conduction band and are donor levels unless indicated by an A for an acceptor level. The band gap at 300 K is 0.67 eV (after Sze and Irvin [140]).

one can readily see that Group IV doping does not give rise to an electronic level in the band gap of Ge, as a Si or Sn atom may perfectly replace a germanium atom by forming four covalent bonds between its four valence electrons and the four nearest Ge neighbors. With respect to metals, one can conclude that the Group I elements Cu, Ag and Au will be triple acceptors, Group II impurities (Zn, Be, Cd, Hg) are double acceptors, while the transition elements (Mn, Fe, Co, Ni) which have two 4s electrons in the outer shell act as double acceptors. In the case of platinum, with one 6s electron, only two acceptor states have been reported, while a third level is expected based on the electronic configuration [139]. The only other established exception to this simple valence rule is the $E_V + 0.05$ eV donor level of Au_s. However, by giving up its one valence electron, Au_s can take up the closed-shell noble gas configuration, which is energetically quite favorable. One can consider that the corresponding donor level for Cu_s and Ag_s coincides with the valence band and is, therefore, unobservable. Finally, group VI elements are expected to behave as double donors, when on a lattice site, which is actually the case for S and Se.

The electronic configuration of a metal atom only gives some indication on the number of levels to be expected in the band gap of germanium. The position of these levels will be dictated to some extent by the atomic radius (size) of the impurities involved. As a rule of thumb, one can assume that the heavier the mass (the larger the atomic number) the further the corresponding level from its respective band will be, as shown in the summary of Figure 5.32 [85,140]. It should also be remarked that the order of the levels is dictated by the rule that the more negative the charge state is, the higher the ionization energy (positive U centers), as can be derived for the case of Au_s in Figure 5.32. It should be remarked that summaries of most of the early work can be found in Refs [4,76,85,139,140]. An overview of the ESR identification of impurities in Ge is given in Ref. [141].

Based on the known energy levels and corresponding charge states one can calculate the contribution of the metal impurities to the free carrier concentration at each desired temperature, provided the degeneracy factors of all the levels and the concentration of the impurity N_T are known. Besides that, ionized metal atoms can cause Coulombic carrier scattering at low temperatures, thereby reducing the mobility. The intensity of the effect will be proportional with $(Zq)^2$ with Zq the ion charge in absolute value. Using this effect, it has been possible to demonstrate with Hall effect measurements that certain substitutional acceptor impurities can assume a double or triple charge state. Combining these two factors enables to estimate the impact of metal doping on the resistivity of germanium.

5.6.2. Optical properties of metals in germanium

Besides thermal ionization, an electron (or hole) can be released from a deep level by absorption of a photon. This gives then rise to the so-called extrinsic or impurity photoconductivity (PC) [72,73], which finds applications in Ge-based IR photodetectors. While the Ge band gap of 0.7 eV corresponds with a wavelength threshold of 1.8 μm, this can be shifted to much longer wavelengths in the case of extrinsic photoconductive Ge. The photoionization threshold shift is inversely proportional with the ionization energy of the impurity. Doping with several metals has been used for that purpose, with Zn and Au among the more popular ones [72]. Figure 5.33 gives an idea of the absorption threshold λ and operation temperature for Au- and Zn-doped germanium [72]. The operation temperature needs to be low enough to populate the deep level with an electron (or hole), so that it can be photoionized to the corresponding band. Since Zn and Au are multiple acceptors, different energy levels (and thresholds) can be achieved, depending on the compensation or counter doping.

The sensitivity of such a detector depends, amongst others, on the concentration of the deep level in question and will be limited by the solubility of the metal. In the case of Zn or Cd quite high concentrations can be reached (up to a few $10^{17}\,cm^{-3}$) [72,73] while for Au_s or Hg_s this is limited to $<10^{16}\,cm^{-3}$ or $\sim 5 \times 10^{14}\,cm^{-3}$, respectively [73]. The impurity absorption coefficient α ($\sim$1–10 cm^{-1} typically) is given by the product of N_T and the photo-ionization cross section σ_{ph} [74]. As an empirical rule of thumb for Ge, one may assume that [74]:

$$\sigma_{ph} = 1.1 \times 10^{-18}\, E_1^{-2} \tag{5.37}$$

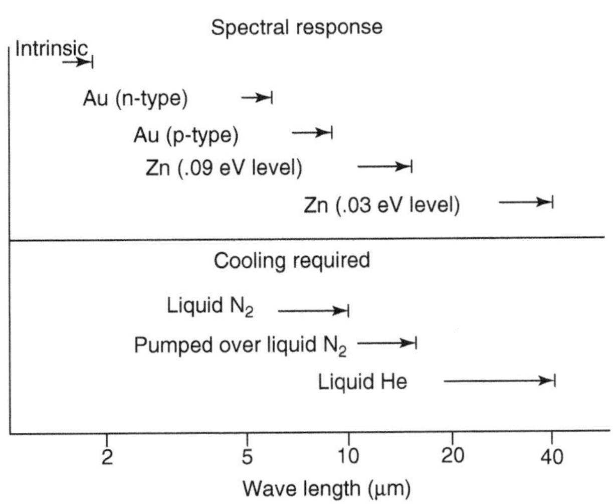

Figure 5.33. Spectral coverage and cooling required for impurity PC in germanium (after Levinstein [72]).

with E_I the ionization energy of the hydrogenic (shallow) impurity. According to Equation (5.37), the deeper the energy of the level, the smaller the photo-ionization cross section and, hence, the sensitivity of the detector. Typical experimental values are given in Table 5.6 [74]. As can be seen, the deeper the acceptor level, the smaller σ_{ph}, although Equation (5.37) predicts a stronger dependence on E_I [74].

5.6.3. Trends in the impact on carrier lifetime in Ge

Perhaps more important than the impact on the resistivity is the effect deep levels have on the carrier recombination (and generation) lifetime. For several reasons, it is much more difficult to distinguish clear trends here. Firstly, the lifetime not only depends on the energy level and concentration, but also on the capture cross section for minority carriers and last but not least on the ambient temperature and the carrier injection level of the measurement. Moreover, in the case of multiple acceptors or donors, each level can act as a recombination (generation) channel, depending on the operation conditions, rendering the interpretation and modeling of the data more complicated. Reviews of the early work can be found in Refs [142,143]; the latter paper gives an exhaustive summary of the capture cross section of impurities and dopant elements in Ge.

Based on the extensive literature overview, an attempt has been made to summarize the impact of TMs (Mn up to Cu) on the minority carrier lifetime in Ge. This parameter is given by:

$$\tau_n = [N_T \sigma_n v_{thn}]^{-1} \quad \text{(p-Ge)} \tag{5.38a}$$

$$\tau_p = [N_T \sigma_p v_{thp}]^{-1} \quad \text{(n-Ge)} \tag{5.38b}$$

and is essential for the operation of solar cells in Ge e.g., as it limits the diffusion length of minority carriers. v_{thn}, v_{thp} is the thermal velocity for electrons and holes. It is taken for convenience as $10^7\,\text{cm s}^{-1}$ at 300 K, here. Figures 5.34 and 5.35 summarize the available literature data on the capture cross sections (or capture rates) at 300 K in p- and n-type Ge, respectively. The capture rate is defined as:

$$c_n = \sigma_n v_{thn} \tag{5.39a}$$

Table 5.6. Photo-ionization cross section for certain impurities in p-type Ge.

Element	E_I (eV)	σ_{ph} (cm^2)
As	0.0127	40×10^{-16}
Zn	0.035	16×10^{-16}
Cu	0.04	8.5×10^{-16}
Au	0.16	$0.1 - 0.9 \times 10^{-16}$
Ni	0.23	2×10^{-16}

After Putley [74].

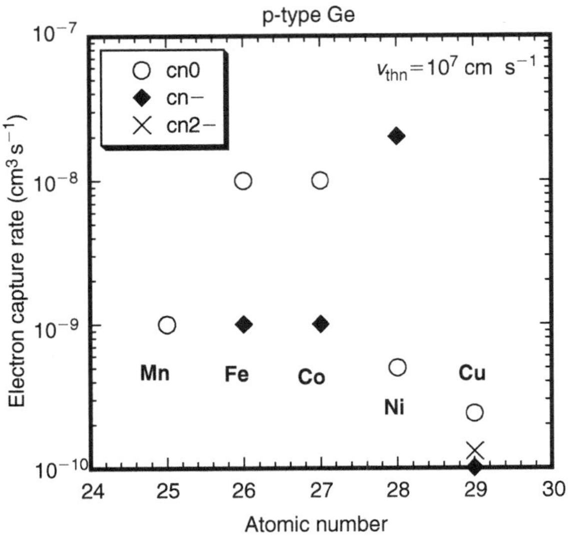

Figure 5.34. Electron capture rate at 300 K versus atomic number, for substitutional TM in p-type Ge.

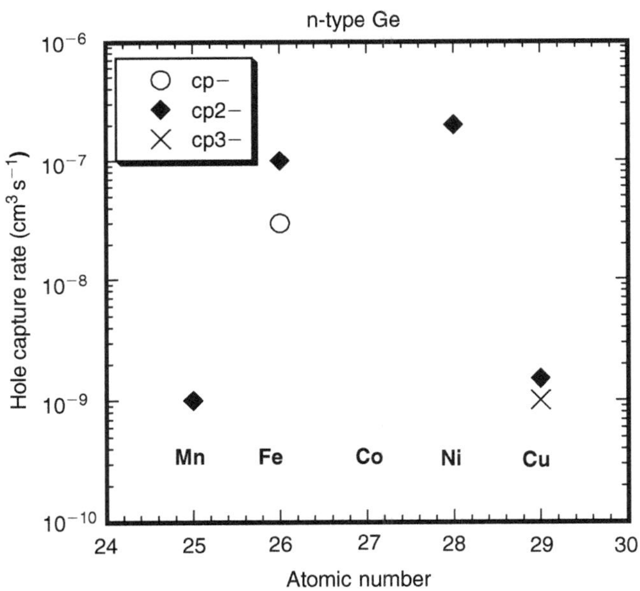

Figure 5.35. Hole capture rate at 300 K versus atomic number, for substitutional TM in n-type Ge.

$$c_p = \sigma_p v_{thp} \tag{5.39b}$$

A few conclusions can be drawn: in both cases, Fe and Ni (in p-Ge also Co) are the most efficient lifetime killers and, secondly, higher lifetimes may be achieved in p- than in n-type

Ge at 300 K, for the same TM concentration. For example, for a Ni concentration of 10^{12} cm^{-3}, one obtains a low-injection lifetime of 50 μs in p-Ge and of 5 μs in n-Ge. This is related to the difference in electron and hole capture cross section. Higher hole capture cross sections are generally observed since most of the TMs give only rise to deep (multiple) acceptor levels, which are attractive for holes and repulsive for electrons.

Some additional comments regard the assignment of the high capture rate of Ni in p-Ge to the single negative charge state (c_n^-). This seems in contradiction to the overall trend and requires a further in-depth analysis of the published data. It is more likely belonging to the neutral state (c_n^0). No data for hole capture by substitutional Co in n-Ge are available in the literature (Figure 5.35) although it could correspond to an efficient recombination center as well.

Besides the recombination lifetime, equally important for device operation is the possible generation current and generation lifetime (τ_{gen}) by the TMs in Ge. These parameters have been calculated based on the following simplifying assumptions: the generation current density is given by:

$$J_{gen} = qn_i W / \tau_{gen} \tag{5.40}$$

with W the depletion width. In Equation (5.40) no field dependence is taken into account. This could arise from:

- PF lowering of the emission barrier for attractive generation centers;
- trap-assisted tunneling;
- a combination of these effects.

This will become important for higher substrate doping levels (junction in a well or a highly doped substrate) for fields starting from 10^3 V cm^{-1} approximately. In practice, this will lead to a stronger bias dependence than predicted by Equation (5.40) (i.e. $\sim V^{1/2}$). In the case of junctions in a silicon p-well, one regularly finds either a power law (V^n with $3 < n < 5$) or even an exponential dependence ($I \sim \exp(qV)$). For the rest of the analysis, we assume no field-enhancement, which could be valid for a lowly doped Ge substrate ($\sim 10^{14}$ cm^{-3}).

The deep-level information is lumped into the generation lifetime τ_{gen}. For the latter, an approach is followed developed by Schroder [144] and further refined in Ref. [145]. Neglecting the effect of the majority capture cross section, one finds [144,145]:

$$\tau_{gen} = 2\tau_{rec}\cosh[(E_T - E_i)/kT] \tag{5.41}$$

with E_T the generation level; E_i the intrinsic Fermi level (taken 0.335 at 300 K for Ge) and τ_{rec} the minority carrier lifetime. The latter is given by Equation (5.38).

Table 5.7. Parameters used to calculate the leakage current density per metal atom.

Atomic number	Element	$c_{min}(cm^3s^{-1})$	E_T (eV)	$\cosh(E_T - E_i)$
25	Mn	10^{-9}	0.37	2.05
26	Fe	10^{-8}	0.34	1.02
27	Co	10^{-9}	0.31	1.5
28	Ni	2×10^{-8}	0.30	2.05
29	Cu	10^{-10}	0.32	1.13
47	Ag	10^{-9}	0.302	1.92
79	Au	2×10^{-9}	0.215	50.5

An estimate of the leakage current density, based on the above formulae has been performed for the TMs in Ge. They can in fact be considered as typical examples of deep levels in Ge and provide an idea about the range of expected generation currents. A further simplification has been made in the sense that all these TMs are multiple acceptors when placed on a lattice site. Here, we assume that only the level closest to mid-gap contributes to electron-hole pair generation. Given the exponential dependence on $E_T - E_i$ in Equation (5.41), this is a valid approximation.

In the calculation, the parameters of Table 5.7 have been used for c_{min}. The leakage current per defect at 300 K is calculated from:

$$I_{defect} = qn_i c_{min}/2\cosh[(E_T - E_i)/kT] \quad \text{(in A)} \tag{5.42}$$

and is represented in Figure 5.36. Fe and Ni stand out as the more efficient leakage current generators. This mainly relates to the high c_{min} (high electron capture cross section).

In Figure 5.37, we have tried to estimate the leakage current density per μm depletion width for the different TMs in Ge. This is given by:

$$J_{gen} = I_{defect} N_T 1\,\mu m \quad (A\,cm^{-2}) \tag{5.43}$$

The true leakage current density then follows from multiplying Equation (5.43) by W in μm. In this way, one obtains a J_{gen} that is independent of the doping density of the substrate.

If one assumes a limit of 10^{-3} A cm^{-2}, which is found typically for state-of-the-art implanted p–n junctions in Ge (see Chapter 8) and for a depletion depth of 1 μm, it is seen that this is reached for Ni and Fe at a concentration in the range of a few 10^{14} cm^{-3}. For the other metals, even higher concentrations have to be introduced. One can derive from this that the trap densities in ion-implanted p–n junctions should be at least of this order of magnitude, neglecting possible electric-field effects. Such concentrations should be easily observable by DLTS for example.

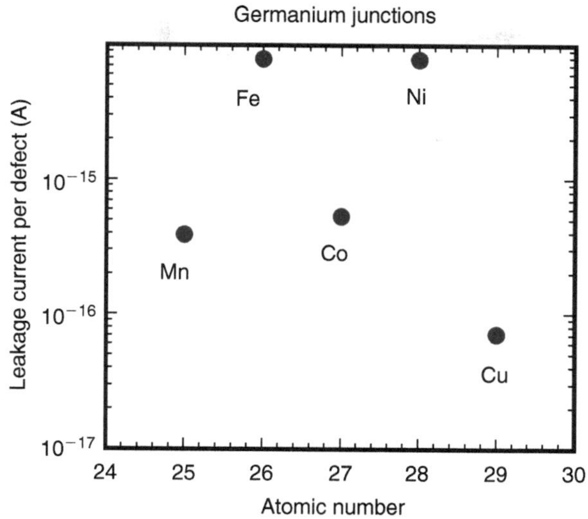

Figure 5.36. Leakage current per defect for the TMs in germanium.

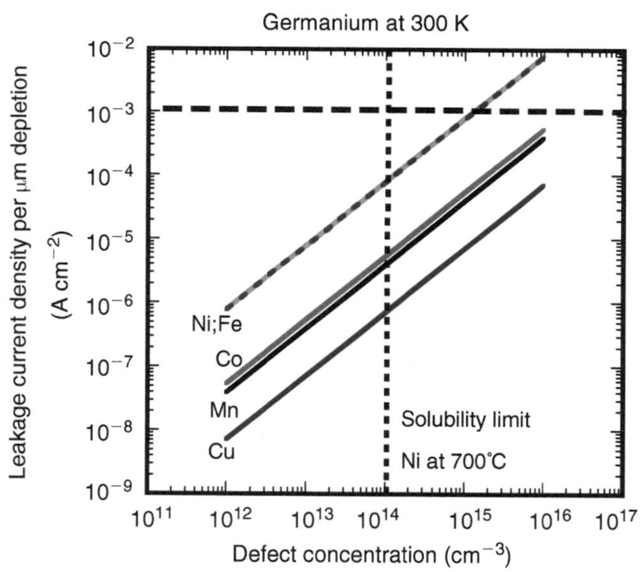

Figure 5.37. Leakage current density as a function of metal concentration in Ge, per μm depletion width.

On the other hand, from the viewpoint of leakage current, one has to take into account another limitation, namely, the solid solubility of the elements. For Ni, this is in the range of $10^{14}\,\text{cm}^{-3}$ at 700°C, so that in practice one can expect a maximum leakage current density (per μm W) in the range $5 \times 10^{-4}\,\text{A}\,\text{cm}^{-2}$.

5.7. Conclusions

While there is a reasonable understanding with respect to the chemical trends for the distribution coefficient, the diffusion coefficient and the maximum solubility of metallic impurities in germanium, this is less the case for the electrical (doping, recombination) properties of some common contaminants like Fe, Cr, Ti, ... The basic reason is the low solubility of these elements combined with difficulties for in-diffusing measurable concentrations in germanium. For some metals currently of interest in microelectronics, like, Zr, Hf, Ti, little if anything is known about even the most elementary properties. This clearly calls for a detailed study. There is also the issue of the missing Pt_s levels in Ge that needs to be resolved.

A related problem is how these metal impurities can be removed form the active device layers. The field of (intrinsic) gettering is largely under-developed in germanium. Although one could rely on similar principles as in silicon, one should consider also the lower thermal budget, which is required for processing on germanium. This places an upper limit on the distance of the gettering sites from the active device layers, as there is less time and thermal energy for the diffusion of the contaminants. The issue is even more crucial for germanium-on-insulator (GOI) approaches leaving little options for implementing (lateral?) gettering sites – perhaps in the highly n^+ or p^+ doped layers.

Alternatively, one could rely on hydrogen passivation of metallic contaminants, which has been studied in the early eighties, particularly with respect to copper in Ge. While it is clear that for some metals hydrogen passivation works, for others no effect has been observed. Moreover, there is the question of the thermal stability of the formed MH_n complexes, which may dissociate at some point.

Finally, there is the search for the missing mobile interstitial metals, like Cu_i, Ni_i, Ag_i, Au_i, ... which should yield a donor level, but so far, only for Cu_i a deep level has been assigned to it. A related question is the formation of pairs (complexes) between the mobile interstitial species and immobile substitutional impurities. Again, this is only documented for Cu_i and Li_i, but no report has been made for other elements. This could be due to the small interstitial solubility at room temperature. If these mobile species and associated complexes can be firmly established (by DLTS for example), one could think of deploying a fast non-destructive identification, using lifetime-based techniques, similar as for the Fe–B or Cr–B case in silicon.

References

[1] A. Seeger and K.P. Chik, *Phys. Stat. Sol.* **29** (1968) 455.

[2] D. Shaw, *Phys. Stat. Sol.* **72** (1975) 11.

[3] F.C. Frank and D. Turnbull, *Phys. Rev.* **104** (1956) 617.

[4] J.A. Burton, *Physica* **20** (1954) 845.

[5] J.A. Burton, E.D. Kolb, W.P. Slichter and J.D. Struthers, *J. Chem. Phys.* **21** (1953) 1991.

[6] C.S. Fuller and J.D. Struthers, *Phys. Rev.* **87** (1952) 526.

[7] W.P. Slichter and E.D. Kolb, *Phys. Rev.* **87** (1952) 527.

[8] D.H. Rank and D.C. Cronemeyer, *Phys. Rev.* **90** (1953) 202.

[9] R.L. Hopkins and E.N. Clarke, *Phys. Rev.* **100** (1955) 1786.

[10] H.H. Woodbury and W.W. Tyler, *Phys. Rev.* **105** (1957) 84.

[11] K. Wolfstirn and C.S. Fuller, *J. Phys. Chem. Solids* **7** (1958) 141.

[12] J.C. Severiens and C.S. Fuller, *Phys. Rev.* **92** (1953) 1322.

[13] C.S. Fuller and J.C. Severiens, *Phys. Rev.* **96** (1954) 21.

[14] R.N. Hall and J.H. Racette, *J. Appl. Phys.* **35** (1964) 379.

[15] K. Weiser, *Phys. Rev.* **126** (1962) 1427.

[16] A. Hiraki, J.W. Cleland and J.H. Crawford Jr., *J. Appl. Phys.* **38** (1967) 3519.

[17] J.C. Pigg and J.H. Crawford Jr., *Phys. Rev.* **135** (1964) A1141.

[18] F.A. Huntley, *Phil. Mag.* **26** (1972) 1047.

[19] A.G. Tweet and C.J. Gallagher, *Phys. Rev.* **103** (1956) 828.

[20] C.S. Fuller and J.A. Ditzenberger, *J. Appl. Phys.* **28** (1957) 40.

[21] F. van der Maesen and J.A. Brenkman, *J. Electrochem. Soc.* **102** (1955) 229.

[22] V.I. Fistul', A. Yakovenko and E.A. Shelonin, *Sov. Phys. Solid State* **20** (1979) 1312.

[23] R.L. Williams, *J. Appl. Phys.* **40** (1969) 2932.

[24] W. Shockley and J.L. Moll, *Phys. Rev.* **119** (1960) 1480.

[25] A.Ya. Potemkin and V.I. Potapov, *Sov. Phys. Solid State* **2** (1960) 1668.

[26] V.I. Fistul', A.G. Yakovenko, A.A. Gvelesiani and G.S. Dmitrieva, *Inorg. Mater.* **9** (1983) 5.

[27] N.A. Stolwijk, W. Frank, J. Hölzl, S.J. Pearton and E.E. Haller, *J. Appl. Phys.* **57** (1985) 5211.

[28] H. Bracht, N.A. Stolwijk and H. Mehrer, *Phys. Rev. B* **43** (1991) 14465.

[29] C.S. Fuller and K.B. Wolfstirn, *J. Phys. Chem. Solids* **26** (1965) 1463.

[30] R.A. Logan, *Phys. Rev.* **100** (1955) 615.

[31] A.G. Tweet, *Phys. Rev.* **106** (1957) 221.

[32] A.G. Tweet, *Phys. Rev.* **111** (1958) 57.

[33] A.G. Tweet, *Phys. Rev.* **111** (1958) 67.

[34] P. Penning, *Phys. Rev.* **110** (1958) 586.

[35] J.F. Battey and R.M. Baum, *Phys. Rev.* **94** (1954) 1393.

[36] W.C. Dunlap Jr., *Phys. Rev.* **96** (1954) 40.

[37] E.E. Haller, P.P. Li, G.S. Hubbard and W.L. Hansen, *IEEE Trans. Nucl. Sci.* **NS-26** (1979) 265.

[38] A.O. Evwaraye, R.N. Hall and T.J. Soltys, *IEEE Trans. Nucl. Sci.* **NS-26** (1979) 271.

[39] E. Simoen, P. Clauws, J. Broeckx, J. Vennik, M. Van Sande and L. De Laet, *IEEE Trans. Nucl. Sci.* **NS-29** (1982) 789.

[40] E. Simoen, P. Clauws, M. Lamon and J. Vennik, *Semicond. Sci. Technol.* **1** (1986) 53.

[41] P. Clauws, G. Huylebroeck, E. Simoen, P. Vermaercke, F. De Smet and J. Vennik, *Semicond. Sci. Technol.* **4** (1989) 910.

[42] I.M. Kotina, V.V. Kuryatkov, S.R. Novikov and T.I. Pirozhkova, *Sov. Phys. Semicond.* **21** (1987) 635.

[43] R.A. Paramonova and A.V. Rzhanov, *Sov. Phys. Solid State* **4** (1963) 1335.

[44] S.J. Pearton, E.E. Haller and J.M. Kahn, *J. Phys. C Solid State Phys.* **17** (1984) 2375.

[45] F.X. Zach, H. Grimmeiss and E.E. Haller, *Mat. Sci. Forum* **83–87** (1992) 245.

[46] E.E. Haller, R.E. McMurray Jr., L.M. Falicov, N.M. Haegel and W.L. Hansen, *Phys. Rev. Lett.* **51** (1983) 1089.

[47] R.E. McMurray Jr., *Solid State Commun.* **53** (1985) 1127.

[48] M. Lax, *Phys. Rev.* **119** (1960) 1502.

[49] V.N. Abakumov, V.I. Perel' and I.N. Yassievich, *Sov. Phys. Semicond.* **12** (1978) 1.

[50] V.N. Abakumov, *Sov. Phys. Semicond.* **13** (1979) 566.

[51] N.G. Zhdanova and S.G. Kalashnikov, *Sov. Phys. Solid State* **6** (1964) 350.

[52] V.G. Alekseeva, N.G. Zhdanova, M.S. Kagan, S.G. Kalashnikov and E.G. Landsberg, *Sov. Phys. Semicond.* **3** (1970) 1179.

[53] V.G. Alekseeva, N.G. Zhdanova, M.S. Kagan, S.G. Kalashnikov and E.G. Landsberg, *Sov. Phys. Semicond.* **6** (1972) 270.

[54] L.S. Darken, P. Sangsingkeow and G.E. Jellison Jr., *J. Electron. Mater.* **19** (1990) 105.

[55] L.S. Darken and G.E. Jellison Jr., *Appl. Phys. Lett.* **55** (1989) 1424.

[56] L.S. Darken, *Phys. Rev. Lett.* **69** (1992) 2839.

[57] V.N. Abakumov and Z.N. Sokolova, *Sov. Phys. Semicond.* **12** (1978) 962.

[58] V.N. Abakumov, *Sov. Phys. Semicond.* **13** (1979) 34.

[59] V.N. Abakumov, *Sov. Phys. Semicond.* **14** (1980) 510.

[60] Y. Kamiura and F. Hashimoto, *Jpn. J. Appl. Phys.* **28** (1989) 763.

[61] R.I. Gloriozova and L.I. Kolesnik, *Semicond.* **27** (1993) 303.

[62] V.I. Fistul', A.G. Yakovenko and E.A. Shelonin, *Sov. Phys. Solid State* **22** (1980) 17.

[63] J.A. Burton, G.W. Hull, F.J. Morin and J.C. Severiens, *J. Phys. Chem.* **57** (1953) 853.

[64] G.S. Kalashnikov, *J. Phys. Chem. Solids* **8** (1959) 52.

[65] R.M. Baum and J.F. Battey, *Phys. Rev.* **98** (1955) 923.

[66] R.G. Shulman and B.J. Wyluda, *Phys. Rev.* **102** (1956) 1455.

[67] G.S. Kalashnikov and A.I. Morozov, *Sov. Phys. Solid State* **1** (1959) 1182.

[68] B.V. Kornilov, *Sov. Phys. Solid State* **4** (1963) 1771.

[69] M.I. Iglitsyn and Yu.A. Kontsevoi, *Sov. Phys. Solid State* **2** (1960) 1039.

[70] B.H. Schulz, *Philips Res. Repts* **16** (1961) 182.

[71] M. Nagae, *J. Phys. Soc. Japan* **18** (1963) 207.

[72] H. Levinstein, *Proc. IRE* **47** (1959) 1478.

[73] R. Newman and W.W. Tyler, *Solid State Phys.* **8** (1959) 49.

[74] E.H. Putley, *Phys. Stat. Sol.* **6** (1964) 571.

[75] W.C. Dunlap Jr., *Phys. Rev.* **100** (1955) 1629.

[76] W. Dunlap Jr., *Progress Semicond.* **2** (1957) 167.

[77] G. Huylebroeck, P. Clauws, E. Simoen, E. Rotsaert and J. Vennik, *Semicond. Sci. Technol.* **4** (1989) 529.

[78] W.C. Dunlap Jr., *Phys. Rev.* **91** (1953) 1282.

[79] W.E. Baker and D.M.J. Compton, *IBM J.* (1960) 296.

[80] H. Kodera, *Jpn. J. Appl. Phys.* **3** (1964) 369.

[81] R.N. Hall, *J. Phys. Chem. Solids* **3** (1957) 63.

[82] A.S. Syed, *Canad. J. Phys.* **40** (1962) 286.

[83] A.A. Bugai, V.E. Kosenko and E.G. Miseliuk, *Sov. Phys. Tech. Phys.* **2** (1957) 1553.

[84] A. Almazouzi, J. Bernardini, E.G. Moya, H. Bracht, N.A. Stolwijk and H. Mehrer, *J. Appl. Phys.* **70** (1991) 1345.

[85] W.W. Tyler, *J. Phys. Chem. Solids* **8** (1959) 59.

[86] M.F. Millea, *J. Phys. Chem. Solids* **27** (1966) 309.

[87] V.E. Kosenko, *Sov. Phys. Solid State* **4** (1962) 42.

[88] E. Suliga and M. Henzler, *J. Phys. C Solid State Phys.* **16** (1983) 1543.

[89] W.C. Dunlap Jr., *Phys. Rev.* **97** (1955) 614.

[90] P.H. Wisbey and B.K. Ridley, *J. Phys. C Solid State Phys.* **3** (1970) 211.

[91] S.J. Pearton, *Solid-State Electron.* **25** (1982) 305.

[92] E. Simoen, P. Clauws, G. Huylebroeck and J. Vennik, *Semicond. Sci. Technol.* **2** (1987) 507.

[93] G. Rupprecht, *J. Phys. Chem. Solids* **22** (1961) 255.

[94] N.G. Zhdanova and V.G. Alekseeva, *Sov. Phys. Solid State* **5** (1963) 397.

[95] B.K. Ridley and T.B. Watkins, *J. Phys. Chem. Solids* **22** (1961) 155.

[96] B.K. Ridley and T.B. Watkins, *Proc. Phys. Soc.* **78** (1961) 710.

[97] B.K. Ridley and R.G. Pratt, *Phys. Lett.* **4** (1963) 300.

[98] B.K. Ridley, *Phys. Lett.* **16** (1965) 105.

[99] M.S. Kagan and S.G. Kalashnikov, *J. Phys. Soc. Japan* **21** (1966) 537.

[100] V.G. Alexeeva, S.G. Kalashnikov, I.V. Karpova and E.G. Landsberg, *J. Phys. Chem. Solids* **22** (1961) 45.

[101] I.V. Karpova, V.G. Alekseeva and S.G. Kalashnikov, *Sov. Phys. Solid State* **4** (1962) 461.

[102] V.G. Alekseeva, I.V. Karpova and S.G. Kalashnikov, *Sov. Phys. Solid State* **3** (1961) 699.

[103] I.G. Aksyanov, V.G. Ivanov, S.K. Novoselov, M.A. Reshchikov and Yu.S. Smetannikova, *Sov. Phys. Semicond.* **21** (1987) 1223.

[104] K.D. Glinchuk, E.G. Miseliuk and N.N. Fortunatova, *Sov. Phys. Solid State* **1** (1959) 1234.

[105] W.K. Schoenmaekers and R. Henck, *IEEE Trans. Nucl. Sci.* **19** (1972) 329.

[106] R. Henck, R. Stuck, P. Siffert, W.K. Schoenmaekers and L.H. De Laet, *IEEE Trans. Nucl. Sci.* **19** (1972) 299.

[107] P. Penning, *Philips Res. Repts* **13** (1958) 17.

[108] A.G. Tweet and W.W. Tyler, *J. Appl. Phys.* **29** (1958) 1578.

[109] F. van der Maesen and J.A. Brenkman, *Philips Res. Rep.* **9** (1954) 225.

[110] P. Penning, *Phys. Rev.* **102** (1956) 1414.

[111] A. Giese, H. Bracht, N.A. Stolwijk and H. Mehrer, *Defect and Diff. Forum* **143–147** (1997) 1059.

[112] A. Giese, N.A. Stolwijk and H. Bracht, *Appl. Phys. Lett.* **77** (2000) 642.

[113] W.W. Tyler, R. Newman and H.H. Woodbury, *Phys. Rev.* **98** (1955) 461.

[114] G. Huylebroeck, P. Clauws, E. Simoen and J. Vennik, *Solid State Commun.* **82** (1992) 367.

[115] S.J. Pearton, J.M. Kahn, W.L. Hansen and E.E. Haller, *J. Appl. Phys.* **55** (1984) 1464.

[116] S.J. Pearton and A.J. Tavendale, *Solid State Electron.* **26** (1983) 1019.

[117] G.K. Wertheim, *Phys. Rev.* **115** (1959) 37.

[118] J.F. Battey and R.M. Baum, *Phys. Rev.* **100** (1955) 1634.

[119] J. Okada, *J. Phys. Soc. Japan* **12** (1957) 741.

[120] J. Okada, *J. Phys. Soc. Japan* **12** (1957) 1338.

[121] F.D. Rosi, *RCA Review* **19** (1958) 349.

[122] P.G. Eliseev and S.G. Kalashnikov, *Sov. Phys. Solid State* **5** (1963) 233.

[123] H.H. Woodbury and G.W. Ludwig, *Phys. Rev. Lett.* **1** (1958) 16.

[124] G.W. Ludwig and H.H. Woodbury, *Phys. Rev.* **113** (1959) 1014.

[125] W.W. Tyler and H.H. Woodbury, *Phys. Rev.* **96** (1954) 874.

[126] L.Y. Wei, *J. Phys. Chem. Solids* **18** (1961) 162.

[127] A.A. Bugai, V.E. Kosenko and E.G. Miseliuk, *Sov. Phys. Tech. Phys.* **2** (1957) 183.

[128] K.D. Glinchuk, E.G. Miseliuk and N.N. Fortunatova, *Sov. Phys. Tech. Phys.* **3** (1958) 980.

[129] K.D. Glinchuk, E.G. Miseliuk and N.N. Fortunatova, *Sov. Phys. Tech. Phys.* **2** (1957) 2283.

[130] A.D. Belyaev and V.G. Malogolovets, *Sov. Phys. Solid State* **5** (1964) 2229.

[131] W.W. Tyler, *Phys. Rev.* **96** (1954) 226.

[132] W.W. Tyler, R. Newman and H.H. Woodbury, *Phys. Rev.* **97** (1955) 669.

[133] M.I. Barnik, B.I. Beglov, D.A. Romanychev and Yu.S. Kharionovskii, *Sov. Phys. Semicond.* **5** (1971) 87.

[134] G. Susila, *J. Phys. Chem. Solids* **31** (1970) 963.

[135] S.J. Pearton and A.J. Tavendale, *J. Appl. Phys.* **54** (1983) 820.

[136] H.H. Woodbury and W.W. Tyler, *Phys. Rev.* **100** (1955) 659.

[137] E.G. Landsberg and S.G. Kalashnikov, *Sov. Phys. Solid State* **3** (1961) 1137.

[138] E.G. Landsberg and S.G. Kalashnikov, *Sov. Phys. Solid State* **5** (1963) 777.

[139] E.M. Conwell, *Proc. IRE* **46** (1958) 1281.

[140] S.M. Sze and J.C. Irvin, *Solid-State Electron.* **11** (1968) 599.

[141] G.W. Ludwig and H.H. Woodbury, *Solid State Phys.* **19** (1962) 223.

[142] G. Bemski, *Proc. I.R.E.* **46** (1958) 990.

[143] V.L. Bonch-Bruevich and E.G. Landsberg, *Phys. Stat. Sol.* **29** (1968) 9.

[144] D.K. Schroder, *IEEE Trans. Electron Devices* **29** (1982) 1336.

[145] E. Simoen and C. Claeys, *IEEE Trans. Electron Devices* **46** (1999) 1487.

CHAPTER 6

Ab-Initio Modeling of Defects in Germanium

Robert Jones and José Coutinho

6.1. Introduction

Point defects possess characteristic optical, electrical and magnetic signatures which can have an impact on Ge devices, e.g. the energy levels, thermal stabilities and activation efficiency of donors or acceptors. Extended defects alter the mechanical properties of the material, and act as sinks for point defects. Advances in modeling techniques have enabled some of these properties to be accurately computed, e.g. defect structures and migration barriers, while less or even no real progress has taken place in the evaluation of some others, such as carrier capture cross-sections or excitonic recombination rates. In this review, we concentrate on the simplest defects in Ge and show how theory has been related to experimental results, particularly deep level transient spectroscopy (DLTS) and Fourier transform infrared (FTIR) spectroscopy.

There are various ways in which defects in materials and especially covalently bonded semiconductors like Ge can be modeled. We can distinguish between quantum mechanical methods like density functional theory (DFT) which attempt to solve the many-body Schrödinger equation, and more approximate methods. Current computational power limits DFT methods to ~500 group-IV atom clusters or supercells. More approximate atomistic methods, such as tight binding schemes, can deal with ~1000 atoms. To extend this especially to problems involving diffusion and atomic motion, molecular dynamics, Monte-Carlo and continuum methods have been used. These require some atomistic potential or functional form giving the binding and migration energies of defects. One limitation is that it is often unclear how accurate the input data are.

Germanium-Based Technologies: From Materials to Devices. C. Claeys and E. Simoen (Editors)

6.2. Quantum Mechanical Methods

To determine accurately the energy of an assembly of atoms, we need to solve the Schrödinger equation for the ions and electrons. This is impossible to do exactly: progress can only be made if a set of approximations are made.

The first approximation is due to Born and Oppenheimer [1] who argue that since the nuclear mass is so much larger than that of the electrons, then we can treat the nuclear motion classically and reduce the Schrödinger equation to one only involving the electrons moving in a potential of fixed nuclear sites denoted by **R**. The solution of this equation is the *structural energy* $E(\mathbf{R})$. To determine dynamical quantities like vibrational modes and diffusion barriers, we need then to solve the classical problem of the nuclei moving in the force field $-\nabla_R E(\mathbf{R})$. The equilibrium structure is given by $-\nabla_R E(\mathbf{R}) = 0$.

We next consider the Schrödinger equation of the electrons. This equation still cannot be solved exactly because of the interaction between every electron pair i and j, described by a Coulomb potential, $\Sigma_{i,j} V(\mathbf{r}_i - \mathbf{r}_j)$. If this term was neglected, then the Schrödinger equation is separable and the exact wavefunction can be written down as a sum of product orbitals $\Pi_i \varphi_i(\mathbf{r}_i)$, where $\varphi_i(\mathbf{r}_i)$ are the eigenfunctions of the Hamiltonian of one electron. Of course, we must require that the total wavefunction is anti-symmetric and this can be achieved by a suitable sum over the product orbitals. For example, in the H_2 molecule, the two electrons are described by an anti-symmetric wavefunction of the form:

$$\varphi_1(\mathbf{r}_1) \times \varphi_2(\mathbf{r}_2) - \varphi_1(\mathbf{r}_2) \times \varphi_2(\mathbf{r}_1) \tag{6.1}$$

Here $\varphi_\lambda(\mathbf{r})$ is a spin-orbital. This type of wavefunction is called a Slater determinant [2] and is specified by an electronic configuration, i.e. a listing of the spin-orbitals λ occupied by N electrons.

The neglect of the electron–electron interaction term is too extreme to make the theory useful and various attempts have been made to take this interaction into account. There are two principle schemes: Hartree-Fock (HF) and DFT. Both of these replace the potential acting on electron i, i.e. $\Sigma_{i \neq j} V(\mathbf{r}_i - \mathbf{r}_j)$ by an average over the positions of the other electrons. Thus this potential becomes $V_{\text{eff}}(\mathbf{r}_i)$. Clearly the Schrödinger equation is then separable and we can write the wavefunction as a Slater determinant. In these theories, the effective potential is composed of two terms: the Hartree potential $V_H(\mathbf{r})$ and the exchange-correlation potential $V_{XC}(\mathbf{r})$. The former is the electrostatic potential due to the density of charge $qn(\mathbf{r})$ at point $\mathbf{r}$ where:

$$n(\mathbf{r}) = \sum_\lambda f_\lambda \left|\varphi_\lambda(\mathbf{r})\right|^2 \tag{6.2}$$

and f_λ is the occupancy of the orbital λ. The exchange-correlation potential differs in the two theories. In HF, it is a complicated function dependent on the orbital $\varphi_\lambda(\mathbf{r})$ and determined by the other orbitals. In DFT, it is rigorously determined only by the electron density [3,4]. This is an important result but the precise dependence on the density is not known except for a particular problem: the homogeneous electron gas. In this case, the Schrödinger equation has been solved numerically and the dependence of the energy on n determined. In this way $V_{XC}(\mathbf{r})$ is known exactly. Now for inhomogeneous problems, e.g. molecules and solids, we assume that $V_{XC}(\mathbf{r})$ at a point $\mathbf{r}$ is given by the homogeneous electron gas value involving the density $n(\mathbf{r})$ at the same point $\mathbf{r}$. This is called the local density functional approximation. To illustrate this, for large electron densities, $V_{XC}(\mathbf{r})$ is proportional to $n^{1/3}(\mathbf{r})$ and thus in local DFT $V_{XC}(\mathbf{r})$ is also proportional to $n^{1/3}(\mathbf{r})$. It is possible to go beyond local DFT using a gradient corrected formalism but the agreement with experiment at least for problems in condensed matter physics, is not uniformly better than the simpler local density theory.

Further approximations are often made. The electrons of say Ge are divided into a set of core electrons, e.g. $1s^2\ 2s^2\ 2p^6\ 3s^2\ 3p^6\ 3d^{10}$, as well as the four valence electrons $4s^2\ 4p^2$. Only the latter are included in the calculation through a specially constructed *pseudopotential* which leads to the same atomic valence energy levels but has no core states. The *pseudowavefunctions* of the valence energy levels are exactly the same as those of the atom outside a small core radius so that the bonding properties of the *pseudoatom* are close to those of the true atom. However, recent studies of Ge–O bonding in the A-centers, shows that the small radius of O leads to short Ge–O bonds and hence an effect of oxygen on the 3d-orbitals of Ge. This can be taken into account by a non-local pseudopotential.

In order to solve the DFT equations for solid state problems, the wavefunctions are represented by basis functions: often plane waves, but localized ones such as Gaussian orbitals have become increasingly popular. The disadvantage with plane waves lies on the large number of functions (typically about 100 per atom) required to converge the properties of interest. This has to be contrasted with about 20 basis functions per atom for localized basis sets. However, programming and computing efficiently the matrix elements of the Hamiltonian between localized basis sets are then important issues. Well known codes using plane waves are VASP [5,6] and ABINIT [7], while SIESTA [8] and AIMPRO [9,10] use localized basis functions.

6.2.1. Clusters and supercells

The structural energy $E(\mathbf{R})$ is of fundamental importance and can be found for a molecule, a cluster of atoms, or indeed for a periodic cell, often called a supercell. The latter is made up of several unit cells of the bulk, but each supercell contains only one defect. Periodic boundary conditions imply that electric multipoles must vanish within the cell to ensure energy convergence. For example, when a defect is charged, a uniform compensating background charge has to be added to ensure that the cell is neutral. However, if no higher order corrections are considered, the self-consistent electron density $n(\mathbf{r})$ is such that the electric dipole and quadrupole within the cell vanish [11]. A problem with the supercell approach which is related to the study of defects is the so called defect-image interaction. We must not forget that we are modeling a periodic array of defects separated by the supercell lattice vectors, and the interaction between a defect and its image in a different supercell may lead to spurious results. This is especially true when dealing with vacancy-related defects which produce long range strain fields, or shallow centers with long range electronic states. Another problem, particularly in modeling defects in Ge, arises from the treatment of exchange and correlation for which local DFT, in the supercell formalism, leads to an almost zero band gap. This is the cause of severe problems in the treatment of charged defects and will be discussed further below.

On the other hand, the difficulty for clusters is that their surfaces must be saturated, e.g. by hydrogen [9], otherwise the gap levels associated with surface dangling bonds could interfere with those of a defect. The Ge-H surface bonds have bonding and anti-bonding levels which lie in the valence and conduction bands, respectively. These couple with levels of Ge resulting in a confinement of the states of the latter, and lead to a larger gap which decreases steadily as the cluster increases in size [9]. When modeling defects in clusters one should use a large enough cluster to avoid surface effects. In practice this is often not possible and a technique has to be found which minimizes the surface influence. The way in which the calculation is done has important effects on the results. For example, whether the surface H and Ge atoms are allowed to move, the lattice parameter used to generate the cluster, the length of the surface Ge–H bonds.

For supercell calculations, one also needs to integrate quantities such as the band structure over the Brillouin zone (BZ). This can be done in several ways but that introduced by Monkhorst and Pack [12] is one of the most common. Accordingly, a uniform mesh of points is generated to lie within the BZ, and *folded* according to the symmetry operations of the zone. The integrand is then sampled on the irreducible points $\mathbf{k}_i$, normally referred as *special* **k**-points, which have a representativity w_i that depend on their site symmetry. This scheme has been particularly successful due to its robustness and easy implementation. Convergence tests are also straightforward by simply producing higher density **k**-point sets.

6.3. Kohn–Sham and Occupancy Levels

The structural energy $E(\mathbf{R})$ can be written in terms of energy levels E_λ associated with each spin-orbital $\varphi_\lambda(\mathbf{r})$ by:

$$E(\mathbf{R}) = \sum_\lambda f_\lambda E_\lambda - \frac{1}{2} E_{\mathrm{H}} + \int d\mathbf{r}\{\varepsilon_{\mathrm{XC}} - \mu_{\mathrm{XC}}\} \cdot n(\mathbf{r}) + E_{\mathrm{ion}}(\mathbf{R}) \tag{6.3}$$

Here f_λ is the occupancy of level λ, E_{H} is the Hartree energy or the electrostatic energy of the electron density $n(\mathbf{r})$ and $\varepsilon_{\mathrm{XC}}$ and μ_{XC} are the exchange-correlation energy and potential densities, respectively, which depend only on $n(\mathbf{r})$. The last term is the electrostatic energy of all the ions. The first term is the band structure energy and in the simplest case dominates the total energy.

Consider now a situation in which an electron is removed from level σ as in an ionization process. In the simplest case when the band structure energy dominates, the ionization energy is simply the change in the band structure energy, i.e. it is just $-E_\sigma$. Similarly, optical absorption from state σ to μ, results in an energy change $E_\mu - E_\sigma$. These results impart a particularly simple interpretation of the Kohn–Sham energy levels E_λ. However, as the charge density changes whenever ionization or optical absorption takes place, the change in energy is not simply the band structure term and the change in all the terms must be considered.

6.4. Formation Energies, Vibrational Modes, Energy Levels

The formation energy E_F of a defect [13] is defined in terms of the structural energy of a cell containing n_i atoms of species i by $E_F = E(\mathbf{R}) - \Sigma_i n_i \mu_i$. Here μ_i is the chemical potential of species i. The chemical potential is determined by the energy per atom from a reservoir in equilibrium with the defective crystal. For instance, for oxygen defects in Ge, the chemical potential of oxygen is determined by the energy of O atoms in a GeO_2 crystal which surrounds Ge when grown by the Czochralski process, i.e. $\mu_O = (\mu_{GeO_2} - \mu_{Ge})/2$, where μ_{GeO_2} is the energy per GeO_2 unit. The Ge chemical potential μ_{Ge} is the energy per Ge atom from an *infinite* Ge crystal.

The formation energy controls the equilibrium density of defects which in the dilute approximation is given by $\rho \exp(-F/KT)$ where ρ is the positional, orientational and spin degeneracy of the defect. Strictly speaking, even at 0 K where entropic terms vanish, the formation energy should include a PV term appropriate for an enthalpy, a zero-point motion sum $\Sigma_i \eta \omega_i$ over all available vibrational frequencies ω_i, and a term $-z_q E_F$ for defects which are charged with z_q electrons. Here E_F is the Fermi-level. However, for low defect concentrations in solids, and when ultra light impurities such as hydrogen or muonium are not involved, we have to consider the last term only.

Localized vibrational modes (LVMs) and their frequencies are quantities which can be accurately predicted by *ab initio* methods. These can be directly compared with infrared absorption, Raman scattering or photoluminescence (PL) data. Most often, two main approximations are assumed, namely (i) the adiabatic (or Born–Oppenheimer) approximation, where the nuclei are regarded as point-like masses surrounded by electrons that adiabatically follow their vibrational movement, and (ii) the harmonic approximation, according to which, the structural energy $E(\mathbf{R})$ of the defective system is expanded to second order of the atomic displacements [14]. Hence, from the second derivatives of the energy $\Phi_{ij} = \partial^2 E / \partial R_i \partial R_j$ we obtain a dynamical matrix $D_{ij} = \Phi_{ij}/\sqrt{m_i m_j}$ (where m_i is the mass of the atom i), whose eigenvalues and eigenvectors give the squared frequencies ω^2 and the corresponding normal coordinates [9].

We define a donor as a defect which is positively charged when ionized, while an acceptor is a defect which is negatively charged when ionized. The donor level relative to the valence band top (E_V) is the energy required to excite an electron from the top of the valence band to the positively charged defect. Alternatively, it is the energy necessary to excite a hole from the positively charged defect to the top of the valence band. On the other hand, an acceptor level relative to the conduction band bottom (E_C) is the energy required to excite an electron, trapped on a negatively charged defect, to the bottom of the conduction band. In terms of total energies the donor level with respect to the valence band is the difference in energies between the final and initial states or $E(\mathbf{R}^0) - E(\mathbf{R}^+) - E_V$ while the acceptor level with respect to the conduction band is $E_C - E(\mathbf{R}^-) + E(\mathbf{R}^0)$. Here, $\mathbf{R}^0$, $\mathbf{R}^{\pm}$ are the stable structures of the defect in each charge state. These electrical levels are thermodynamic quantities. They tell us the charge state taken by a defect with respect to the Fermi level location. In this sense they should not be confused with one-electron or the Kohn–Sham levels E_λ discussed above.

There is no standard procedure for the calculation of the energy levels of defects. One common way [13] is to model the defect in a supercell and to evaluate the formation energy $E_F(z_q)$ of the defect with z_q additional electrons with respect to the neutral case ($z_q = \ldots, -1, 0, +1, \ldots$),

$$E_F(z^q) = E(\mathbf{R}^z{}_q) - \Sigma_i n_i \mu_i - z^q E_F \tag{6.4}$$

where E_F is the Fermi-level position with respect to E_V. However, problems arise in accurately evaluating these energies. For example, the calculation of E_V is not straightforward as the highest occupied crystalline state shifts when the defect is introduced in the supercell. Moreover, and as mentioned above, total energies within the supercell approximation are affected by multiple interactions due to the 3D replication of defects. This is graver when charged states are involved in the calculations. Several corrections have been proposed, and the most popular comes from Makov and Payne, which accounts for the energy of an array of point-like charges [15]. For instance, to correct the formation energy from the dominant monopole interaction between charged defects one has to add $\alpha_M z^2{}_q / L_c \varepsilon$ to $F(z_q)$. Here α_M is the Madelung constant associated with a supercell of characteristic length L_c, and ε is the permittivity of the host crystal. Given that $\varepsilon = 16.2$ in Ge, and considering a standard characteristic supercell size $L_c = 11$–17 Å, we end with corrections of the order of 0.1 eV for first ionized states, and ~0.4 eV for second ionized states. The latter is unacceptable when dealing with a 0.7 eV gap material.

An alternative procedure [16,17] is to compare the calculated level positions with those of a standard defect whose donor or acceptor levels are known. We shall discuss this in more detail below.

6.5. Defect Modeling in Ge

Density functional calculations give excellent values of the lattice parameter, at 5.568 Å and a bulk modulus of 74.2 GPa, for germanium. These are to be compared with observed values of 5.658 Å and 77.2 GPa, respectively [18]. However, the one-electron (Kohn–Sham) band structure diagram shown in Figure 6.1 is in poor agreement because of the underestimate of the gap. The calculations give a 0.2 eV direct gap at Γand a 0.3 eV indirect gap with the conduction band minimum at L. These results are to be compared with the measured gap at Γof 0.90 eV, and with the indirect gap with the conduction band minimum at L of 0.74 eV at 1.5 K [19].

There have been relatively few modeling studies of defects in germanium using quantum mechanical methods. This is partly because attention has been focused on Si and other

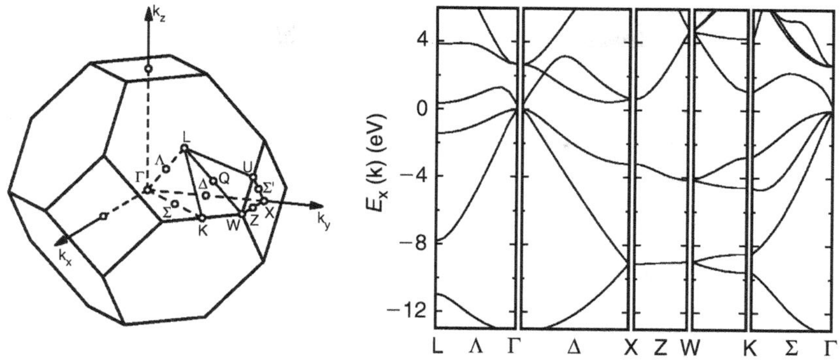

Figure 6.1. BZ of a face centered cubic crystal indicating high symmetry **k**-points (left), and Kohn–Sham band structure along high symmetry directions (right). This is from a pseudopotential density functional calculation within the local density approximation. The band gap is only 0.2 eV.

materials where there are many good and reliable experimental results and has permitted a realistic assessment of the likely error in the calculations. In addition however, the problem with the theoretical band gap of germanium causes concern about the accuracy of modeling studies. This means that defects can hardly be charged and predicted properties can then be quite wrong. For example, the A-center or vacancy-oxygen (VO) defect, where an oxygen interstitial atom is trapped by a vacancy, has an acceptor level at $E_V + 0.32$ eV [20]. Infrared absorption studies have assigned LVM bands at 620 cm^{-1} and 669 cm^{-1} to vibrating Ge–O–Ge units in VO^0 and VO$^-$ [21,22], respectively. Recent density functional supercell calculations [18] on the other hand yield an acceptor state resonant with the conduction band, and the additional electron in VO$^-$ actually falls in a conduction band state, resulting in a negligible difference between the vibrational frequencies of neutral and negatively charged centers. These findings on the VO defect in Ge have to be contrasted with those in Si where LVMs at 839 cm^{-1} and 872 cm^{-1} are calculated for VO^0 and VO$^-$, and can be favorably compared with respective experimental band frequencies of 836 cm^{-1} and 885 cm^{-1} [18]. In this case, the density functional band gap for Si is about 0.54 eV, which is sufficient to contain an empty Kohn–Sham level of the VO defect. Inserting an electron into this level perturbs the O–Si bond lengths and causes the change in vibrational mode.

6.6. Defects in Germanium

Point defects in semiconductors have been classified by several schemes and one of the most common refers to their chemical composition. This separates defects in two main classes, namely (i) intrinsic defects such as vacancies and self-interstitials which result in a lack or excess of Ge atoms within a confined volume, respectively, and (ii) extrinsic defects containing chemical species which are *alien* to the host material. Ge crystals grown by the Czochralski process contain oxygen and carbon, introduced from the SiO_2 crucible and graphite puller components. Hence oxygen and carbon can be important impurities. Likewise Ge when grown by epitaxial processes contains appreciable amounts of hydrogen arising from the GeH_4 source. During processing, transition metal impurities can be introduced.

Radiation, especially during ion implantation, introduces many defects. In Si, vacancies and interstitials become mobile below 200 K, and cannot be detected at room temperature. Those that escape annihilation quickly complex with impurities such as dopants, oxygen, carbon and hydrogen. For high energy radiation, or when the material is annealed at temperatures where vacancy-impurity and self-interstitial-impurity complexes anneal out, multi-vacancy and multi-interstitial complexes can be formed, especially in ion-implanted or neutron irradiated material. Our understanding of the details of the evolution of defects in Si has largely come from the success of techniques like electron paramagnetic resonance (EPR) [23–25], FTIR [26,27] PL [28], DLTS [29–31] and others. It is known that at cryogenic temperatures vacancy production by 1–3 MeV electrons is 100 times greater in p-Si than n-Si, an effect attributed to the trapping of interstitials by boron, preventing Frenkel pair annihilation [23]. In n-Si, Frenkel pair annihilation dominates and relatively few vacancies and interstitials remain to form thermally stable complexes which can degrade devices.

Not all of the above techniques have been successfully applied to Ge. Accordingly, in Ge, it is not completely clear as to the evolution of radiation damage. It seems likely that again the vacancy and self-interstitial are mobile by room temperature and hence many annihilate through pairing. The few that escape form complexes with dopants, oxygen, hydrogen and with each other. Of special interest is the prediction that Ge vacancies have a much lower formation energy than the self-interstitial [32,33]. This could account for the belief that self-diffusion is dominated by a vacancy mechanism.

Unfortunately, there is a sparsity of both calculations and experiments dealing with the properties of vacancies and interstitials, as well as their reaction with impurities. A recent experimental study of n-Ge [34] has compared defects introduced by different types of radiation at room temperature. γ-rays, electrons and heavy ions were used. DLTS experiments then revealed $(2-/-)$ levels of VSb, VP and VO at 0.377, 0.293 and 0.281 eV below the conduction band bottom. The formation of such complexes around room temperature is a consequence of the rapid migration of negatively charged vacancies. This conclusion is in line with perturbed angular correlation experiments, to be discussed below, which suggest that both the vacancy and interstitial are mobile below room temperature. In ion or proton-implanted Ge, additional defects are found such as one with [AQ1] a level at $E_C - 0.29$ eV and assigned to V_2 [35] Other defects with shallower levels, possibly interstitial related, have also been detected. The dominant defects anneal by 100–200°C but V_2 persists to 400°C [34].

For heavy ion irradiation, there are suggestions that more stable multi-vacancy defects are formed [36]. Implantation of 3 MeV Si ions into Ge-rich SiGe alloys (99.2%), with a dose low enough to prevent amorphization, revealed defects which persisted to at least 500°C. Higher doses, similar to those used to generate {113} defects in Si, led to broad DLTS peaks possibly arising from extended vacancy or interstitial clusters.

Full boron activation following implantation can be achieved with annealing to 150°C [37] or rapid thermal annealing treatments in range of 400–650°C. Boron appears immobile even for heat treatments at 800°C [38]. This strongly contrasts with boron implanted Si where the transformation of interstitial clusters leads to a transient enhanced diffusion of boron. The absence of this in Ge raises questions about the stability of interstitial clusters and the mechanism for boron diffusion. In contrast, P, and to a lesser extent As, implants exhibit considerable diffusion and loss through to the surface [39]. Further, a considerable fraction of the dopant is inactive, at least to 600–850°C [40].

6.6.1. Vacancies and divacancies in Ge

Experimental information on the electronic structure of vacancies in Ge has largely been derived from perturbed angular correlation spectroscopy (PACS) studies [41]. The technique is sensitive to an anisotropic electric field at the ^{111}In site. Here $10^{13}\,cm^{-3}$ ^{111}In atoms were implanted and the damage subsequently eliminated through a 600°C anneal leaving the In atoms at substitutional sites. n- and p-Ge samples were then electron irradiated at 77 K and subsequently investigated by PACS at various temperatures. Two axial centers are formed around ~200 K and have been attributed to In–*V* and In–I complexes. Around this temperature, it appears that V and I become mobile and become trapped at In defects. The resulting complexes are stable up to about 120°C. The In–V complex has only been detected in p-Ge, for $p > 10^{15}\,cm^{-3}$, while In–I was most abundant in n-Ge, provided $n < 10^{18}\,cm^{-3}$, and not found in p-Ge with $p > 10^{16}\,cm^{-3}$. For the first defect, the argument was made that the Fermi-level has to be below $E_V + 0.2\,eV$ for the In–V defect to form. This level was assigned to the $(-/0)$ transition of V so that for higher concentrations of the p-dopant, the vacancy is neutral which then diffuses and is hypothesized to be preferentially bound to In^- rather than to the more numerous Ga p-dopant. Thus the $V(-/0)$ level is placed at $E_V + 0.2\,eV$. This result would not be affected if V also possessed a $(2-/-)$ level lying above $E_V + 0.2\,eV$ (see below). It was also argued that the second signal arises from the trapping of I^+ by In^-. Clearly, this requires the Fermi-level to be below the donor level of *I*, and leads to an estimate of the $I(0/+)$ level at $E_C - 0.04\,eV$. For $n > 10^{18}\,cm^{-3}$, I would be neutral and it is hypothesized that I–In^- defects would not form, while for $p > 10^{16}\,cm^{-3}$, I^+ would be trapped by the more numerous Ga^- dopants. The low temperatures, $T \approx 200\,K$, where both I–In and V–In defects form, and well above the irradiation temperature, indicate a fast diffusion mechanism for both *V* and *I*.

Other experiments have however suggested that V^- is unstable in conflict with the PACS interpretation. These studies suggest that V^- is a negative-*U* center [42]. This is based on a study of the diffusion of As^+ mediated by V^{2-}, V^- and V^0 and carried out in p-Ge. Acceptable agreement could only be obtained if V^- did not exist or was immobile. The former implies that V^- would be a negative-*U* center.

Density functional calculations of the vacancy have led to results in partial agreement with the PACS experiments. Early supercell and cluster calculations revealed that the lattice distortions near the vacancy are similar, although smaller in magnitude, to those calculated for the vacancy in Si [32,43]. Consequently, the vacancy in Ge, unlike that in Si [44], is a double donor defect with a positive-*U* ordering of levels [32]. The supercell calculations were carried out in 128 atom cells where one atom was removed, and energy was calculated using a Γ-point BZ sampling scheme, which is now seen as inadequate [32]. The formation energies and structures of the vacancy were found for the $2+$, $+$, 0, $-$ and $2-$ charge states. The symmetries, readily understood from Watkins analysis of the same defect in Si [23], are found to be T_d, D_{2d}, D_{2d}, C_{2v} and C_{2v}, respectively. There are large volume changes by the atoms surrounding the vacancy moving inwards leading to volumetric contractions of up to 40% in the double negative charge state. For tetrahedral V^{2+}, the bonds across the vacancy are shifted from 3.96 Å in the ideal vacancy, to 3.57 Å for the relaxed case resulting in a 26% decrease in volume. In the positively charged and neutral cases, two of these bonds strengthen to 3.44 Å with a volume reduction of about 40%. For the single and double negative charge states, one bond is reduced to about 3.1 Å with between 50% and 55% volume decrease. The formation energies lie between about 1.7–1.9 eV, suggesting equilibrium vacancy concentrations between 5–50 $\times$ 10^{13} cm^{-3} at 800°C. This is considerably less than the formation energy of the vacancy in Si calculated to be about 4.36 eV [45]. The (2−/−), (−/0), (0/+) and (+/2+) levels are given as $E_V + 0.4$, $E_V + 0.38$, $E_V + 0.2$ and $E_V + 0.0$ eV. The (−/0) level has to be contrasted with an experimental estimate at $E_V + 0.2$ eV.

Cluster DFT calculations with about 400 atoms have also been carried out on vacancies and divacancies [43]. The strongest pairing has been found for V^- with bonds of length 3.36 Å while V^{2+} does not reconstruct and has T_d symmetry with bonds of length 3.54 Å. No details of energy levels are given. Interestingly, the Jahn-Teller distortion magnitudes and directions were shown to significantly depend on the lattice parameter used to generate the cluster. The DFT lattice parameter resulted in much larger relaxations. This lattice parameter is 1.2% smaller than the experimental one. The calculated Jahn-Teller relaxation energies are about 0.05, 0.32 and 0.29 eV for V^+, V^0 and V^- in Si and about factors between 2 and 3 smaller in Ge. The Si results contrast with experimental values of 0.4, 1.5, and 2.4 eV for V^+, V^0 and V^- respectively [47].

For V_2, a strong pairing distortion is found for Si in agreement with EPR data, but in Ge the distortion is weak and exceptional. These cluster calculations predict the Jahn-Teller relaxation energies of 0.10, 0.22 and 0.15 eV for +, 0 and −charge states in Si, respectively [43]. Experimentally, the energies are 1.3 and 2.4 eV for the + and − charge states, again exceeding the theoretical values [48]. However, these early experimental findings have been recently revised and are now believed to be about 0.16 and 0.54 eV for V_2^+ and V_2^-, respectively [49]. For Ge, a slight outward expansion of V_2 is predicted [43]. This distortion, called resonating valence bond [50], was obtained with a slight energy preference with respect to the pairing distortion.

Very recent density functional studies, carried out on large ~500 atom clusters, have suggested a distinct electrical activity for the vacancy in Ge when compared with Si [46]. The symmetries of V^{2+}, V^+, V^0, V^- and V^{2-}, were found to be T_d, D_{2d}, D_{2d}, D_2 and D_{2d}, respectively. The Kohn–Sham levels associated with these charge states are shown schematically in Figure 6.2 [46]. In V^{2+}, the t_2 gap level is empty and there is no driving force for a change to the tetrahedral symmetry. However, the four Ge atoms move outwards leading to a small volume increase of about 3%. In V^+ and V^0, the t_2 level is partially occupied and a Jahn-Teller distortion occurs, leading to a splitting of the t_2 level into b_2 and e levels. This reduces the ab (cd) lengths by 6% and 17%, together with a volume shrinkage of 12% and 29% with respect to the perfect unrelaxed vacancy. The negatively charged vacancy adopts a D_2 symmetry and a volume decrease by 34% with respect to the ideal one. In the double minus charge state (V^{2-}), the vacancy suffers a tetragonal Jahn-Teller distortion, leading to D_{2d} symmetry.

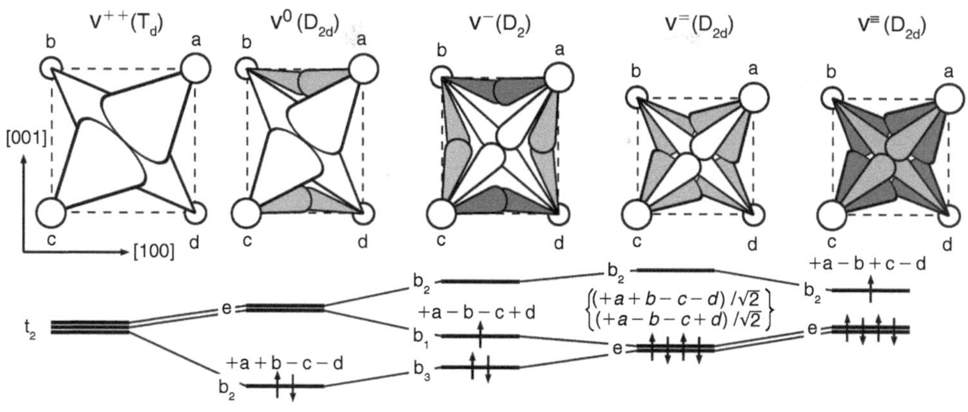

Figure 6.2. Scheme of the structural distortions in a Ge vacancy for several charge states of interest (up) after Coutinho *et al.* [46]. The symmetry of each structure is shown within parentheses. Unoccupied, highest occupied and lower (fully occupied) levels are represented in white, light gray and dark gray bonds, respectively. The Jahn-Teller splitting pattern of the triplet t_2 gap level is also shown (down), along with the symmetry representation of each level and the Linear Combination of Atomic Orbitals (LCAO) description of the highest occupied level.

However, its electronic and atomic structures differ from that of V^0 and V^+. In V^{2-} the four available electrons are accommodated in a $e^{\uparrow\downarrow\uparrow\downarrow} b_2^0$ configuration (see Figure 6.2), and the vacancy volume is about 60% of the original perfect vacancy. Finally, the triple minus charge state (V^{3-}) also shows a Jahn-Teller effect with a similar distortion (although weaker) to that in V^{2-}. An inward movement occurs along the [010] principal C_2 axis.

Of special interest is the prediction that unlike the Si vacancy, the vacancy in Ge does not possess a donor or double donor level [46]. However, the defect possesses three acceptor levels at $V(-/0) = E_V + 0.17\,\text{eV}$, $V(2-/-) = E_C - 0.54\,\text{eV}$ and $V(3-/2-) = E_C - 0.27\,\text{eV}$. The acceptor states are consistent with the PACS result. The relatively large Jahn-Teller distortions in V^0 and V^{2-} when compared to V^- result in almost superimposed $(-/0)$ and $(2-/-)$ levels. This indicates that the Ge vacancy could be a negative-U center with a metastable negative charge state and consistent with the results of diffusion studies discussed above [42]. The very large differences in the properties of the vacancy in Si and Ge are striking.

There does not seem to have been any first-principles calculations of the diffusion energy of the vacancy. Preliminary results indicate a barrier of about 0.3 eV which suggests motion occurs around 120 K assuming a normal pre-exponential factor. This is reasonably close to that found in the PACS experiments. However, studies of vacancy-mediated self-diffusion in Ge obtained at elevated temperatures indicate an activation barrier of about 3 eV [51]. This figure includes the formation energy of a vacancy and is in line with the activation barrier for the vacancy assisted diffusivity of copper in Ge [52,53]. Taking into account the 2 eV from the *ab initio* formation energy of a single vacancy [32], we can estimate the migration barrier of V at about 1 eV. This is of course in conflict not only with the PACS experiments, but also with the observed growth of VO defects which start to appear at

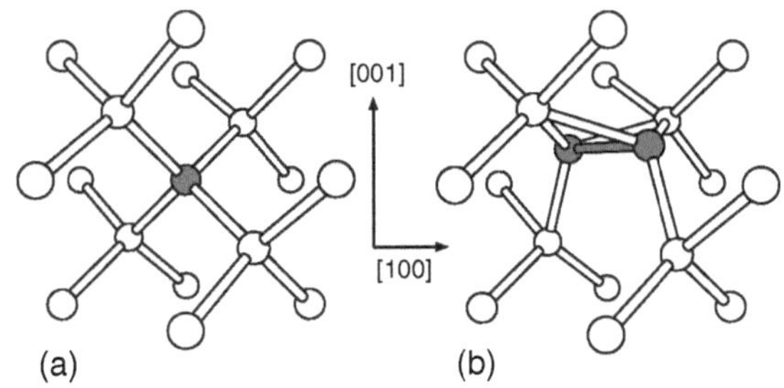

Figure 6.3. Cluster of Ge atoms as in bulk (a), and in a 110-split interstitial in Ge (b). The defect has C_{2v} symmetry with its principal axis along [001], and the Ge–Ge split-unit (shown as gray atoms) is aligned along the [110] direction.

temperatures as low as 60 K [54] in Ge electron irradiated at cryogenic temperatures. The *high temperature form* of the vacancy in Ge seems distinct from the *low temperature form*. The same appears to be the case for the vacancy in Si [55].

6.6.2. The self-interstitial

DFT calculations [33,56] have found that the [110] split interstitial arrangement shown in Figure 6.3b, where two Ge atoms aligned along [110] share a lattice site is the most stable in the −, 0 and + charge states.

This structure is similar to that found for the neutral self-interstitial in Si. A donor level (0/+) was placed at 0.08 eV below the valence band top, i.e. the positively charged defect is thermodynamically unstable in conspicuous disagreement with the experimental assignment at $E_C - 0.3$ eV found from PACS studies discussed above [41]. Instead, an acceptor level was placed close to $E_C - 0.3$ eV. It was recognized that understanding the PAC signal, attributed to (In^-–I^+) for n-type samples is then problematic. The neutral self-interstitial was found to have a formation energy of 3.55 eV, i.e. lower than for the analogous defect in Si, but well above the 1.9 eV formation energy for the single vacancy [32]. This corroborates early ^{71}Ge tracer-diffusion measurements which conclude that self-diffusion in Ge is dominated by a vacancy mechanism [51] as the thermal concentration of vacancies should be well above that of interstitials.

6.6.3. Nitrogen defects

FTIR studies on implanted nitrogen in Ge [57] following 400°C anneals to remove implantation damage, revealed a number of vibrational bands including a pair at 659 and 825 cm^{-1}. Isotopic studies using mixtures of ^{14}N and ^{15}N demonstrated that these modes can be attributed to an interstitial nitrogen dimer. This defect anneals around 700°C. Seven other nitrogen related modes probably arising from implantation damage were also reported. Density functional calculations of an interstitial nitrogen dimer gave infrared active modes at 632 and 754 cm^{-1} in reasonable agreement with the 659 and 825 cm^{-1} bands. The dimer is believed to be electrically inert. The substitutional defect has modes calculated to lie at $574\,cm^{-1}$ in good agreement with its experimental band frequency at $577\,cm^{-1}$ [58]. Its structure comprises a $\langle 111 \rangle$ off-centered N_s atom, probably with deep electrical activity due to its unsaturated Ge radical along the trigonal axis [57–59]. A mode observed at $590\,cm^{-1}$ has been assigned to the single nitrogen interstitial and although this is close to one calculated at $617\,cm^{-1}$, the analogous defect in Si is found to be mobile at room temperature [60] and hence this assignment should be reconsidered.

6.6.4. *Carbon in germanium*

Carbon in Si is an abundant contaminant, usually with a concentration of the order of 10^{15}–$10^{16}\,cm^{-3}$. After irradiating Si with MeV electrons, carbon atoms are responsible for a multitude of electrically active centers due to its ejection to an interstitial lattice location [28]. In Ge, the scenario is rather different as the carbon solubility limit is believed to be very much lower, at around 10^{8}–$10^{10}\,cm^{-3}$ [61]. Carbon has been implanted into Ge to doses of $\approx 10^{20}\,cm^{-3}$ and annealed at 450°C to remove damage [62], a single mode was revealed at $531\,cm^{-1}$ by FTIR. Channeling measurements indicated that one third of the implanted carbon atoms take a substitutional site in the Ge lattice, and the annealing behavior of the channeling and FTIR data are similar. Annealing studies show the loss of substitutional carbon above 700°C, possibly via surface segregation. Theoretical investigation carried out in 131 atom clusters showed that the T_2 vibrational mode of substitutional carbon lies around $516\,cm^{-1}$ [62] in good agreement with the data and confirming the substitutional nature of the defect.

6.6.5. *Oxygen in germanium*

Experimental data on oxygen in germanium have been in depth discussed in Chapter 4. Interstitial oxygen has been modeled by several groups using DFT in supercells [18,63]. The oxygen atom lies near a bond centered site with a Ge–O–Ge angle of $\approx 140°$ and a Ge–O length of 1.73 Å. The ground state structure is shown in Figure 6.4a. The formation energy of dissolved oxygen in equilibrium with GeO_2 is 1.2 eV, suggesting that high concentrations of oxygen can dissolve into Ge. The experimental solubility [64] is $2 \times 10^{18}\,cm^{-3}$ with a heat of solution of 1.2 eV in good agreement with the theory [18]. LVMs for interstitial oxygen are found at 847 and $401\,cm^{-1}$ and can be compared with experimental modes at 863 and $412\,cm^{-1}$ [65]. The diffusion energy of oxygen is calculated to be 1.7 eV and 0.3 eV less than the experimental barrier – the same underestimate found for the barrier of oxygen diffusion in Si [18]. In the saddle point the O atom is three-fold coordinated, producing a donor state high in the gap. It is possible that the error in the migration barrier arises from the underestimation of the gap when using supercells and the local density approximation. However, the structure of interstitial oxygen, having no gap levels, is clearly correctly found using supercells as shown by the agreement between the calculated and observed local mode.

Oxygen dimers, shown in Figure 6.4b, are bound by 0.6 eV and diffuse with lower barriers than single oxygen defects. The migration energy for O_2 in Ge is about 0.6 eV lower than

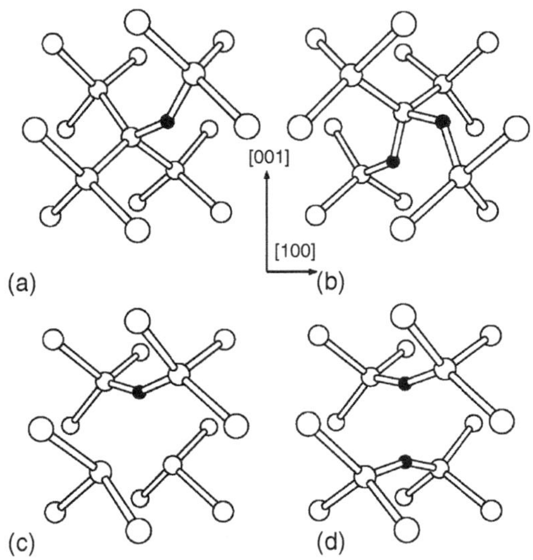

Figure 6.4. Oxygen and di-oxygen centers in Ge. Interstitial oxygen (a), interstitial oxygen dimer (b), VO (c) and vacancy-di-oxygen (d) centers according to Coutinho *et al.* [18]. Ge and O atoms are shown in white and black, respectively.

the barrier for the diffusion of the single O_i center [18]. Vibrational modes of the dimer have been found and compare favorably with experimental results [21].

As in Si, interstitial oxygen in Ge is an efficient trap for a mobile vacancy. The resulting VO complex, shown in Figure 6.4c, is a common product of radiation damage [54]. The binding energy of O with V was calculated by about 0.4 eV and considerably less than the 1.6 eV for the same defect in Si [18]. This small value could arise from an underestimate of the formation energy of the vacancy and should be recalculated using a more accurate method. The LVM of VO is calculated to be 642 cm^{-1} and the same for VO^- in conflict with experiment. This reflects the difficulty of treating charged defects in germanium using supercells. The experimental band frequencies of VO^0 and VO^- are 620 and 669 cm^{-1}, respectively [22,54,66], and a mode at 716 cm^{-1} has been assigned to VO^{2-} [20]. The binding energy between neutral VO and another O_i to form VO_2 has been estimated by 1.3 eV [18]. This value is considerably greater than the binding energy of V and O_i. The VO_2 complex has IR-active modes at 690 and 363 cm^{-1}, and the highest is not far from an absorption band at 731 cm^{-1} that has been assigned to V_nO or VO_2 [66].

The electrical levels of VO in Ge have recently been found experimentally using DLTS and Laplace-DLTS studies [20,67]. In irradiated oxygen-rich Ge, prominent electron and minority hole traps with levels at $\Delta H = 0.26$ eV and $\Delta H = 0.32$ eV have been assigned to VO(2−/−) and VO (−/0), respectively. The high resolution Laplace-DLTS technique, combined with uniaxial stress, has also been used to investigate these traps [67]. Unfortunately, only the splitting of the (2−/−) level could be resolved, indicating that VO^{2-} has orthorhombic-I (C_{2v}) symmetry. The lack of a stress-splitting effect on the (−/0) transition

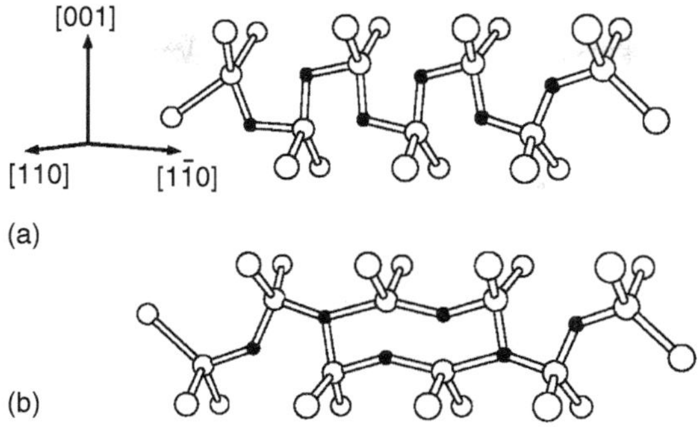

Figure 6.5. Small oxygen interstitial aggregate with six oxygen atoms (a) and thermal double donor model in Ge from Coutinho *et al.* [70]. Ge and O atoms are shown in white and black, respectively.

was interpreted as indicating a fast thermal reorientation of the neutral defect at the temperature of the measurement ($T \approx 140$–200 K), leading to an apparent tetrahedral symmetry signature [67]. The symmetry of the negative charge state has also been established after EPR measurements on quasi-monoisotopic ^{74}Ge samples. Here, a clear orthorhombic-I (C_{2v}) signal was assigned to a spin-1/2 VO^- center. This EPR center has the same annealing behavior as the 669-cm^{-1} absorption band attributed to VO^- [66].

It must be remarked that DLTS studies on oxygen implanted into phosphorous doped Ge, followed by anneals at 500°C, report levels around $E_V + 0.39$ eV [68] which are passivated by exposure to atomic hydrogen. Although this is close to the VO double acceptor level at $E_C - 0.28$ eV, the defects are unlikely to be related. This is because the anneal temperature of VO is 135°C [20]. Instead of exposure to atomic hydrogen, molecular hydrogen leads to a level at $E_C - 0.34$ eV.

6.6.6. Thermal donors

Thermal donors (TDs) are formed when O doped Ge is heated to around 370°C [69]. They comprise a family of shallow double donors and are believed to be composed of long chains of oxygen atoms. It is possible that as in Si, the first donor TD1 contains 5 oxygen atoms and TD_n contains $n + 4$ oxygen atoms [70,71]. The first three members of the family are bi-stable and possess an electrically inactive form which is stable in the dark. According to recent modeling studies, oxygen clustering is energetically favored in Ge, and for small O complexes (less than 6–7 oxygen atoms), the aggregates take the staggered form [69,71] shown in Figure 6.5a. These are believed to be the electrically inactive species. For larger clusters, an electrically active species becomes the ground state and its structural appearance is shown in Figure 6.5b. Here the inner oxygen atoms of the chain bridge second-nearest neighboring Ge atoms lying along the [1 $\bar{1}$ 0] direction and compress the lattice along [001], while the atoms near the end are over-coordinated [70,71].

The electrically activity comes from the over-coordinated oxygen atoms but their levels lie above those of strain induced states arising from the inner parts of the chain. Hence the donor electrons actually occupy these strain induced states which have very little overlap with Ge. Such a model accounts for the small oxygen hyperfine parameters for TDs in Si [72], as well for the [001] compressive piezospectroscopic stress-energy tensor [73,74].

The vibrational properties of the donors have been investigated [69] and although the donors contain a large number of O atoms, only two O related vibrational modes are observed for each donor in the FTIR spectrum. These modes have been associated with the end O atoms in a chain. Vibrational mode frequencies of 723, 732, 740, 747 cm^{-1} for $O_n + 4$ defects with $n = 1, \ldots, 4$, associated with the first four TD members, compare favorably with experimental values of 769, 774, 778 and 781 cm^{-1} for TDn, with $n = 1, \ldots, 4$. EPR measurements have been carried out on Ge [75] and indicate that the donor has C_{2v} symmetry similar to donors in Si.

6.6.7. Hydrogen in germanium

Early modeling studies [76] suggested that H^+ in Ge, unlike in Si, was barely stable and hence passivation of acceptors less likely to occur. Photo-thermal ionization spectroscopy identified several shallow acceptors in Ge grown in a H atmosphere and rapidly quenched. Two were assigned to H trapped by Si and C impurities [77,78]. Stress experiments on optical transitions showed that the ground state did not split and this isotropic state was attributed to a tunneling of H around the substitutional impurities. Similar defects are Zn–H and Be–H all having single acceptor levels around 10 meV and shallower than the substitutional dopants Zn and Be. The concentrations of all these H-related centers are low and less than $10^{11}\,cm^{-3}$. Shallow donor defects, e.g. H and Li bound to O, have been reported but the tunneling model has not been entirely accepted: a static model has been discussed by Ham [79].

LVMs of H in Ge have been experimentally found and reported [80,81]. Here, a mode at 1794 cm^{-1} is attributed to a bond centered proton (H^+_{BC}) and a second defect with a mode at 745 cm^{-1} attributed to anti-bonded hydride ion (H^-_{AB}) suggesting that H is a negative-U center as it is in Si. Further support for this picture comes from DLTS studies [82] assigning a (0/+) level for hydrogen at $E_C - 0.11$ eV while (−/0) should lie close to E_V rather than near mid-gap as in Si. This highlights a difference between the two elements and suggests that in Ge shallow donors may not be easily passivated by H. Nevertheless, H has been shown to displace and passivate transition metal donor states such as Cr, Co and Cd [83], as well as deep acceptors like Cu [84,85]. There have been a number of modeling studies of hydrogen centers in Ge [86–88]. As in Si, the calculations predict a bi-stable center with the H atom sitting at BC and AB sites for the positive and negative charge states, respectively. The H^+_{BC} center gives rise to a stretch mode at 1897 cm^{-1}, whereas H^-_{AB} has stretch and bend modes at 1239 and 704 cm^{-1}, respectively [88]. These values are close to the FTIR data reported above.

Hydrogen in Si doped Ge crystals, i.e. dilute GeSi alloys, has also been studied by FTIR and first-principles calculations. From an analysis of the FTIR data, it was concluded that H^+_{BC} does not form preferential Ge–H–Ge or Si–H–Ge units [81]. This is in line with the minute energy differences between Ge–H–Ge, Si–H–Ge, Si–H–Si units from the calculations [88]. On the other hand, the calculations find that H^-_{AB} binds preferentially to Si impurities in Si doped Ge [88]. These results were recently confirmed by FTIR measurements [89]. In these experiments, GeSi samples were proton-implanted at $T < 20\,\text{K}$, and *in situ* optical measurements show that besides the presence of the usual bands that appear in low-T H^+-implanted Ge, new bands at 815.8, 1430.2 and $1630.7\,\text{cm}^{-1}$ developed. Based on isotope and annealing data, as well as the dependence of line intensity on the concentration of Si, these bands were assigned to wag, wag-overtone and stretch modes of a Si–H_{AB}^- defect in Ge-rich GeSi alloys. These frequencies are well accounted by calculations which predict wag and stretch modes at 768 and $1449\,\text{cm}^{-1}$ [88].

Other hydrogen related defects have been investigated. Local modes of H_2^* and IH_2 defects have been calculated and compared with experimental results [87,90–92]. Hydrogen related platelets in Ge seem similar to those in Si, with molecular hydrogen stabilizing the platelets which do however form at lower temperatures than in Si [93]. Plasma treated H samples show a reduction of deep-level defects associated with Ni, Au, Cd, Zr, Ti and Cu [83,84,94] but not for donors like Te, S and Se [83]. However, there have been no modeling studies of these defects.

Vacancy-hydrogen defects have been investigated theoretically [95] and their vibrational modes compared with experimental values. IR-active modes of VH_4 lie at $2107\,\text{cm}^{-1}$ compared with an experimental value of $2062\,\text{cm}^{-1}$, while those of VH_2 lie at 2102 and $2078\,\text{cm}^{-1}$ compared with observed modes at 1993 and $1980\,\text{cm}^{-1}$. Other defects VH and VH_3 are considered along with D isotopic shifts. Hydrogen seems able to passivate implantation damage from oxygen [68]. H interactions with He introduced bubbles has been reported in Ref. [96].

6.7. Electrical Levels of Defects

Recently, the problem of calculating electrical levels of defects in Ge has been considered using H-terminated clusters containing 297 atoms [97]. The Ge–H bond length was fixed at a length corresponding to a band gap of 0.7 eV. The energies of relaxed clusters in charge states ranging from +2 to −2 were then found for both defect-free and defective clusters. We define $E(2-/-)$ and $E(-/0)$ to be the energy differences $E(\mathbf{R}^{2-}) - E(\mathbf{R}^-)$ and $E(\mathbf{R}^-) - E(\mathbf{R}^0)$, respectively. For a defect-free and *infinite* cluster, $E(2-/-)$ and $E(-/0)$ would be both equal to the energy of the conduction band bottom (E_C). Similarly, $E(0/+) = E(\mathbf{R}^0) - E(\mathbf{R}^+)$ and $E(+/2+) = E(\mathbf{R}^+) - E(\mathbf{R}^{2+})$ would both be equal to the energy of the top of the valence band (E_V). Here $E(\mathbf{R}^-)$, $E(\mathbf{R}^{2+})$, etc, stand for the structural energies of a cluster with one, two, etc, additional electrons, respectively. To determine the levels of a defect with respect to E_C and E_V, the corresponding energies $E(2-/-)$, $E(-/0)$, etc, for a relaxed defect were found and compared with the corresponding quantities in the defect-free cluster. This gave the acceptor energy levels with respect to E_C and the donor levels with respect to E_V. To refer an acceptor level to E_V, or a donor level from E_C, the experimental gap is subtracted from these energies.

Table 6.1 shows that, for substitutional P, $E(0/+)$ is $-6.25\,\text{eV}$ or $0.63\,\text{eV}$ higher than $E(0/+)$ for the defect-free cluster. Since the latter is identified with E_V, the donor level of P_s was placed at $E_V + 0.63\,\text{eV}$, or at $E_C - 0.07\,\text{eV}$ if we take the band gap to be 0.7 eV.

Table 6.1. Transition energies (eV) for relaxed defect-free Ge cluster (Ge), and defective clusters (P_s and B_s) in various charge states. In the latter the central Ge atom was replaced by B and P species.

Defect	E(2+/+)	E(0/+)	E(−/0)	E(2−/−)
Ge	−8.17	−6.88	−4.94	−3.59
P_s		−6.25		
B_s			−5.60	

Table 6.2. Calculated energy levels (eV) of defects in Ge with respect to E_V. The experimental gap is taken to be 0.70 eV being the average of the low and room temperature gaps. Experimental values, in brackets, taken from Sze [98] and Grimmeiss *et al.* [99].

Defect	(0/+)	(−/0)	(2−/−)
B_s		0.04 (0.01)	
P_s	0.63 (0.70)		
As_s	0.63 (0.70)		
S_s	0.61 (0.42)		
Se_s	0.61 (0.43)		
Te_s	0.66 (0.61)		
Be_s		0.14 (0.02)	0.14 (0.06)
Zn_s		0.26 (0.04)	0.27 (0.10)

This can be compared with an experimental value of $E_C - 0.01$ eV or $E_V + 0.70$ eV when using the same band gap of 0.7 eV. Clearly, the method describes well this shallow donor.

Turning to substitutional boron, we see that $E(-/0)$ is lower by 0.66 eV than E_C placing the acceptor level at $E_V + 0.04$ eV, which is to be compared with an experimental value of 0.01 eV above the valence band top. Similarly the first donor levels of the substitutional chalcogen defects S, Se and Te can be evaluated. These are given in Table 6.2 and again they are in good agreement with the experiments (shown in brackets). Second donor levels are however not described accurately. Experimentally, they differ from the single donor levels by about 0.3 eV in both Ge and Si but the calculated differences is only about 0.1 eV. As previously reported [17], this is likely to arise from deficiencies in the description of the exchange-correlation energy.

As_s and Be_s acceptor defects appear to give levels in reasonable agreement with experiment. However, with Zn_s larger deviations from the experimental data are obtained. To understand Zn with an atomic configuration $3d^{10}4s^2$, we recall that the vacancy model regards four of its valence electrons to bond with Ge neighbors leaving a configuration $3d^8$, with the defect acting as a double acceptor. The d-orbitals hybridize with the t_2 orbitals arising from the four surrounding Ge atoms, and the resulting $t_2^6e^4t_2^4$ configuration is sensitive to a Jahn-Teller

Table 6.3. Energy levels (eV) of deep-level defects in Ge with respect to E_V and calculated using Zn_s as a marker. Experimental values [20,100,101] are shown in brackets.

Defect	**(0/+)**	**(−/0)**
Zn_s	0.26 (0.04)	0.27 (0.09)
Au_s	0.36 (0.15)	0.47 (0.50)
VP	0.33 (0.31)	0.39 (0.41)
VO	0.42 (0.27)	0.47 (0.49)

distortion, splitting the t_2 manifold. The calculated acceptor levels are close to $E_V + 0.26\,eV$ and deeper than the experimental ones by about 0.2 eV. It was suggested that the gap states of Zn_s are unlike those of other defects investigated, which were largely effective-mass like. This suggests that the method cannot really treat defects with deep vacancy like levels and that it would be better to use a marker method. Here the acceptor level of a defect is found from the electron affinity of a marker M, $E_M(-/0)$, together with its experimental acceptor level $E_M^{expt}(-/0)$, using $E(-/0) - E_M(-/0) + E_M^{expt}(-/0)$. The marker was taken to be substitutional Zn.

To investigate the levels using the marker method, Au_s and the radiation defects VO and VP were considered. The structure of all these centers shows Jahn-Teller distortions. The substitutional Au is slightly distorted from T_d symmetry, the Au atom being displaced along the $\langle 100 \rangle$ direction. In the VO defect, the oxygen atom is strongly displaced from the substitutional site along the same direction, forming bonds with two Ge atoms, and the two remaining Ge atoms form a reconstructed bond. This defect has C_{2v} symmetry. The VP defect is formed when a mobile vacancy is trapped by a substitutional P atom, and it has C_{1h} symmetry. Electrical levels of these defects have been measured in Refs. [20,35, 102–104]. Table 6.3 shows the calculated acceptor levels of VP to be in good agreement with the experimental values with errors around 0.02 eV. Those of Au and VO however, depart more with errors of 0.1–0.2 eV.

6.8. Summary

In this chapter we have attempted to describe the problems and successes in applying density functional methods to defects in Ge. It is noted that the very large error in the band gap found from using local DFT can create problems. For example, if a defect possesses an electrical level in the band gap and the structure depends on the charge state, then a supercell approach is unlikely to capture the structural change on charging the defect. This can be overcome but only at the expense of using a method which produces a realistic band gap. One simple approach may be in using hydrogen terminated clusters as these lead to a substantial band gap. However, it remains to be seen whether this method will not introduce some new problem and the convergence of the defect properties with cluster size needs to be carefully considered. At this stage there are relatively few results on defects in Ge but this may change in the future if Ge proves to become, once again, an important semiconductor.

References

[1] M. Born and K. Huang, *Dynamical Theory of Crystal Lattices*, Oxford University Press, London, 1954.

[2] J.C. Slater, *Quantum Theory of Atomic Structure*, Vol. 2, McGraw Hill, New York, 1960.

[3] P. Hohenberg and W. Kohn, *Phys. Rev.* **136** (1964) B864.

[4] W. Kohn and L.J. Sham, *Phys. Rev.* **140** (1965) A1133.

[5] J.P. Perdew, J.A. Chevary, S.H. Vosko, K.A. Jackson, M.R. Pederson, D.J. Singh and C. Fiolhais, *Phys. Rev. B* **46** (1992) 6671.

[6] G. Kresse and J. Furthmuller, *Phys. Rev.* B **54** (1996) 11169.

[7] X. Gonze, J.-M. Beuken, R. Caracas, F. Detraux, M. Fuchs, G.-M. Rignanese, L. Sindic, M. Verstraete, G. Zerah, F. Jollet, M. Torrent, A. Roy, M. Mikami, Ph. Ghosez, J.-Y. Raty and D.C. Allan, *Comp. Mat. Sci.* **25** (2002) 478.

[8] J.M. Soler, E. Artacho, J.D. Gale, A. García, J. Junquera, P. Ordejón and D. Sánchez-Portal, *J. Phys. Condens. Matter* **14** (2002) 2745.

[9] R. Jones and P.R. Briddon, *Identification of defects in semiconductors*, Chapter 6, edited by M. Stavola, *Semiconductors and Semimetals*, Vol. 51A Academic Press, Boston 1998, p. 287.

[10] P.R. Briddon and R. Jones, *phys. stat. sol. (b)* **217** (2000) 131.

[11] L. Kleinman, *Phys. Rev. B* **24** (1981) 7412.

[12] H.J. Monkhorst and J.D. Pack, *Phys. Rev. B* **13** (1976) 5188.

[13] J.E. Northrup and S.B. Zhang, *Phys. Rev. B* **47** (1993) 6791.

[14] A.M. Stoneham, *Defects in Solids*, Oxford University Press, London 1975.

[15] G. Makov and M.C. Payne, *Phys. Rev. B* **51** (1995) 4014.

[16] A. Resende, R. Jones, S. Öberg and P.R. Briddon, *Phys. Rev. Lett.* **82** (1999) 2111.

[17] J. Coutinho, V.J.B. Torres, R. Jones and P.R. Briddon, *Phys. Rev. B* **67** (2003) 035205.

[18] J. Coutinho, R. Jones, P.R. Briddon and S. Öberg, *Phys. Rev. B* **62** (2000) 10824.

[19] S. Zwerdling, B. Lax, L.M. Roth and K.J. Button, *Phys. Rev.* **114** (1959) 80.

[20] V.P. Markevich, I.D. Hawkins, A.R. Peaker, V.V. Litvinov, L.I. Murin, L. Dobaczewski and J.L. Lindström, *Appl. Phys. Lett.* **81** (2002) 1821.

[21] P. Vanmeerbeek and P. Clauws, *Phys. Rev. B* **64** (2001) 245201.

[22] V.V. Litvinov, L.I. Murin, J.L. Lindström, V.P. Markevich, and A.N. Petuhg, *Fiz. Tekh. Poluprovodn.* **36** (2002) 658 [*Semiconductors* **36** (2002) 621].

[23] G.D. Watkins, *Radiation Damage in Semiconductors*, Dunod, Paris 1967, p. 97.

[24] G.D. Watkins, *Identification of defects in semiconductors*, Chapter 1 edited by M. Stavola, *Semiconductors and Semimetals*, Vol. 51A Academic Press, Boston 1998, p. 1.

[25] R. van Kemp, E.G. Sieverts and C.A.J. Ammerlaan, *Phys. Rev. B* **40** (1989) 4037.

[26] E.V. Lavrov, L. Hoffmann and B. Bech Nielsen, *Phys. Rev. B* **60** (1999) 8081.

[27] R.C. Newman, *Infra-Red Studies of Crystal Defects*, Taylor & Francis Ltd, London 1973.

[28] G. Davies and R.C. Newman, *Handbook on Semiconductors*, edited by S. Mahajan, Vol. **3b**, Elsevier Science, Amsterdam 1994, p. 1557.

[29] L. Dobaczewski, A.R. Peaker and K. Bonde Nielsen, *J. Appl. Phys.* **96** (2004) 4689.

[30] P. Pellegrino, P. Lévêque, J. Lalita, A. Hallén, C. Jagadish and B.G. Svensson, *Phys. Rev. B* **64** (2001) 195211.

[31] P.M. Mooney, *Identification of defects in semiconductors*, Chapter 1, edited by M. Stavola, *Semiconductors and Semimetals*, Vol. 51B, Academic Press, Boston 1998 p. 93.

[32] A. Fazzio, A. Janotti, A.J.R. da Silva and R. Mota, *Phys. Rev. B* **61** (2000) R2401.

[33] M.D. Moreira, R.H. Miwa and P. Venezuela, *Phys. Rev. B* **70** (2004) 115215.

[34] V.P. Markevich, A.R. Peaker, L.I. Murin, V.V. Emtsev, V.V. Litvinov, N.V. Abrosimov and L. Dobaczewski, *Sol. St. Phen.* **108–109** (2005) 253.

[35] J. Fage-Pedersen, A. Nylandsted Larsen and A. Mesli, *Phys. Rev. B* **62** (2000) 10116.

[36] A. Colder, M. Levalois and P. Marie, *Eur. Phys. J. AP* **13** (2001) 89.

[37] P.J. MacDonald and D.W. Palmer, *Inst. Phys. Conf. Ser.* **23** (1975) 404.

[38] S. Uppal, A.F. Willoughby, J.M. Bonar, A.G.R. Evans and N.E.B. Cowern, *J. Appl. Phys.* **90** (2001) 4293.

[39] E. Simoen, A. Satta, M. Meuris, T. Janssens, T. Clarysse, A. Benedetti, C. Demeurisse, B. Brijs, I. Hoflijk, W. Vandervorst and C. Claeys, *Sol. St. Phen.* **108–109** (2005) 691.

[40] C.O. Chui, K. Gopalakrishnan, P.B. Griffin, J.D. Plummer and K.C. Saraswat, *Appl. Phys. Lett.* **83** (2003) 3275.

[41] H. Haesslein, R. Sielemann and C. Zistl, *Phys. Rev. Lett.* **80** (1998) 2626.

[42] E. Vainonen-Ahlgren, T. Ahlgren, J. Likonen, S. Lehto, J. Keinonen, W. Li and J. Haapamaa, *Appl. Phys. Lett.* **77** (2000) 690.

[43] S. Ögüt and J.R. Chelikowsky, *Phys. Rev. B* **64** (2000) 245206.

[44] G.A. Baraff, E.O. Kane and M. Schlüter, *Phys. Rev. B* **21** (2000) 5662.

[45] M.I.J. Probert and M.C. Payne, *Phys. Rev. B* **67** (2003) 075204.

[46] J. Coutinho, R. Jones, V.J.B. Torres, M. Barroso, S. Öberg and P.R. Briddon, *J. Phys: Cond. Matter.* **17** (2005) L521.

[47] G.D. Watkins, *Deep Centers in Semiconductors* 2nd edition, edited by S.T. Pantelides, Gordon and Breach, Yverdon 1996, p. 177.

[48] G.D. Watkins and J.W. Corbett, *Phys. Rev.* **138** (1965) A543.

[49] G.D. Watkins, *Physica* B **376** (2006) 50.

[50] O. Sugino and A. Oshiyama, *Phys. Rev. B* **42** (1990) 11869.

[51] M. Werner, H. Mehrer and H.D. Hochheimer, *Phys. Rev. B* **32** (1985) 3930.

[52] S. Mayburg, *Phys. Rev.* **95** (1954) 38.

[53] A. Giese, N.A. Stolwijk and H. Bracht, *Appl. Phys. Lett.* **77** (2000) 642.

[54] R.E. Whan, *Phys. Rev.* **140** (1965) A690.

[55] H. Bracht, J. Fage Pedersen, N. Zangenberg, A. Nylandsted Larsen, E.E. Haller, G. Lulli and M. Posselt, *Phys. Rev. Lett.* **91** (2003) 245502.

[56] A.J.R. da Silva, A. Janotti, A. Fazzio, R.J. Baierle and R. Mota, *Phys. Rev. B* **62** (2000) 9903.

[57] F. Berg Rasmussen, R. Jones and S. Öberg, *Phys. Rev. B* **50** (1994) 4378.

[58] H.J. Stein, *Appl. Phys. Lett.* **52** (1988) 153.

[59] C. Cunha, S. Canuto and A. Fazzio, *Phys. Rev. B* **48** (1993) 17806.

[60] J.P. Goss, I. Hahn, R. Jones, P.R. Briddon and S. Öberg, *Phys. Rev. B* **67** (2003) 045206.

[61] R.I. Scace and G.A. Slack, *J. Chem. Phys.* **30** (1959) 1551.

[62] L. Hoffmann, J.C. Bach, B. Bech Nielsen, P. Leary, R. Jones and S. Öberg, *Phys. Rev. B* **55** (1997) 11167.

[63] E. Artacho, F. Yndurain, B. Pajot, R. Ramirez, C.P. Herrero, L.I. Khirunenko and K.M. Itoh, *Phys. Rev. B* **56** (1997) 3820.

[64] W. Kaiser and C.D. Thurmond, *J. Appl. Phys.* **32** (1961) 115.

[65] A.J. Mayur, M.D. Sciacca, M.K. Udo, A.K. Ramdas, K. Itoh, J. Wolk and E.E. Haller, *Phys. Rev. B* **49** (1994) 16293.

[66] P. Vanmeerbeek, P. Clauws, H. Vrielinck, B. Pajot, L. Van Hoorebeke and A. Nylandsted Larsen, *Phys. Rev. B* **70** (2004) 035203.

[67] V.P. Markevich, V.V. Litvinov, L. Dobaczewski, J.L. Lindström, L.I. Murin, S.V. Vetrov, I.D. Hawkins and A.R. Peaker, *Physica B* **340–342** (2003) 844.

[68] A.J. Tavendale and S.J. Pearton, *J. Appl. Phys.* **54** (1983) 3213.

[69] L.I. Murin, J.L. Lindström, V.P. Markevich, T. Hallberg, V.V. Litvinov, J. Coutinho, R. Jones, P.R. Briddon and S. Öberg, *Physica B* **308–310** (2001) 290.

[70] J. Coutinho, R. Jones, L.I. Murin, V.P. Markevich, L. Lindström, S. Öberg and P.R. Briddon, *Phys. Rev. Lett.* **87** (2001) 235501.

[71] R. Jones, J. Coutinho, S. Öberg and P.R. Briddon, *Physica B* **308–310** (2001) 8.

[72] J. Michel, J.R. Niklas and J.-M. Spaeth, *Phys. Rev. B* **40** (1989) 1732.

[73] J.M. Trombetta, G.D. Watkins, J. Hage and P. Wagner, *J. Appl. Phys.* **81** (1997) 1109.

[74] G.D. Watkins, *Early Stages of Oxygen Precipitation in Silicon,* Vol. **17** in the series NATO ASI, edited by R. Jones, Kluwer Academic Publishers, Dordrecht (1996) p. 1.

[75] F. Callens, P. Clauws, P. Matthys, E. Boesman and J. Vennik, *Phys. Rev. B* **39** (1989) 11175.

[76] S.K. Estreicher and Dj.M. Maric, *Phys. Rev. Lett.* **70** (1993) 3963.

[77] E.E. Haller, B. Joos and L.M. Falicov, *Phys. Rev. B* **21** (1980) 4729.

[78] J.M. Kahn, R.E. McMurray, E.E. Haller and L.M. Falicov, *Phys. Rev. B* **36** (1987) 8001.

[79] F.S. Ham, *Phys. Rev. B* **38** (1988) 5474.

[80] M. Budde, B. Bech Nielsen, C. Parks Cheney, N.H. Tolk and L.C. Feldman, *Phys. Rev. Lett.* **85** (2000) 2965.

[81] R.N. Pereira, B. Bech Nielsen, L. Dobaczewski, A.R. Peaker and N.V. Abrosimov, *Phys. Rev B* **71** (2005) 195201.

[82] L. Dobaczewski, K. Bonde Nielson, N. Zangenberg, B. Bech Nielson, A.R. Peaker and V.P. Markevich, *Phys. Rev. B* **69** (2004) 245207.

[83] S.J. Pearton and A.J. Tavendale, *J. Appl. Phys.* **54** (1983) 820.

[84] S.J. Pearton, *Appl. Phys. Lett.* **40** (1982) 253.

[85] S.J. Pearton and A.J. Tavendale, *J. Appl. Phys.* **54** (1983) 440.

[86] P.J.H. Denteneer, C.G. Van de Walle and S.T. Pantelides, *Phys. Rev. Lett.* **62** (1989) 1884.

[87] S.K. Estreicher, M.A. Roberson and Dj. M. Maric, *Phys. Rev.* B **50** (1994) 17018.

[88] A. Balsas, V.J.B. Torres, J. Coutinho, R. Jones, B. Hourahine, P.R. Briddon and M. Barroso, *J. Phys. Condens. Matter* **17** (2005) S2155.

[89] R.N. Pereira, B. Bech Nielsen and J. Coutinho, *Physica B* **376–377** (2006) 22.

[90] M. Budde, B. Bech Nielsen, R. Jones, J. Goss and S. Öberg, *Phys. Rev. B* **54** (1996) 5485.

[91] M. Budde, B. Bech Nielsen, P. Leary, J. Goss, R. Jones, P.R. Briddon, S. Öberg and S.J. Breuer, *Phys. Rev. B* **57** (1998) 4397.

[92] J. Coutinho, V.J.B. Torres, R.N. Pereira, R. Jones, S. Öberg and P.R. Briddon, *Mat. Sci. Eng. B* **124–125** (2005) 363.

[93] M. Hiller, E.V. Lavrov and J. Weber, *Phys. Rev. B* **71** (2005) 045208.

[94] S.J. Pearton, J.M. Kahn, W.L. Hansen and E.E. Haller, *J. Appl. Phys.* **55** (1984) 1464.

[95] B.J. Coomer, P. Leary, M. Budde, B. Bech Nielsen, R. Jones, S. Oberg and P.R. Briddon, *Mat. Sci. Eng. B* **58** (1999) 36.

[96] S.M. Myers, H.J. Stein and D.M. Follstaedt, *Phys. Rev. B* **51** (1999) 9742.

[97] R. Jones, A. Carvalho, J. Coutinho, V.J.B. Torres, S. Öberg and P.R. Briddon, *Sol. St. Phen.* **108–109** (2005) 697.

[98] S.M. Sze, *Physics of Semiconductor Devices*, 2nd edition, Wiley, New York (1981).

[99] H.G. Grimmeiss, L. Montelius and K. Larsson, *Phys. Rev. B* **37** (1988) 6916.

[100] V.V. Ostroborodova, *Sov. Phys. – Sol. State* **7** (1965) 484.

[101] V.P. Markevich, A.R. Peaker, V.V. Litvinov, V.V. Emtsev and L.I. Murin, *J. Appl. Phys.* **95** (2004) 4078.

[102] M.K. Nissen, A.G. Steele and M.L.W. Thewalt, *Phys. Rev. B* **41** (1990) 7926.

[103] V.P. Markevich, A.R. Peaker, V.V. Litvinov, V.V. Emtsev and L.I. Murin, *J. Appl. Phys.* **95** (2004) 4078.

[104] V.P. Markevich, I.D Hawkins, A.R. Peaker, K.V. Emtsev, V.V. Emtsev, V.V. Litvinov, L.I. Murin and L. Dobaczewski, *Phys. Rev. B* **70** (2004) 235213.

CHAPTER 7

Radiation Performance of Ge Technologies

Vladimir Markevich, Anthony R Peaker and Arne Nylandsted Larsen

7.1. Introduction

This chapter discusses radiation and particle damage in germanium, the identification of the defects produced, their removal by annealing and the effects they have on material and device properties. We focus primarily on radiation that is capable of depositing sufficient energy in germanium crystals to effect a lattice displacement or displacements. In the simplest case this will constitute removing a germanium atom from its normal substitutional site into an interstitial position leaving behind a vacancy, in other words creating a Frenkel-pair. Both the vacancy and the interstitial are sufficiently mobile at room temperature to move in the lattice and react with other species so that additional defect reactions involving the formation of complexes also occur. Energetic electrons, gamma rays, neutrons and energetic ions of any atomic species are capable of initiating such displacements and reactions. The latter mechanism is especially important in relation to semiconductor technology as it is the basis for device fabrication using ion implantation.

The chapter is structured so that generic issues associated with radiation damage and implantation in semiconductors are considered initially, gradually focusing onto the specifics associated with germanium. In this first part (Section 7.2) the fundamental difference between radiation which usually creates isolated Frenkel-pairs (gamma and electron) and radiation events creating clusters (alpha, neutron and ion implantation) are contrasted.

In the second part (Section 7.3) defects and defect reactions reported for germanium are critically reviewed and catalogued. In the discussion part of this section comparisons are made with silicon and silicon germanium.

Germanium-Based Technologies: From Materials to Devices. C. Claeys and E. Simoen (Editors)

In the final section (Section 7.4) the impact of radiation damage on the properties of germanium materials and devices is considered. Here again we concentrate on the impact of damage which has produced atomic displacements but will also consider the direct effect of ionizing radiation on devices where the energy deposited produces an ionization event which results in a hole-electron pair rather than a displacement.

In this latter case we need to distinguish between transient ionization and longer term effects. In the case of transient ionization the hole and electron will migrate and if created in the active region of a device may cause an error or noise. In insulators such as the gate oxide in some metal-oxide-semiconductor (MOS) devices the electrons and holes produced are not free as in the semiconductor bulk and charged regions are induced which may induce a permanent change of characteristics.

A further case which needs to be mentioned is that where an ionization event, which in itself has insufficient energy to produce a displacement on its own, can contribute energy to the migration or formation of a defect species. One example of this is radiation enhanced defect motion which is extensively documented in wider band gap materials. In such materials the energy available from the electron-hole recombination is much greater than in germanium so as a general rule recombination enhanced defect motion is more important in wider gap materials and is not considered to be an important mechanism in germanium. However, in general, defect migration is dependent on the charge state of the defect and this may be affected by electron-hole generation and capture of a carrier. This process may be important in germanium and is discussed later.

Also in the final section contrasts are made between the behaviour of germanium devices when subject to irradiation and the much more extensive literature on silicon technologies. Predictions are made of the likely impact of radiation on germanium MOS devices. In this chapter the implications of residual implant damage are also considered in the context of their device significance.

7.2. Interaction of Radiation with Solids

7.2.1. Damage processes

For sufficiently high particle energies the energy transferred during an elastic collision can be large enough to displace an atom from its lattice site. The primary particle hits an atom in the lattice, transferring energy to effect the displacement. Multiple atom displacements can occur when the recoil atom has sufficient energy to displace other atoms. The energy transfer process has been studied in considerable detail as a general issue and a comprehensive understanding was achieved in the 1960s which has not changed conceptually since. Very clear early reviews of the basic processes and the specific case of germanium were published by Corbett [1] and by Brown and Augustyniak [2].

The probability for such atomic displacements is higher for higher energy and higher mass of the incident particle so that such multiple displacements are seen in high energy electron

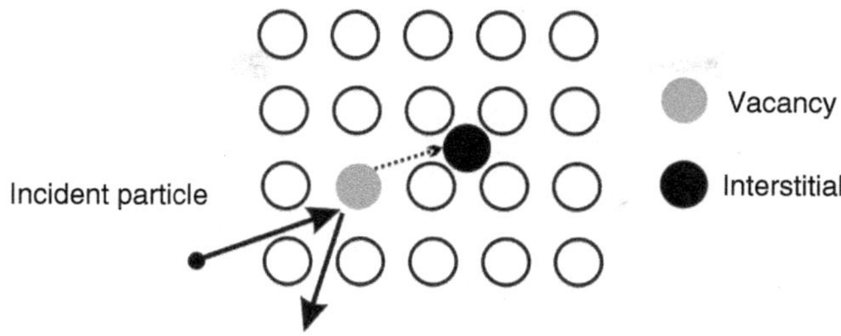

Figure 7.1. Schematic of a Frenkel-pair.

irradiation, in irradiation with high-energy neutrons and protons, and ion implantation. These cases are different primarily because of the different localization of damage as will be discussed in the next section. In general a highly localized generation of defects is more likely to produce a more complex defect structure. However, it is universally true that as a result of these displacements interstitials and vacancies appear, the vacancy-interstitial pair being referred to as a Frenkel defect (Figure 7.1). If one of the components is lost to the surface it is referred to as a Schottky defect.

As will be seen in more detail below, these primary radiation defects are highly mobile at room temperature and can move some distance through the lattice. They can either disappear from the material, by vacancy-interstitial recombination or by accumulation at the surface in the case of interstitials to form an extra plane or annihilation in the case of vacancies.

Alternatively, interstitials can accumulate in the bulk of the material to form extrinsic stacking faults, for example at the end of range in ion implantation or in the case of vacancies to produce voids. However, more commonly the migrating vacancies and interstitials become trapped by impurity atoms, giving rise to stable defects or to defect complexes.

It is evident that an important parameter for describing radiation damage is the threshold energy for displacement. This is the minimum kinetic energy transferred to an atom in the lattice from a particle necessary to displace permanently the atom from its lattice site, thus generating a stable defect, such as the Frenkel-pair referred to above. In the case of particles with a much higher kinetic energy than that necessary to transfer the minimum displacement energy, the initial displacement gives rise to a series of cascades which stop when the highest-energy particle has a kinetic energy smaller than the threshold. For ion damage in particular the mean free path between two successive collisions decreases towards the end of the range, so that the defects produced can interact. In consequence the threshold displacement energies are critical parameters for both low- and high-energy irradiation conditions.

The cascade process is illustrated in Figure 7.2 and is important in that it is the basis for damage clusters which tend to result in defect complexes such as is seen in ion implantation and neutron irradiation. The figure is derived from a Monte Carlo calculation of a 1 MeV arsenic implant into germanium using the software SRIM [3]. A threshold energy

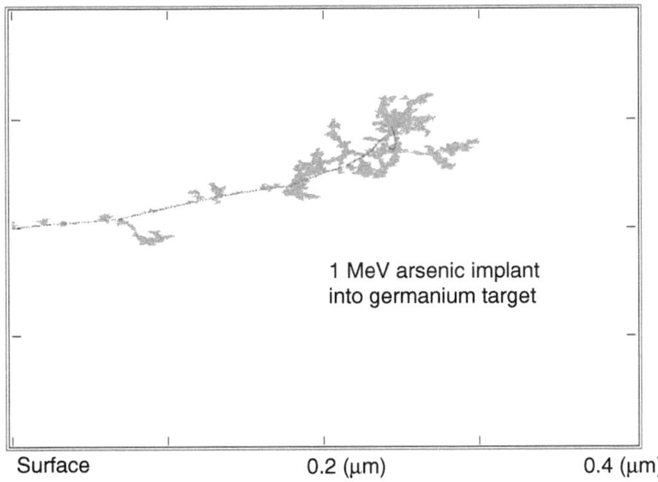

Figure 7.2. Monte Carlo simulation using SRIM [3] of damage clusters resulting from the implantation of a single arsenic ion. This shows a high concentration of displacements near the end of range of the incident and energetic displaced atoms.

for germanium displacement of 15 eV has been assumed. The energy has been transferred to the germanium from the implanted particle. The detailed physics of displacement energies and the energy transfer process is rather complex and is not dealt with here. Comprehensive treatments have been published elsewhere but it is important to state two well established generalizations. Firstly, the higher the energy the greater the probability of displacement and, secondly, that the displacement threshold energy is not isotropic so that the magnitude and character of displacements can depend on orientation.

The observation that damage is dependent on the total kinetic energy lost in the material is sometimes called "the NIEL hypothesis." NIEL stands for "Non Ionizing Energy Loss," and is usually expressed in units of $\mathrm{keV\,cm^{-2}\,g^{-1}}$ although other units are used so care must be taken in interpreting absolute values of this parameter. NIEL provides a convenient if somewhat simplistic way of comparing damage from different radiation sources. NIEL can be calculated for electrons, protons, neutrons, etc., using analytic expressions that sum the elastic and inelastic contributions.

In view of the fact that the NIEL concept does not consider the type of damage produced and ignores recombination of vacancies and interstitials and reactions with other defects it has proved surprisingly valuable to the device community in testing radiation hardness of devices. It is widely used to compare the damage in a practical testing regime (e.g. mono-energetic electrons) with the effects in a real radiation environment such as that experienced by satellites in near space where a spectrum of electron and proton irradiation is present. The basis for NIEL calculations is:

$$\mathrm{NIEL} = \left[\frac{N}{A}\right] \times [\sigma_{\mathrm{e}} T_{\mathrm{e}} + \sigma_{\mathrm{i}} T_{\mathrm{i}}] \tag{7.1}$$

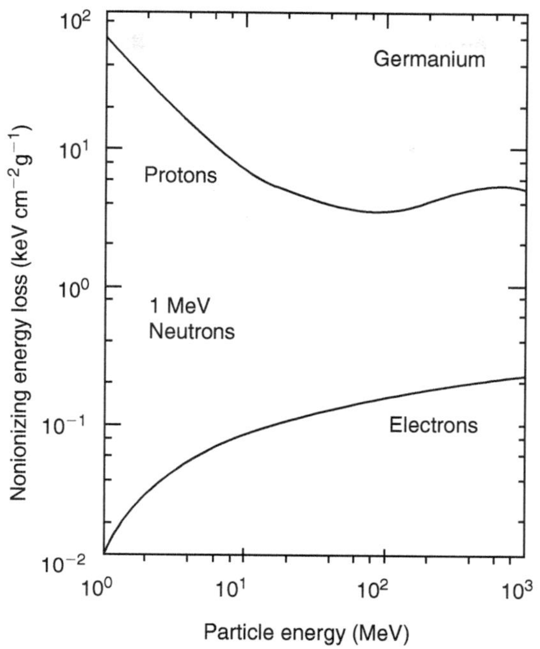

Figure 7.3. NIEL calculation for protons, neutron and electrons in germanium (after Marshall *et al.* [5]).

where σ_e and σ_i are the total elastic and inelastic cross sections, respectively, T_e and T_i are elastic and inelastic effective average recoil energies corrected for ionization loss, respectively, N is Avogadro's number, and A is the gram atomic weight of the target material [4]. The NIEL calculation has undergone considerable refinement since it was first proposed. The details of this are presented elsewhere [5] but the concept will be used in the discussions of damage later in this chapter.

7.2.2. Comparison of electron, gamma ray, neutron and proton damage

The NIEL concept is of use in comparing different types of radiation damage and is easy to use but it does not consider the defect physics behind such damage. It takes no account of the defect reactions occurring in semiconductors subsequent to damage events such as clustering and annealing, and as a result does not provide a comprehensive picture of radiation damage in germanium. In this section we reproduce earlier work on NIEL calculations and then consider the general issues differentiating different radiation damage resulting from different sources. Figure 7.3 results from NIEL calculations by Marshall *et al.* [4,5]. It can be related to Equation (7.1) by considering the difference in mass between electrons and protons and in consequence the much greater deposition of energy by protons. The neutron case is intermediate between the electron and proton because of the similar mass of the neutron and proton but modified by the reduced electronic interaction compared to the charged particles.

The complex shape of the energy loss curve for the proton is due to the different dependencies of elastic and inelastic scattering on energy. It is interesting to note that the energy loss for

20-MeV protons is the same as for 200-MeV protons but the relative contributions of elastic and inelastic scattering are very different. The shape of the proton energy loss is very different to the silicon case [6] due to the higher maximum displacement energy in germanium. These seemingly esoteric issues are of great importance in the development of radiation detectors and in recent years in relation to devices for the Large Hadron Collider at CERN.

Although the NIEL calculation has proved to be surprisingly valuable in silicon (and what evidence there is suggests that the same is true in silicon–germanium and germanium) in order to understand annealing effects and ion implantation it is necessary to look more deeply at the damage mechanisms. In general, electron damage produces well spaced displacements in which the majority of reactions only involve a single Frenkel-pair. As a result the point defects tend to consist of an intrinsic defect complexed with an impurity. In n-type germanium the most obvious manifestation is the E centre which is a donor-vacancy complex.

Although high-energy electrons produce displacements, high-energy photons, e.g. γ-rays cannot displace atoms directly. However, for photon energies greater than about 200 eV the energetic electrons produced by ionization events are backscattered and referred to as Compton electrons. These can interact with the lattice to produce displacements. The end result is that the damage resulting from γ and electron irradiations is essentially identical for comparable energy irradiations. A practical example of this is the γ rays from ^{60}Co which have energies of 1.33 and 1.17 MeV. As the photon never loses the whole of its energy in any one collision the energies of the Compton electrons have a spectrum which is always less than the γ ray energy. Most practical sources of γ rays produce a relatively weak flux of Compton electrons and so the rate of damage production tends to be low compared to electron irradiation or particle damage.

Fast neutrons, i.e. neutrons with energies greater than 0.5 MeV, produce displacements and, as the primary knock-on atom (PKA) energy increases, displacement cascades. This is a function of the target material as well as the neutron energy and so the cascade effect is expected to be more significant in germanium than in silicon.

Low-energy neutrons, e.g. neutrons in thermal equilibrium with the reactor core, have energies of the order of 0.5 eV. Such neutrons tend to undergo inelastic collisions with the semiconductor matrix atoms. In this case nuclear reactions occur, producing transmutation and secondary radiation. In addition the reaction products from the target atom may migrate through the lattice causing displacements. In the germanium case the reactions are importantly different for different naturally occurring isotopes. ^{74}Ge is the most abundant isotope ($\sim$37%) which reacts with the neutron to produce a donor while ^{70}Ge present naturally at $\sim$21% produces an acceptor.

Table 7.1 gives the natural abundance of the isotopes of germanium and the subsequent equations indicate the reactions that occur during irradiation with thermal neutrons which produce atoms other than alternative isotopes of germanium and hence introduce impurities into the germanium lattice.

Table 7.1. The isotopic abundance of natural Ge and reaction products after nuclear reactions with thermal neutrons.

Isotope	**Natural abundance**	**Reaction product**
^{70}Ge	20.6%	^{71}Ga
^{72}Ge	27.4%	^{73}Ge
^{73}Ge	7.6%	^{74}Ge
^{74}Ge	36.7%	^{75}As
^{76}Ge	7.7%	^{77}Se

The main nuclear reactions which result from irradiation with thermal neutrons are:

$$
\begin{aligned}
&{}^{76}_{32}\text{Ge} + n_{\text{th}} \rightarrow {}^{77}_{32}\text{Ge} \xrightarrow[\tau\,^{1}\!/_{2} = 11.3\,\text{h}]{\beta^-} {}^{77}_{33}\text{As} \xrightarrow[\tau\,^{1}\!/_{2} = 38.8\,\text{h}]{\beta^-} {}^{77}_{34}\text{Se} \\
&{}^{70}_{32}\text{Ge} + n_{\text{th}} \rightarrow {}^{71}_{32}\text{Ge} \xrightarrow[\tau\,^{1}\!/_{2} = 11.4\ \text{days}]{\text{EC}} {}^{71}_{31}\text{Ga} \\
&{}^{74}_{32}\text{Ge} + n_{\text{th}} \rightarrow {}^{75}_{32}\text{Ge} \xrightarrow[\tau\,^{1}\!/_{2} = 83\ \text{min}]{\beta^-} {}^{75}_{32}\text{As}
\end{aligned}
\tag{7.2}
$$

Protons carry charge and so the interaction with the semiconductor lattice is rather different to the neutron case. The most obvious manifestation is in the electronic interaction which is much greater than for the neutral particle and hence for similar energies the NIEL is much greater. Very high-energy protons will pass through a semiconductor device and so any permanent damage due to displacements results primarily from the electronic stopping; such displacement events are well spaced along the proton track. The calculation of stopping of ions in general was reviewed comprehensively by Ziegler *et al.* [7]. The issue of energetic light ions is rather complex but has been revisited by Ziegler relatively recently [8].

However, for lower-energy protons which come to rest in the semiconductor and possibly in the active region of the semiconductor device, the proton radiation produces cluster damage near the end of range.

The third differentiating feature of proton irradiation is that for low-energy protons the hydrogen ion remains in the semiconductor and is extremely reactive with the possibility of passivating dopants or, ironically, radiation defects. Figure 7.4 provides data on the penetration depth and longitudinal straggle of protons of various energies.

7.2.3. Ion-implantation damage

Although the effects of protons on semiconductors are traditionally classed as radiation damage it exemplifies some of the issues associated with ion implantation. Ion-implantation

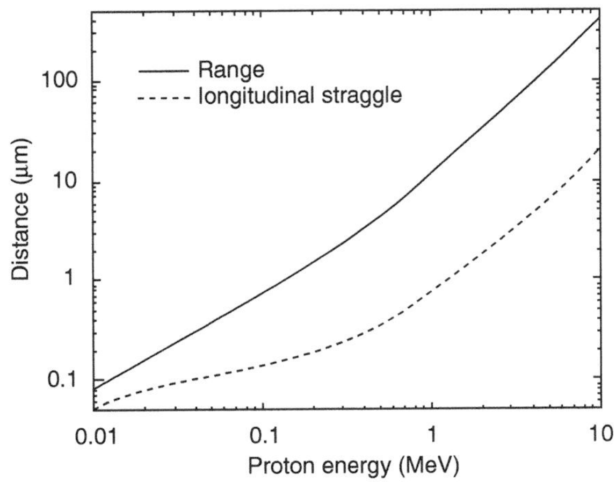

Figure 7.4. Range and longitudinal straggle of protons implanted into germanium calculated using SRIM [3].

damage differs semantically from radiation damage in the sense that it is a process technology for devices but, in reality, for most potential applications of germanium it is the most important form of damage, and has similarities to other forms of radiation damage. However, it is much more complex; an ion striking germanium (or any other semiconductor) experiences interaction with electrons surrounding the matrix atoms as electronic stopping and then as the ion slows down the predominant damage is due to elastic scattering which produces displacements in the germanium matrix. The complication is that the displaced germanium atoms have themselves sufficient kinetic energy to cause more displacements and such damage is characterized by cascades as was shown in Figure 7.2. Early literature on implantation referred to these clusters of damage as thermal spikes and associated the notional lattice temperature with their formation. A formidable amount of work has been done in silicon in trying to elucidate the nature of these regions. Even at relatively modest damage levels the local properties of the semiconductor are modified but neither theoretical models nor experimental techniques have succeeded in elucidating the detailed atomic structure of such regions. It is known that during annealing they form better defined defects consisting in the silicon case of, among other things, interstitial clusters and several unidentified defects that have point defect like properties. However, it is widely reported that such point defects although quite similar to those observed in electron damage do tend to have distinctly different characteristics both in terms of their electron trapping behaviour as measured by deep level transient spectroscopy (DLTS) and their annealing characteristics.

In the case of germanium much less data appears in the literature relating to these issues but the limited work that has been done seems indicative of the fact that phenomenologically there is similar behaviour. The relative paucity of literature on ion implantation in germanium results from the original germanium devices being made exclusively by diffusion and alloying to introduce impurities. Although a number of experimental systems for ion implantation were in operation during the sixties the technique only achieved practical application in the semiconductor field during the mid-seventies by which time germanium had been relegated to niche applications. The upsurge of interest in germanium as a mainstream semiconductor has stimulated much work recently into ion-implantation damage in germanium and some of this is reported elsewhere in this book. However, it is important to point out that there is no reason to believe that a rigorous understanding will be any simpler than for the silicon case, which after 20 years of intensive study is still not rigorously understood.

7.3. Primary Radiation-Induced Defects and their Interactions with Impurities in Crystalline Ge

7.3.1. Frenkel-pairs, the lattice vacancy, divacancy, and self-interstitial atom in Ge

In contrast to silicon, reliable information on the properties of intrinsic point defects in crystalline germanium is rather limited in spite of numerous studies devoted to obtaining results on the properties of vacancies and self-interstitials in Ge. The results of early studies were summarized in a monograph by Corbett [1]. Further developments in this field and the progress in understanding the properties of primary radiation-induced defects in Ge were reviewed later by Mashovets [9]. The latest progress and current understanding of the properties of intrinsic defects in Ge have been described recently in a monograph by Claeys and Simoen [10].

It is believed that both the vacancy and the self-interstitial are mobile at room temperature in Ge, so, to study the properties and stability of the primary displacement defects a number of particle irradiations were originally performed at low temperature, 4.2 K [11–15], 25 K [16] and 77 K [16–19]. A general and well established result from the early electrical studies is that independent of initial doping Ge becomes highly resistive and p-type after high irradiation fluencies.

It has been firmly established that the irradiation with 1–3-MeV electrons at 4.2 K results in the effective introduction of close Frenkel-pairs in Ge [9,15,20]. Capture cross section of their creation coincides with the value calculated under an assumption that the threshold energy of formation of a Frenkel-pair in Ge is 30 eV, and the introduction rate is in the range of 1 cm^{-1} for irradiation with 1-MeV electrons. It should be noted that only very small changes in electrical properties of p-type Ge are observed after electron irradiation at 4.2 K. On the basis of an analysis of changes in electrical properties of Ge crystals after the low-temperature irradiations it was argued that the close Frenkel-pairs possess acceptor properties in Ge, they most likely have two acceptor levels in the gap [9]. However, the exact positions of energy levels of the pairs are not known. The close Frenkel-pairs anneal out in the temperature range 20–65 K and there is a strong experimental evidence that the majority of the pairs annihilate [15]. The annealing temperature varies slightly with the energy of bombardment particles, being lower for the particles with lower energies. It is likely that these variations are related to the difference in spatial distribution between the components of the Frenkel-pairs after irradiations with particles of different energies. There is no consensus in relation to which component of the Frenkel-pair becomes mobile in the temperature range 20–65 K and is responsible for the annihilation of the pair.

There are a lot of controversial results regarding the positions of energy levels and activation energies for the migration of the isolated vacancy and self-interstitial in Ge. Some important results regarding charge states of the vacancy and self-interstitial were obtained from early studies of the introduction rate of stable secondary radiation-induced defects versus concentration of group-V impurities. It was found that the cross section of the formation of the stable defects became five to six time higher when the concentration of group-V impurities increased by three or four orders of magnitude [21]. The observed dependence was explained by Mashovets [9] in terms of a charge-state-dependent probability of the vacancy-interstitial separation. It was argued that the self-interstitials possess a donor level in the upper half of the band gap and the separation probability of the Frenkel-pair could decrease if the charge state of the interstitial changes from 0 to +1 upon movement of the Fermi level away from the conduction band, the vacancy being negatively charged. This picture has been proved recently in perturbed angular correlation spectroscopy (PACS) studies [19,22,23]. The results of these studies are described briefly in the Chapter 3 (Section 3.3.2). It has been established with a high degree of certainty that there is an acceptor level of the single vacancy in Ge at about $E_V + 0.20\ (\pm 0.04)$ eV.

Ab-initio modelling studies have shown that the structure and electronic properties of the vacancy in Ge are similar to those of the vacancy in Si: the vacancy in Ge having five charge states: (2+), (+), (0), (−), and (2−) [24,25]. However, these results have been questioned in a recent work [26], where it has been argued that the donor levels of the vacancy in Ge are resonant states in the valence band but there are three vacancy-related acceptor levels in the gap. The results of this study are described in detail in Chapter 6. It should be noted that there is a consensus in the modelling studies that the vacancy formation energies in germanium (from 1.7 to about 2.5 eV) are significantly smaller than in silicon, for all charge states.

Recent *ab-initio* modelling studies have not indicated the existence of a donor level of the Ge self-interstitial in the upper half of the gap [27,28]. According to these studies there is a single acceptor level of the self-interstitial in the lower half of the gap and for both the neutral and singly negatively charged states the $\langle 110 \rangle$-split dumbbell configuration is found to be the energetically favoured one [28]. The formation energy for the neutral Ge interstitial (3.55 eV) was found to be comparable with the corresponding value for Si (3.5 eV) according to the calculations [28].

It has been proved experimentally that the diffusivity of the vacancy in Si depends on its charge states [29]. One can expect the same behaviour of the vacancy in Ge. Indeed, some experimental indications of different mobility of the Ge vacancy at different locations of the Fermi level in Ge crystals have been obtained [9,14,20]. Furthermore, values of the activation energy for the diffusivity of the Ge vacancy in different charge states were given [9,19,20], e.g. a value of 0.42 eV was found for the singly negatively charged state of the vacancy (n-type Ge) and for p-type or intrinsic Ge a migration energy of 0.52 eV was suggested for V^0 [20]. These values are higher than in silicon [29] and suggest annealing stages at temperatures much higher than 100 K, i.e. in the range of 200 K. However, it was found that in oxygen-rich Ge samples, which were irradiated at 25 K, vacancy-oxygen (VO) related local vibration mode lines emerged upon annealing in the temperature range 45–80 K [16].

The above observation was considered as evidence for vacancy diffusion in Ge at low (~50 K) temperatures.

Divacancies (V_2) can be created in semiconductors when the energy of irradiation species is high enough or by the association of two single vacancies. The vacancy in Ge is negatively charged over a broad range of Fermi level position in the gap, therefore, the latter mechanism of V_2 formation in Ge is suppressed because of Coulombic repulsion of vacancies. The electronic properties of V_2 in Ge are not well understood and there are large discrepancies in the positions of electronic levels which have been assigned to the divacancy, a short review of the results obtained has been given in Ref. [30]. It is very likely that an acceptor level with the activation energy of 0.29 eV for the electron emission is related to V_2 in Ge. This level anneals out in the temperature range 150–200°C [30].

Some annealing stages occurring in irradiated Ge crystals at temperatures lower than 200 K were linked to the mobility of self-interstitials [14,15,19], however, the proposed arguments for such assignments are not very solid.

7.3.2. Interaction of the intrinsic points defects with impurities in Ge

As in the case of silicon, there are some strong indications that the vacancy in Ge can interact with oxygen and group-V donor atoms (P, As, Sb, Bi) with the formation of electrically active complexes stable at room temperature.

It has been argued recently [30–33] that the vacancy-group V impurity atom pair or E centre is the dominant vacancy-related defect produced in donor-doped oxygen-lean Ge crystals by electron or gamma irradiation at room temperature. In oxygen-rich material the interstitial oxygen competes for vacancies, producing the vacancy-oxygen complex or A centre with a consequent reduction in the generation rate of the E centres. As it was mentioned above there are strong indications that in n-type or intrinsic Ge at $T \sim 300$ K the single vacancy is negatively charged. The positively charged donor provides a much more attractive capture site for the negatively charged vacancy than neutral interstitial oxygen by almost an order of magnitude. However, the overall branching ratio depends also on the relative concentration of O_i and the group-V ions.

Figures 7.5 and 7.6 show DLTS spectra of gamma-irradiation induced majority and minority carrier traps in n-type Ge crystals doped with different donor species. The dominant DLTS peaks in the spectra were argued to be related to the E centres. It was inferred from the analysis of the data obtained that the E centres in Ge have at least three charge states: double negative, single negative and neutral, and introduce two energy levels into the gap [30–32]. The first acceptor level is in the lower half of the gap and the second acceptor level is in the upper half of the gap. It has been shown that the directly measured capture cross sections of electrons at the singly negatively charged E centres are temperature dependent and can be described by the multi-phonon-assisted capture model [34]. From the combined analysis of the carrier emission and capture data the values of enthalpy and entropy of ionization for the second acceptor level of the E centres have been determined. The carrier

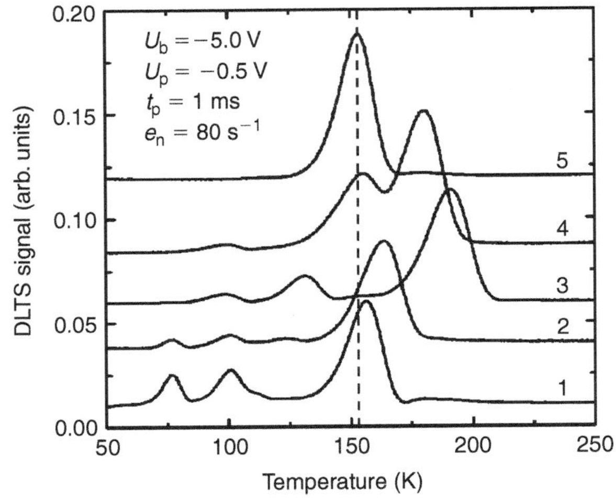

Figure 7.5. DLTS spectra for gamma-irradiated oxygen-lean Ge crystals doped with (1) P, (2) As, (3) Sb, and (4) Bi. Spectrum 5 was recorded on a γ-irradiated oxygen-rich Ge sample. The spectra have been vertically displaced for clarity. Measurement settings were $e_n = 80\,s^{-1}$, bias $-5 \rightarrow -0.5$ V, and pulse duration 1 ms (after Markevich *et al.* [32]).

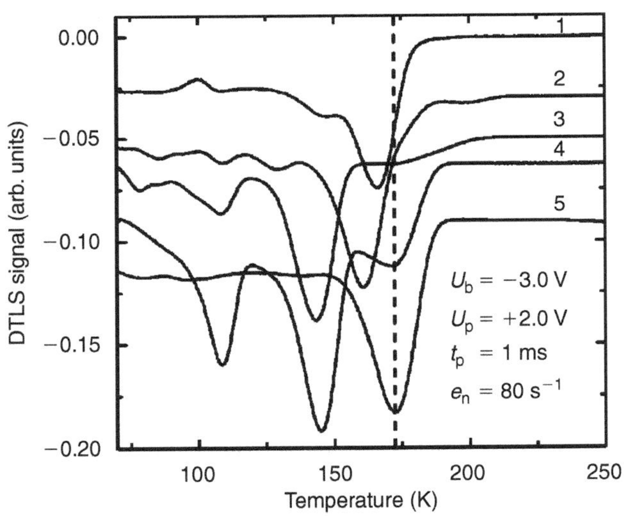

Figure 7.6. "Injection" DLTS spectra for γ-irradiated oxygen-lean Ge crystals doped with (1) P, (2) As, (3) Sb, and (4) Bi. Spectrum 5 was recorded on a γ-irradiated oxygen-rich Ge sample. The spectra have been vertically displaced for clarity. Measurement settings were $e_n = 80\,s^{-1}$, bias $-3 \rightarrow +2.0$ V, and pulse duration 1 ms (after Markevich *et al.* [32]).

Table 7.2. Electronic parameters for the dominant radiation-induced electron traps in Ge crystals doped with different donor impurities. The values of the activation energy for electron emission (ΔE_{ne}), pre-exponential factor (A_{ne}), and apparent capture cross section (σ_{na}) have been derived from Arrhenius plots of T^2-corrected emission rates determined from Laplace DLTS measurements. The values of the activation energy for electron capture ($\Delta E_{\sigma n}$) and capture cross section extrapolated to $1/T = 0$ (σ_{n0}) have been determined from direct capture measurements using LDLTS technique. The values of enthalpy (ΔH), entropy (ΔS), and free energy of ionization (ΔG) have been derived from the combined analysis of the emission and capture data. The cited values of ΔG are at 220 K (after Markevich *et al.* [32]).

Material	ΔE_{ne} eV	A_{ne} $s^{-1}K^{-2}$	σ_{na} cm^2	$\Delta E_{\sigma n}$ eV	σ_{n0} cm^2	ΔH eV	ΔS k	ΔG eV
Ge:P	0.293	9.6×10^6	2.7×10^{-15}	0.07	3.8×10^{-16}	0.225	1.95	0.19
Ge:As	0.310	1.1×10^7	2.9×10^{-15}	0.065	1.6×10^{-16}	0.245	2.9	0.19
Ge:Sb	0.377	2.1×10^7	5.9×10^{-15}	0.083	9.2×10^{-17}	0.295	4.2	0.215
Ge:Bi	0.349	1.5×10^7	4.2×10^{-15}	0.085	1.6×10^{-16}	0.265	3.3	0.20

Table 7.3. Electronic parameters for the dominant radiation-induced hole traps in Ge crystals doped with different donor impurities. The values of the activation energy for hole emission (ΔE_{pe}), pre-exponential factor (A_{pe}), and apparent capture cross section (σ_{pa}) have been derived from Arrhenius plots of T^2-corrected emission rates determined from Laplace DLTS measurements (after Markevich *et al.* [32]).

Material	ΔE_{pe} eV	A_{pe} $s^{-1}K^{-2}$	σ_{pa} cm^2
Ge:P	0.348	1.1×10^8	9.2×10^{-14}
Ge:As	0.334	1.05×10^8	8.75×10^{-14}
Ge:Sb	0.307	2.2×10^8	1.8×10^{-13}
Ge:Bi	0.305	1.5×10^8	1.25×10^{-13}

emission and capture parameters for both acceptor levels of the E centres are summarized in Tables 7.2 and 7.3. The emission of an electron from the doubly negatively charged state of the centres is accompanied by rather large changes in entropy (ΔS), so, the free energy of the electron ionization, $\Delta G(2-/-) = \Delta H(2-/-) - T\Delta S(2-/-)$, changes significantly with temperature. Consequently, the position of the second acceptor level of the centres $\{E_C - E(2-/-) = \Delta G(2-/-)\}$ is temperature dependent. This is an indication of a phonon-assisted ionization process.

Very recently, Lindberg *et al.* [33] have succeeded in making good n^+p-mesa diodes from $2\,\Omega$ cm p-type Ge to study irradiation-induced levels in the lower half of the band gap. Because of the way the n^+-top layer was produced (molecular-beam epitaxially grown layer doped with Sb) some Sb had diffused into the p-type layer rendering it practicable to study Sb-related defects in the lower half of the band gap. This investigation not only confirmed the existence of the single acceptor state with an enthalpy of ionization of $\Delta H_p = 0.309$ eV but also revealed a new Sb-related line with an apparent enthalpy of ionization of

$\Delta H_p = 0.095$ eV. This new line had the same intensity as the line of the single acceptor and annealed at the same temperature. Based on these observations it was concluded that this shallow level is the single donor-replica of the Sb-V complex. Thus, the Sb-V complex in Ge has four charge states. It can be expected that the E centres incorporating P, As and Bi atoms also possess a single donor level close to the valence band edge but exact information on the positions of this level is not available at the moment.

An optical absorption line at 620 cm^{-1} in irradiated oxygen-rich Ge was originally assigned to the vacancy-oxygen centre by Whan [16]. Recently, by using different oxygen isotopes and Ge samples with different free carrier concentrations, it was shown that absorption lines at 621.4, 669.1 and 716.2 cm^{-1} correspond to stretching vibrations of VO^0, VO^- and VO^{2-} respectively [35–38]. It has been found in DLTS studies that the annealing rates of a hole trap and an electron trap with activation energies for carrier emission of 0.33 and 0.28 eV, respectively, are identical with those of the absorption lines. Together with the correlation of the magnitude of these traps with the oxygen content in Ge samples these findings provide a very reliable identification of the two deep states as related to the first and second acceptor levels of the VO center [35,36]. Thus, the A centre in Ge has three different charge states.

Figure 7.7 shows the available experimental data on electronic properties of the dominant defects in n-type Ge crystals induced by irradiation with gamma-rays and electrons ($E < 5$ MeV). The data on annealing temperature of the traps upon 30 min isochronal are also presented. There is no reliable information available on interactions of the Ge interstitial with impurities in crystalline germanium.

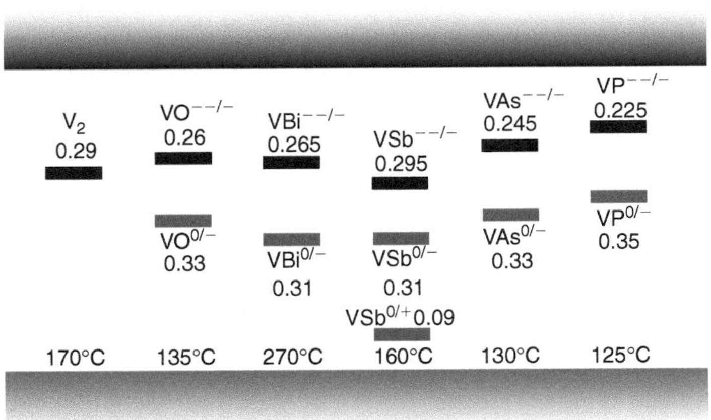

Figure 7.7. Representation of irradiation-induced defects in germanium. The figures adjacent to the levels indicate the activation energies for hole emission and enthalpies for electron emission from the state. In the case of the double acceptor states of the E centres, it is known that the defects possess a large entropy factor, so that the values given are significantly different from the true energy level of the defect in the gap. The temperatures given at the bottom of the diagram are derived from isochronal annealing studies and represent the temperature at which the concentration of the defect is reduced to half its original value after a 30 min anneal with the defect in its negative charge state.

7.3.3. *Ion-implantation-induced damage: multi-vacancy and multi-self-interstitial complexes in Ge*

It is evident that in silicon there are profound differences in the defect populations produced by electron irradiation compared to ion implantation [39]. The principal reason for this is the high damage rate which near the end of range of the implanted particle is likely to result in clustering of self-interstitials and nearer the surface in clusters of vacancies. In the silicon case reverse modelling of annealing and *ab-initio* calculations [40–42] predict stable clusters of small numbers of vacancies and/or interstitials. These anneal out at temperatures around those at which dopant diffusion is important and hence play an important role in mediating the diffusion. In germanium it is likely that similar effects occur but as yet there is no strong evidence. However, it is important to consider that defect diffusion in germanium is much more dependent on vacancy mediation than in silicon. In consequence it is vacancy clusters rather than interstitial clusters that are much more likely to be of technological significance.

The LERMAT group have undertaken quite detailed studies of ion implantation into germanium [43–45] and have catalogued defect introduction rates for high-energy heavy ions in Ge:Sb [46]. Lower energy Si implantation which is more likely to be representative of the kinetics of dopant implantation damage related to device fabrication has been studied in Refs [47,48].

Figure 7.8 shows the damage resulting from 3 MeV silicon implantation into n-type oxygen-lean $Ge_{0.9992}Si_{0.0008}$ compared with the spectrum resulting from gamma-rays irradiation of

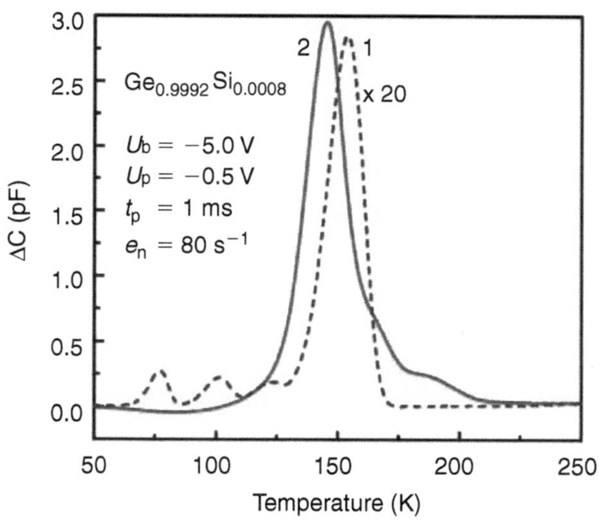

Figure 7.8. DLTS spectra for oxygen-lean $Ge_{0.9992}Si_{0.0008}$:P samples after (1) irradiation with gamma-rays from a ^{60}Co source ($\Phi = 2.0 \times 10^{17} cm^{-2}$) and (2) implantation with Si ions ($\Phi = 1.0 \times 10^{10} cm^{-2}$). Measurement settings were $e_n = 80 s^{-1}$, bias $-5 \rightarrow -0.5 V$, and pulse duration 1 ms. Spectrum 1 is multiplied by 20 (after Peaker *et al.* [47]).

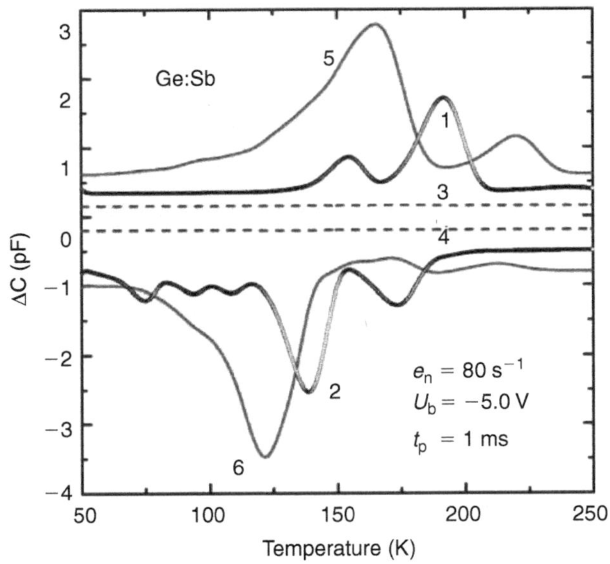

Figure 7.9. DLTS spectra for oxygen-lean Ge:Sb crystals after (1–4) irradiation with 3-MeV electrons ($\Phi = 1.5 \times 10^{15}\,\text{cm}^{-2}$) and (5,6) implantation with Si ions ($\Phi = 1.0 \times 10^{12}\,\text{cm}^{-2}$). The spectra 3–6 were recorded after annealing the samples at 400°C for 30 min. For all the spectra the emission rate window was $e_n = 80\,\text{s}^{-1}$ and pulse duration was 1 ms. Bias sequences were $-4.0 \rightarrow 0$ V for the spectra 1, 3 and 5 and were $-4.0 \rightarrow +2.0$ V for the spectra 2, 4 and 6. The spectra have been vertically displaced for clarity (after Peaker *et al.* [47]).

the same material. No annealing has taken place in either case. It can be seen that the spectra are distinctly different. The concentration of defects is substantially higher in the ion-implanted sample, and the E centre is not the dominant defect after ion implantation.

It is valuable to compare the germanium data with that of self-implanted silicon, for identical doses the total defect concentration in n-type germanium is much higher, in some cases by almost an order of magnitude, prior to annealing.

In Figure 7.9 the effects of annealing are shown. Here the DLTS spectra for Ge:Sb after 1 MeV Si implantation with a dose of $10^{12}\,\text{cm}^{-2}$ and following annealing at 400°C (5,6) are compared with those for an electron irradiated sample (3 MeV) after an identical anneal. Spectra of the electron irradiated sample prior to annealing are also shown. As expected from the results described in the previous paragraph the annealed electron irradiated Ge shows no defects whatsoever while the Si ion implanted material contains a high concentration of defects. These would result in high excess generation currents in Ge devices. It is apparent that very different species of defects are produced by the ion implantation compared to the electron irradiation. It has been found that a substantial concentration of electrically active defects remain even after the 500°C 30 min anneals in Ge crystals, which were implanted with relatively high doses ($\geq 10^{12}\,\text{cm}^{-2}$) of Si ions. There is as yet no clear evidence as to the identity of these defects but it is likely that multi-vacancy or multi-interstitial

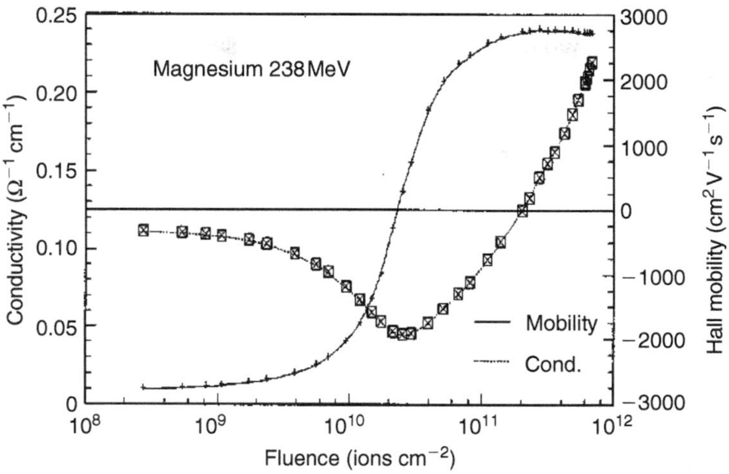

Figure 7.10. Conductivity (square symbols) and Hall mobility as a function of ion fluence for n-type germanium. The initial carrier concentration was $2.4 \times 10^{14}\,cm^{-3}$. Irradiation and measurement were at room temperature (after Marie *et al.* [46]).

complexes are involved which as for the silicon case seem to be stable up to relatively high temperatures.

7.4. Effects on Devices

The most obvious effect on the macroscopic properties of germanium as a result of radiation displacement damage is the change in free carrier concentration. In n-type material this is manifested as carrier loss and is predominantly due to the formation of the vacancy donor complex or E centre as discussed above. The formation of an E centre by incident radiation converts the donor via its reaction with the vacancy into an acceptor, or in cases where the material is significantly below room temperature a double acceptor. As a result generation of a single vacancy removes two carriers at normal temperatures or three carriers in a cooled radiation detector or spacecraft. Continued exposure to radiation results in type conversion and the formation of p-type material, this is illustrated in Figure 7.10 which shows the carrier loss in n-type germanium and the classic behaviour of hole mobility in the presence of two carrier types indicative of type conversion at high fluence. Although these results are for energetic ions (238 MeV) the behaviour in relation to conductivity change and type conversion is quite typical of n-type germanium under irradiation although as discussed earlier the magnitude of the effects differ with different energies and ion mass. In the case of p-type material a small increase in the hole population is seen.

The effects of these changes in conductivity are device specific. Perhaps the most sensitive to radiation damage are Ge:Ga photo-detectors which are used in scientific satellites to detect long wavelength infrared radiation. Less than 4 rads of gamma rays increase the conductivity changing the responsivity of the detector by decades [49].

These devices are particularly susceptible as they operate in space at very low temperatures (about 4 K). Two other types of devices for which there is now considerable experience of radiation damage are the tandem solar cell, currently finding favour in satellites because of its extremely high efficiency due to utilization of a large part of the solar spectrum [50,51]. These devices currently have commercial efficiencies of 28% and efficiencies of almost 40% are possible in principle.

Protons are a particular potent source of damage to solar cells in near earth orbit. In particular the earth's magnetic field concentrates energetic particles at two altitudes known as the inner and outer van Allen belts so making the radiation environment very variable with altitude. The inner radiation belt extends over altitudes of 1000–6000 km. The radiation density is highest in the Earth's equatorial plane and consists mostly of protons with energies of the order of 10–50 MeV. However, the inner belt also contains electrons, low-energy protons, and oxygen ions with energies of 1–100 keV. The lower energy protons are very damaging to solar cells which need to be protected by high density glass covers. This has relevance to germanium technologies because of the use of germanium substrates for gallium arsenide solar cells, but more significantly for high efficiency tandem cell structures referred to earlier. These use germanium as one of the active layers to capture more of the infrared element of the sun's radiation.

The outer van Allen belt ranges from 10.000 to 65.000 km radius. Here the proton density is much less and the dominant radiation originates from electrons with energies less than 10 MeV. This is of particular significance for geo-stationary satellites which have an orbit radius of 42,000 km.

The germanium part of tandem solar cells is relatively radiation hard. Firstly it is the lower layer and so receives less proton damage than the other constituents of the cell but most importantly the germanium is usually quite heavily doped p-type and shows little conductivity change on irradiation in the space environment. Studies of the minority carrier lifetime and radiation damage coefficients of germanium [52] show that the diffusion length damage coefficients of germanium are considerably less than those of other common solar cell materials.

As mentioned previously, the other germanium device for which there is a great deal of data on radiation damage is not surprisingly radiation detectors. Here the starting material is of extremely high resistivity and the effects are associated with carrier lifetimes in the material and to a lesser extent the active volume if additional acceptors are created by the radiation process.

A number of early studies were undertaken on the radiation resistance of bipolar transistors [e.g. 53]. The issues here apart from carrier removal are the creation by displacement events of generation recombination centres. Historically germanium transistors were made with a very wide base region and so radiation damage affected the base transport factor. In reality such devices are no longer relevant to the mainstream electronics industry and so will not be considered here. Present day bipolar transistors in silicon or in germanium or indeed in other materials have very narrow base regions and so the effect of recombination centres on the base transport factor is something of an irrelevancy with the exception of power transistors. The fundamental issues associated with the creation of generation recombination centres and their effect on device parameters are discussed in detail in Chapter 8 of this book.

If we move on to predictions of the possible effect of radiation damage on germanium integrated circuits then the situation is unlikely to be very different to the silicon case. Some work has been done on radiation hardness of SiGe hetero-junction bipolar transistors [54]. The SiGe devices were very radiation hard compared to silicon devices with similar functionality. However, the radiation hardness was not attributed to the presence of germanium but to the differences in transistor structure and in particular the heavier doping of the SiGe device. This underlines the considerable difficulties in conducting systematic comparisons of devices which are based on different materials and require very different structures for optimal design.

In the MOS case germanium is only ever likely to be used beyond the 45 nm node in conjunction with high-κ dielectrics. There is a vast amount of literature in relation to conventional silicon complementary-metal-oxide-semiconductor (CMOS) both in terms of displacement damage and ionization caused by radiation in relation to single event upsets. As a generality reference should be made to the relevant reviews on silicon [e.g. 3,10,55,56] until such time as hard data becomes available for the germanium case.

7.5. Conclusions

Well-defined defect states are produced in germanium by gamma or electron irradiation. In all cases this damage can be removed by low temperature (<400°C) anneals leaving no detectable electronic states in the gap. In most cases the physical and chemical identity of the defects, including the charge state, is known. In ion-implanted material defects are produced which are distinctly different to the electron irradiation case and which are much more difficult to remove by annealing. The states remaining after 400–500°C anneals are powerful generation centres but at the present time their physical and chemical identities are unknown.

References

[1] J.W. Corbett, Electron Radiation Damage in Semiconductors and Metals, Vol. **7** of Solid State Physics, *Advances in Research and Applications*, Academic, New York, 1966.

[2] W.L. Brown and W.M. Augustyniak, *J. Appl. Phys.* **30** (1959) 1300.

[3] J.R. Srour, C.J. Marshall and P.W. Marshall, *IEEE Trans. Nucl. Sci.* **50** (2003) 653.

[4] P.W. Marshall, C.J. Dale, G.P. Summers, E.A. Wolicki and E.A. Burke, *IEEE Trans. Nucl. Sci.* **36** (1989) 1882.

[5] C.J. Dale, P.W. Marshall, E.A. Burke, G.P. Summers and E.A. Wolicki, *IEEE Trans. Nucl. Sci.* **35** (1988) 1208.

[6] The Monte Carlo simulation software SRIM used for calculating ion ranges is available from the website: www.srim.org

[7] J.F. Ziegler, J.P. Biersack and U. Littmark, The Stopping and Range of Ions in Solids, Pergamon Press, New York, 1985.

[8] J.F. Ziegler, *J. Appl. Phys/Rev. Appl. Phys.* **85** (1999) 1249.

[9] T.V. Mashovets, In *Radiation Effects in Semiconductors 1976*, edited by N.B. Urli and J.W. Corbett, Inst. Phys. Conf Ser. No. 31, Bristol and London, 1977, 30.

[10] C. Claeys and E. Simoen, Radiation effects in advanced semiconductor materials and devices, *Springer Series in Materials Science*, Springer, Berlin, 2002.

[11] G.W. Gobeli, *Phys. Rev.* **112** (1958) 732.

[12] J.W. MacKay and E.E. Klontz, *J. Appl. Phys.* **30** (1959) 1269.

[13] T.A. Callcot and J.W. MacKay, *Phys. Rev.* **161** (1967) 698.

[14] J. Bourgoin and F. Mollot, *Phys. Stat. Solidi* (b) **43** (1971) 343.

[15] P. Ehrhart and H. Zillgen, *J. Appl. Phys.* **85** (1999) 3503.

[16] R.E. Whan, *Phys. Rev.* **140** (1965) A690.

[17] N. Fukuoka and H. Saito, *Jpn. J. Appl. Phys.* **13** (1974) 1524.

[18] N. Fukuoka and H. Saito, *Jpn. J. Appl. Phys.* **15** (1976) 237.

[19] H. Haesslein, R. Sielemann and C. Zistl, *Phys. Rev. Lett.* **80** (1998) 2626.

[20] V.V. Emtsev, T.V. Mashovets and V.V. Michnovich, *Sov. Phys. Semicond.* **26** (1992) 12.

[21] V.V. Emtsev, T.V. Mashovets and S.M. Ryvkin, in *Radiation Damage in Semiconductors*, Inst. Phys. Conf. Ser. No. **16**, London and Bristol, 1973, p. 17.

[22] R. Sielemann, H. Haesslein, L. Wende and Ch. Zistl, *Physica B* **273–274** (1999) 565.

[23] R. Sielemann, H. Haesslein, M. Müller, L. Stadler and V.V. Emtsev, *Physica B* **308–310** (2001) 529.

[24] A. Fazzio, A. Janotti, A.J.R. da Silva and R. Mota, *Phys. Rev. B* **61** (2000) R2401.

[25] S. Ögüt and R. Chelikowsky, *Phys. Rev. B* **64** (2001) 245206.

[26] J. Coutinho, R. Jones, V.J.B. Torres, M. Barroso, S. Öberg and P.R. Briddon, *J. Phys.: Condens. Matter.* **17** (2005) L521.

[27] A.J.R. da Silva, A. Janotti, A. Fazzio, R.J. Baierle and R. Mota, *Phys. Rev. B* **62** (2000) 9903.

[28] M.D. Moreira, R.H. Miwa and P. Venezuela, Phys. *Rev. B* **70** (2004) 115215.

[29] G.D. Watkins, in *Properties of Crystalline Silicon*, edited by R. Hull, (EMIS Data Review Series No. 20, INSPEC, London, 1999), Chapter 11.1, p. 643.

[30] J. Fage-Pedersen, A. Mesli and A. Nylandsted Larsen, *Phys. Rev. B* **62** (2000) 10116.

[31] V.P. Markevich, A.R. Peaker, V.V. Litvinov, V.V. Emtsev and L.I. Murin, *J. Appl. Phys.* **95** (2004) 4078.

[32] V.P. Markevich, I.D. Hawkins, A.R. Peaker, K.V. Emtsev, V.V. Emtsev, V.V. Litvinov, L.I. Murin and L. Dobaczewski, *Phys. Rev. B* **70** (2004) 235213.

[33] C.E. Lindberg, J. Lundsgaard Hansen, P. Bomholt, A. Mesli, K. Bonde Nielsen, A. Nylandsted Larsen and L. Dobaczewski, *Appl. Phys. Lett.* **87** (2005) 172103.

[34] C.H. Henry and D.V. Lang, *Phys. Rev. B* **15** (1977) 989.

[35] V.P. Markevich, I.D. Hawkins, A.R. Peaker, V.V. Litvinov, L.I. Murin, L. Dobaczewski and J.L. Lindström, *Appl. Phys. Lett.* **81** (2002) 1821.

[36] V.P. Markevich, V.V. Litvinov, L. Dobaczewski, J.L. Lindström, L.I. Murin and A.R. Peaker, *Phys. Status Solidi (c)* **0** (2003) 702.

[37] P. Vanmeerbeek and P. Clauws, *Phys. Rev. B* **64** (2001) 245201.

[38] P. Vanmeerbeek, P. Clauws, H. Vrielinck, B. Pajot, L. Van Hoorebeke and A. Nylandsted Larsen, *Phys. Rev. B* **70** (2004) 035203.

[39] B.G. Svensson, B. Mohadjeri, A. Hallen, J.H. Svensson and J.W. Corbett, *Phys. Rev. B* **43** (1991) 2292.

[40] N.E.B. Cowern, G. Mannino, P.A. Stolk, F. Roozeboom, H.G.A. Huizing, J.G.M. van Berkum, F. Cristiano, A. Claverie and M. Jaraiz, *Phys. Rev. Lett.* **82** (1999) 4460.

[41] T.E.M. Staab, A. Sieck, M. Haugk, M.J. Puska, Th. Fraunheim and H.S. Leipner, *Phys. Rev. B* **65** (2002) 115210.

[42] J.L. Hastings, S.K. Estreicher and P.A. Fedders, *Phys. Rev. B* **56** (1997) 10215.

[43] P. Marie, M. Levalois and P. Bogdanski, *J. Appl. Phys.* **74** (1993) 868.

[44] P. Marie and M. Levalois, *J. Appl. Phys.* **75** (1994) 1852.

[45] A. Colder, M. Levalois and P. Marie, *J. Appl. Phys.* **88** (2000) 3082.

[46] P. Marie, M. Levalois and E. Paumier, *J. Appl. Phys.* **79** (1996) 7555.

[47] A.R. Peaker, V.P. Markevich, L.I. Murin, N.V. Abrosimov and V.V. Litvinov, *Mater. Sci. Eng. B* **124–125** (2005) 166.

[48] V.P. Markevich, A.R. Peaker, L.I. Murin, V.V. Emtsev, V.V. Litvinov, N.V. Abrosimov and L. Dobaczewski, *Solid State Phen.* **108–109** (2005) 253.

[49] M. Patrashin, B. Fouks, U. Grözinger, D. Lemke and J. Wolf, *J. Appl. Phys.* **82** (1997) 1450.

[50] D.C. Senft, *Proc. Thirty-First IEEE Photovoltaic Specialist Conference* (IEEE Cat. No. 05CH37608), 2005, p. 536.

[51] C. Baur, M. Meusel, F. Dimroth, A.W. Belt, M. Nell, G. Strobl, S. Taylor and C. Signorini, *Proc. Thirty-First IEEE Photovoltaic Specialist Conference* (IEEE Cat. No. 05CH37608) 2005, p. 548.

[52] H. Yoon, K.M. Edmondson, G.S. Kinsey, R.R. King, P. Heberter, R.K. Ahrenkiel, B.T. Cavicchi and N.H. Karam, *Proc. Thirty-First IEEE Photovoltaic Specialists Conference* 2005, p. 842.

[53] L. Taylor, *I.R.E. Trans. Nucl. Sci.* **9** (1962) 280.

[54] J.D. Cressler, *Proc. IEEE* **93** (2005) 1559.

[55] A.H. Johnston, *IEEE Trans. Nucl. Sci.* **45** (1998) 1339.

[56] S.C. Witczak, R.C. Lacoe, J.V. Osborn and J.M. Hutson, *IEEE Trans. Nucl. Sci.* **52** (2005) 2602.

CHAPTER 8

Electrical Performance of Ge Devices

Michel Houssa, Alessandra Satta, Eddy Simoen, Brice De Jaeger, Marc Meuris, Matty Caymax and Marc Heyns

8.1. Introduction

As the scaling of advanced metal-oxide-semiconductor (MOS) devices is approaching its technological and fundamental limits, new materials are needed to further improve the performances of integrated circuits (IC). As an example, tremendous research efforts have focused recently on the replacement of ultra-thin SiO_2 or SiON gate insulator by high-κ gate dielectrics [1,2]. The use of physically thicker high-κ materials, with equivalent capacitance lower than one nanometer thick SiON layers, results in a large decrease of the gate leakage current flowing through the device, allowing to reduce static power consumption. In addition, high-κ dielectrics will be most likely implemented in advanced devices in combination with metal gates, instead of the "classical" poly-Si gates.

Another material of much interest for high-performance IC applications is germanium, which is currently considered as a potential replacement for silicon [3–10]. The main advantage of Ge over Si is its superior electron and hole mobility, allowing for boosting the drive current of the transistors. The lower band gap of Ge (~0.66 eV) compared to Si (~1.1 eV) should also allow further V_{DD} scaling. Finally, the reduced thermal budget needed for dopant activation of a Ge-based MOSFET (metal-oxide-semiconductor field-effect transistor), i.e. around 500–600°C, compared to Si (1000–1100°C), also makes this material particularly interesting with respect to high-κ/metal gate integration.

From a historical point of view, the first (bipolar) transistors and integrated circuits were built using Ge substrates. However, Si supplanted Ge quickly, mainly due to the remarkable

Germanium-Based Technologies: From Materials to Devices. C. Claeys and E. Simoen (Editors)

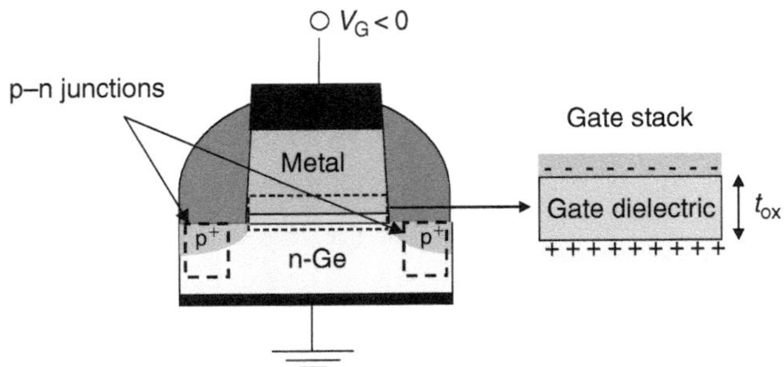

Figure 8.1. Schematic picture of a MOSFET, showing the two fundamental building blocks of the structure, i.e. the p–n (source/substrate and drain/substrate) junction and the semiconductor/insulator/metal gate stack.

properties of silicon dioxide, used as a Si surface passivation layer, high-quality gate insulator as well as field isolation between adjacent devices. Unlike Si, Ge does not have a stable natural oxide than can passivate its surface. GeO_2 is indeed known to be thermodynamically unstable at around 400°C, due to the formation of volatile GeO at these temperatures [11]. The incorporation of nitrogen in the GeO layer has been shown to result in improved thermal stability of the insulator, as well as to lead to better Ge surface passivation [6,12,13].

More recently, high-κ gate dielectrics have been studied as possible gate insulators for Ge-based MOSFETs [3–10]. So far, HfO_2 and ZrO_2 have been widely studied, mainly because these materials are serious candidates for the replacement of SiO_2 in Si-based MOSFETs. Good device characteristics, mainly on p-channel devices, have been recently demonstrated by combining these high-κ gate dielectrics with Ge, together with proper Ge surface passivation, either using GeON interlayer [8], Si capping layer [14] or PH_3 surface treatment [15]. For completeness, let us remind that sulfur passivation of the Ge surface has also been studied in the past [16,17].

In this chapter, the two major building blocks of the Ge-based MOSFETs will be reviewed, namely the p–n junction and the Ge/insulator/metal gate stack (see Figure 8.1). The basic principles of operation of these devices will be first outlined. Next, the most important results and issues reported so far for these Ge-based devices will be presented and discussed.

8.2. Germanium p–n Junctions

A p–n junction is made by a p- and an n-type semiconductor in direct contact with each other, as illustrated in Figure 8.1. When the two sides are of the same material, it is called a homo-junction. There are several ways to fabricate such a structure, relying either on dopant diffusion or on ion implantation, followed by an activation anneal. Due to the different doping density and carrier type, a built-in potential V_{bi} will be established in thermal equilibrium between the p- and the n-side, which will block the diffusion of excess carriers from one side of the barrier to the other. This diffusion is driven by the thermal potential

kT/q. The net result is a Fermi level which is constant throughout the junction and zero net current flow. Current will flow when an external potential drop is applied between the two electrodes.

In this section, the basic equations governing the charge transport across a p–n junction will be briefly recalled. A more detailed discussion can be found in textbooks on device physics (e.g. Refs [18,19]). First, the ideal diode equation will be given, followed by the model for the non-ideal case, where the current I_{pn} is governed by generation and recombination of charge carriers by deep levels in the band gap of the semiconductor material. This is also called the Shockley-Read-Hall (SRH) theory [20,21]. For planar diodes, typical for Complementary MOS (CMOS) transistors (Figure 8.1), one has to consider two-dimensional (geometrical) effects in the charge transport, whereby charge carriers are not only generated or recombined in the bulk of the material, but also at its surface. This leads to a separation of the current into a volume (area) and a peripheral (perimeter) component. As will be seen, for planar Ge p–n junctions this is often the dominant source of leakage current. Some attention will be also paid to specific types of germanium junctions, which are used for special applications. Finally, the current–voltage (I_{pn}–V_{pn}) characteristics of advanced shallow junctions in germanium intended as highly doped drain (HDD) junctions are discussed and the implications for CMOS on Ge will be outlined.

8.2.1. Theory of a large-area p–n junction

The current through an ideal, defect-free p–n junction is described by the Shockley equation, given by [18,19]:

$$I_{pn} = I_{pn0}\left[\exp\left(\frac{qV_{pn}}{kT}\right) - 1\right] \tag{8.1}$$

whereby V_{pn} is the externally applied bias and the saturation current I_{pn0} obeys:

$$I_{pn0} = qn_i^2 A_J\left(\frac{D_n}{L_n N_A} + \frac{D_p}{L_p N_D}\right) \tag{8.2}$$

In Equation (8.2), A_J is the junction area, L_n and L_p the diffusion length for electrons or holes and D_n and D_p, the diffusivity of electrons or holes, respectively. Finally, N_A and N_D are the doping density (assumed uniform) at the p- and n-side of the junction. The saturation current I_{pn0} is a pure diffusion current, which for an asymmetrically doped junction, with $N_A \ll N_D$ or $N_D \ll N_A$ is governed by the diffusion of the minority carriers in the lowly doped material. In other words:

$$I_{pn0} = qn_i^2 A_J\left(\frac{D_n}{L_n N_A}\right) \quad \text{or} \quad I_{pn0} = qn_i^2 A_J\left(\frac{D_p}{L_p N_D}\right) \tag{8.3}$$

In reverse operation, where $V_{pn}/kT < 0$, Equation (8.1) reduces to $I_{pn} \approx I_{pn0}$, so that the leakage current is constant and independent of the bias. At the same time, the temperature dependence will be governed by the n_i^2 factor in Equation (8.2) or (8.3), yielding an activation energy equal to E_G, the band gap of the semiconductor material.

In forward operation, when $qV_{pn} \gg kT$, the exponential term in Equation (8.1) will dominate, leading to an exponential bias dependence with ideality factor $m_{pn} = 1$. In practice, the forward current will become limited by the parasitic series resistance R_s, once the potential drop $I_{pn}R_s$ is no longer negligible with respect to V_{pn}.

The presence of a deep level E_T in the band gap will have a strong impact on the charge transport across a junction, both in forward and reverse operation, as illustrated by the band gap scheme of Figures 8.2 and 8.3, respectively. For an indirect band gap material like germanium,

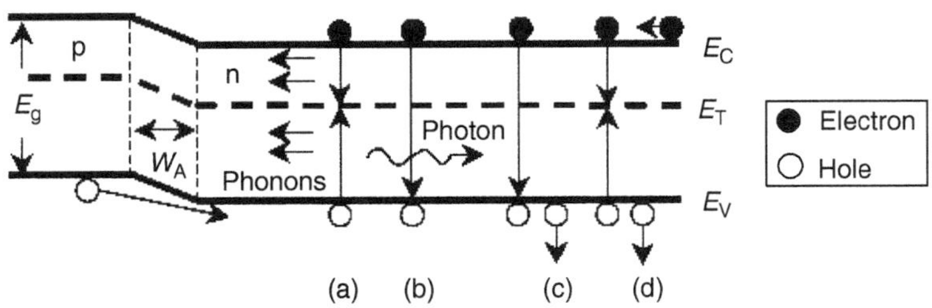

Figure 8.2. Forward biased junction with recombination mechanisms: (a) trap-assisted multi-phonon or SRH recombination; (b) band-to-band or radiative recombination; (c) direct Auger recombination and (d) trap-assisted Auger recombination.

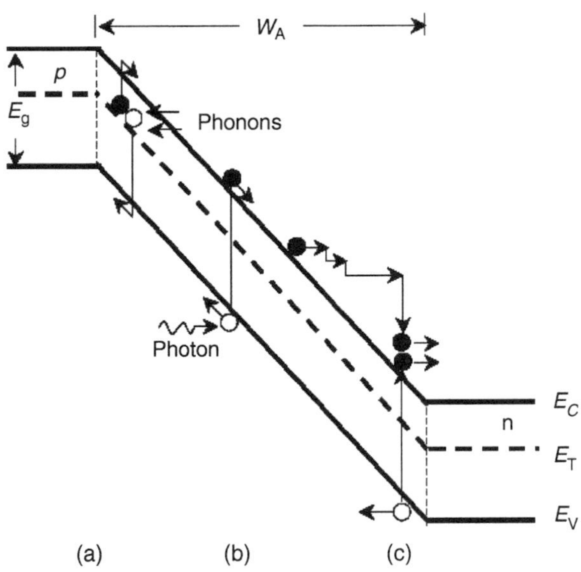

Figure 8.3. Reversely biased junction with generation mechanisms: (a) multi-phonon generation via a deep-level center; (b) optical electron-hole pair generation and (c) impact ionization.

the optical and direct Auger transitions can be in the first instance neglected, while the trap-related Auger transitions will become important at high-carrier densities (high injection level). Considering only thermal transitions, i.e. phonon-assisted ones, the SRH time constant for recombination in forward operation can be approximated by [18–21]:

$$\frac{1}{\tau_{rec}} = \frac{1}{\tau_n} + \frac{1}{\tau_p} \tag{8.4}$$

In Equation (8.4), τ_n, τ_p are given by:

$$\tau_n = \frac{1}{\sigma_n v_{th} N_T} \quad \text{and} \quad \tau_p = \frac{1}{\sigma_p v_{th} N_T} \tag{8.5}$$

with N_T the deep-level concentration, v_{th} the thermal velocity of the respective carriers and σ_n, σ_p the capture cross section for electrons and holes, respectively.

Similarly, for carrier generation in the depletion region W_A of a reversely biased junction, one can show that the generation lifetime is approximated by [18–21]:

$$\tau_{gen} = \tau_p \exp\left(\frac{E_T - E_i}{kT}\right) + \tau_n \exp\left(-\frac{E_T - E_i}{kT}\right) \tag{8.6}$$

In Equation (8.6), E_i is the intrinsic Fermi level position. For $\sigma_n = \sigma_p$ this can be further simplified to [22,23]:

$$\tau_{gen} = 2\tau_{rec}\cosh\left(\frac{E_T - E_i}{kT}\right) \approx \tau_{rec}\exp\left(\frac{|E_T - E_i|}{kT}\right) \quad \text{for } E_T \neq E_i \tag{8.7}$$

The leakage current of a non-ideal p–n diode is then often represented by [18–21]:

$$I_{pn} \approx I_{pn0} + I_{gen} = I_{pn0} + \frac{A_J q n_i W}{\tau_{gen}} \tag{8.8}$$

whereby the depletion region width W_A is given by [18–21]:

$$W = \sqrt{\frac{2\varepsilon_S}{q}\left(\frac{N_A + N_D}{N_A N_D}\right)\left(V_{bi} - V_{pn}\right)} \tag{8.9}$$

and:

$$V_{bi} = \frac{kT}{q} \ln\left(\frac{N_A N_D}{n_i^2}\right) \tag{8.10}$$

From Equation (8.8), it follows that the reverse current of a p–n junction is in many cases a linear function of the depletion width W, which can be found from the measured capacitance C_{pn} through $W = \varepsilon_s A_J / C_{pn}$, ($\varepsilon_s$ is the permittivity of the semiconductor). The slope of the linear fit is proportional with τ_{gen}^{-1} and the intercept with the saturation diffusion current I_{pn0} [24]. Combining Equations (8.7) and (8.8), one expects an activation energy of the reverse current that is given by $E_G/2$ for lower T where the generation current is dominant and by E_G for the higher-temperature range where the diffusion current takes over.

The non-ideal forward characteristic is usually represented by:

$$I_{pnF} = I_{pnsat}\left\{\exp\left[\frac{q\left(V_{pnF} - I_{pn}R_s\right)}{m_{pn}kT}\right] - 1\right\} \tag{8.11}$$

with m_{pn} the ideality factor, ranging between $m_{pn} = 1$ (pure diffusion current – Equation (8.1)) and $m_{pn} = 2$ (pure recombination current). The saturation current at small forward bias V_{pnF} is approximated by:

$$I_{pnsat} = \frac{A_J q n_i W_A}{2\tau_{rec}\cosh\left(\frac{E_T - E_i}{kT}\right)} \tag{8.12}$$

Finally, the net free carrier concentration in the depletion region of a one-sided, asymmetric p^+–n or n^+–p junction can be found from the reverse C_{pn}–V_{pn} characteristic, since based on Equation (8.9), N_A ($\ll N_D$) or N_D ($\ll N_A$) is found from the slope of a C_{pn}^{-2} versus V_{pn} graph. In case of a non-uniformly doped substrate, the local carrier density profile $n(x)$ can be found from [25]:

$$n(x) = \frac{2}{A_J^2 \varepsilon_S q \frac{dC_{pn}^{-2}}{dV_{pn}}} \tag{8.13}$$

whereby x corresponds to the depletion region at a certain reverse bias $W_A(x)$. The spatial resolution of the C_{pn}–V_{pn} profiling technique is limited by the Debye length given by [25]:

$$\lambda_D = \sqrt{\frac{kT\varepsilon_S}{q^2 n(x)}} \tag{8.14}$$

8.2.2. *Theory of a planar p–n junction*

The theory presented so far, only considers one-dimensional effects, i.e, charge transport into the bulk of the material. This is a valid approximation for junctions with a large area compared with the perimeter P_J. However, for the source and drain junctions of a MOS transistor, a significant fraction of the current flows in the lateral direction, parallel to the surface of a wafer. This follows from the planar technology, whereby the active diode region is opened in an insulating layer – SiO_2 in the case of silicon CMOS. For shallow state-of-the-art junctions, the lateral dimensions are comparable or even larger than the vertical ones, so that one has to consider recombination and generation effects in the lateral depletion region, as represented in Figure 8.4.

Besides carrier generation/recombination (GR) through bulk deep levels in the depletion region, one has to account for GR events by traps at the semiconductor-insulator interface in the lateral depletion region with extent W_P. The corresponding surface-state generation (S_g) or recombination velocity (S_r) is given, respectively, by:

$$S_g = \frac{S_r}{2\cosh\left(\frac{E_{Ts} - E_i}{kT}\right)}. \quad \text{(in cm s}^{-1}\text{)} \tag{8.15a}$$

$$S_r = \sigma_s v_{th} N_{Ts} \quad \text{(in cm s}^{-1}\text{)} \tag{8.15b}$$

with E_{Ts}, σ_s and N_{Ts} (in cm^{-2}) the energy level, capture cross section and surface density of the interface traps. The corresponding peripheral surface generation current is then given by:

$$I_{gsP} = A_S q n_i S_g \tag{8.16}$$

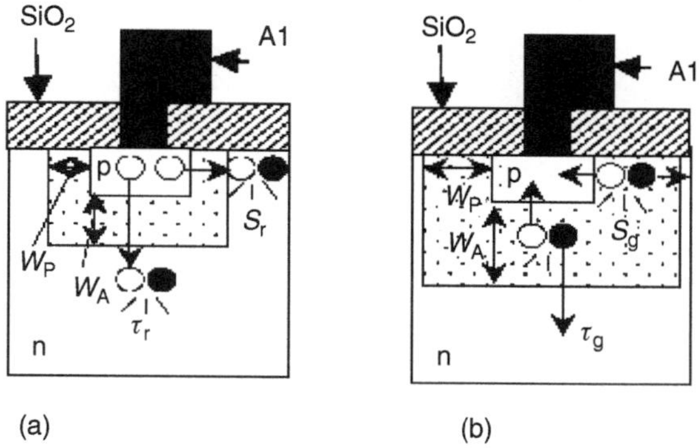

Figure 8.4. Schematic representation of the recombination (a) and generation (b) components in a planar silicon junction diode.

while the corresponding recombination current obeys:

$$I_{rsP} = \frac{A_s q n_i S_r}{2\exp\left(\frac{qV_{pn}}{2kT}\right)}\left[\exp\left(\frac{qV_{pn}}{kT}\right) - 1\right] \tag{8.17}$$

The total current of a planar p–n junction is thus composed of the area (or bulk), the peripheral and, in the case of a corner intensive structure a corner component, with densities represented by J_A (A cm^{-2}), J_P (A cm^{-1}) and J_C (A/corner). In turn, each of these geometrical components contains an ideal (diffusion) and a defect-assisted (non-ideal) contribution. To separate the geometrical components for a rectangular p–n junction, neglecting corner and parasitic currents, one can combine the I_{pn}–V_{pn} characteristics of at least two junctions with different area (A_{Ji}, i = 1,2) and perimeter (P_{Ji}, i = 1,2). This yields the set of equations:

$$I_{pn1} = A_{J1}J_A + P_{J1}J_P \tag{8.18a}$$

$$I_{pn2} = A_{J2}J_A + P_{J2}J_P \tag{8.18b}$$

from which the current densities can be solved. In practice, one usually designs more types of junctions with different P_J/A_J ratios. Representing I_{pn}/A_J versus P_J/A_J at the same V_{pn} should yield a straight line with slope J_P and intercept J_A.

A more direct way to separate the geometrical current components relies on the so-called gated diode structure [26,27], represented in Figure 8.5 [28]. In that case, the gate voltage

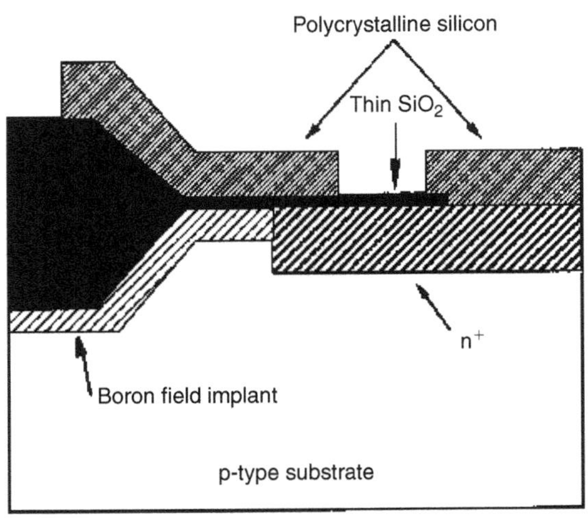

Figure 8.5. Schematic representation of a gated silicon n^+–p junction diode, representing the peripheral gated region. The gate voltage across the MOS structure, defined by the polycrystalline gate electrode, the thin SiO_2 and the p-type substrate controls the surface potential and hence the surface generation current (after Czerwinski *et al.* [28]).

V_G across the MOS capacitor at the periphery of the p–n junction controls the surface potential and, hence, the peripheral current component. The latter can be completely suppressed when the interface is in accumulation or inversion, while its contribution is maximal for some intermediate V_G, depleting the underlying surface. This means that studying I_{pn} at fixed V_{pn} in function of V_G enables to separate J_A from J_P, using only one structure [26,27].

8.2.3. Theory of an ideal germanium p–n junction

One of the concerns for the development of CMOS on germanium substrates is the intrinsic carrier concentration that is significantly higher than in silicon, owing to the smaller band gap. One can calculate n_i from:

$$n_i^2 = N_V N_C \exp\left(-\frac{E_G}{kT}\right) \tag{8.19}$$

with the density of states in the valence and conduction band respectively given by:

$$N_V = 1.98 \cdot 10^{15} T^{3/2} \tag{8.20a}$$

$$N_C = 1.00 \cdot 10^{15} T^{3/2} \tag{8.20b}$$

and the temperature dependence of the band gap [29]:

$$E_G = 0.741 - \frac{4.56 \cdot 10^{-4} T^2}{(T + 210)} \text{(eV)} \tag{8.21}$$

From Equations (8.3) or (8.8) it follows that the leakage current of a p–n junction is proportional to n_i^2 (diffusion current) or n_i (generation current). Since n_i at room temperature is about a factor of 1000 higher in Ge than in Si, a significant increase of the reverse current density is to be expected for a Ge p–n junction. One can for example calculate the diffusion current density according to Equation (8.3), assuming a single-sided junction, a realistic recombination lifetime of 100 μs and a diffusivity of 50 cm^2/s^{-1}. The fundamental limit of the reverse current at 298 K for an ideal defect-free Ge substrate derived from Figure 8.6 is in the range 10^{-4}–10^{-7} A cm^{-2}, depending on the substrate doping density and minority carrier lifetime. This is significantly larger than for silicon p–n junctions where for high-quality diodes J_A values below 1 nA cm^{-2} can be achieved [30].

Another concern regarding planar Ge p–n junctions is the impact of the peripheral reverse current on the total leakage current, due to the high surface-state carrier generation. According to Equations (8.15) and (8.16) besides the higher n_i one should also take into consideration the density of interface states which is usually much larger in germanium compared with silicon,

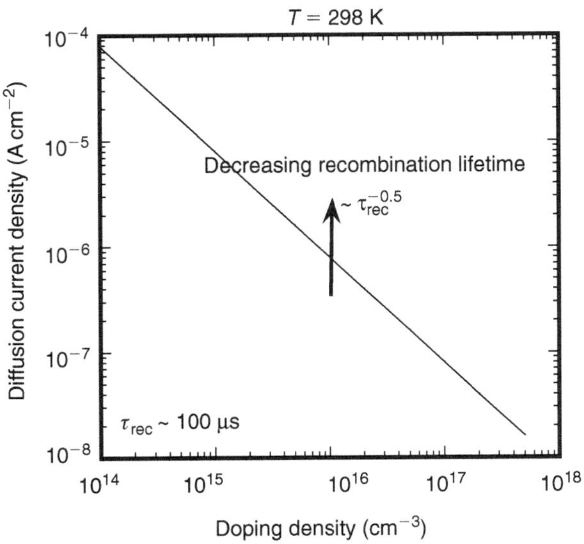

Figure 8.6. Diffusion current density at 298 K for a Ge p–n junction. A lifetime of 100 μs was assumed and a diffusivity $D = 50\,cm^2 s^{-1}$. So, for a doping density of $10^{16}\,cm^{-3}$ and a defect-free substrate, the fundamental leakage current density limit at 298 K will be $10^{-6}\,A\,cm^{-2}$.

irrespective of the type of insulating dielectric used [31]. Typical values for the surface generation velocities are in the range of a few $100\,cm\,s^{-1}$ and corresponding surface current densities are found to be around 10^{-4}–$10^{-3}\,A\,cm^{-2}$ [31], clearly higher than the theoretical diffusion current limit.

From Equation (8.16) and assuming $n_i = 10^{13}\,cm^{-3}$ and $S_g = 100\,cm\,s^{-1}$ one arrives at a peripheral surface-state current density of $1.602 \cdot 10^{-4}\,A\,cm^{-2}$. It implies that in many cases, this component will dominate the total leakage current of a planar Ge junction, as will be illustrated below. Only for large-area diodes (a few mm^2) or well-passivated interfaces, one may expect the volume leakage current to become dominant at room temperature.

8.2.4. Germanium bulk p–n junction diodes

Historically, several fabrication techniques have been utilized to make germanium p–n junctions. One of the first proposed methods is based on the segregation behaviour of the most relevant doping elements during the crystal growth of Ge (see Chapters 2 and 3). A high-purity Ge ingot is usually p-type at the seed end and n-type at the tail with a p–n junction in between [32]. Doping elements of the appropriate type can also be added during the pulling, so that a p–n junction can be realized during Ge crystal growth [33,34].

It was soon recognized that a better impurity control could be achieved by using diffusion at high temperatures from a gas or solid source [35]. In the latter case, the desired doping atoms can be in-diffused from an alloy with germanium [36,37]. The introduction of ion

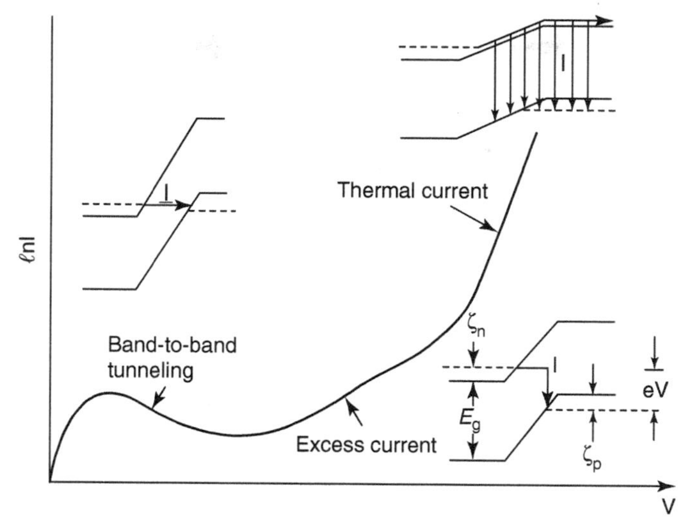

Figure 8.7. Current–voltage characteristic of a degenerate p–n junction. Insets show the relative positions of the bands and the paths of electron flow for three principal components of current. ζ_n and ζ_p are the penetrations of the Fermi levels into the conduction and valence bands (after Meyerhofer *et al.* [44]).

implantation followed by thermal annealing has revolutionized the p–n junction technology, as it allows a more precise control of the doping profiles, especially for shallow junctions. More details regarding the processing of advanced Ge junctions will be provided in Chapters 10 and 11.

Ion implantation has also been applied to the fabrication of high-purity Ge nuclear-radiation detectors [38,39]. Before, p-i-n detectors were made by Li drifting (diffusion of Li under application of an electric field) [40], but due to the instability of the Li-donors at room temperature this approach has been abandoned. A Ge nuclear-radiation detector is usually operated at liquid nitrogen temperature, since the material is intrinsic at room temperature. Moreover, cooling to 77 K reduces significantly the thermally activated leakage current processes. This is important as the noise of the system is partly due to the shot noise of the dark current. The leakage current also contributes to the broadening of the spectral peak and, therefore, limits the energy resolution. Such a p–n junction is working under a large reverse bias as to fully deplete the bulk of the material, which is utilized as sensitive volume. In order to achieve a depletion width of several mm to cm for a reasonable V_{pn} of a few 100 V, one has to use high-purity Ge material with a doping density $<10^{11}\,cm^{-3}$. As the volume of such a detector is quite large, another issue is the surface generation by interface traps. It has been demonstrated that a good passivation of the surface states can be achieved by employing a hydrogenated amorphous germanium layer [41].

At the other end of the spectrum, a special kind of junction was discovered by Esaki [42] – also called Esaki diode – consisting of a degenerately doped n region in contact with a degenerately doped p region. The corresponding forward characteristic has the shape depicted in Figure 8.7, exhibiting a negative differential resistance regime. The peak at low forward V_{pn} is due to band-to-band tunneling [42–44]; at higher voltages, the classical thermal current dominates. In between, excess current generated by trap-assisted tunneling may be observed [44].

The smaller band gap of Ge is not always a drawback as it offers the opportunity for fabricating photodiodes in the infrared regime, in the wavelength range of 1–1.5 μm which is of high interest for optical telecommunication and fiber optics. In the past, discrete germanium p–n junctions have been developed for that purpose (see e.g. Refs [45–48]). Nowadays, however, the goal is to integrate germanium photodetectors on a silicon substrate. For a detailed coverage of this and related topics, the Reader is referred to Chapter 12.

8.2.5. State-of-the-art shallow germanium p–n junctions

Typical implantation conditions for state-of-the-art shallow junctions are: for p^+–n, a B dose of 4×10^{15} at. cm^{-2} at 13 keV and for n^+–p: P at 50 keV and 3.3×10^{14} at. cm^{-2}. This is followed by an activation and damage anneal at 500°C for 5 min, for example. Using the single-mask lay-out of Figure 8.8, one can obtain a gated diode, using the same gate dielectric as for the fabrication of MOS capacitors present on the same mask. The resulting p–n junctions exhibit good characteristics, as can be judged from Figure 8.9, which scale in this case with the perimeter. In other words, normalizing the diode current for the perimeter P_J yields a nearly identical reverse current density for diodes with different A_J and P_J, demonstrating the dominance of this component over the bulk leakage current.

The role of the thermal anneal is demonstrated in Figure 8.10. For annealing at 300°C, insufficient damage removal is obtained, so that a large density of implantation defects is still present in the depletion region. This creates a high generation current which overwhelms the peripheral component. On the other hand, annealing at 500°C leads to the result of Figure 8.9, where the impact of the residual ion-implantation defects is marginal compared with the perimeter current.

Another important consideration for Ge CMOS junctions is that the real operation temperature is expected to be around 100°C. It implies that the fundamental diffusion components,

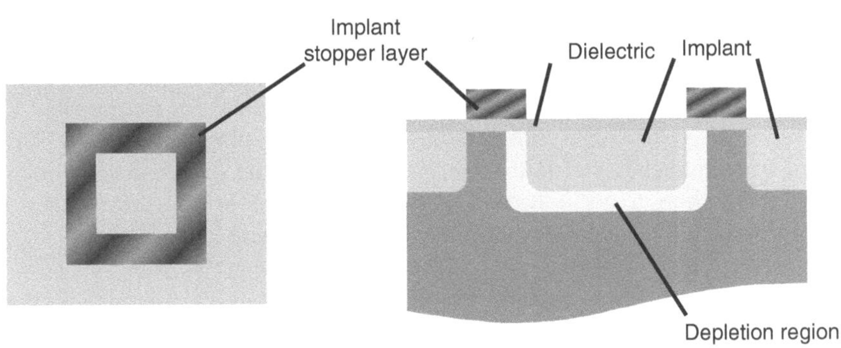

Figure 8.8. Single-mask processing lay-out of a gated p–n junction diode on a germanium substrate.

associated with the diode perimeter and bulk will become dominant compared with deep-level-assisted generation currents. In other words, it will be more difficult to control or lower the reverse current by an appropriate processing step. Fortunately, in the case of a germanium-on-insulator scheme, expected to be a more realistic solution for Ge CMOS than on bulk wafers, the impact of the volume leakage current is suppressed by the insulator below the thin Ge layer. This implies that the optimization of the peripheral leakage current component, through an appropriate passivation of the germanium surface is key for the successful implementation of shallow p–n junctions.

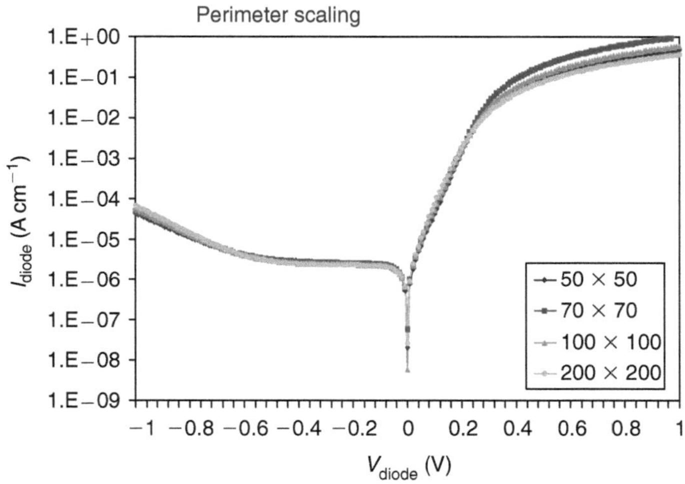

Figure 8.9. Current normalized by the perimeter of square p^+–n junctions with different size, fabricated on a germanium substrate and 4 nm HfO_2 as a gate dielectric.

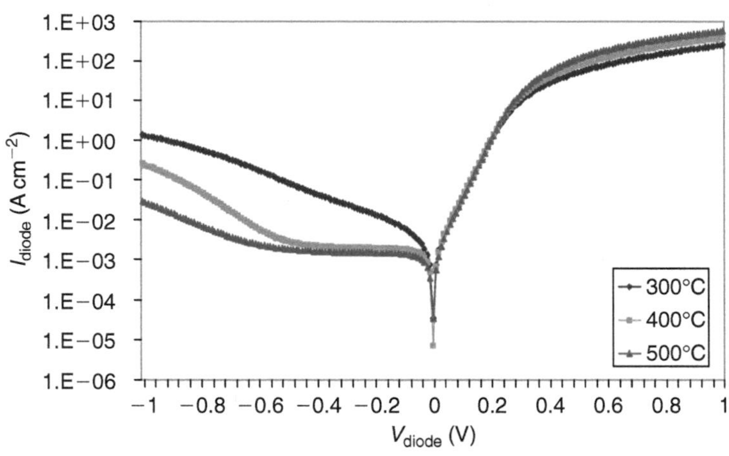

Figure 8.10. Current normalized by the area of a 70 × 70 μm^2 square p^+–n junction, fabricated on a germanium substrate and 4 nm HfO_2 as a gate dielectric and corresponding with different annealing temperatures.

8.3. Germanium-Based Gate Stacks

8.3.1. Equivalent oxide thickness

The stack formed by the semiconductor, the gate insulator and the metal gate constitutes the other fundamental building block of the MOSFET. From an electrical point of view, this system behaves like a parallel plate capacitor, as illustrated in Figure 8.1: when a gate voltage V_G is applied to the gate, charges on the metal are compensated by opposite charges in the semiconductor, these latter charges forming the channel connecting the source and the drain of the transistor. The capacitance C of this parallel plate capacitor is given by (in accumulation and inversion, neglecting poly-Si depletion effects)

$$C = \frac{A\varepsilon_r\varepsilon_0}{t_{ox}} \tag{8.22}$$

where A is the capacitor area, ε_r the relative dielectric constant of the material (3.9 for SiO_2), ε_0 the permittivity of free space (8.85×10^{-12} F/m) and t_{ox} the gate oxide thickness. From Equation (8.22), it appears that decreasing t_{ox} allows to increase the capacitance of the structure, and hence to increase the number of charges in the channel for a fixed value of V_G. However, as far as Si-based technology is concerned, the SiO_2 gate layer thickness is approaching its fundamental limits [2]. An alternative way of increasing the capacitance is to use an insulator with a higher relative dielectric constant than SiO_2. One could then use a physically thicker gate layer and, hopefully, reduce the leakage current flowing through the structure and also improve the device reliability.

The EOT of a gate insulator is defined as the thickness of the SiO_2 layer that would be required to achieve the same capacitance density than the insulator in consideration. According to Equation (8.22), EOT is thus given by

$$\mathrm{EOT} = \frac{\varepsilon_{r,SiO_2}}{\varepsilon_{r,high\text{-}\kappa}} t_{high\text{-}\kappa} \tag{8.23}$$

where $t_{high\text{-}\kappa}$ and $\varepsilon_{r,high\text{-}\kappa}$ are the thickness and relative dielectric constant of the high-κ material, respectively. As an example, using ZrO_2 as gate dielectric ($\varepsilon_r \approx 20$) would allow us to use a 5 nm thick layer in order to achieve a capacitance equivalent to about 1 nm thick SiO_2 layer; the EOT of this ZrO_2 layer is thus ~1 nm.

Actually, when a high-κ metal oxide like ZrO_2 or HfO_2 is deposited on a Si substrate, an ultra-thin (lower-κ) interfacial layer, either SiO_x or SiM_yO_x, (where M is Zr or Hf) forms at the silicon interface, or is intentionally deposited as a passivation layer, as illustrated in Figure 8.11.

The capacitance of the gate stack, C_{tot}, then results from the combination in series of the low-κ and high-κ dielectric layer capacitances, i.e.

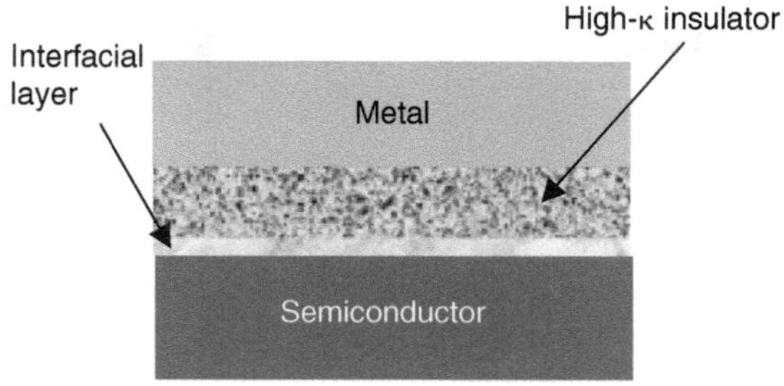

Figure 8.11. Schematic illustration of a MOS structure with a gate stack, formed by a low-κ interfacial layer and a high-κ dielectric layer.

$$\frac{1}{C_{\text{tot}}} = \frac{1}{C_{\text{low-}\kappa}} + \frac{1}{C_{\text{high-}\kappa}} \tag{8.24}$$

The EOT then reads

$$t_{\text{eq}} = \left(\frac{\varepsilon_{\text{r,SiO}_2}}{\varepsilon_{\text{r,low-}\kappa}}\right) t_{\text{low-}\kappa} + \left(\frac{\varepsilon_{\text{r,SiO}_2}}{\varepsilon_{\text{r,high-}\kappa}}\right) t_{\text{high-}\kappa} \tag{8.25}$$

The presence of the low-κ interfacial layer increases the EOT of the gate stack, and should be thus as thin as possible to achieve the EOT required by the International Technology Roadmap for Semiconductors (ITRS). For example, a gate stack formed by a 0.7 nm SiO_x interfacial layer and a 5 nm ZrO_2 layer results in an EOT of about 1.7 nm (assuming $\varepsilon_{\text{r,low-}\kappa} = 3.9$), as compared to an EOT of ~1 nm if the interfacial layer was not present.

As mentioned previously, HfO_2 and ZrO_2 are considered as the most promising candidates to replace SiO_2 in high-performance Si-based technologies (sub-45 nm nodes). For this reason, these two materials have been in detail studied as possible gate insulators for Ge-based MOSFETs. These results are reviewed in the next section and also discussed in Chapter 10.

8.3.2. Ge/HfO_2 gate stacks

Since GeO_2 is unstable, one would expect to be able to deposit HfO_2 or ZrO_2 directly on Ge, without the formation of an interfacial layer. A typical transmission electron microscopy (TEM) picture of a 9 nm HfO_2 layer deposited by atomic layer deposition (ALD) on a HF-last treated Ge surface is shown in Figure 8.12. One observes the presence of an ultra-thin GeO_x interfacial layer, typically less than 3 Å. This GeO_x layer is actually present at the Ge surface after the HF treatment. This interfacial layer is much thinner than the typical SiO_2 interfacial layer formed during high-temperature processing of Si devices, which is typically about 1 nm. From a scaling point of view, the Ge/HfO_2 system thus presents a potential advantage over its Si/HfO_2 counterpart.

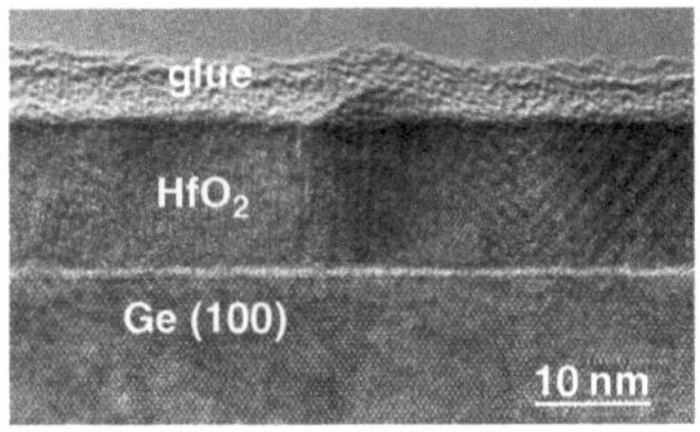

Figure 8.12. High-resolution cross-sectional TEM image of a Ge/GeO_x/HfO_2 gate stack.

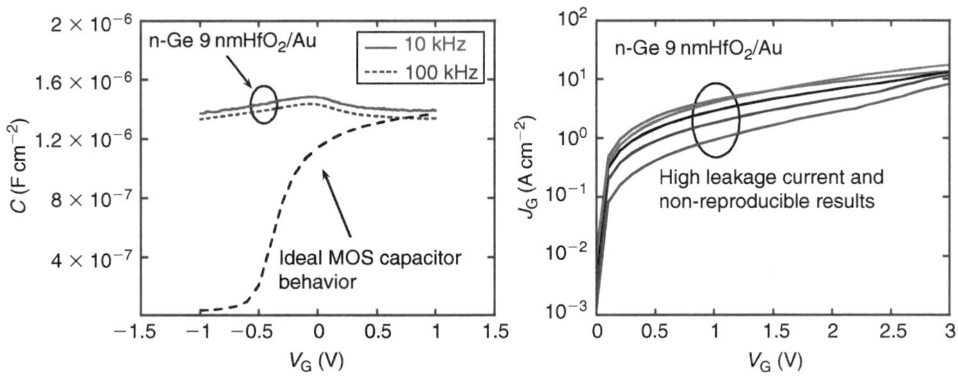

Figure 8.13. High-frequency *C–V* characteristics (left) and *I–V* characteristics (right) of Ge 9 nm HfO_2/Au capacitors.

Unfortunately, the electrical properties of capacitors formed by the direct deposition of HfO_2 on Ge are extremely bad, as illustrated in Figure 8.13. The high-frequency capacitance–voltage (*C–V*) characteristics of the structure do not present the typical transition (capacitance drop) from accumulation to depletion observed in MOS capacitors [18]. The flat *C–V* characteristics rather suggest that the Fermi level at the Ge surface is pinned at a fixed value, due to the presence of a very high density of interface states (in the 10^{13}–$10^{14}\,cm^{-2}$ range). The leakage current flowing through the gate dielectric is also very high, and current–voltage characteristics are not reproducible from one device to the other. These poor electrical results are likely caused by the formation of Ge–Hf bonds at the interface as well as to the diffusion of Ge across the HfO_2 layer [49], either during the deposition of the HfO_2 layer or after post-deposition anneals (PDA).

Similar results were reported for HfO_2 layers deposited by metal-organic chemical vapor deposition (MOCVD) [49] or atomic beam deposition [50]. Proper passivation of the Ge surface is thus required before the deposition of the high-κ layer. The major surface treatments prior to high-κ deposition on Ge are reviewed in the next sections.

8.3.3. Passivation by an ultra-thin GeON interlayer

The presence of an ultra-thin (0.3–0.5 nm) nitrided germanium oxide interlayer has been shown to improve much the electrical properties of the devices. This interlayer was either

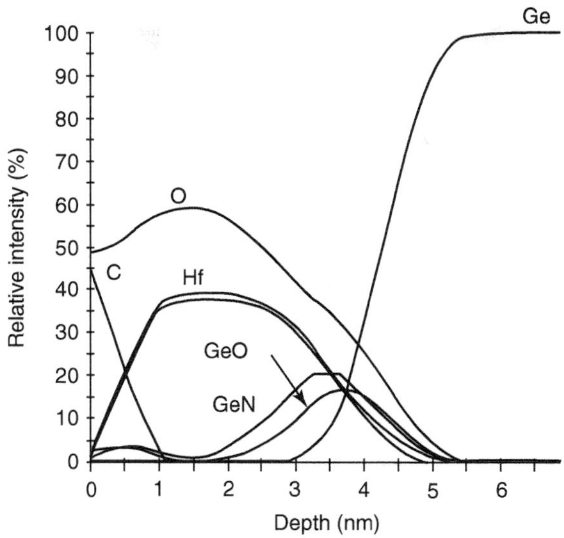

Figure 8.14. Elemental depth profiles of a Ge/GeON/HfO_2 stack obtained from angle-resolved XPS.

formed by thermal nitridation of the Ge surface in NH_3 [8,12,49], or by deposition of a GeON layer by atomic beam deposition [50]. Since both methods lead to very similar results, we only discuss here the data obtained on GeON/HfO_2 stacks formed during atomic beam deposition. Experimental details about the deposition of these stacks can be found elsewhere [50].

Elemental depth profiles of a 0.5 nm GeON/3 nm HfO_2 stack are shown in Figure 8.14. These profiles were reconstructed from angle-resolved X-ray photoemission spectroscopy (XPS) measurements. A well defined stack is observed, with about 50 at.% N content. These profiles also indicate the absence of Ge (Hf) diffusion in the HfO_2 layer (GeON interlayer and Ge substrate), respectively.

The *C–V* characteristics of a n-Ge/GeON/HfO_2/TiN structure are shown in Figure 8.15. Compared to the data shown in Figure 8.13, a clear improvement in the characteristics is observed when an ultra-thin GeON interlayer is present. A bump near the flat-band voltage (V_{FB}) is observed, which amplitude depends on the frequency of the AC signal. This bump is the signature of slow interface states that contribute to the capacitance of the structure at low frequency [51–53]; this bump is also observed in Ge surfaces exposed to NH_3 [54]. Simulated *C–V* curves, taking into account the presence of a broad density of acceptor-like interface states near the conduction band edge of Ge, are shown in Figure 8.16. Note that a very similar energetic distribution of interface states was recently derived from conductance versus frequency measurements by Afanas'ev *et al.* [55]. Comparison between the data and simulations allows us to estimate the average density of acceptor-like interface states to be about $5 \times 10^{12} cm^{-2}$.

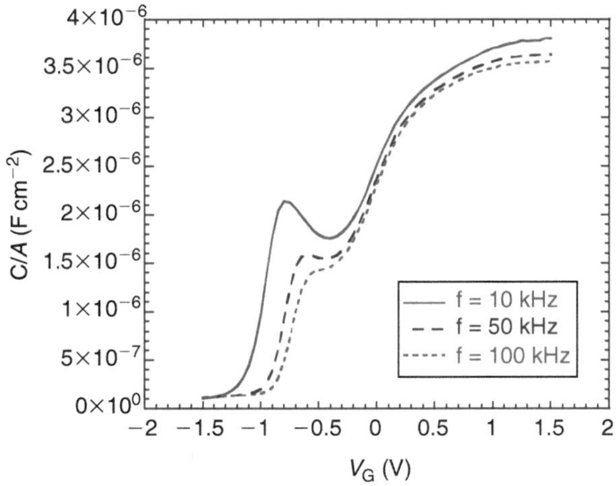

Figure 8.15. *C–V* characteristics of a n-Ge/GeO(N)/3 nm HfO_2/TiN structure measured at different frequencies.

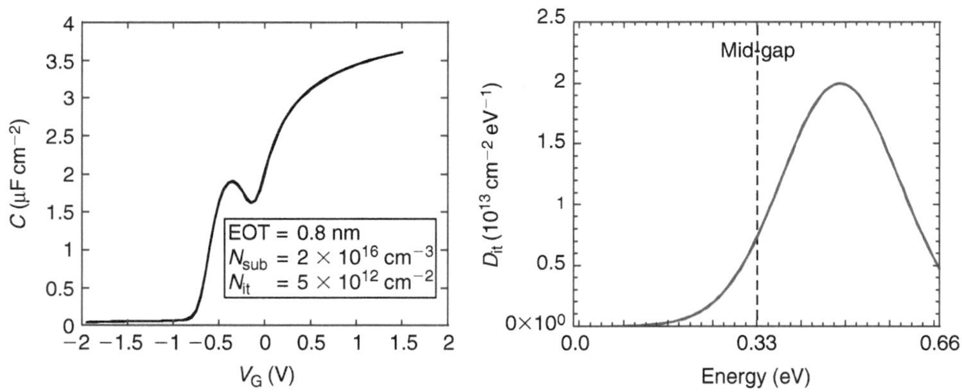

Figure 8.16. Simulated *C–V* characteristics of a n-Ge/high-κ/TiN structure with 0.8 nm EOT (left) with a broad density of acceptor-like interface states in the upper part of the Ge band gap (right).

It should be pointed out that these interface defects are most likely not Ge dangling bonds at the Ge/GeO(N) interface, as recently demonstrated by comparing electrical measurements and electron-spin resonance experiments on similar gate stacks [55].

The current–voltage characteristics of the capacitors are also much improved when a GeON interlayer is present, as illustrated in Figure 8.17. First of all, the uniformity of the characteristics measured across the wafer is much better, compared with the data shown in Figure 8.13 (right). Next, the leakage current is also much reduced in the devices with nitrided interlayer, as shown in Figure 8.17 (right), where the average leakage current of capacitors with GeO and GeON interfacial layer at a fixed (V_G–V_{FB}) value is compared; V_{FB} is the flat-band potential. In addition, the leakage current of the GeON/HfO_2 structure is not altered by

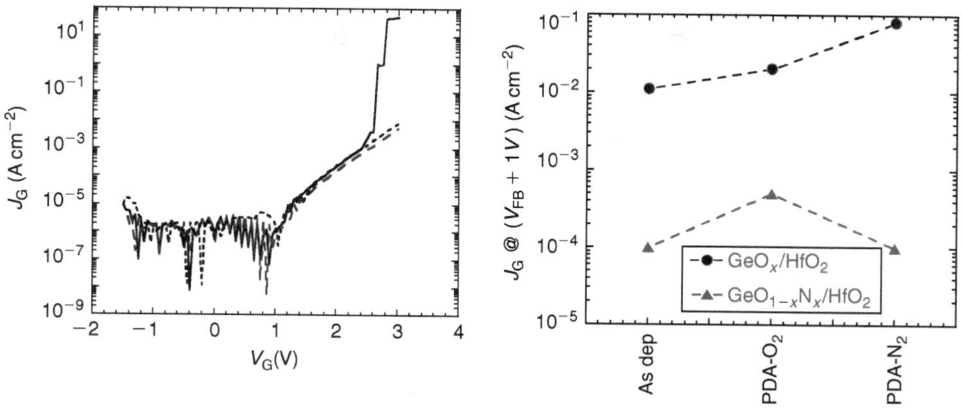

Figure 8.17. J_G–V_G characteristics of n-Ge/GeON/3 nm HfO_2/TiN capacitors (left). Gate leakage current density at V_{FB} + 1 V, for as-deposited stacks, and stacks annealed in O_2 during 1 min at 400°C or in N_2 at 500°C during 1 min.

a PDA at 500°C in N_2, which is the typical thermal budget required for dopant activation in MOSFETs. On the contrary, the leakage current of the GeO_x/HfO_2 structures is increased by an order of magnitude after the annealing. These results indicate a much better thermal stability of the GeON/HfO_2 stack. The improved thermal stability, together with the inhibition of Ge and Hf diffusion, leads to the improved device characteristics observed on the Ge surface passivated by an ultra-thin GeON interlayer.

It has to be noted that the GeON interlayer has a slightly higher dielectric constant than GeO_x, allowing further reduction of the EOT of the gate stack. The lowest-EOT achieved so far with such gate stacks is about 0.7–0.8 nm [52,56].

On the other hand, the incorporation of nitrogen near the Ge interface results in the presence of a large density of fixed positive charges. As shown in Figure 8.18, the incorporation of N leads indeed to a substantial shift of V_{FB} towards more negative values, indicating the presence of up to $10^{13}\,\mathrm{cm}^{-2}$ positive charges. This density can be reduced by about a factor 2 after PDA. However, such positive charge densities are harmful for the control of the threshold voltage of pMOSFETs, and they can potentially act as hole scattering centers, limiting the channel mobility of the devices.

Ring-type Ge p- and nFETs have been fabricated on bulk Ge wafers, using an molecular beam epitaxial (MBE) GeON layer as the Ge surface passivation prior to the deposition of the high-κ layer. The gate stack consists of a 4 nm HfO_2 dielectric and a 50 nm TiN metal gate. Figure 8.19 shows typical I_S–V_G and I_S–V_D characteristics of a 3.5 µm long and 100 µm wide Ge pFET. The EOT for these devices is estimated at 1.1 nm from the accumulation C–V, corrected for quantum-mechanical effects. The effective hole mobility μ_{eff} extracted from the I_S–V_G characteristics is about $90\,\mathrm{cm}^2\mathrm{V}^{-1}\mathrm{s}^{-1}$, which is comparable with state-of-the-art Si pFETs hole mobility with HfO_2 dielectric.

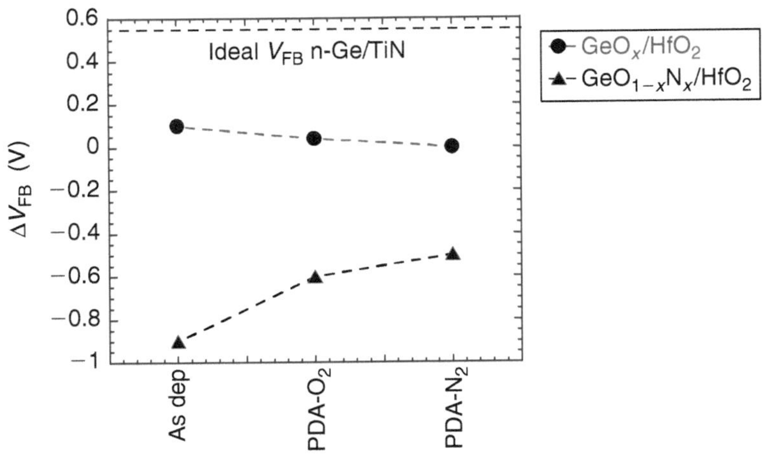

Figure 8.18. Flat-band voltage of n-Ge/GeO(N)/3 nm HfO_2/TiN structures, for as-deposited stacks, and stacks annealed during 1 min in O_2 at 400°C or in N_2 at 500°C.

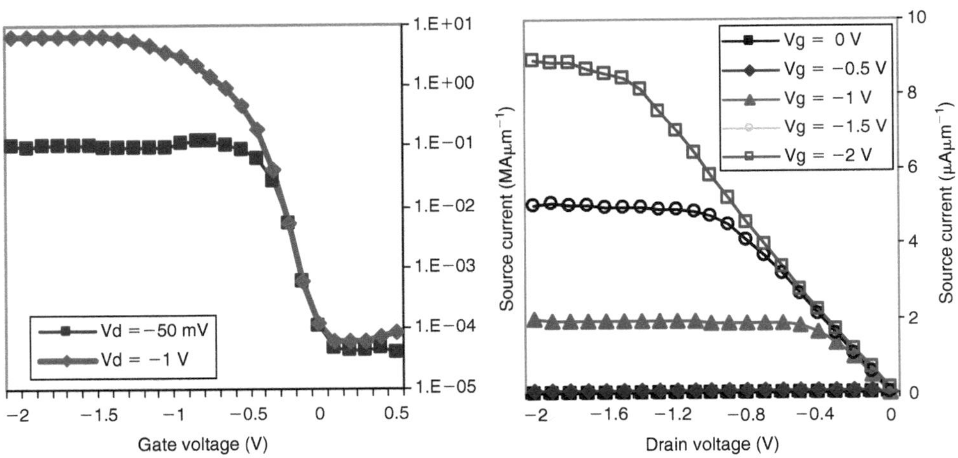

Figure 8.19. Typical I_S–V_G and I_S–V_D characteristics of a 3.5 μm long by 100 μm wide Ge pFET, using an MBE GeON layer as the Ge surface passivation.

The Ge nFETs with GeON as the Ge surface passivation typically have a very low performance [7,56,57], with drive current and μ_{eff} values well below the corresponding pFET values. The worse nFET behavior has been tentatively ascribed to an asymmetric and peaked distribution of (the high density) of interface states in the Ge band gap (illustrated in Figure 8.16), preventing the nFET to become inverted [58,59].

8.3.4. Si surface passivation

Another promising Ge surface passivation treatment consists in depositing an ultra-thin Si epitaxial interlayer, followed by its (partial) chemical or thermal oxidation [14,60]. The Ge surface

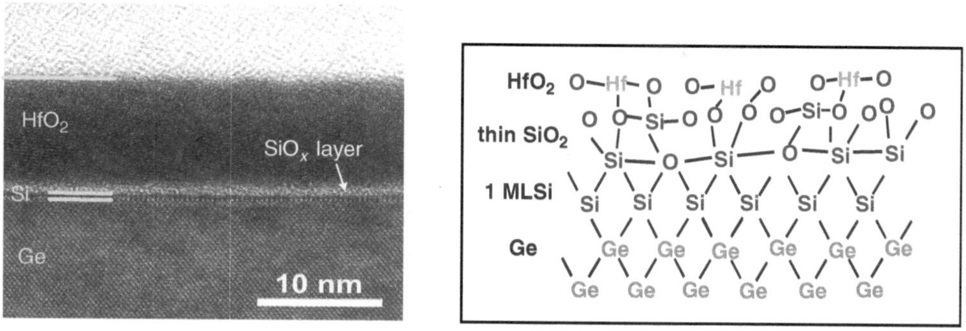

Figure 8.20. High resolution cross-sectional TEM picture of a gate stack formed by a Ge substrate, an ultra-thin epi–Si layer, partly oxidized into SiO_x, and a HfO_2 layer deposited by ALD (left). Schematic drawing of the Ge/Si/SiO_x/HfO_2 gate stack (right).

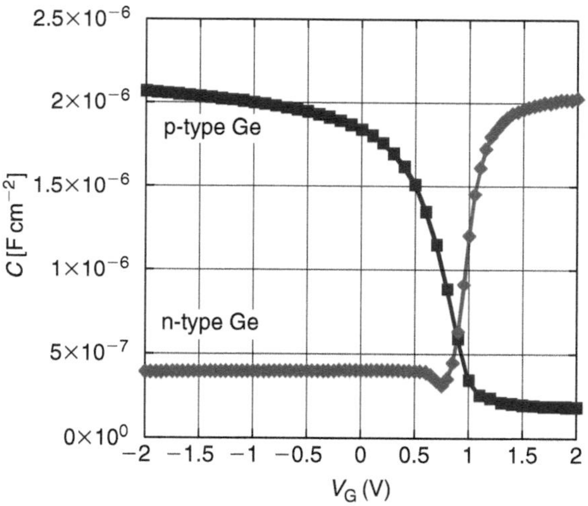

Figure 8.21. 100 kHz *C–V* characteristics of p- and n-Ge/Si/SiO_x/4 nm HfO_2/TiN gate stacks.

is first etched in an HF solution, to remove most of the native GeO_x layer. The wafer is then transferred to an epitaxial reactor, where it is baked in H_2 at a typical temperature of 600–650°C, to obtain a clean Ge surface. An ultra-thin Si epitaxial layer is deposited on the Ge surface, using SiH_4 at a typical temperature of 500°C. This interlayer is next partly oxidized in, e.g. an O_3-based solution, to form an ultra-thin SiO_x interlayer, typically 0.5–1.0 nm thick.

A high-resolution cross-sectional TEM picture of a Ge/Si/SiO_x/HfO_2 stack is shown in Figure 8.20. The HfO_2 layer was deposited by ALD (atomic layer deposition) at 300°C, using $HfCl_4$ and H_2O as precursors. A well defined gate stack, with sharp and smooth interfaces is observed from the TEM picture.

Capacitor structures on p- and n-type Ge with such a Si/SiO_x passivation can result in much improved *C–V* characteristics as compared to the GeON interlayer case, see Figure 8.21.

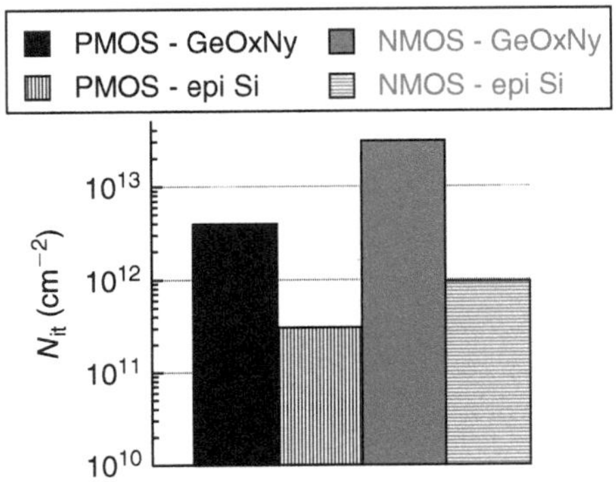

Figure 8.22. N_{it} from charge pumping measurements on Ge n- and pFETs passivated with GeON (NH_3 anneal) or Si/SiO_x.

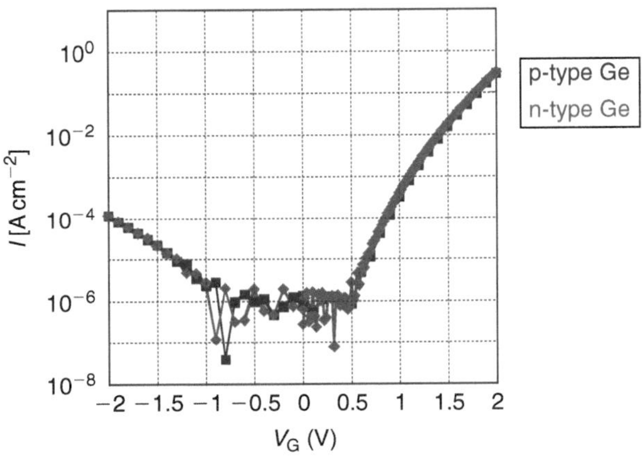

Figure 8.23. *I–V* characteristics of p- and n-Ge/Si/SiO_x/4 nm HfO_2/TiN gate stacks.

The steeper transition between accumulation and depletion and the absence of bumps near V_{FB} both indicate a lower density of interface states in the Ge band gap. Charge pumping measurements on long-channel p- and nFETs confirm the reduced number of interface states (N_{it}) as compared to the GeON interlayer case, as shown in Figure 8.22. The reduction in N_{it} is about one decade, yielding values of $3 \times 10^{11}\,cm^{-2}$ for pFET and $9 \times 10^{11}\,cm^{-2}$ for nFET. These values remain however significantly higher than the corresponding Si values of about $2 \times 10^{10}\,cm^{-2}$, especially for the nFET. From the V_{FB} value it can be deduced that the use of the Si/SiO_x passivation results in the presence of a fixed negative charge with a significantly lower density than the fixed positive charge of about $10^{13}\,cm^{-2}$ observed in the GeON case.

The corresponding *I–V* characteristics show very low-leakage currents, as illustrated in Figure 8.23. Moreover, low-leakage currents are also obtained when scaling the HfO_2 thickness down to 2 nm, as shown in Figure 8.24: the leakage current at +1 V is about

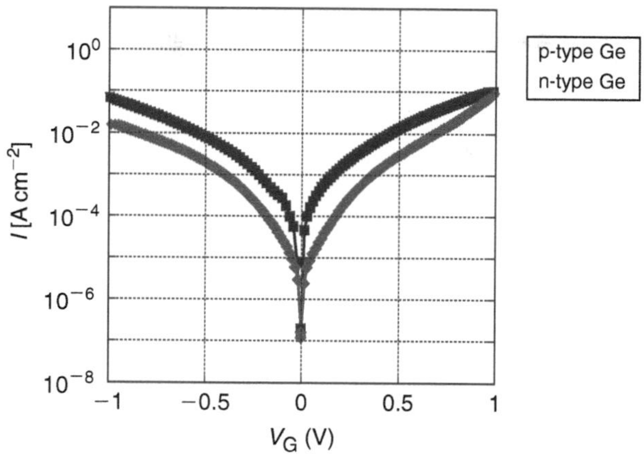

Figure 8.24. I–V characteristics of p- and n-Ge/Si/SiO_x/2 nm HfO_2/TiN gate stacks.

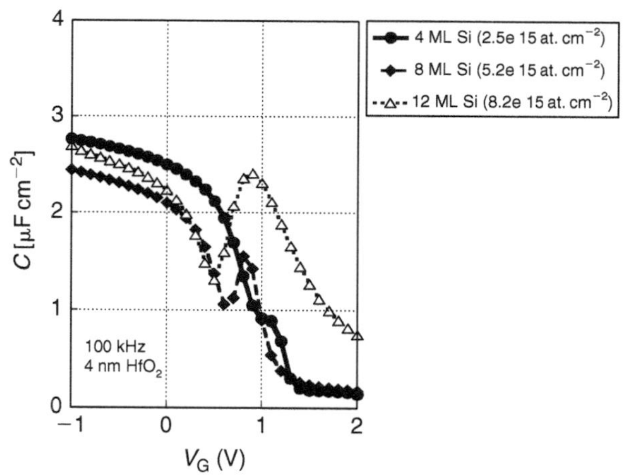

Figure 8.25. 100 kHz C–V characteristics for 4, 8 and 12 monolayers (ML) of Si. The legend mentions the number of Si at. cm^{-2} as measured by TXRF.

0.1 A cm^{-2}, which is typically 5 orders of magnitude less compared to SiO_2/poly-Si-based devices with a comparable EOT ~1 nm; this value is also lower than the specification of the ITRS roadmap for high-performance logic applications, which is between 10 and 100 A cm^{-2}.

It must be noted that the Si layer thickness has a serious impact on the C–V characteristics [61]. Si layers with a thickness of 4, 8 and 12 monolayers have been compared. The deposited Si thickness is determined by Total X-ray reflection fluorescence (total number of Si at. cm^{-2}, the values are indicated in Figure 8.25). For four monolayers of Si, the bump visible in all C–V's is minimized, indicating the lowest N_{it}.

A detailed TEM analysis (Figure 8.26) allows interpreting these results. The four monolayers of Si results in a smooth Ge–Si interface without visible defects, as evidenced from Figure 8.26a. The Si layer appears to stay epitaxial with the Ge substrate. On the contrary, the eight monolayers of Si results in periodical contrast variations at the interface, cf. Figure 8.26b. These contrast variations are probably related to the presence of stress induced by the silicon layer. In the stress distribution, roughly a 10 nm periodicity can be observed, which corresponds with the mismatch between the Si lattice parameter ($a = 0.5431$ nm) and the Ge lattice parameter ($a = 0.56575$ nm), with $25d_{\mathrm{Ge}(110)} \sim = 26d_{\mathrm{Si}(110)} = 9.97$ nm. Additionally, the 12 monolayers of Si show not only stress but also dislocations at the interface, see Figure 8.26c. This relatively thick layer of Si apparently relaxes through the formation of a regular array of misfit dislocations, again with a periodicity of about 10 nm.

Germanium p- and nFETs have been fabricated on Ge-On-Si wafers, using a Si/SiO_x layer as the Ge surface passivation prior to the deposition of the high-κ layer. The silicon-like process flow used to fabricate the Ge devices is outlined in [62]. The gate dielectric is 4 nm HfO_2 and the metal gate is a 10 nm PVD TaN/70 nm PVD TiN stack. Figure 8.27 shows

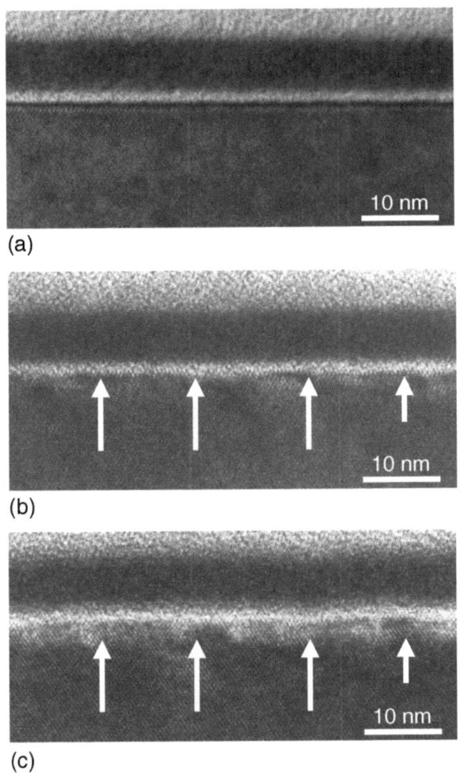

Figure 8.26. a, b and c: TEM analysis for 4, 8 and 12 monolayers of Si. In a), the slightly darker interfacial layer is the Si layer while the brighter interfacial layer on top is the SiO_2 layer. The arrows emphasize the periodicity of the contrast variations observed on the TEM images.

typical I_S–V_G characteristics of long- (10 and 1 μm) and short-channel (0.25 μm) Ge p- and nFETs. The EOT is estimated at 1.45 nm from the accumulation *C–V*, corrected for quantum-mechanical effects. Figure 8.28 shows the effective mobility μ_{eff}, extracted from the I_S–V_G characteristics of the long-channel devices.

The pFET mobility of ~250 $cm^2 V^{-1} s^{-1}$, obtained from these conventional long-channel (10 μm) devices, is superior to the best values obtained with NH_3 [7] or PH_3 passivation [15] (see also next paragraph) on ring-shaped long channel devices, see Table 8.1. The pFET

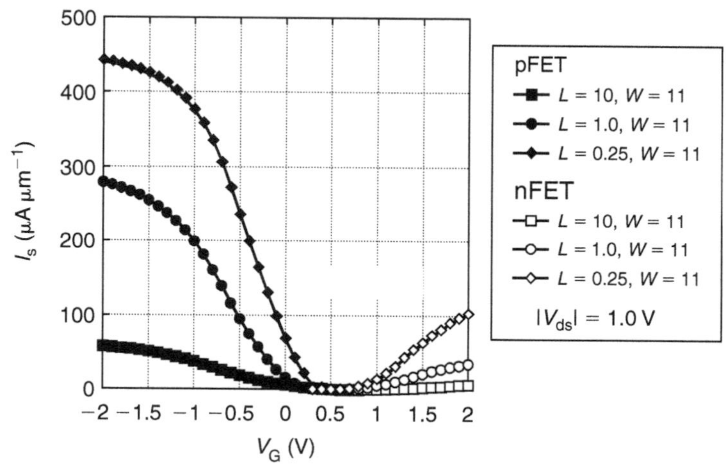

Figure 8.27. I_S–V_G characteristics of long- and short-channel p- and nFETs, using a $Ge/Si/SiO_x$/4 nm HfO_2/TaN/TiN gate stack.

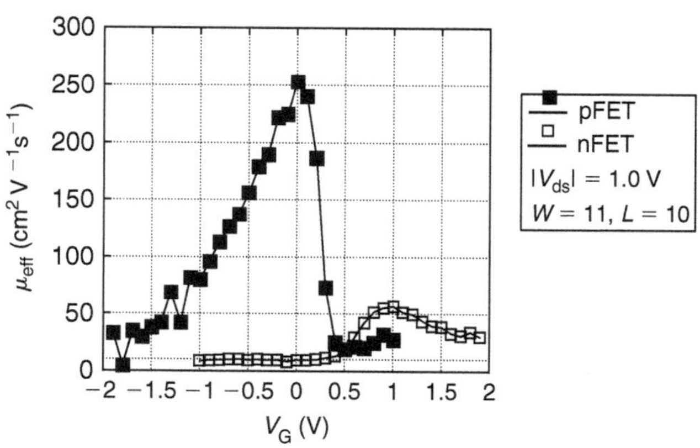

Figure 8.28. The effective mobility μ_{eff} extracted from the I_S–V_G characteristics of the long-channel (10 μm) p- and nFETs shown in Figure 8.25.

Table 8.1. Comparison of the low-field mobility μ_{eff} and the EOT between Si/SiO_x, NH_3 and PH_3 passivations.

	pFET		nFET	
	μ_{eff} ($cm^2V^{-1}s^{-1}$)	EOT (nm)	μ_{eff} ($cm^2V^{-1}s^{-1}$)	EOT (nm)
Si/SiO_x (this work)	250	1.45	50	1.45
NH_3 [7]	210	1.6	~0.2	Not mentioned
PH_3 [15]	125	⩾1.7	300	⩾1.7

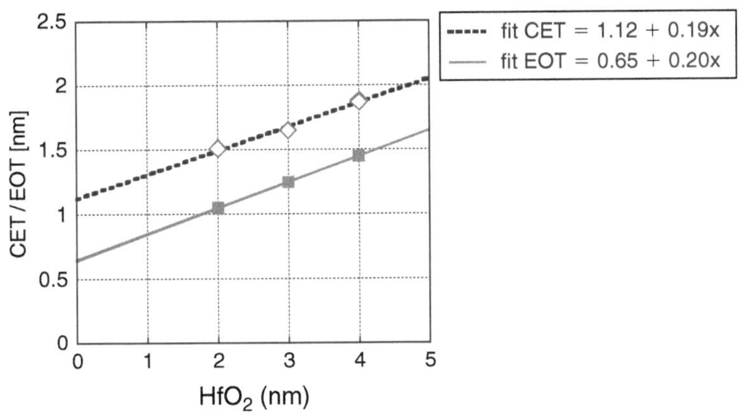

Figure 8.29. EOT as a function of HfO_2 physical thickness of p-Ge/Si/SiO_x/HfO_2 gate stacks. Solid line is a linear fit to the data according to Equation (8.25).

drive current of the 0.25 μm deep sub-micron device moreover exceeds 300 μ*A* μm^{-1} for $|V_{GS}-V_T| = |V_{DS}| = 1$ V, which is similar to state-of-the-art Si pFET. The latter is a very promising result that can likely be significantly improved. Indeed, up till now little specific device architecture engineering has been performed on the deep sub-micron Ge pFET, as the attention has been focused mostly on developing an optimized surface passivation for Ge, with long-channel mobility as the main performance parameter.

The nFET mobility of ~50 $cm^2V^{-1}s^{-1}$ is still low, though significantly superior to the data published for NH_3 passivation. Good nFET mobility has only been reported for the PH_3 passivation [15].

Apart from the poor nFET performances, EOT scaling is also an issue for devices with the Si/SiO_x layer as the Ge surface passivation. The EOT of Ge/Si/SiO_x/HfO_2 stacks, extracted from the analysis of *C–V* characteristics (accounting for quantization of charge carriers in the accumulation layer) are shown in Figure 8.29. Fitting these data with Equation (8.25) allows estimating the interfacial SiO_x thickness and the dielectric constant of HfO_2 to be 0.65 nm and 19.5, respectively. This result is very similar to the results obtained on Si substrates with a

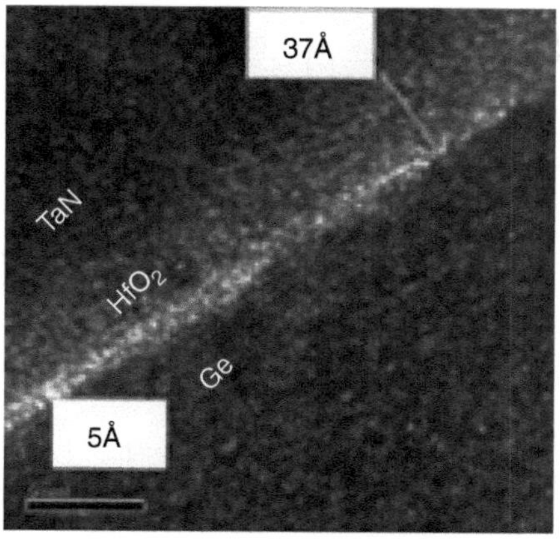

Figure 8.30. High-resolution cross-sectional TEM picture of a Ge/GeO/HfO_2 gate stack with a Ge surface exposed to a PH_3 plasma at 400°C (after Whang *et al.* [15]).

HfO_2 high-κ gate dielectric. In other words, the EOT scaling possibilities seem not better than in the Si substrate case with a HfO_2 high-κ gate dielectric. This means that the Si/SiO_x passivation technique will probably not allow fabricating Ge devices with EOT values as low as 0.5 nm, which are however a requirement for aggressively scaled high-performances devices in the sub-45 nm nodes.

The poor nFET performance and the limited EOT scaling possibilities motivate the ongoing search for alternative surface passivation techniques and/or gate dielectrics. Recent findings related to these topics are briefly reviewed in the next sections.

8.3.5. PH_3 surface passivation

HfO_2-based MOS devices were fabricated on Ge surfaces exposed to a PH_3 plasma [15]. This surface treatment was shown to reduce the formation of the GeO interlayer, as well as to prevent Ge and Hf out-diffusion. A typical TEM picture of such a gate stack is shown in Figure 8.30. An extremely thin interlayer is observed, typically 0.3–0.4 nm thick. Reduced interlayer thickness, which is attributed to formation of P–O bonds during plasma treatment, enables to scale EOT down to about 0.7 nm, with low-leakage current of the order of $1\,mA\,cm^{-2}$ at V_G–V_{FB} = 1 V [15], as shown in Figure 8.31.

As shown in Figure 8.32, gate stacks with Ge surfaces exposed to a PH_3 plasma also exhibit very good *C–V* characteristics, both on n- and p-type substrates.

Well-behaved n- and p-channel FETs were fabricated using the Ge surface exposure to a PH_3 plasma [15]. The peak hole and electron mobilities were reported to be about 125 and $300\,cm^2V^{-1}s^{-1}$, respectively, corresponding to an improvement in channel mobilities by a factor 1.6 (electrons) and 1.8 (holes) compared to Si-based FETs.

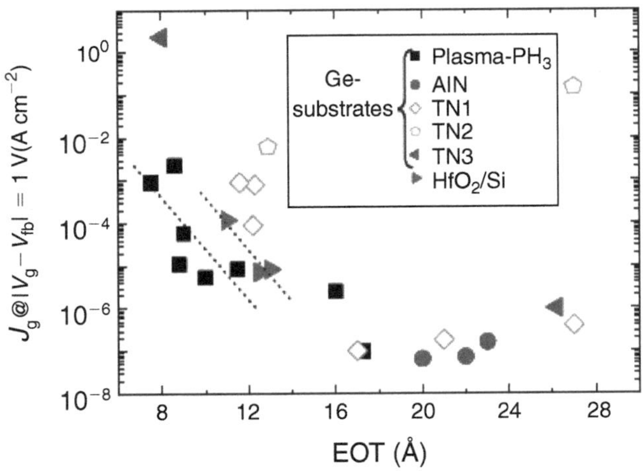

Figure 8.31. Gate-leakage current at $|V_G–V_{FB}| = 1$ V as a function of EOT of Ge/HfO_2 gate stacks with different Ge surface preparations. TN stands for thermal nitridation of the surface. Data for AlN layers deposited on the Ge surface prior to HfO_2 deposition are also shown (after Whang *et al.* [15]).

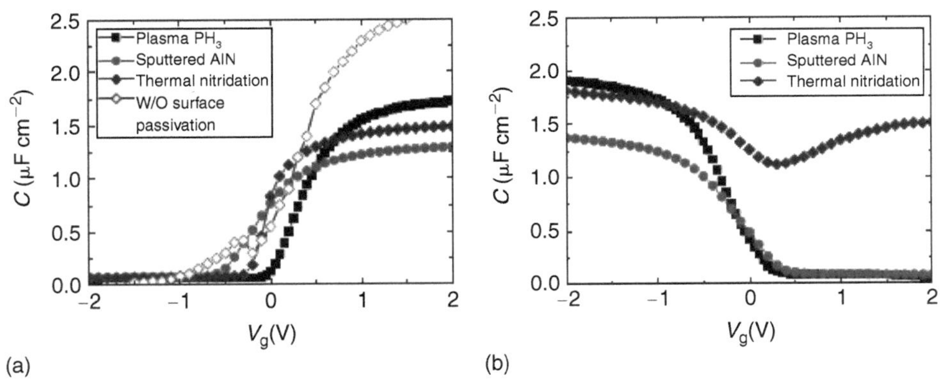

Figure 8.32. High-frequency *C–V* characteristics of n-type (a) and p-type (b) Ge/HfO_2/TaN gate stacks with different Ge surface preparations (after Whang *et al.* [15]).

8.3.6. Alternative high-κ on Ge

Alternative high-κ dielectrics to HfO_2 and ZrO_2 have been investigated recently for their potential use in Ge-based MOSFETs. Among these materials, Al_2O_3 and $LaAlO_3$ have attracted much attention, because they can be deposited directly on Ge. These materials have been successfully integrated into Ge-based MOSFETs, with very promising electrical characteristics [9,63,64]. As an illustration, the I_D–V_D characteristics and channel mobilities of $LaAlO_3$-based n and pMOSFETs are shown in Figures 8.33 and 8.34, respectively [9]. Both types of devices are functioning, with carrier mobilities over those observed in Si-based devices.

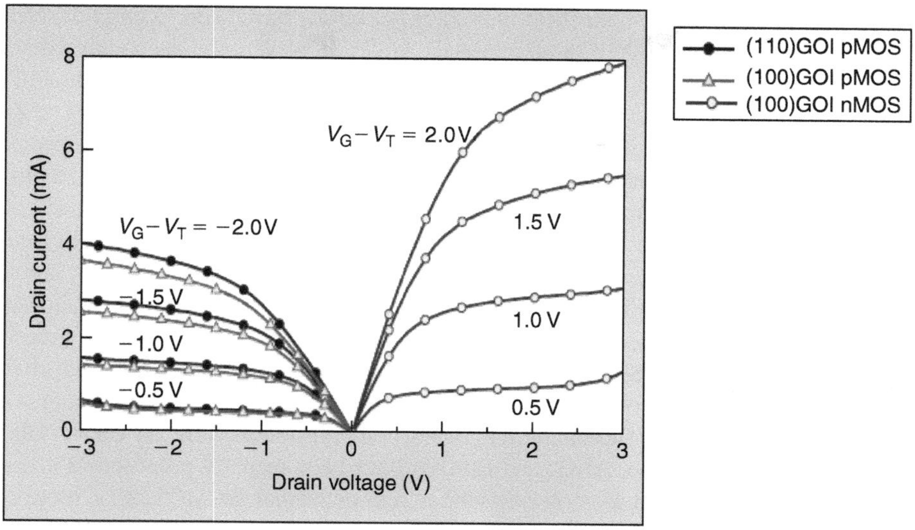

Figure 8.33. I_D–V_D characteristics of n- and pMOSFETs with Ge/$LaAlO_3$/metal gate stacks. Devices were fabricated on germanium-on-insulator (GOI) substrates (after Chin and McAlister [9]).

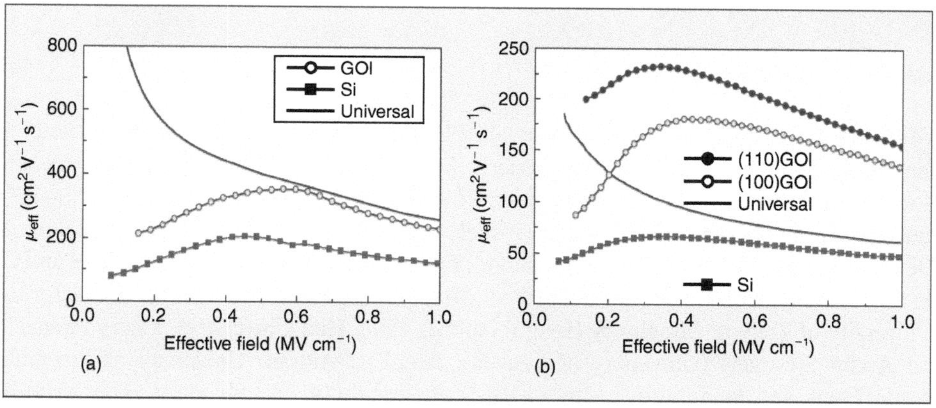

Figure 8.34. Electron (a) and hole (b) mobilities of Ge/$LaAlO_3$/metal gate-based MOSFETs. Devices were fabricated on germanium-on-insulator (GOI) substrates. Data for Si-based devices and the universal mobility curves for Si/SiO_2-based devices are also shown for comparison (after Chin and McAlister [9]).

8.4. Conclusion

The need for high-κ gate dielectrics and metal gates into advanced CMOS technologies has re-opened the door to germanium, as a potential replacement for silicon. Recent achievements on the fundamental building blocks of MOSFET, i.e. the p–n junctions and the Ge/high-κ/metal gate stacks have been presented and discussed. One of the fundamental issues related to the operation of these devices consists in the proper passivation of the Ge surface, since this material lacks a good thermal oxide. Since HfO_2 and ZrO_2 are the most likely candidate high-κ gate dielectrics on Si, a lot of efforts have been recently devoted to the deposition and integration of these materials on Ge. However, non-functioning devices are obtained when attempting to deposit them directly on the Ge surface. Passivation of the surface by growing or depositing a GeON interlayer, in combination with HfO_2 or ZrO_2 high-κ layers, produces decent pMOSFET characteristics, but is hampered by the presence of a large density of fixed positive charges, as well as a large density of interface states in the upper part of the Ge band gap, resulting in non-functioning nMOSFETs. An alternative surface preparation method, based on the deposition of an ultra-thin Si epi layer, followed by its partial oxidation, leads to much improved pMOSFET performances with peak mobility values well above Si pMOSFET ones. The nMOSFET characteristics are still poor, however. Another passivation scheme, using Ge surface exposure to a PH_3 plasma, has been recently reported to give good nMOSFET characteristics while retaining decent pMOSFET characteristics. On the other hand, other high-κ materials than HfO_2 and ZrO_2 are also being investigated, with the hope that such materials could be deposited directly on Ge, forming a properly passivated interface. Al_2O_3 and $LaAlO_3$ have been recently proposed as possible candidates. First results lead to good behaving n- and p-channel Ge-MOSFETs, more confirmation of these results being required. Finally, it should be pointed out that further optimization of the high-κ deposition process, fundamental understanding of the high-κ/Ge interaction and progress towards the integration of these materials into Ge-based devices is necessary to fulfill the requirements of sub-45 nm CMOS generations.

Acknowledgements

We are grateful to our colleagues for their valuable contribution to the work presented here: Jan Van Steenbergen, Renaud Bonzom, Frederik Leys, Sven Van Elshocht, Annelies Delabie, Florence Bellenger, Gareth Nicholas, Thierry Conard, Roger Loo, and Ben Kaczer (IMEC), Paul Zimmerman and David Brunco (Intel Assignees at IMEC), Thanasis Dimoulas (Demokritos) and Geoffroy Raskin (Umicore). Fruitful discussions with Wilman Tsai and Chi On Chui (Intel), Krishna Saraswat and Paul McIntyre (Stanford University), Akira Toriumi (University of Tokyo), Minghwey Hong (National Tsing Hua University), Valery Afanas'ev and Andre Stesmans (University of Leuven), Jean-Luc Autran (University of Provence), Martin Frank and Jean-Pierre Locquet (IBM) are gratefully acknowledged. This work has been financially supported by the IMEC industrial affiliation program on Ge and III-V devices and the European Project IST-ET4US-2048 "Epitaxial Technologies for Ultimate Scaling".

References

[1] G.D. Wilk, R.M. Wallace and J.M. Anthony, *J. Appl. Phys.* **89** (2001) 5243.

[2] For a recent review, see *High-κ Gate Dielectrics*, edited by M. Houssa, Institute of Physics Publishing, London, 2004.

[3] D.P. Norton, J.D. Budai and M.F. Chisholm, *Appl. Phys. Lett.* **76** (2000) 1677.

[4] C.O. Chui, H. Kim, D. Chi, B.B. Triplett, P.C. McIntyre and K.C. Saraswat, *IEDM Technical Digest IEEE,* Piscataway NJ, 2002, p. 437.

[5] H. Kim, C.O. Chui, K. Saraswat and P.C. McIntyre, *Appl. Phys. Lett.* **83** (2003) 2647.

[6] H. Shang, H. Okorn-Schmidt, J. Ott, P. Kozlowski, S. Steen, E.C. Jones, H.S. Wong and W. Hanesch, *IEEE Electron Dev. Lett.* **24** (2003) 242.

[7] A. Ritenour, S. Yu, M.L. Lee, N. Lu, W. Bai, A. Pitera, E.A. Fitzgerald, D.L. Kwong and D. Antoniadis, *IEDM Technical Digest IEEE,* Piscataway NJ, 2003, p. 433.

[8] W.P. Bai, N. Lu, J. Liu, A. Ramirez, D.L. Kwong, D. Wristers, A. Ritenour, L. Lee and D. Antoniadis, *VLSI Technical Digest IEEE,* Piscataway NJ, 2003, p. 121.

[9] A. Chin and S.P. McAlister, *IEEE Circuit Dev. Mag.* (2005) 27.

[10] E.J. Preisler, S. Guha, B.R. Perkins, D. Kazazis and A. Zaslavsky, *Appl. Phys. Lett.* **86** (2005) 223504.

[11] K. Prabhakaran, F. Maeda, Y. Watanabe and T. Ogino, *Appl. Phys. Lett.* **76** (2000) 2244.

[12] D.J. Hymes and J.J. Rosenberg, *J. Electrochem. Soc.* **135** (1988) 961.

[13] C.O. Chui, F. Ito and K.C. Saraswat, *IEEE Electron Dev. Lett.* **25** (2004) 613.

[14] N. Wu, Q. Zhang, C. Zhu, D.S. Chan, M.F. Li, N. Balasubramanian, A. Chin and D.L. Kwong, *Appl. Phys. Lett.* **85** (2004) 4127.

[15] S.J. Whang, S.J. Lee, F. Gao, N. Wu, C.X. Zhu, J.S. Pan, L.J. Tang and D.L. Kwong, *IEDM Technical Digest IEEE,* Piscataway NJ, 2004, p. 307.

[16] T. Weser, A. Bogen, B. Konrad, R.D. Schnell, C.A. Schug and W. Steinmann, *Phys. Rev.* B **35** (1987) 8184.

[17] P. Kruger and J. Pollmann, *Phys. Rev. Lett.* **64** (1990) 1808.

[18] S.M. Sze, 'Physics of semiconductor devices', John Wiley & Sons, New York, 1981.

[19] G.W. Neudeck, 'The pn junction diode', In: 2nd *Modular Series on Solid State Devices*, edited by G.W. Neudeck and R.F. Pierret, Addison-Wesley Publishing Company 1989.

[20] W. Shockley and W.T. Read Jr., *Phys. Rev.* **87** (1952) 835.

[21] R.N. Hall, *Phys. Rev.* **87** (1952) 387.

[22] D.K. Schroder, *IEEE Trans. Electron Dev.* **29** (1982) 1336.

[23] E. Simoen and C. Claeys, *IEEE Trans. Electron Dev.* **46** (1999) 1487.

[24] Y. Murakami and T. Shingyouji, *J. Appl. Phys.* **75** (1994) 3548.

[25] W.C. Johnson and P.T. Panousis, *IEEE Trans. Electron Dev.* **18** (1971) 965.

[26] A.S. Grove and D.J. Fitzgerald, *Solid-St. Electron.* **9** (1966) 783.

[27] D.J. Fitzgerald and A.S. Grove, *Surface Sci.* **9** (1968) 347.

[28] A. Czerwinski, E. Simoen and C. Claeys, *Appl. Phys. Lett.* **72** (1998) 3503.

[29] Y.P. Varshni, *Physica* **34** (1967) 149.

[30] J. Vanhellemont, E. Simoen, A. Kaniava, M. Libezny and C. Claeys, *J. Appl. Phys.* **77** (1995) 5669.

[31] M. Randolph and L.G. Meiners, *J. Electrochem. Soc.* **136** (1989) 2699.

[32] E.E. Haller, W.L. Hansen and F.S. Goulding, *Adv. Phys.* **30** (1981) 93.

[33] F.S. Goucher, G.L. Pearson, M. Sparks, G.K. Teal and W. Shockley, *Phys. Rev.* **81** (1951) 637.

[34] G.K. Teal, M. Sparks and E. Buehler, *Proc. I.R.E.* **40** (1952) 906.

[35] R.N. Hall and W.C. Dunlap, *Phys. Rev.* **80** (1950) 467.

[36] J.S. Saby and W.C. Dunlap Jr., *Phys. Rev.* **90** (1953) 630.

[37] D.R. Muss and R.F. Greene, *J. Appl. Phys.* **29** (1958) 1534.

[38] G.S. Hubbard, E.E. Haller and W.L. Hansen, *IEEE Trans. Nucl. Sci.* **24** (1977) 161.

[39] G. Riepe and D. Protić, *Nucl. Instrum. Meth.* **165** (1979) 31.

[40] A.J. Tavendale, *IEEE Trans. Nucl. Sci.* **12** (1965) 255.

[41] W.L. Hansen, E.E. Haller and G.S. Hubbard, *IEEE Trans. Nucl. Sci.* **27** (1980) 247.

[42] L. Esaki, *Phys. Rev.* **109** (1958) 603.

[43] R.S. Claassen, *J. Appl. Phys.* **32** (1961) 2372.

[44] D. Meyerhofer, G.A. Brown and H.S. Sommers Jr., *Phys. Rev.* **126** (1962) 1329.

[45] D.E. Sawyer and R.H. Rediker, *Proc. I.R.E.* **46** (1958) 1122.

[46] H. Ando, H. Kanbe, T. Kimura, T. Yamaoka and T. Kaneda, *IEEE J. Quant. Electron.* **14** (1978) 804.

[47] T. Mikawa, S. Kagawa, T. Kaneda, T. Sakurai, H. Ando and O. Mikami, *IEEE J. Quant. Electron.* **17** (1981) 210.

[48] M.C. Brain, *IEEE J. Quant. Electron.* **18** (1982) 219.

[49] S. Van Elshocht, B. Brijs, M. Caymax, T. Conard, S. De Gendt, S. Kubicek, M. Meuris, B. Onsia, O. Richard, I. Teerlinck, J. Van Steenbergen, C. Zhao and M. Heyns, *Mat. Res. Soc. Symp., Proc.* **811** (2004) 163.

[50] A. Dimoulas, G. Mavrou, G. Vellianitis, E. Evangelou, N. Boukos, M. Houssa and M. Caymax, *Appl. Phys. Lett.* **86** (2005) 032908.

[51] M. Houssa, J.L. Autran, A. Stesmans and M.M. Heyns, *Appl. Phys. Lett.* **81** (2002) 709.

[52] P. Masson, J.L. Autran, M. Houssa, X. Garros and Ch. Leroux, *Appl. Phys. Lett.* **81** (2002) 3392.

[53] G. Apostolopoulos, G. Vellianitis, A. Dimoulas, J.C. Hooker and T. Conard, *Appl. Phys. Lett.* **84** (2004) 260.

[54] M. Houssa, B. De Jaeger, A. Delabie, S. Van Elshocht, V.V. Afanas'ev, J.L. Autran, A. Stesmans, M. Meuris and M.M. Heyns, *J. Non-Crystalline Sol.* **351** (2005) 1902.

[55] V.V. Afanas'ev, Y.G. Fedorenko and A. Stesmans, *Appl. Phys. Lett.* **87** (2005) 032107.

[56] M. Houssa, T. Conard, J. Van Steenbergen, G. Mavrou, Y. Panayiotatos, A. Dimoulas, M. Meuris, M. Caymax and M.M. Heyns, G. Nicholas, *ECS Trans.* Vol. 1, issue 5, p. 9 (2006).

[57] C.O. Chui, H. Kim, P.C. McIntyre and K.C. Saraswat, *IEDM Technical Digest IEEE, Piscataway* NJ, 2003 p.437.

[58] J. Croon, B. Kaczer, G.S. Lujan, S. Kubicek, G. Groeseneken and M. Meuris, *Proc. ICMTS* **18** (2005) 191.

[59] K. Martens, B. Kaczer, P. Roussel, G. Groeseneken, and H. Maes, *ECS Trans.* Vol. 1, issue 5, p. 27 (2006).

[60] R. Bonzom, F. Leys, R. Loo, O. Richard, D. Vanhaeren, J. Rip, J. Van Steenbergen, B. De Jaeger, H. Bender, W. Vandervorst, M. Caymax, and M. Meuris, Presented at the Symposium on *In Situ Studies of Gas/Solid Surface Reaction Dynamics*, MRS Spring Meeting, 2005.

[61] B. De Jaeger, R. Bonzom, F. Leys, O. Richard, J. Van Steenbergen, G. Winderickx, E. Van Moorhem, G. Raskin, F. Letertre, T. Billon, M. Meuris and M. Heyns, *Microel. Eng.* **80** (2005) 82.

[62] B. De Jaeger, M. Houssa, A. Satta, S. Kubicek, P. Verheyen, J. Van Steenbergen, J. Croon, B. Kaczer, S. Van Elshocht, A. Delabie, E. Kunnen, E. Sleeckx, I. Teerlinck, R. Lindsay, T. Schram, T. Chiarella, R. Degraeve, T. Conard, J. Poortmans, G. Winderickx, W. Boullart, M. Schaekers, P.W. Mertens, M. Caymax, W. Vandervorst, E. Van Moorhem, S. Biesemans, K. De Meyer, L. Ragnarsson, S. Lee, G. Kota, G. Raskin, P. Mijlemans, J.-L. Autran, V. Afanas'ev, A. Stesmans, M. Meuris and M. Heyns, *Proc. ESSDERC* (2004) 189.

[63] D.S. Yu, C.H. Huang, A. Chin, C. Zhu, M.F. Li, B.J. Cho and D.L. Kwong, *IEEE Electron Dev. Lett.* **25** 138 (2004) 138.

[64] J.J.-H. Chen, N.A. Bojarczuk, J.H. Shang, M. Copel, J.B. Hannon, J. Karasinski, E. Preisler, S.K. Banerjee and S. Guha, *IEEE Trans. Electron Dev.* **51** (2004) 1441.

CHAPTER 9

Device Modeling

David Esseni, Pierpaolo Palestri and Enrico Sangiorgi

9.1. Introduction

In this chapter we describe the modeling approaches developed for the simulation of germanium devices. The main focus will be on metal-insulator-semiconductor (MIS) devices, with particular attention on germanium-on-insulator (GOI) structures. Most of the approaches were originally developed for silicon devices, and thus we will describe them briefly, stressing the differences between Si and Ge and how they translate in different modeling requirements.

The chapter is organized according to a "bottom-up" structure, beginning with Section 9.2 which describes the main differences between the basic properties of Ge and Si, and then spanning from band-structure calculation up to the determination of the drain current in MIS devices for both n-channel and p-channel transistors. However, so far, most of the models have been developed for n-channel devices and cannot be easily extended to p-channel transistors. This is reflected also in the material presented in this chapter, which is more complete and richer of examples for n-channel than it is for p-channel transistors.

Section 9.3 is devoted to band-structure calculation. Since the main interest is in MIS devices, where carriers are quantized in a 2D inversion layer, a relevant fraction of the section is focused on the calculation of the energy states in 2D systems.

In the semi-classical physical framework that we will consider hereafter, the transport modeling is essentially based on the Boltzmann transport equation (BTE), whose general solution is very complex because the scattering integrals make the stationary BTE an integral–differential equation in a multi-dimensional space of the phases (which has six dimensions for a bulk

Germanium-Based Technologies: From Materials to Devices. C. Claeys and E. Simoen (Editors)

semiconductor and four dimensions for a 2D inversion layer). A dramatic simplification is obtained by neglecting the scattering term, which leads to the ballistic transport regime. This simplified approach is mostly useful to investigate the upper-limits of the performance that can be attained with a device structure, so that it can be used for a preliminary investigation of the possible benefits related to new device structures, such as Ge channel Metal-oxide-semiconductor field effect transistor (MOSFETs). This aspect is described in Section 9.4, where comparisons between Si and Ge devices are provided.

The solution of the BTE beyond the ballistic approximation is considered in Section 9.5, starting from approximate solutions such as the Drift-Diffusion approach, and then considering more accurate approaches, such as the Monte-Carlo (MC) method.

Finally, in Section 9.6 we will draw our conclusion and propose a "roadmap" for the forthcoming activities in the field of the simulation of advanced Ge devices.

9.2. Modeling Germanium versus Silicon

In this section we will summarize the main differences between Si and Ge that are relevant in order to extend the simulation tools, originally developed for Si devices, to the analysis of Ge devices. Some points will be described extensively in the following sections, so that they will be only briefly mentioned here.

The main differences between semiconductor materials (e.g. Si and Ge) can be traced back to the different band structure. As it will be shown in the next section, the conduction band of Ge exhibits relevant differences with respect to the Si one: the absolute energy minima are placed in different positions of the First-Brillouin-Zone (1BZ) (L instead of Δ) and secondary minima exist, which are very close in energy to the lowest minima. The implications of these aspects on the modeling approaches used for carrier quantization and drain current computation will be described in Sections 9.3, 9.4 and 9.5.

On the other hand, the valence band structure of Ge is similar to the one of Si, only the parameters that describe the energy minima are different, so that the simulation of Ge pMOSFETs can follow the methodologies developed for Si devices.

Another important feature of the band structure is the band gap. Since the band gap of Ge is $E_G = 0.66\,\text{V}$ [1], this material has a lower threshold energy for impact ionization than Si has. Thus, the generation of electron-hole pairs by impact ionization is expected to play an important role also for Ge devices operating with a low supply voltage (e.g. 1.1 V for the 65-nm technology node of the International Technology Road map for semiconductors). The impact ionization scattering rate for electrons in Si and Ge, as obtained in [2,3], are reported in Figure 9.1. However, in conventional simulators, based on the Drift-Diffusion approach (see Section 9.5.3), the impact ionization is defined only in terms of ionization coefficients, i.e. the number of electron-hole pairs generated per unit length in the presence of a given electric field. The ionization coefficients for electrons and holes in Ge are reported in Figure 9.2.

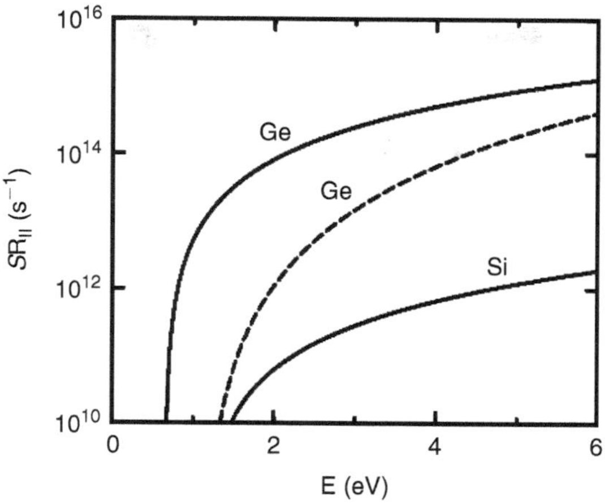

Figure 9.1. Impact ionization scattering rate for electrons in Si and Ge as a function of the energy of the primary electron. Solid lines are from [3], dashed line is from [2]. In all cases, the scattering rates have been calculated by assuming the Keldysh expression $SR_{II} = P_{II}((E - E_{th})/E_{th})^{\eta}$, adjusting the parameters in order to reproduce the measured ionization coefficients. The difference between the scattering rates in [2,3] is due to the fact that different combinations of parameters can provide the same ionization coefficients, in particular if the phonon models are different.

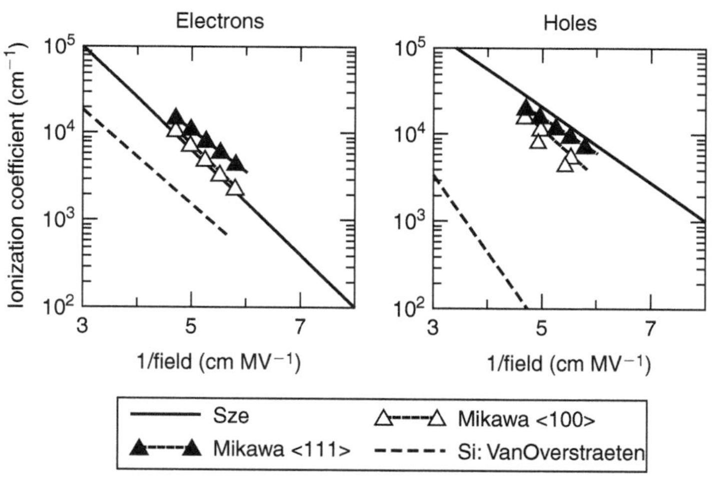

Figure 9.2. Impact ionization coefficient (ionization rates) for electrons (left plot) and holes (right plot). Data for Ge in [1,4] are compared with data for Si from [5].

The small band gap E_G of Ge also implies a high corresponding intrinsic concentration $n_i = 2 \times 10^{13}\text{cm}^{-3}$ at 300 K, see Figure 9.3, resulting in large junction leakage. The correct evaluation of junction leakage is thus essential in order to assess the possible advantages of Ge structures. The leakage can be drastically reduced by using GOI structures, since the area of the junctions is reduced.

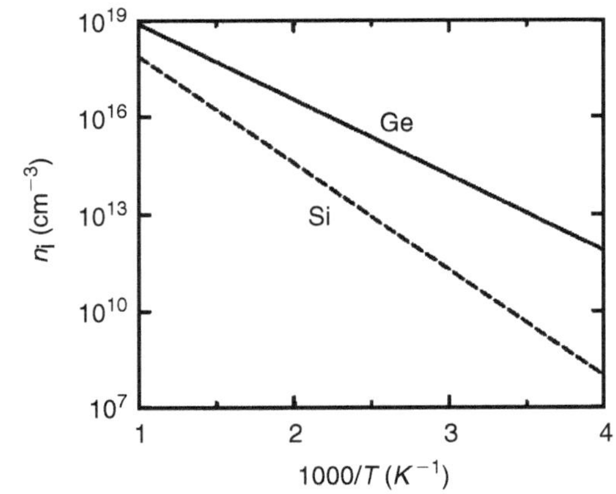

Figure 9.3. Intrinsic carrier density as a function of the inverse of the temperature. Data from Ref. [1].

Due to the lower thermal conductivity of Ge (0.6 W cm^{-1} K^{-1} instead of 1.48 W cm^{-1} K^{-1} in Si), self-heating effects play an important role in determining the performance of Ge devices; this is particularly true in the case of GOI structures, where the effects are emphasized by the very low thermal conductivity of the buried insulator layer. A comparison between Si and Ge ultra-scaled devices in the presence of self-heating is presented in [6].

Alternative dielectrics must be used in Ge based MIS devices, due to the low stability of GeO_2. This will be discussed more in detail in Chapter 10. Most of the high-κ dielectrics lead to reduced mobility of the inversion layer because of the scattering mechanisms associated to remote polar phonons (see Section 9.5.6), charge trapped in the dielectric and poor quality of the semiconductor–dielectric interface. All these effects must be taken into account when simulating Ge based MIS devices.

9.3. Band Structure

The relationship between the energy of the carriers and their wave-vector, referred as band structure, is one of the main ingredients of physically based modeling approaches.

From the band structure it is possible to evaluate important parameters such as the density-of-states (DoS) and the group velocity. In this section we discuss the structure of the conduction and valence bands of germanium. We will start considering the bulk material (i.e. without any confining potential) and then we will extend our analysis to the case of 2D germanium inversion layers, where quantization effects play an important role.

9.3.1. Conduction band of bulk germanium

The band structure of bulk materials can be calculated using different approaches: Tight-Binding [7], Non-local pseudo-potentials [8] and the k.p method (this latter is essentially used for the hole band structure) [9]. In this section we analyze the conduction band structure as obtained using pseudo-potentials. A brief explanation of the method is provided in the following.

The Schrödinger equation for electrons in a crystal can be written as:

$$\left[-\frac{\eta^2}{2m_0}\nabla^2 + V_P(\vec{r})\right]\psi(\vec{r}) = E\psi(\vec{r}), \qquad (9.1)$$

where the crystal potential V_P is periodic in the real space.

Since the solution of the Schrödinger equation are Bloch function, they can be written as:

$$\psi_{\vec{K},n}(\vec{r}) = \sum_{\vec{G}} f_{\vec{K},n}(\vec{G})\exp\left[j(\vec{K}+\vec{G})\bullet\vec{r}\right] \qquad (9.2)$$

where the sum extends over *all the vectors* of the reciprocal space. For a given $\vec{\kappa}$, different solutions refer to different bands (index n).

By substituting Equation (9.2) in Equation (9.1) one obtains:

$$\frac{\eta^2}{2m_0}\left|\vec{K}+\vec{G}\right|^2 f_{\vec{K},n}(\vec{G}) + \sum_{\vec{G}'} f_{\vec{K},n}(\vec{G}')V\left(\vec{G}-\vec{G}'\right) = E_{FB}(\vec{K},n)f_{\vec{K},n}(\vec{G}) \qquad (9.3)$$

The energy dispersion relationship E_{FB} as a function of the wave-vector $\vec{\kappa}$ can thus be obtained by solving for the eigenvalues of Equation (9.3), where $\vec{\kappa}$ is a parameter of the equation. The terms $V|\vec{G}–\vec{G}|$ are the Fourier components of the crystal potential. In the empirical pseudo-potential method, considering group IV semiconductors (as Si and Ge), these coefficients are assumed to depend only on $|\vec{G}–\vec{G}'|$ and they are used as fitting parameters to reproduce reflectivity and photoemission experiments. Usually, only the terms $V(\sqrt{3})$, $V(\sqrt{8})$, $V(\sqrt{11})$ are considered (where the vectors of the reciprocal space are indicated in units of $2\pi/a_0$, with a_0 the lattice parameter).

The approach just described is the so-called "Local-Pseudo-Potential" method. The Non-Local method also accounts for a non-local term in the potential $V_P(\vec{r})$ of Equation (9.1) [8]. Furthermore, the spin-orbit coupling can be included, as described in [10].

The conduction band profile obtained by using our non-local-pseudo-potential solver is reported in Figure 9.4. We have used the same parameters as [11] except for the value of the potential $V(\sqrt{8}) = 0.019$Ry, instead of the value 0.0019Ry reported in [11], that appears to be a typing error. Spin-orbit coupling has been neglected.

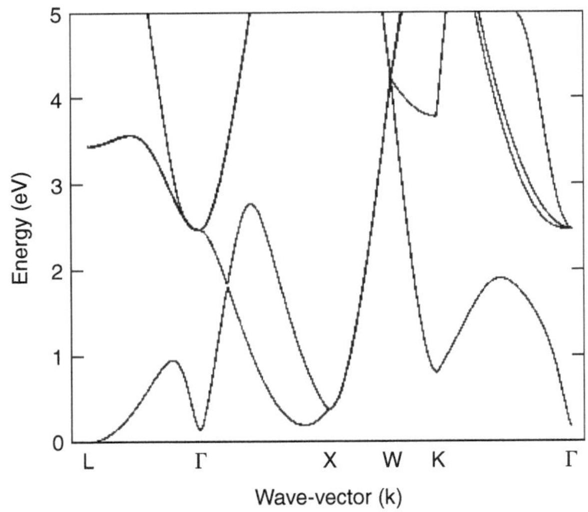

Figure 9.4. Full-band structure of the conduction band of Ge as obtained using the non-local pseudo-potential approach.

We observe that the minimum energy is located at the L-point. Since there are 8 of these minima in the 1BZ, but only *a half of each minima* lies inside the 1BZ, the correct multiplicity to be used for the calculation of the DoS is 4. Other 6 minima (referred as Δ) are placed between the Γ and the X points, as in silicon. Finally, a third minimum is found at the Γ point. The secondary minima are relatively close to the minima in L and are relevant for high-field transport. For example, as shown in [12], electron transfer to the Δ minima can produce a negative differential conductivity.

Many practical modeling approaches employed to simulate the electrical characteristics of semiconductor devices cannot handle the semiconductor full-band structure, but only its approximation in the proximity of the energy minima with a non-parabolic dispersion relationship:

$$E(1+\alpha E) = \frac{\eta^2 k_l^2}{2m_l} + \frac{\eta^2 k_t^2}{2m_t} \tag{9.4}$$

where α is the non-parabolicity coefficient. The constant energy surfaces are ellipsoids. The subscript l refers to the longitudinal axis of the ellipsoid, whereas t refers to the components of the $\vec{\kappa}$ normal to the principal axis. The symbols m_l and m_t indicate the longitudinal and transversal effective masses, respectively, with m_l larger than m_t. The parameters for the minima of Ge are reported in Table 9.1, which includes the values extracted from Figure 9.4 and those collected from the literature.

9.3.2. Valence band of bulk germanium

The calculation of the valence band structure using non-local-pseudo-potentials is made difficult by the presence of strong spin-orbit coupling [10]. An alternative, simpler

Table 9.1. Parameters of the minima of the conduction band.

	From Figure 9.4	Fischetti [11]	Jacoboni [13]	Rahman [14]
m_l (L) $[m_0]$	1.58	1.57	1.59	1.64
m_t (L) $[m_0]$	0.093	0.094	0.080	0.08
α (L) $[eV^{-1}]$		0.33	0.3	
m_l (Δ) $[m_0]$	0.89	1.85	1.35	0.95
m_t (Δ) $[m_0]$	0.194	0.195	0.29	0.20
$E_{\Delta\to L}$ [eV]	0.19	0.173	0.18	
m (Γ) $[m_0]$	0.049	0.049	0.04	0.049
$E_{\Gamma\to L}$ [eV]	0.145	0.135	0.14	

approach for the valence band calculation in proximity of the band maximum is the k.p method, that has been applied to germanium in [9]. One of the main advantages of the k.p method is that it is easy to include the spin-orbit coupling and the strain. The dispersion relationship is obtained by solving the eigenvalue problem [15]:

$$[H_{KP}(\vec{K}) + H_{SO}]\psi_{\vec{K}} = \left[E_{FB}(\vec{K}) - \frac{\eta^2\,|\vec{K}|^2}{2m_0}\right]\psi_{\vec{K}} \tag{9.5}$$

The 6×6 H_{KP} matrix is typically written as:

$$H_{KP}^{6\times6} = \begin{pmatrix} H_{KP}^{3\times3} & 0^{3\times3} \\ 0^{3\times3} & H_{KP}^{3\times3} \end{pmatrix} \tag{9.6}$$

where:

$$H_{KP}^{3\times3} = \begin{pmatrix} Lk_x^2 + M\left(k_y^2 + k_z^2\right) & Nk_xk_y & Nk_xk_z \\ Nk_xk_y & Lk_y^2 + M\left(k_x^2 + k_z^2\right) & Nk_yk_z \\ Nk_xk_z & Nk_yk_z & Lk_z^2 + M\left(k_x^2 + k_y^2\right) \end{pmatrix} \tag{9.7}$$

The parameters L, M and N are adjusted to reproduce the cyclotron resonance experiments [9] and the calculations obtained with the more sophisticated method of the pseudo-potentials [16].

The eigenvectors ψ are six-element vectors representing the wave-function in the base of the Bloch functions of the unstrained crystal corresponding to $K = 0$. The matrix H_{SO}

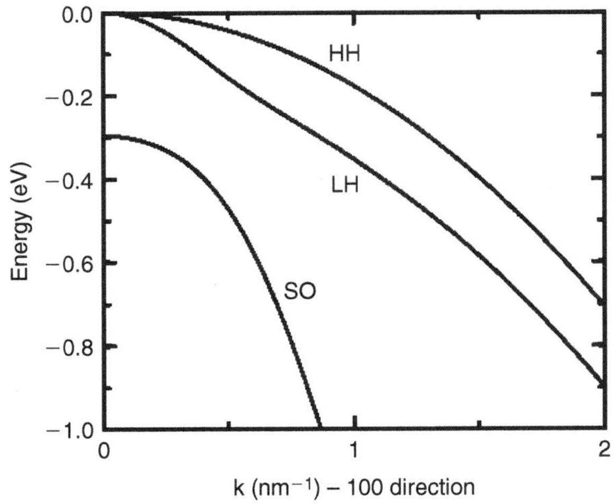

Figure 9.5. Valence band of unstrained Ge calculated using the k.p method. Parameters from [16] have been modified by adding $-\eta^2/2m_0$ to L and M, since in [16] the left-hand-side of the Schrödinger equation does not include the term $-\eta^2/2m_0$. $L = -120.13\,\mathrm{eV\AA^{-2}}$, $M = -21.49\,\mathrm{eV\AA^{-2}}$, $N = -128.17\,\mathrm{eV\AA^{-2}}$, $\Delta_{SO} = 0.296\,\mathrm{eV}$.

accounts for the spin-orbit coupling and for a 6 × 6 Hamiltonian, it is given by [15]:

$$H_{SO}^{6\times 6} = -\frac{\Delta_{SO}}{3}\begin{pmatrix} 0 & j & 0 & 0 & 0 & -1 \\ -j & 0 & 0 & 0 & 0 & j \\ 0 & 0 & 0 & 1 & -j & 0 \\ 0 & 0 & 1 & 0 & -j & 0 \\ 0 & 0 & j & j & 0 & 0 \\ -1 & -j & 0 & 0 & 0 & 0 \end{pmatrix} \tag{9.8}$$

where Δ_{SO} is the spin-orbit splitting in $K = 0$ and j is the imaginary unit. Figure 9.5 reports the k.p calculation of the valence band of unstrained Ge. As in silicon, we observe a Heavy-Hole, a Light-Hole and a Split-Off band.

In the proximity of the valence band maxima, the full-band structure can be approximated with a warped relationship:

$$E_{FB}\left(\vec{K} = \left(k_x, k_y, k_z\right)\right) = -\frac{\eta^2}{2m_0}\left(A\left|\vec{K}\right|^2 \pm \sqrt{B^2\left|\vec{K}\right|^4 + C^2\left(k_x^2k_y^2 + k_y^2k_z^2 + k_x^2k_z^2\right)}\right) \tag{9.9}$$

The signs + and − refer to light-and heavy-holes, respectively. The parameters for Ge [9] are $A = 13.0$, $B = 8.9$, and $C = 10.3$. The above expression does not take into account the non-parabolicity. Extension to the non-parabolic case have been proposed in [17] for the valence band of Si.

9.3.3. *Energy dispersion in germanium inversion layers: electrons*

In the 2D inversion layer of MIS devices, the confining potential varies over distances comparable with the electron wave-length. As a result, electrons do not behave as free carriers in the quantization direction normal to the semiconductor–oxide interface (z direction) and a so-called 2D electron gas (2Deg) is obtained. The band structure of the 2Deg consists of different subbands with its own dispersion relationship $E_\mu(k_x,k_y)$, where μ is the subband index.

In principle, $E_\mu(k_x,k_y)$ should be obtained by solving the Schrödinger equation considering both the crystal potential and the confining potential. When considering thin semiconductor films, this can be implemented using the Tight-Binding method applied at the atomistic structure of the film. The band structure of thin Ge films has been calculated using the Tight-Binding method in [18]. As an alternative, the Linear Combination of Bulk Bands (LCBB) method can be used [19–21]. In the LCBB method the unknown wave-function of the 2Deg is written as linear combination of the Bloch functions of the bulk crystal (that can be computed using the non-local pseudo-potential method):

$$\psi^{2D}_{k_x,k_y} = \sum_{n,k_z} \mathrm{A}(k_x,k_y,k_z,n)\psi k_x,k_y,k_z,n \tag{9.10}$$

As explained in [19], the range of k_z that must be included in the expansion is equal to the periodicity length in the 3D reciprocal lattice space, hence it depends on the quantization direction. The expansion volume is thus defined as a prism having the same volume as the 1BZ in the 3D reciprocal lattice space of the bulk crystal and having as a base the 1BZ of the 2Deg. By taking the appropriate expansion volume the Schrödinger equation becomes a separate eigenvalue problem for each in-plane vector (k_x,k_y).

Results for Ge inversion layers [19] are reported in Figure 9.6, considering different quantization directions.

The calculation procedures based on either the Tight-Binding or the pseudo-potential approach are time consuming. For example, the approach requires the solution of the Schrödinger equation for each vector in the k-plane. The DoS and the electron concentration have to be calculated numerically, so that the self-consistent calculation of the potential profile through the Poisson equation becomes almost computationally prohibitive even at equilibrium, that is even when we know *a priori* that the electron states are occupied according to the Fermi–Dirac distribution function. The problem can be simplified by exploiting the effective-mass-approximation (EMA) [22]: in proximity of the energy minima of the bulk band structure, we can write the Schrödinger equation as:

$$\left[E_{\mathrm{FB}}\left(k_x,k_y,-\mathrm{j}\frac{\mathrm{d}}{\mathrm{d}z}\right)+V(z)\right]\psi = E_\mu(k_x,k_y)\psi \tag{9.11}$$

where the kinetic operator has been replaced by the dispersion relationship of the bulk crystal. Since k_z has been replaced by an operator, we need an analytical expression for E_{FB}. The expression given by Equation (9.4) is most frequently used by setting $\alpha = 0$.

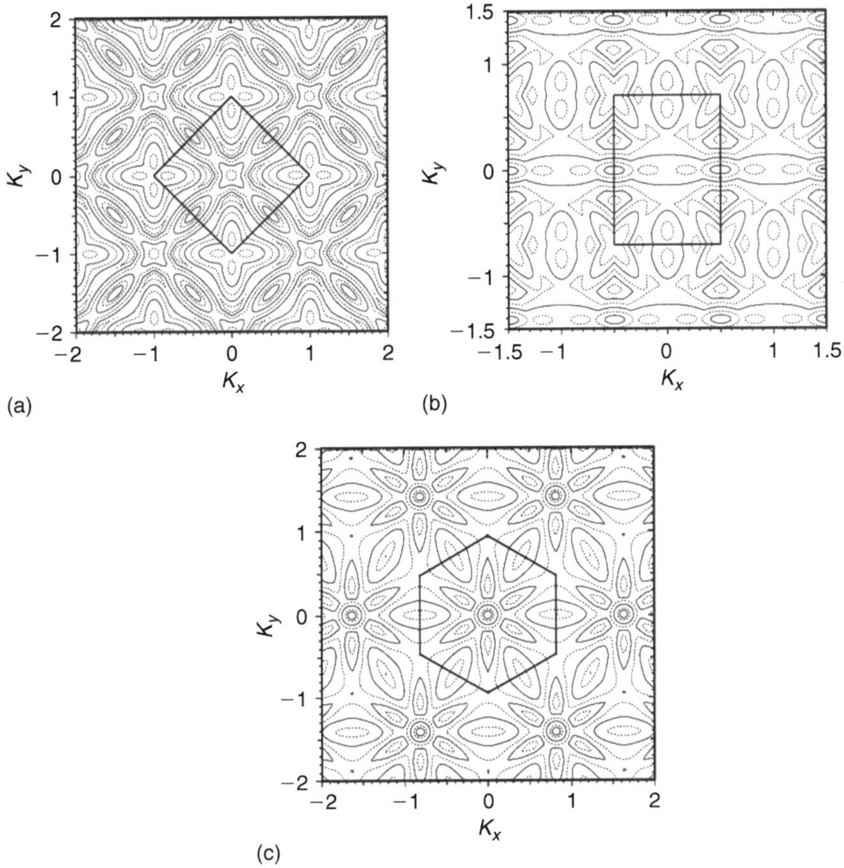

Figure 9.6. Energy dispersion relationship of the lowest subband in Ge inversion layers calculated with the linear combination of bulk states method described in [19]. Different plots refer to different quantization directions: 100 in plot (a), 110 in plot (b) and 111 in plot (c). The confining potential is a quantum well with a thickness of 5 nm and a barrier height of 3 eV. The solid lines indicate the 1BZ of the 2D system.

In the case of silicon with (100) quantization, k_z coincides with one of the principal axis of the constant energy ellipsoids, and the Schrödinger equation reads:

$$-\frac{\eta^2}{2m_z}\frac{d^2\psi}{dz^2} + V(z)\psi = E_{\mu 0}\psi \tag{9.12}$$

where m_z is either m_l or m_t, depending on what minimum we are considering. The dispersion relationship for the μ-th subband is:

$$E_\mu = E_{\mu 0} + \frac{\eta^2 k_x^2}{2m_x} + \frac{\eta^2 k_y^2}{2m_y} \tag{9.13}$$

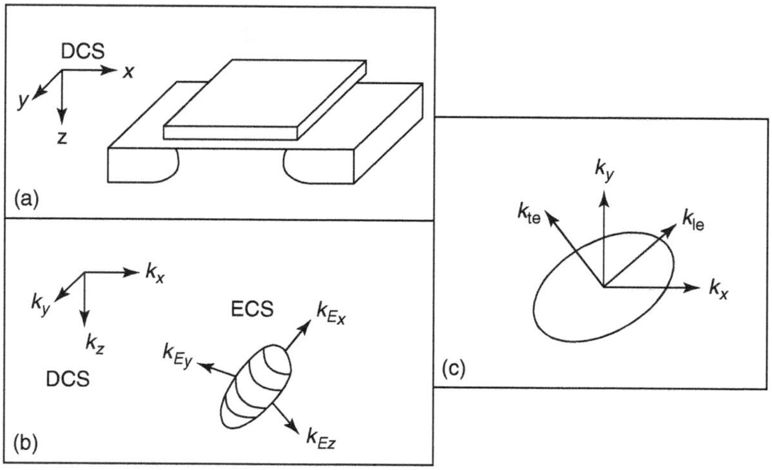

Figure 9.7. Plot (a): definition of the DCS. The quantization direction is z. Plot (b): the ECS differs from the DCS. Plot (c): the constant energy curves in the transport plane are ellipses with principal axis in the direction k_{le} and k_{te}. This latter system of coordinates is indicated as ECS.

The effective masses m_x and m_y are equal to either m_l or m_t. The problem becomes more complex when the quantization direction does not coincide with one of the principal axis of the constant energy ellipsoids. This happens for (111) and (110) silicon (and for the Δ-valleys of Ge) as well as for the L-valleys of Ge. The problem of generalizing the EMA to arbitrary orientation of the ellipsoids has been analyzed first by Stern and Howard [23], and more recently in [14].

With reference to Figure 9.7, the system of coordinates used to describe the device (device coordinate system, DCS), does not coincide with the coordinate system corresponding to the principal axis of the constant energy ellipsoid (ellipsoid coordinate system, ECS). We can define a linear transformation relating the wave-vector in the DCS to the wave-vector in the ECS:

$$\vec{K}_D = \vec{\mathrm{A}}\, \vec{K}_E \tag{9.14}$$

As demonstrated in [23], if we define a matrix:

$$\hat{\mathrm{W}} = \begin{pmatrix} w_{11} & w_{12} & w_{13} \\ w_{21} & w_{22} & w_{23} \\ w_{31} & w_{32} & w_{33} \end{pmatrix} = \hat{\mathrm{A}} \begin{pmatrix} 1/m_l & 0 & 0 \\ 0 & 1/m_t & 0 \\ 0 & 0 & 1/m_t \end{pmatrix} \hat{\mathrm{A}}^{\mathrm{T}} \tag{9.15}$$

the Schrödinger equation becomes:

$$-\frac{\eta^2}{2m_{\mathrm{QM}}} \frac{\mathrm{d}^2\zeta}{\mathrm{d}z^2} + V(z)\zeta(z) = E_{\mu 0}\zeta(z), \tag{9.16}$$

where $m_{QM} = 1/w_{33}$. The dispersion relationship is:

$$E_{\mu} = E_{\mu 0} + \frac{\eta^2}{2}\left[\left(w_{11} - \frac{w_{13}^2}{w_{33}}\right)k_x^2 + \left(w_{22} - \frac{w_{23}^2}{w_{33}}\right)k_y^2 + 2\left(w_{12} - \frac{w_{13}w_{23}}{w_{33}}\right)k_x k_y\right] \tag{9.17}$$

The constant energy curves in the transport plane are ellipses (see Figure 9.7c). By defining a new system of coordinates corresponding to the principal axis of the ellipse (ECS):

$$E_{\mu} = E_{\mu 0} + \frac{\eta^2 k_{\mathrm{le}}^2}{2m_{\mathrm{le}}} + \frac{\eta^2 k_{\mathrm{te}}^2}{2m_{\mathrm{te}}} \tag{9.18}$$

where m_{le} and m_{te} are the longitudinal and transverse mass in the ECS. For example, the ellipses corresponding to subbands originated by the L minima in (100) inversion layers are rotated by 45 degrees with respect to the DCS (see Figure 9.6).

Notice that the wave-function obtained with this approach is given by:

$$\psi(z) = \zeta(z)\exp\left[-\mathrm{j}z\left(w_{13}k_x + w_{23}k_y\right)/w_{33}\right] \tag{9.19}$$

where $\zeta(z)$ is the wave-function defined in Equation (9.15). The values of m_{le}, m_{te} and m_{QM} corresponding to the L and Δ minima for different quantization directions are reported in Table 9.2 (from [23]).

9.3.4. *Energy dispersion in germanium inversion layers: holes*

The hole inversion layer can be studied by using the Tight-Binding as well as the linear combination of bulk states described in the previous section for the electrons. The same issues concerning the computational complexity of the above methods clearly apply even to

Table 9.2. Value of the effective masses for the L and Δ minima considering different quantization directions. The multiplicity of the subbands is indicated with n_ν.

	Δ-valleys				**L-valleys**			
Orientation	m_{te}	m_{le}	m_{QM}	n_ν	m_{le}	m_{te}	m_{QM}	n_ν
100	m_t	m_t	m_l	2	m_t	$(m_t + 2m_l)/3$	$(3m_t m_l)/(m_t + 2m_l)$	4
	m_t	m_l	m_t	4				
110	m_t	$(m_t + m_l)/2$	$(2m_t m_l)/(m_t + m_l)$	4	m_t	$(m_l + 2m_t)/3$	$(3m_t m_l)/(m_l + 2m_t)$	2
	m_t	m_l	m_t	2	m_t	m_l	m_t	2
111	m_t	$(m_t + 2m_l)/3$	$(3m_t m_l)/(m_t + 2m_l)$	6	m_t	m_t	m_l	1
					m_t	$(m_t + 8m_l)/9$	$(9m_t m_l)/(m_t + 8m_l)$	3

the case of the holes. Furthermore, the approach based on the EMA described in [23] cannot be easily extended to the holes because of the strong warpage of the hole energy dispersion. However, as shown by Fischetti *et al.* [24], it is possible to use the k.p method for the hole inversion layers. In fact, in the k.p. method, the dependence on k_z is analytical, and thus we can replace k_z with $-\mathrm{j}\,\mathrm{d/dz}$. The Schrödinger equation becomes:

$$\left[H_{\mathrm{KP}}\left(k_x, k_y, -\mathrm{j}\,{}^{\mathrm{d}}\!/\!{}_{\mathrm{dz}}\right) + H_{\mathrm{SO}} + H_{\mathrm{STR}} + V(z) - \frac{\eta^2}{2m_0}\,{}^{\mathrm{d}^2}\!/\!{}_{\mathrm{dz}^2}\right]\psi_{\mu,k_x,k_y}(z)$$
$$= \left[E_\mu\left(k_x, k_y\right) - \frac{\eta^2\left(k_x^2 + k_y^2\right)}{2m_0}\right]\psi_{\mu,k_x,k_y}(z)$$

where $V(z)$ is the confining potential. By discretisizing the wave-function over a spatial mesh in z, consisting of N points, the solution of the Schrödinger equation consists in finding the eigenvalues of a $6N\times 6N$ matrix. One main disadvantage of this approach is that the Schrödinger equation must be solved for each vector in the transport plane.

9.4. Performance Limit

Recently, a few papers have addressed the issue of the maximum currents and switching performance that can be achieved using semiconductors other than silicon for the transistor channel [18,25–27]. All the above papers are based on a ballistic transport model. In particular, the studies presented in [25,26] provide an extensive investigation of the quantization and transport directions that provide the highest on-current in Si and Ge MOSFETs. The paper by Rahman *et al.* [18] is mainly focused on the possible effects of band-structure effects in ultra thin (UT) semiconductor films using a Tight-Binding approach that goes beyond the conventional EMA method. In the following we will present simple analytical expressions for the on-current as a function of the channel material and device parameters that can be very useful to gain an intuitive insight on the impact that the different material and design options have on the on-current magnitude. The discussion is based on the approach introduced in [27].

9.4.1. Analytical expression for the ballistic current

The calculation of the ballistic current I_{BL} produced by the L-valleys of germanium is more complex than the case for the Δ-valleys of silicon; in fact the main axes of the equi-energy ellipses are not aligned with the transport direction. The calculation of I_{BL} is based on the assumption that, for a truly ballistic channel, only the states with a positive component v_x of the velocity in the DCS are filled and thus contribute to the inversion density and the ballistic transport, see Figure 9.8.

Consequently the current per unit width produced by each subband can be written as:

$$\frac{I_{\mathrm{BL}}}{W} = \frac{q}{A_{\mathrm{D}}} \sum_{(k'_x, k'_y) v_x > 0} v_x F(E, E_{\mathrm{Fs}}) \tag{9.20}$$

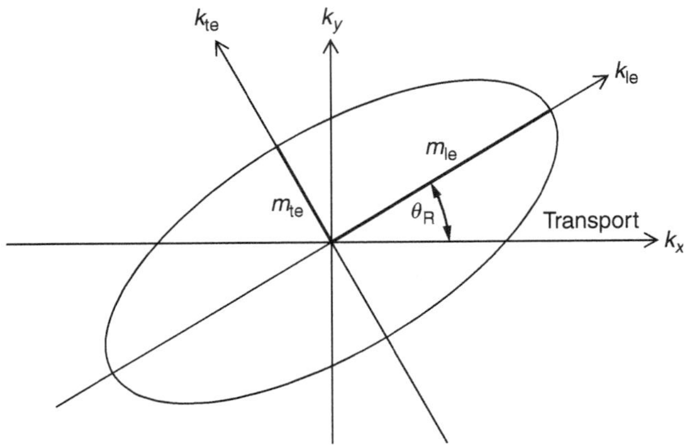

Figure 9.8. Equi-energy ellipsoid in the transport plane stemming from one of the L-valleys of germanium. (k_x,k_y) indicate the DCS whereas (k_{le},k_{te}) indicate the ECS. m_{le} and m_{te} are the longitudinal and transverse mass in the ECS, respectively.

where the sum is taken over the states (k'_x,k'_y) in the ECS which have a positive v_x in the DCS. The function $F(E,E_{Fs})$ denotes the Fermi–Dirac occupation function at the virtual source, which is governed by the Fermi Level E_{Fs} at the source end of the channel and A_D is the area of the device.

Hence, in order to calculate the sum, we must express v_x in the ECS, so that the domains where v_x is positive can be identified. If we denote with θ_R the angle between the DCS and the ECS, then v_x in the ECS is given by:

$$v_x = \frac{\eta}{m_{le}} k_{le} \cos(\vartheta_R) - \frac{\eta}{m_{te}} k_{te} \sin(\vartheta_R) \tag{9.21}$$

From the above expression of v_x in the ECS, we can calculate the range of angle θ_R where v_x is positive, hence the range that must be included in the calculation of I_{BL}. By converting the sum in an appropriate integral, one finally obtains:

$$\frac{I_{BL}}{W} = \frac{q}{\eta^2\sqrt{2}} \left(\frac{kT}{\pi}\right)^{3/2} F_{1/2}\left(\frac{E_{fs} - E_i}{kT}\right)\sqrt{m_{ID}} \tag{9.22}$$

which has been deliberately written in a form similar to the ballistic current for an ellipse aligned to the transport direction, except for the fact that the effective mass m_{ID} is defined as:

$$\boldsymbol{m}_{\mathbf{ID}} = m_{te} \cos^2 \vartheta_R + m_{le} \sin^2 \vartheta_R \tag{9.23}$$

In Equation (9.22) the symbol $F_{1/2}$ denotes the Fermi Integral of order 1/2 as defined in [28].

Clearly, for a given occupation factor $F(E,E_{fS})$ (i.e. a given distance between the eigenvalue and the source Fermi Level), the current is maximum when the effective mass m_{ID} is the largest. The effective mass m_{ID}, in turn, depends on both the longitudinal and transverse masses and also on the angle θ_R, hence on the orientation of the transport direction with respect to the ellipse longitudinal axis. Since m_{le} is larger than m_{te} by definition, the transport direction that yields the maximum current is the one along the transverse axis of the ellipse. Hence, according to the definition of the angle θ_R given in Figure 9.8, the transport direction that results in the maximum current corresponds to $\theta_R = \pi/2$. In fact, it can be easily verified from Equation (9.22) that m_{ID} coincides with m_{le} for $\theta_R = \pi/2$.

The dependence of the ballistic current on m_{ID} discussed so far has been based on Equation (9.22), hence by considering a given value of the occupation factor $F_{1/2}(E,E_{Fs})$. It should be realized, however, that for a given value of the gate bias, the occupation factor $F_{1/2}(E,E_{Fs})$ and the inversion density depend in turn on the masses m_{le} and m_{te}.

An analysis based on approximated, analytical expressions for the occupation factor $F_{1/2}(E,E_{fs})$ shows that, if we take the best transport direction so that $m_{ID} = m_{le}$, then the ballistic current always increases with decreasing of the transverse mass m_{te}, whereas the dependence on m_{le} is not monotonic. The reason for this non-monotonic dependence on the longitudinal mass stems from the behavior of the DoS of the 2Deg.

In fact, the masses m_{le} and m_{te} govern the DoS of the elliptical valley, to which the capacitance of the inversion layer is proportional. More precisely, if we suppose that the 2Deg is degenerate and we embrace the so-called quantum limit approximation (that assumes that the entire inversion density is in the lowest subband), then the capacitance of the inversion layer can be approximated with the quantum capacitance C_{QM}:

$$C_{QM} \cong \frac{q^2 D_{2D}}{2} = \frac{q^2 n_v \sqrt{m_{le} m_{te}}}{2\pi\eta^2} \tag{9.24}$$

Equation (9.24) indicates that semiconductors with very light in-plane masses can result in relatively small C_{QM} values, especially when the degeneracy of the valleys n_v is small. When the gate oxide of the MOSFETs is aggressively scaled, the capacitance of the inversion layer becomes progressively more critical for the overall drive capacitance at the gate of the transistor. Hence, when m_{le} becomes too small, the degradation of the drive capacitance tends to reduce the current I_{BL}. For relatively large m_{le} values, instead, the reduction of the injection velocity for increasing m_{le} prevails, so that I_{BL} starts decreasing for increasing m_{le}.

This tradeoff between the improvement of the injection velocity and the degradation of the drive capacitance can be quantitatively important for alternative channel materials, such as GaAs, whose quantum capacitance is critically low for the realization of MIS transistors having very small equivalent oxide thicknesses.

9.4.2. Results: Ge versus Si MOSFETs

Figure 9.9 shows I_{BL} versus the gate voltage V_G for different silicon and germanium MOSFETs, in the ballistic limit. The devices are double-gate transistors with a semiconductor

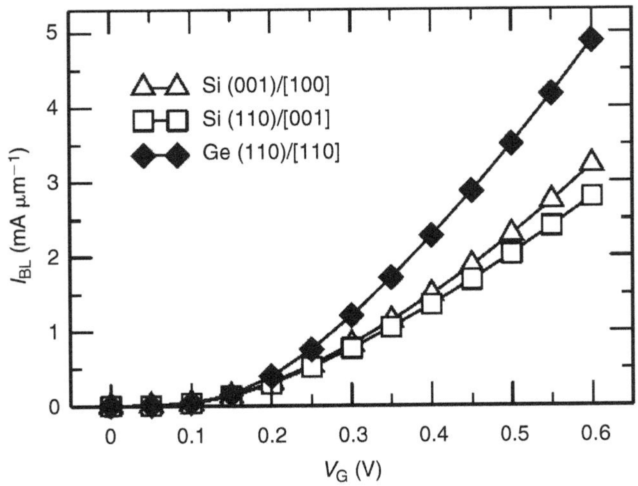

Figure 9.9. Simulated drain current in the ballistic transport regime versus V_G for silicon and germanium MOSFETs with different quantization and transport directions. Equivalent oxide thickness $t_{ox} = 0.6\,\text{nm}$. The off-current $I_{OFF} = 1\,\text{mA}\,\mu\text{m}^{-1}$.

film thickness $t_{SCT} = 7\,\text{nm}$. The current provided by each subband has been calculated according to the ballistic current reported in Equation (9.22), then the overall on-current is obtained by summing over the different subbands and valleys by accounting for the appropriate degeneracy. The calculations include all the relevant valleys for germanium, namely the L, Δ- and Γ-valley. These I_{ON} characteristics have been obtained by calculating the eigenvalues with a self-consistent solver of the 1D Schrödinger and Poisson equations, where the ballistic transport condition is accounted for by considering only the states with a positive v_x in the calculation of the inversion density N_{INV}. The notation Si(001)/[100] denotes a silicon MOSFET with (001) quantization direction and [100] transport direction. In the case of the germanium MOSFETs the best transport direction has been chosen for both quantization directions.

As it can be seen the Ge(110) exhibits a ballistic current larger than silicon. Consistently with previous studies [25,26], we can see that the Ge(110) can achieve the best on-current, with an improvement with respect to silicon that seems to increase with the gate voltage, for a given equivalent oxide thickness.

The comparison between different channel materials for MOSFETs is illustrated in Figure 9.10 in the context of an aggressive scaling of the equivalent oxide thickness t_{ox}; obviously the t_{ox} values reported in the figure implicitly assume the use of high-κ dielectrics. For the sake of simplicity, the supply voltage has been kept to a constant value $V_{DD} = 0.6\,\text{V}$ for the different t_{ox}.

It should be noticed that, if the gate capacitance of the transistor is simply taken as the oxide capacitance, then, for a given V_{DD}, the scaling of t_{ox} should imply an increase of the inversion density N_{INV} proportional to $1/t_{ox}$. For a given material, the increase of N_{INV} typically

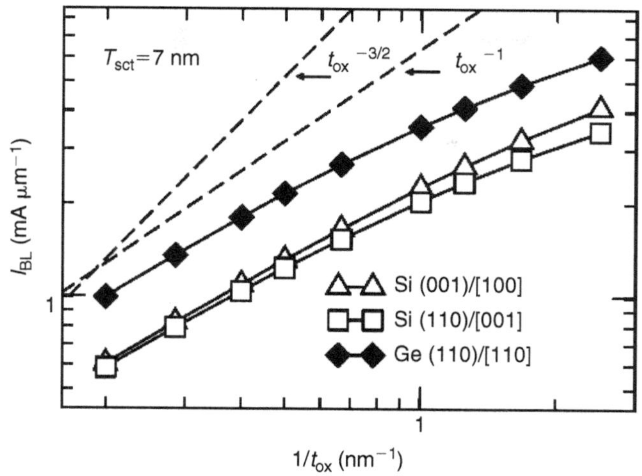

Figure 9.10. Simulated drain current in the ballistic transport regime versus the equivalent oxide thickness T_{OX} for silicon and germanium MOSFETs with different quantization and transport directions. The dashed lines indicates the $1/t_{ox}$ and $(1/t_{ox})^{1.5}$ curves. $I_{OFF} = 1\,\text{mA}\,\mu\text{m}^{-1}$ and $V_{DD} = 0.6\,\text{V}$.

results in an increase of the degeneracy of the lowest subband that should even result in increased injection velocity at the source for a ballistic MOSFET. Consequently, when the t_{ox} value is scaled, the on-current I_{ON} should increase as $(1/t_{ox})^r$ with r between 1 and 1.5.

Figure 9.10 shows that, instead, the I_{ON} improvements with oxide scaling are smaller than expected. This dependence of the I_{ON} on t_{ox} is due to the finite capacitance of the inversion layer, that limits the increase of the overall gate capacitance of the transistor when t_{ox} is reduced. In the range of t_{ox} considered in Figure 9.10 the largest drain current is always achieved by the Ge(110) transistor.

9.5. Semi-classical Transport

In the previous section we have evaluated the performance limits of Ge devices by assuming ballistic transport. However, scattering inside the channel plays an important role even in ultra-scaled MOSFET devices. This has been demonstrated both experimentally, since the measured current is lower than the prediction by the ballistic models, and by means of MC simulations [29,30].

An effective way to model carrier transport in the presence of scattering is to follow the semi-classical approach, that is based on the BTE. In this approach, transport is modeled according to classical mechanics (thus neglecting effects such as tunneling), whereas the calculation of the scattering rates (i.e. the transition rates due to interaction between carriers and phonons, ionized impurities, etc.) is based on quantum-mechanics. The effect of carrier quantization in the direction normal to the transport plane can be taken into account using corrections to the potential, or writing a system of coupled BTEs for each subband.

In this section the BTE in bulk semiconductors is described in 9.5.1, whereas the system of coupled BTEs to be used in inversion layers is described in 9.5.2. The solution of the BTE based on its moments is addressed in 9.5.3. The rest of the section is devoted to the full solution of the BTE using the MC method. The MC method in bulk semiconductors is described in 9.5.4. The introduction of quantum corrections, and the modifications introduced in bulk MC to extend them to the study of the inversion layer are described in 9.5.5. Finally, the use of the MC method for the solution of the multi-subband BTE is described in 9.5.6.

It is worth mentioning that full-quantum approaches for the study of carrier transport in MOSFETs in the presence of scattering (such as NEGF) have been proposed [30] but are very computationally demanding, so that full-quantum approaches are usually applied to the case of ballistic transport.

Finally the reader should be aware that this section is general and not specific to Ge devices. Very few simulation studies on Ge have been presented so far. Most of the approaches described here (e.g. multi-subband Monte-Carlo (MSMC)) have been presented only for Si, but it is possible to extend them to Ge.

9.5.1. BTE: bulk semiconductor

The base of semi-classical transport is the BTE. Considering the distribution function $f(\vec{r}, \vec{K}, t)$, where the reciprocal space is 3D, continuity in the position/momentum-space yields [31]:

$$\frac{\partial f}{\partial t} + \vec{v} \cdot \nabla_{\vec{r}} f + \frac{q}{\eta} \vec{F} \cdot \nabla_{\vec{K}} f = \sum_{\vec{K}} f(\vec{r}, \vec{K}', t)\left[1 - f(\vec{r}, \vec{K}, t)\right] S(\vec{K}', \vec{K}) - f(\vec{r}, \vec{K}', t)\left[1 - f(\vec{r}, \vec{K}', t)\right] S(\vec{K}, \vec{K}') \quad (9.25)$$

where $\vec{v} = \nabla \vec{K} E(\vec{K})/\eta$ is the group velocity, $S(\vec{K}, \vec{K}')$ is the scattering rate (probability per unit time) between states K and K'.

The left-hand-side of the BTE represents the drift and the diffusion of the carriers due to the electric field and their random distribution. The right-hand-side is given by the scattering terms that describe the scattering events as changes of the energy and the wave-vector $\vec{K}$ of the carriers that take place instantaneously in time and without any change of the carrier position in the real space.

The solution of the BTE is the distribution function $f(\vec{r}, \vec{K}, t)$. All macroscopic quantities can be calculated from the distribution function. The carrier concentration is given by:

$$n(\vec{r}, t) = \frac{1}{\Omega} \sum_{\vec{K}} f(\vec{r}, \vec{K}, t) \quad (9.26)$$

where Ω is the volume of the considered cell in real space. The current density is given by:

$$\vec{J}_n(\vec{r},t) = \frac{q}{\Omega}\sum_{\vec{K}} \vec{v}(\vec{K}) f(\vec{r},\vec{K},t) \tag{9.27}$$

and so on.

9.5.2. BTE: 2D inversion layers

In confined systems, the band structure consists of different subbands μ, each of them having its dispersion relationship $E_\mu(\vec{K})$. The wave-vector is defined only in the transport plane. We need a distribution function for each subband $f_\mu(\vec{r}, \vec{K}, t)$.

The BTE for the μ-th subband reads [32]:

$$\begin{aligned}\frac{\partial f_\mu}{\partial t} + \vec{v}_\mu \cdot \nabla_{\vec{r}} f_\mu + \frac{q}{\eta}\vec{F}_\mu \cdot \nabla_{\vec{K}} f_\mu = \sum_{\mu',\vec{K}'} f_{\mu'}(\vec{r},\vec{K}',t)\left[1 - f_\mu(\vec{r},\vec{K},t)\right] \\ S_{\mu'\mu}(\vec{K}',\vec{K}) - f_\mu(\vec{r},\vec{K},t)\sum_{\mu',K'}\left[1 - f_{\mu'}(\vec{r},\vec{K}',t)\right] S_{\mu\mu'}(\vec{K},\vec{K},')\end{aligned} \tag{9.28}$$

$S_{\mu\mu'}(\vec{K}, \vec{K})$ is the scattering rate from the state K in the μ-th subband to the state K' in the μ'-th subband. Due to inter-subband scattering, the BTEs of the different subbands are coupled. On the basis of the energy conservation, for each subband μ the electric field has to be replaced by a driving field equal to the gradient in the real space of the bottom energy of the subband:

$$\vec{F}_\mu = \frac{1}{q}\nabla_{\vec{r}} E_{\mu 0} \tag{9.29}$$

9.5.3. Solution of the BTE: methods based on the moments

The exact solution of the BTE is a difficult task, in particular when realistic potential profiles and complex device structures are to be accounted for. Most of the commercial tools for device simulation are based on the Drift-Diffusion model [33,34]. This model is based on the moments of the BTE, i.e. both sides of the BTE are multiplied by appropriate quantities and then integrated over the momentum space [31]. In particular, the Drift-Diffusion model is based on the first two moments: in the first momentum both sides of the BTE are simply integrated as they are, whereas in the second momentum both sides are multiplied by the group velocity before integrating. However, each momentum shows terms that depend on the next momentum, so that a closure relation must be introduced. In particular, in the Drift-Diffusion model, it is assumed that the energy distribution of the carriers is

Maxwellian, with an effective temperature equal to the lattice temperature. If we consider free electrons (no quantization), the Drift-Diffusion model consists of the two equations:

$$\begin{aligned} \frac{\partial n}{\partial t} + \frac{1}{q}\nabla\cdot\vec{J}_n &= G - R \\ \vec{J}_n &= qn\mu_n\vec{F} + qD_n\nabla n \end{aligned} \tag{9.30}$$

Similar equations hold for the holes.

The parameters of the model are: the mobility μ_n, the diffusion coefficient D_n, the generation and recombination rates G and R, respectively. These parameters depend on the carrier and doping concentrations, on the electric field and on the lattice temperature. Such dependences can be measured experimentally. Based on this approach, the Drift-Diffusion simulators developed for Si devices can be used even for Ge devices by simply replacing the parameters of Si with those of Ge. This approach has been followed in [35].

The Drift-Diffusion model is strictly local, that is the current density at a given point in the device depends only on the carrier concentrations and the electric field at the same point, not on the whole potential profile experienced by the carriers. In nano-MOSFETs, non-local effects and quasi-ballistic transport become relevant, so that the accuracy of the Drift-Diffusion model is largely questionable. One possible extension consists in the inclusion of more than two moments of the BTE. A model based on six moments has been presented in [36]. In this case the number of parameters involved in the model is larger and the interpretation of the results becomes progressively more difficult.

Quantum effects can be included in conventional Drift-Diffusion tools using the Density Gradient model [37–39], i.e. by correcting the classical potential in order to reproduce quantum effects such as carrier repulsion from the semiconductor/dielectric interfaces.

The conventional Drift-Diffusion is derived by applying the method of the moments to the BTE for the free-electron gas. By applying this method to the multi-subband BTE it is possible to develop Multi-Subband-Drift-Diffusion models [40].

9.5.4. Solution of the BTE: MC for bulk Ge

The MC method [41] is a powerful tool for the solution of the BTE without any simplifying assumption. The motion of carriers is divided in ballistic free-flight, where the carrier wave-vector changes due to the electric field, interrupted by scattering events. The ingredients of the MC are the band structure (used to compute the group velocity) and the scattering rate. In this subsection we describe the use of the MC method to analyze the transport properties of bulk Ge. We have already described the band structure and the parameters of the minima. The main scattering mechanisms in bulk materials are phonon scattering and ionized impurity scattering.

The formulas for the scattering rates of these mechanisms can be found in several textbooks, e.g. [31]. Due to the relevance of phonons, in particular at low doping levels, we

Table 9.3. Parameters for phonons in Ge. Data from Ref. [13].

Valley	Type	Energy	Coupling constant
L	Acoustic	n.a.	11 eV
Δ	Acoustic	n.a.	9 eV
Γ	Acoustic	n.a.	5 eV
L	Optical intravalley	37 meV	5.5×10^8 eV cm^{-1}
L	Intervalley f-type	27.6 meV	3.0×10^8 eV cm^{-1}
L	Intervalley f-type	10 meV	2.0×10^7 eV cm^{-1}
Δ	Intervalley g-type	8.6 meV	7.89×10^7 eV cm^{-1}
Δ	Intervalley g-type	37 meV	9.46×10^8 eV cm^{-1}
L↔Δ	Intervalley	27.6 meV	4.06×10^8 eV cm^{-1}
L↔Γ	Intervalley	27.6 meV	2.0×10^8 eV cm^{-1}
Δ↔Γ	Intervalley	27.6 meV	1.0×10^9 eV cm^{-1}

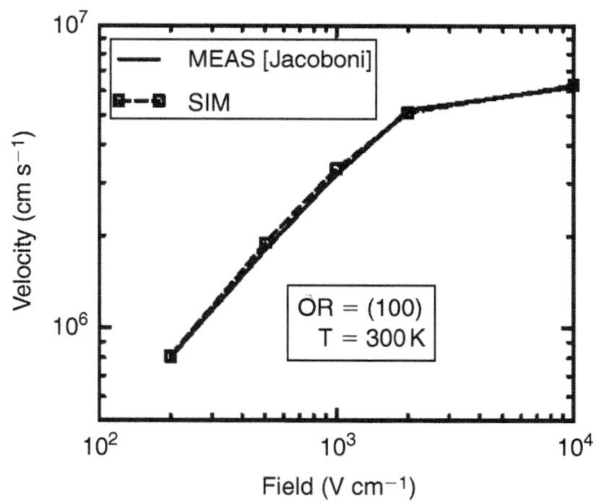

Figure 9.11. Drift velocity as a function of the electric field in bulk Ge. Experimental data from [13] are compared with MC simulations using the band structure described in Table 9.1 and the phonons of Table 9.3.

describe the phonon model in Ge with more detail. The dispersion relationship of phonons in Ge has been measured in [42]. These data can be used to evaluate the energy of the phonons involved in the transitions between the different minima (valleys) in the 1BZ (valleys). Another important ingredient are the coupling constants which cannot be obtained directly from experiments and are usually calibrated by comparing the MC simulations to the measured transport properties. The coupling constants and the phonon energies of Ge are reported in Table 9.3.

Figure 9.11 reports the results of MC simulations obtained using the valley parameters of Table 9.1 and the phonons of Table 9.3. The results indicate that with the appropriate choice of parameters it is possible to reproduce the experimental results of [13]. By inspection of

Figure 9.11, it is evident that the low-field electron mobility in bulk Ge is larger than in Si, thanks to the low effective mass of the L-valleys. On the other hand, the saturation velocity is lower. In fact, at high effective field, carriers are transferred to the Δ-valleys (due to the small energy separation between L- and Δ-valleys, see Table 9.1), that feature a much higher effective mass. The same effect is responsible for negative differential resistance at low temperature [12].

The MC method has been also used to study hole transport in Ge [43,44]. However the simulation task is much more challenging than for electrons due to the warped nature of the valence band.

9.5.5. MC with quantum corrections

When considering MOSFETs we have to account for the quantization in the inversion layer, that requires the solution of the BTE for 2D systems. MC simulators based on the multi-subband BTE will be described in the following subsection. However, many of the MC tools currently available consider a free-electron gas. When applied to MOSFETs, these tools must be modified in order to account for quantization and for the influence of the roughness of the semiconductor–dielectric interface (surface roughness scattering). Surface roughness can be implemented following the specular-diffusive approach: particles hitting the interface have a finite probability to be scattered randomly [45].

Quantum corrections to the potential are introduced in conventional MC tools to mimic carrier repulsion from the interface. Among several different approaches, it is worth to mention the effective potential [46], based on the convolution between the classical potential and a Gaussian function representing the electron wave-packet, and the approach described in [47], where the potential is modified in order to reproduce the carrier distribution provided by the solution of the Schrödinger equation in the vertical direction in the different sections along the channel.

Since in the presence of quantum corrections the carriers are displaced from the interface, the usual specular-diffusive approach for surface roughness cannot be used. Surface roughness must be implemented as an additional scattering mechanism (see e.g. [48,49] for the case of electrons in Si inversion layers).

A full-band MC with quantum corrections has been applied to transport in Ge MOSFETs in [2]. Surface roughness has been included as an additional scattering mechanism. Impact ionization has been included too. It has been found that the on-current improvement in short-channel Ge MOSFET compared to their Si counterpart is much lower than the differences in the bulk mobility between Ge and Si.

9.5.6. Multi-subband MC

Solution of the coupled BTE for the different subbands described in 9.5.3 can be accomplished by using the MC method [50,51]. Simulators solving the multi-subband BTE are referred to as MSMC.

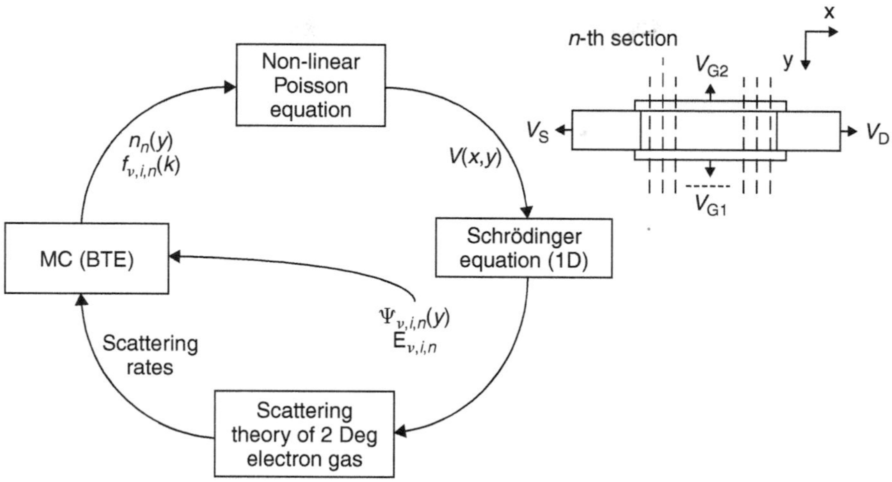

Figure 9.12. Flowchart of a MSMC tool described in [51]. The potential is indicated as $V(x,y)$, $E_{\nu,i,n}$ is the subband energy (ν is the valley index, i the subband index and n the index of the section), $\psi_{\nu,i,n}$ the wave-function, $f_{\nu,i,n}$ is the subband occupation. The symbol $n_n(y)$ represents the carrier concentration profile in the n-th section.

A typical flowchart is reported in Figure 9.12. Starting with a first guess for the 2D potential, the device cross-section is partitioned into N sections along the channel direction x, and the Schrödinger equation in the y direction is solved in all the sections, thus obtaining the subband profile along the channel. The corresponding wave-functions are then used to compute at each section the scattering rates. Carrier transport is simulated using the MC method, where, as discussed in Section 9.5.2 the driving field is the derivative of the subband energy bottom along x. The MC simulation provides the occupation probability at each section and for each wave-vector in the transport plane. After each MC step, the electron concentration is used to re-calculate the potential by solving the Poisson equation (the non-linear scheme proposed in [52] has been used in [51]).

The most important scattering mechanisms to be considered in inversion layers are: phonons, ionized impurities, surface roughness, film thickness fluctuations and surface optical (SO) phonons [53]. The scattering due to the surface roughness and the fluctuations of the semiconductor film thickness are not present in a bulk material. Both the scattering mechanisms are due to the modification of the quantum well that confines the carriers produced by the non-uniformities of the semiconductor–oxide interface [54]. In particular, the film thickness fluctuation scattering is specific to GOI devices and it is due to the change of the bottom energy of the subband produced by the thickness variation of the semiconductor film. SO phonons are due to the polar nature of the insulator and are especially relevant in devices featuring high-κ dielectric [55]. They are thus relevant in Ge based MOSFETs, where the insulator is often a high-κ material.

So far, most of the efforts in developing MSMC models have been devoted to Si. However, some of the models are based on the generalized EMA [51] that can be applied even to Ge, provided the scattering models are calibrated. Starting with phonons, the study of the mobility in bulk Ge has made possible to evaluate the phonon energies and the coupling constants for electron–phonon interaction. However, it is not guaranteed that the same phonon parameters used for bulk germanium will correctly describe the transport in germanium inversion layers. In the case of silicon e.g. in order to reproduce the inversion layer mobility many references report increased coupling constant values for acoustic phonons ($\sim$14 eV), compared to the $\sim$9 eV value used in bulk Si [53,56]. A similar recalibration might be necessary for the Ge inversion layers. Unfortunately, reliable universal mobility curves for Ge are not available yet. Furthermore, the picture is made even more complex by the use of high-κ dielectrics, thus adding the parameters of the SO phonons to the set of model parameters to be adjusted to fit the experimental data. In addition, at high effective field values, surface roughness is the dominant scattering mechanism, but the parameters of the roughness are not known a priori, and need to be adjusted based on the experimental data. Again, in Si the experimentally observed existence of a universal mobility curve suggests that the quality of the semiconductor–oxide interface does not vary much with the fabrication process, at least as far as its impact on the charge transport is concerned. A similar reproducibility of the interface in Ge devices should absolutely not be taken for granted.

Finally it is worth mentioning that so far all efforts in MSMC have been devoted to the modeling of the electron inversion layer. The development of MSMC simulators for pMOSFETs should be now pursued because large advantages in hole mobility are foreseen in germanium MOSFETs.

9.6. Conclusions

In this chapter we have described the main modeling approaches for Ge devices, focusing on MIS type of devices. We have compared recently analyzed band structure and transport properties of Ge MIS inversion layers with those of bulk Ge, that have been extensively studied in the 60's and in the 70's. So far, most of the recent Ge modeling effort has been devoted to performance limit evaluation (i.e. assuming pure ballistic transport). However, since carrier scattering is still expected to play a role in very short-channel devices, models for dissipative transport must be developed, conveniently extending tools developed for Si to the study of Ge.

As a first step, models based on the moments of the BTE can be used. In these models, the material properties are expressed in terms of parameters, e.g. carrier mobility, generation rates, etc., which can be directly measured, making feasible to calibrate the models on the behavior of the real devices, at least for large enough geometries. However, this is not the case for the nanometric MOSFETs fabricated in today and near future CMOS technologies, where the use of the moments of the BTE is at least questionable.

The full solution of the BTE using the MC method is probably the most powerful approach to study dissipative transport in short-channel devices. In particular, MSMC are especially useful, since they can be used to include on a physical basis effects such as size-induced carrier quantization and all the related effects on carrier transport and device electrostatics, non-conventional crystal orientations and remote phonons, these latter important in devices featuring high-κ dielectrics. Using MSMC simulators it is possible to investigate how mobility improvement of Ge over Si in bulk material translates in increase of the transistors on-current and to optimize the device design in order to exploit the benefits of Ge at their best.

MSMC tools have been already developed for n-channel Si devices. However, their extension to Ge devices implies the calibration of the physical parameters, a process that requires experimental data, mainly inversion channel mobility data. So far, universal inversion channel mobility data for Ge inversion layers have not been presented in the literature due to poor and unstable interface quality and strong remote phonons scattering. Hence, a sound experimental analysis of mobility in Ge MOSFETs is needed to produce reliable data for the calibration of both MSMC and commercial simulators based on the moments of the BTE.

A challenging yet important milestone is the development of simulation tools for pMOSFETs, since hole transport is expected to be particularly advantageous in Ge.

Finally it is important to mention that the modeling of Ge devices must include self-heating effects and generation of electron-hole pairs due to impact ionization because of the lower thermal conductance and reduced band gap of germanium with respect to Silicon.

References

[1] S.M. Sze, *Physics of Semiconductor Devices*, 2nd edition, Wiley, New York.

[2] B. Ghosh, X. Wang, X.-F. Fan, L.F. Register and S.K. Banerjee, *IEEE Trans. Electron Dev.* **52** (2005) 547.

[3] M.V. Fischetti, *IEEE Trans. Electron Dev.* **38** (1991) 634.

[4] T. Mikawa, S. Kagawa, T. Kaneda and Y. Toyama, *Appl. Phys. Lett.* **37** (1980) 387.

[5] R. Van Overstraeten and H. De Man, *Solid State Electron.* **13** (1970) 583.

[6] E. Pop, C.O. Chui, S. Sinha, R. Dutton and K. Goodson, *IEEE IEDM Technical Digest* (2004) 411.

[7] T.B. Boykin, G. Klimeck and F. Oyafuso, *Phys. Rev. B* **69** (2004) 115201.

[8] J.R. Chelikowsky and M.L. Cohen, *Phys. Rev. B* **14** (1976) 556.

[9] G. Dresselhaus, A.F. Kip and C. Kittel, *Phys. Rev.* **98** (1955) 368.

[10] L.R. Saravia and D. Brust, *Phys. Rev.* **176** (1968) 915.

[11] M.V. Fischetti and J.M. Higman, Theory and calculation of the deformation potential electron–phonon scattering rates in semiconductors, in Monte Carlo device simulations: full band and beyond, edited by K. Hess, Kluwer, Norwell (Mass.) 1991 p. 123.

[12] E.G.S. Paige, *IBM J. Res. Dev.* **13** (1969) 562.

[13] C. Jacoboni, F. Nava, C. Canali and G. Ottaviani, *Phys. Rev. B* **24** (1981) 1014.

[14] A. Rahman, M.S. Lundstrom and A.W. Ghosh, *J. Appl. Phys.* **97** (2005) 053702.

[15] T. Manku and A. Nathan, *J. Appl. Phys.* **73** (1993) 1205.

[16] M.V. Fischetti and S.E. Laux, *J. Appl. Phys.* **80** (1996) 2234.

[17] S. Rodriguez-Bolivar, F.M. Gomez-Campos and J.E. Carceller, *Semicond. Sci. Technol.* **20** (2005) 16–22.

[18] A. Rahman, G. Klimeck, T.B. Boykin and M. Lundstrom, *IEEE IEDM Technical Digest* (2004) 139.

[19] D. Esseni and P. Palestri, *Phys. Rev. B* **72** (2005) 165342.

[20] L.W. Wang and A. Zunger, *Phys. Rev. B* **59** (1999) 15806.

[21] F. Chirico, A. Di Carlo and P. Lugli, *Phys. Rev. B* **64** (2001) 045314.

[22] J.M. Luttinger and W. Kohn, *Phys. Rev.* **97** (1955) 869.

[23] F. Stern and W.E. Howard, *Phys. Rev.* **163** (1967) 816.

[24] M.V. Fischetti, Z. Ren, P.M. Solomon, M. Yang and K. Rim, *J. Appl. Phys.* **94** (2003) 1079.

[25] S.E. Laux, *IEEE IEDM Technical Digest* (2004) 135.

[26] T. Low, Y.T. Hou, M.F. Li, C. Zhu, A. Chin, G. Samudra, L. Chan and D.-L. Kwong, *IEEE IEDM Technical Digest* (2003) 691.

[27] M. De Michielis, D. Esseni and F. Driussi, *Proc. ESSDERC* (2005) 165.

[28] J.S. Blakemore, *Solid State Electron.* **25** (1082) 1067.

[29] P. Palestri, D. Esseni, S. Eminente, C. Fiegna, E. Sangiorgi and L. Selmi, *IEEE Trans. Electron Dev.* **52** (2005) 2727.

[30] A. Svizhenko and M.P. Anantram, *IEEE Trans. Electron Dev.* **50** (2003) 1459.

[31] M.S. Lundstrom, *Fundamentals of Carrier Transport*, Addison Wesley, New York 1990.

[32] D.K. Ferry and S.M. Goodnick, *Transport in Nanostructures*, Cambridge University Press, Cambridge, 1997.

[33] Silvaco ATLAS User Manual.

[34] Integrated Systems Engineering, Dessis 8.0 User Manual, 2002

[35] A. Chin, H.L. Kao, Y.Y. Tseng, D.S. Yu, C.C. Chen, S.P. McAlister, C.C. Chi, *Proc. ESSDERC* (2005) 285.

[36] T. Grasser, C. Jungemann, H. Kosina, B. Meinerzhagen and S. Selberherr, *Proc. SISPAD* (2004)1.

[37] A. Wettstein, A. Schenk and W. Fichtner, *IEEE Trans. Electron Dev.* **48** (2001) 279.

[38] M.G. Ancona, *Superlatt. Microst.* **7** (1990) 119.

[39] M.G. Ancona, Z. Yu, W.-C. Lee, R.W. Dutton and P.V. Voorde, *Proc. SISPAD* (1997) 97.

[40] G. Curatola, G. Doornbos, J. Loo, Y.V. Ponomarev and G. Iannaccone, *IEEE Trans. Electron Dev.* **52** (2005) 1851.

[41] C. Jacoboni and L. Reggiani, *Rev. Mod. Phys.* **55** (1983) 645.

[42] B.N. Brockhouse and P.K. Iyengar, *Phys. Rev. B* **111** (1958) 747.

[43] S. Bosi, C. Jacoboni and L. Reggiani, *J. Phys. C. Solid State Phys* **12** (1979) 1525.

[44] L. Reggiani, C. Canali, F. Nava and G. Ottaviani, *Phys. Rev. B* **12** (1977) 2781.

[45] E. Sangiorgi and M.R. Pinto, *IEEE Trans. Electron Dev.* **39** (1992) 356.

[46] D.K. Ferry, R. Akis and D. Vasileska, *IEEE IEDM Technical Digest* (2000) 287.

[47] B. Winstead and U. Ravaioli, *IEEE Trans. Electron Dev.* **50** (2003) 440.

[48] P. Palestri, S. Eminente, D. Esseni, C. Fiegna, E. Sangiorgi and L. Selmi, *Solid State Electron.* **49** (2005) 727.

[49] Chu-Hao, J. Zimmerman, M. Charef, R. Fauquembergue and E. Constant *Solid State Electron.* **28** (1985) 733.

[50] M.V. Fischetti and S.E. Laux, *Phys. Rev. B* **48** (1993) 2244.

[51] L. Lucci, P. Palestri, D. Esseni and L. Selmi, *IEEE IEDM Technical Digest* (2005) 631.

[52] F. Venturi, R.K. Smith, E. Sangiorgi, M.R. Pinto and B. Ricco, *IEEE Trans. Comput-Aided De.* **8** (1989) 360.

[53] D. Esseni, A. Abramo, L. Selmi and E. Sangiorgi, *IEEE Trans. Electron Dev.* **50** (2003) 2445.

[54] D. Esseni, *IEEE Trans. Electron Dev.* **51** (2004) 394.

[55] M.V. Fischetti, D.A. Neumayer and E.A. Cartier, *J. Appl. Phys.* **90** (2001) 4587.

[56] C. Jungemann, A. Edmunds and W.L. Engl, *Solid State Electron.* **36** (1993) 1529.

CHAPTER 10

Nanoscale Germanium MOS Dielectrics and Junctions

Chi On Chui and Krishna C. Saraswat

10.1. Introduction

This Chapter focuses on advanced germanium metal-oxide-semiconductor device technology including nanoscale gate dielectrics and shallow junctions. The Chapter is built up as follows: Section 10.2 investigates the synthesis and physical properties of germanium oxynitride dielectrics. Basic electrical and dielectric–substrate interface characteristics on fabricated germanium oxynitride metal-oxide-semiconductor capacitors are also discussed. In Section 10.3, high-κ dielectric candidates for germanium metal-oxide-semiconductor device applications are first identified and their deposition techniques are introduced. The importance of various interface engineering schemes are examined and their resultant effects on the finished capacitors are explored. Besides, the leakage current densities extracted from these nanoscale dielectrics are benchmarked against each others. In Section 10.4, the activation of ion implanted and solid source diffused p-type and n-type dopants in germanium are thoroughly studied, and the related activation dependences are revealed. The deactivation of pre-activated dopants is observed, which has not been reported before, and a possible mechanism for such a deactivation process is proposed. Finally, some recent works on metal germanide contacts are reviewed.

10.2. Germanium Oxynitride Dielectrics

Silicon has been the most important semiconductor material in the modern electronic industry since 1960s primarily due to its very high quality native oxide for surface passivation.

Germanium-Based Technologies: From Materials to Devices. C. Claeys and E. Simoen (Editors)

After over 40 years of miniaturization, the scaling of classical bulk Si metal-oxide-semiconductor field-effect transistors (MOSFETs) is approaching many fundamental limits, which mandates integration of novel materials and innovative device structures [1] to continue the historic progress in information technology.

Among various high mobility channel candidates, germanium has recently been suggested [1–4] to alleviate the problem of MOSFET drain current saturation by providing a higher source injection velocity [5]. Historically, Ge had been one of the most important semiconductors, and the instability of native Ge oxides (GeO_x) was one of the key enablers in the first (point-contact) transistor demonstration [6]. Conversely, its inferior properties, especially when compared to silicon dioxide (SiO_2), make this dielectric unsuitable for Ge MOSFET gate insulation and integrated-circuit (IC) field isolation, and has therefore prevented very large-scale integration (VLSI) of Ge devices in the past.

Over the last four decades, a variety of grown and deposited dielectric approaches have been suggested for Ge MOS applications. Gate quality germanium dioxide (GeO_2) dielectrics have been grown by wet chemical oxidation [7], thermal oxidation [8–10], electron cyclotron resonance plasma oxidation [10], vacuum ultraviolet-assisted oxidation [11], as well as remote plasma oxidation [12]. In order to improve their thermal and chemical stability, either thermal nitridation [13–19] or plasma anodic nitridation [20] was applied to transform the Ge oxides to Ge oxynitrides (GeO_xN_y). Alternatively, several different deposited dielectrics have been attempted including SiO_2 [21–30], SiO_2 on a thin Si cap [2,27,30–35], silicon nitride (Si_3N_4) [26,28], GeO_2 [28], germanium nitride (Ge_3N_4) [28, 36], aluminum oxide (Al_2O_3) [37–39], and aluminum–phosphorus oxide (AlP_xO_y) [40]. Nevertheless, none of these approaches has been previously shown to hold promises as a dielectric for nanoscale Ge MOSFETs.

Before the interest in Ge as a channel material for decanano-scale MOSFETs could be facilitated, a stable as well as scalable gate dielectric technology has to be developed. In this section, we investigate the grown GeO_xN_y as nanoscale Ge MOS dielectric candidates. We first present the GeO_xN_y synthesis and their physical properties. After which, we report their basic electrical characteristics on the fabricated metal-gated MOS capacitors. Finally, the GeO_xN_y–Ge interface properties and the general gate leakage behavior from these MOS capacitors are analyzed.

10.2.1. Germanium oxynitride synthesis and properties

Amongst all the grown dielectric approaches listed above, thermal nitridation of the intentionally grown Ge oxides is apparently the most promising and accessible technique. As the first step towards GeO_xN_y synthesis, different methods for Ge surface oxidation including thermal oxidation, native oxidation, and chemical oxidation were studied as listed in Table 10.1. Before either thermal or chemical oxidation, the surface native GeO_x was first removed in dilute hydrofluoric acid (HF).

Core-level Ge(3d) and O(1s) X-ray photoemission spectra (XPS) were taken on different Ge oxides as plotted in Figure 10.1. In order to analyze the chemical bonding nature and

Table 10.1. Different Ge pre-oxidation treatments and surface oxidations.

Sample number	**(i)**	**(ii)**	**(iii)**	**(iv)**
Pre-oxidation treatment				
Reagent	50:1 HF	–	50:1 HF	50:1 HF
Duration (s)	30	–	30	30
Surface oxidation				
Type	Thermal	Native	Chemical	Chemical
Reagent	Dry O_2	Air	30% H_2O_2	conc. HNO_3
Temperature (°C)	500	20	20	20
Duration (min)	8	–	1	1

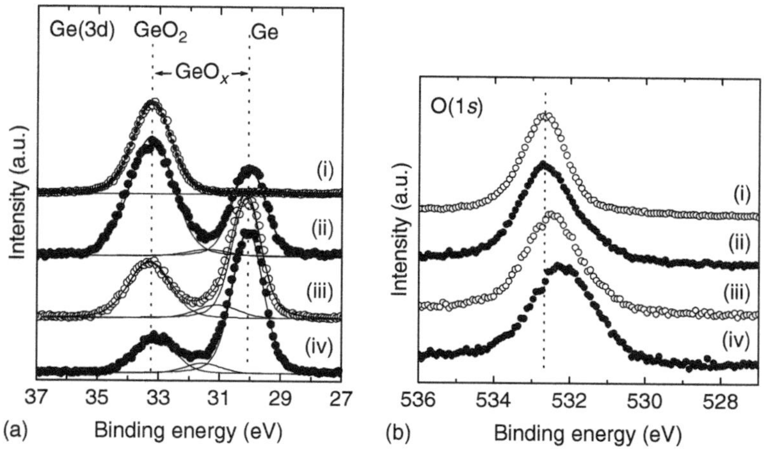

Figure 10.1. Core-level Ge(3d) (a) and O(1s) (b) XPS spectra on differently prepared Ge surface oxides: (i) thermal oxide, (ii) native oxide, (iii) H_2O_2 chemical oxide, and (iv) HNO_3 oxide. The Ge surfaces were HF cleaned prior to oxidation (except on sample (ii)).

stoichiometry of these oxides, the Ge(3d) signals were further peak-fitted assuming the existence of Ge, GeO_x, and/or GeO_2 configurations. Since the thermal oxide on sample (i) was relatively thick, the substrate elemental Ge peak (with binding energy ~30 eV) was absent given its small photoelectron escape depth. On the other hand, the elemental Ge peak was observed on samples (ii)–(iv) as these native and chemical oxidations are usually self-limiting. Nonetheless, from both the Ge(3d) and O(1s) spectra, the bulk of the thermal oxide (i) was mainly composed of stoichiometric GeO_2 while the other samples consisted of a mixture of GeO_2 and GeO_x. Moreover, the relative intensity ratio of GeO_2 to GeO_x is lower for the chemical oxides prepared with either hydrogen peroxide (H_2O_2) (iii) or nitric acid (HNO_3) (iv), whose sub-stoichiometry may generate high density of trap states at the dielectric–substrate interface. These interface states could eventually induce charge trapping and de-trapping (hysteretic current–voltage behavior) and MOSFET channel carrier scattering (mobility and drive current degradation). The thermal oxide was thereby chosen for subsequent nitridation to engineer the high quality GeO_xN_y, with both processes carried out in a commercial rapid thermal processing (RTP) system.

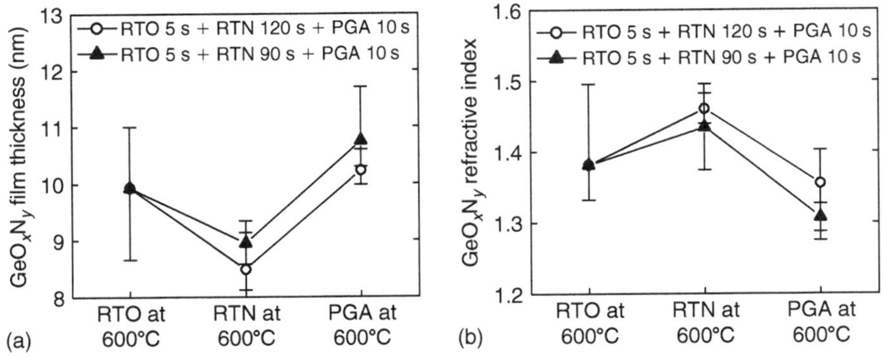

Figure 10.2. (a) GeO_xN_y film thickness and (b) refractive index estimated using optical ellipsometry after various rapid thermal processes.

The starting substrates used to synthesize GeO_xN_y were (100) oriented n-type Ge wafers with a net background doping concentration of $\sim 7 \times 10^{15}\,cm^{-3}$. These substrates were first cleaned by cyclic rinsing between deionized (DI) water and HF [41] to remove native GeO_x. The HF concentration employed was 50:1 and the rinse time was 15 s in each chemical for a total duration of 150 s. The cyclic HF (CHF) clean was finished with a 10 s DI water rinse for chemical safety reasons. The cleaned Ge substrates were then blown dry with nitrogen gas (N_2) and immediately loaded into the RTP system.

The GeO_xN_y formation experiments [42] consisted of an initial rapid thermal oxidation (RTO) at 500–600°C with 5–120 s soak time in dry oxygen (O_2) to form GeO_2 followed by *in-situ* rapid thermal nitridation (RTN) at 600°C with 1–5 min soak time in ammonia (NH_3) ambient to convert the oxides into oxynitrides. NH_3 was chosen as the nitriding agent due to its ability to incorporate more nitrogen (N) into the oxynitride film over other species like nitrous oxide (N_2O) or nitric oxide (NO), as inferred from their behavior on Si oxides [43]. During the oxynitridation, the rapid thermal ramp rates were set to $100°C\,s^{-1}$ for both RTO and RTN and the wafer temperature was brought down to room temperature prior to RTN. On selected samples, an *in-situ* post-growth anneal (PGA) at 600°C for 10 s in dry O_2 was performed. The change in GeO_xN_y film thickness and refractive index during the entire process was monitored using optical ellipsometry. For example, as shown in Figure 10.2, the GeO_xN_y film thickness decreased while the refractive index increased after the RTN treatment following the initial 5 s RTO step. Both phenomena could be explained by the increased coordination between Ge and O atoms through the N incorporation, with longer RTN soak time giving denser coordination that resembles the Si oxynitridation experience [44]. The optional (PGA) thickened the GeO_xN_y by additional surface Ge oxidation, diluted the N concentration within the GeO_xN_y film, and thereby reduced the Ge and O atom coordination.

Lastly, angle-resolved XPS (AR-XPS) was employed to qualitatively depth profile the N content within the GeO_xN_y layer [42]. The XPS signals were first peak-fitted around the

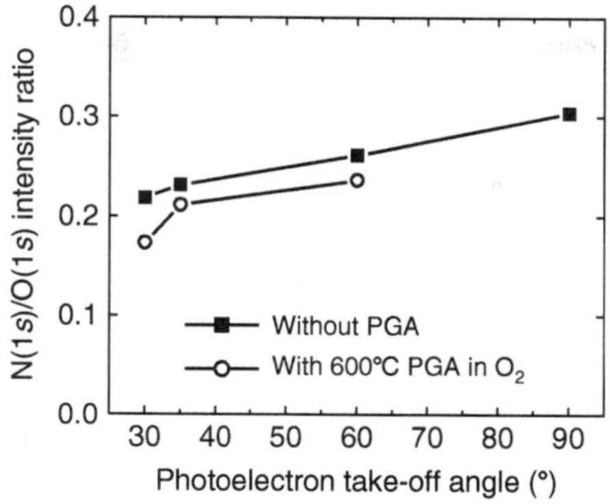

Figure 10.3. XPS intensity ratios of N(1s) to O(1s) signals measured by angle-resolved XPS on two typical GeO_xN_y films (grown by RTO at 600°C for 5 s followed by RTN at 600°C for 120 s) with and without PGA.

N(1s) and O(1s) binding energies and the extracted intensities were normalized by their corresponding sensitivity factors. Illustrated in Figure 10.3 is the intensity ratio of N(1s) to O(1s) signals measured on two typical GeO_xN_y films (grown by RTO at 600°C for 5 s followed by RTN at 600°C for 120 s) with and without PGA. On both films, the N(1s)/O(1s) intensity ratio increases gradually with increasing photoelectron take-off angle; in other words, more N is piled up near the bottom GeO_xN_y–Ge interface (rather than near the top surface) and raises the local dielectric permittivity (κ). Besides, the oxygen PGA at 600°C lowered the N concentration and caused a slight change in the N depth profile within the GeO_xN_y film. In general, the scalability of GeO_xN_y can be improved by incorporating more N. One such strategy is to reduce the initial thermal oxide thickness so as to trim down the resultant lower-κ top surface portion of GeO_xN_y; an alternative scheme could be enhancing the degree of nitridation (e.g. lengthening RTN soak time and raising NH_3 flow rate) and eliminating the optional oxygen PGA.

10.2.2. Basic MOS electrical characterizations

MOS capacitors were fabricated using the GeO_xN_y films to examine their dielectric bulk and GeO_xN_y–Ge interface electrical qualities. Utilizing the same surface pre-cleaning and oxynitridation recipe described in Section 10.2.1, GeO_xN_y films were grown on both (100) and (111) oriented p- and n-type Ge substrates. No *in-situ* oxygen PGA was performed to obtain a higher resultant dielectric permittivity. Next, gate electrodes of various sizes were formed by electron beam evaporation of about 70 nm of tungsten (W) through a shadow mask. Aluminum (Al) was then evaporated on the wafer backside to reduce the sample contact (and series) resistance. Finally, on the completed W/GeO_xN_y/Ge capacitor stacks, a forming gas (H_2/N_2) anneal was carried out at 300°C for 30 min. This forming gas anneal treatment was selected to rectify the abnormal flat-band voltage (V_{FB}) shift originated from the radiation charging during W evaporation [42].

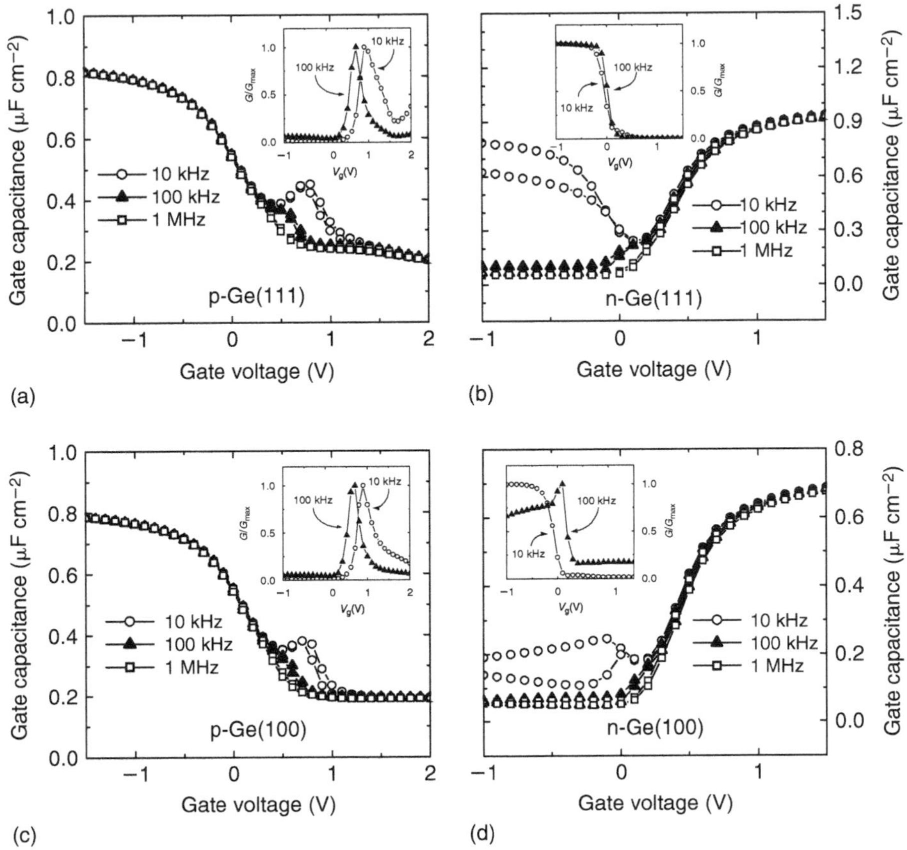

Figure 10.4. Multi-frequency bi-directional *C–V* characteristics measured from forming gas annealed W/GeO_xN_y/Ge capacitors fabricated on Ge substrates with different doping type and orientation. The insets capture the corresponding *G–V* measurements.

Capacitance–voltage (*C–V*) and conductance–voltage (*G–V*) characteristics were measured on these GeO_xN_y MOS capacitors using a HP4275A multi-frequency LCR meter. Since the amount of accumulation capacitance dispersion versus measurement frequency was negligible for most of the samples, the equivalent capacitance-based SiO_2 thickness (EOT) was extracted from the 100 kHz *C–V* data. Figure 10.4 shows the bi-directional *C–V* measurements on the W/GeO_xN_y/Ge capacitors with the GeO_xN_y films grown by RTO at 600°C for 15 s followed by RTN at 600°C for 5 min. Although an identical surface cleaning and oxynitridation condition was exerted, the extracted electrical parameters from various substrates indicated some differences as well as similarities as summarized in Table 10.2.

On one hand, the EOT difference between the (111) and (100) oriented samples is very small for p-type substrates but this mismatch becomes substantial on n-type substrates given their comparable doping levels. Moreover, the (111) samples deliver a lower EOT value than the (100) samples even though the oxidation rate should be higher on the (111)

Table 10.2. Electrical parameters extracted from GeO_xN_y MOS capacitors.

Doping type & orientation	**p-(111)**	**p-(100)**	**n-(111)**	**n-(100)**
Doping concentration (cm^{-3})	5×10^{17}	3×10^{17}	1×10^{16}	1×10^{16}
EOT from 100 kHz *C–V* (nm)	4.24	4.40	3.72	5.06
Flat-band voltage (V)	−0.05	0.01	0.27	0.32
C–V hysteresis (mV)	15	16	45	35
Interface trap level density ($cm^{-2}eV^{-1}$)	5×10^{12}	4×10^{12}	5×10^{12}	3×10^{12}

surfaces [45]. Therefore, this counter-intuitive observation should be accompanied with an enhanced nitridation of the (111) oxides, which may suggest either a better penetration of N atoms into the (111) oxides for oxygen replacement or a higher level of nitridation due to the larger number of available surface atoms than on the (100) surfaces. Also, the slightly smaller V_{FB} observed on the (111) substrates for both doping types could be related to a lower amount of negative fixed charges within the dielectric. Besides, the n-type substrates tend to produce relatively more *C–V* hysteresis than the p-type counterparts. As a result, any Ge p-channel MOSFETs with GeO_xN_y dielectric might be slightly more prone to channel carrier loss due to interface charge trapping. Additionally, the interface trap level densities (D_{it}) on the order of 3×10^{12} to $5 \times 10^{12} cm^{-2}eV^{-1}$ were preliminary determined using the Terman method [46] from the 1 MHz curves showing a higher D_{it} value on the (111) surfaces than the (100) surfaces. These GeO_xN_y–Ge interface trap level properties are further analyzed in Section 10.2.3.

On the other hand, the major sources of AC small-signal energy loss (increase in equivalent conductance) in these GeO_xN_y MOS capacitors during *C–V* measurements can be identified from the corresponding *G–V* measurements [47]. Usually, the shape of these *G–V* curves could also be used to explain any deviations from the ideal *C–V* behavior. As shown in the insets of Figure 10.4 for instance, the equivalent parallel conductance captured on the p-type Ge substrates goes through a peak in weak inversion and drops to a low value in strong inversion. These conductance peaks are indicative of small-signal energy loss dominated by generation and recombination through interface trap levels, which coincide with the humps observed in the *C–V* traces at the same applied gate voltage. On the n-type Ge substrates, however, the small-signal energy loss constitutes mainly of generation and recombination through bulk trap levels as the *G–V* curves do not peak up as a function of gate bias in weak inversion. The variation of the measured conductance with gate bias is mainly due to the corresponding variation of capacitance [47]. This gate bias independent loss process also matches well with the characteristic "low-frequency" behavior observed in the high-frequency *C–V* traces.

After acquisition of the baseline electrical characteristics on the GeO_xN_y dielectrics, we have further examined the effects and degree of nitridation on their MOS capacitor behavior. Figure 10.5 depicts the normalized bi-directional *C–V* characteristics measured at 10 kHz from as-grown W/GeO_2/Ge and W/GeO_xN_y/Ge capacitors on (100) oriented Ge substrates. The GeO_2 film was grown by RTO at 600°C for 30 s and the GeO_xN_y layers were formed by RTO at 600°C for 5 s followed by RTN at 600°C for 2–5 min. The *C–V*

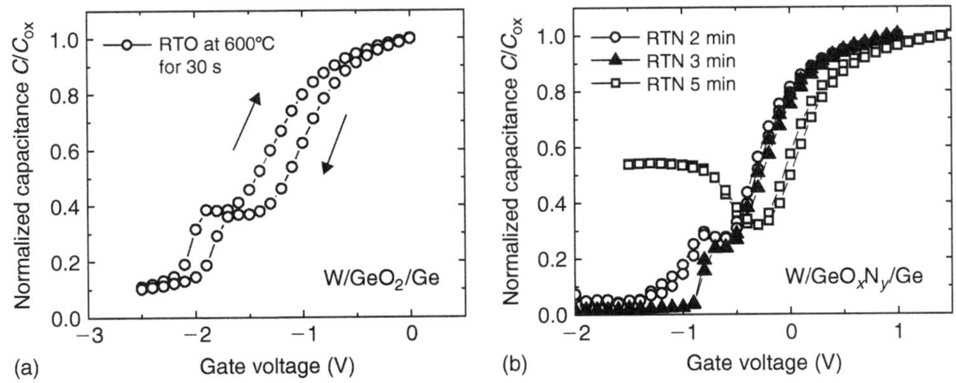

Figure 10.5. Normalized bi-directional capacitance–voltage characteristics measured at 10 kHz from as-grown (a) W/GeO_2/Ge and (b) W/GeO_xN_y/Ge capacitors.

data were normalized with respect to their individual maximum accumulation capacitances to simplify the comparisons of V_{FB} value and bi-directional hysteresis. Only the 10 kHz data are displayed for easy exposure of any abnormal *C–V* behavior.

After nitridation of the GeO_2 to form GeO_xN_y (Figure 10.5b), a drastic reduction in *C–V* hysteresis is obtained which proves its effectiveness to lower the amount of interface charge trapping over the inferior quality GeO_2 (Figure 10.5a). This conclusion is still valid even after normalizing these hysteresis amounts to their corresponding EOT values. Indeed, the D_{it} values extracted from the 1 MHz *C–V* curves (not shown) also decreased after nitridation. However, the degree of nitridation in GeO_xN_y should be well optimized before the real benefit can be discerned. For instance, Figure 10.5b captures the *C–V* characteristics of different GeO_xN_y with identical RTO conditions and varying RTN soak time, while keeping the same RTN temperature and NH_3 flow rate. When the RTN soak time was increased from 2 to 3 min, the kink existed near inversion shrank and the V_{FB} shifted slightly towards the positive voltage direction. Once the soak time was further extended to 5 min the inversion capacitance shot up to about half of the accumulation capacitance (C_{ox}), indicating an efficient carrier generation near the excessively nitrided GeO_xN_y–Ge interfaces. Although, the appearance of this "low-frequency" behavior is not unexpected from the Ge substrate which has a high intrinsic carrier concentration and short minority carrier response time [48], this over-nitridation additionally led to an increased interface charge trapping and negative fixed charge generation within the dielectric. These adverse effects may ultimately degrade the channel carrier mobility due to scattering.

10.2.3. Dielectric–substrate interface analyses

Of crucial importance to the MOS device performance, the GeO_xN_y–Ge interface trap level properties could be characterized using a variety of techniques. The Terman method was previously employed to calculate D_{it}, which however necessitates a theoretical computation of the semiconductor surface capacitance and a measurement of the device doping profile. The second well-known technique is the conductance method in the depletion or weak inversion biasing [47] but requires a broader frequency range beyond our measurement system capability (10 kHz–10 MHz). Amongst other classical techniques, the combined low-high frequency capacitance method [49] is particularly simple and accurate for our GeO_xN_y–Ge interface analyses.

In order to illustrate the D_{it} extraction procedure, we synthesized W/GeO_xN_y/Ge capacitors with the GeO_xN_y films grown by RTO at 600°C for 10 s followed by RTN at 600°C for 5 min and without PGA on (100) oriented p- and n-type Ge substrates. On these capacitors, quasi-static *C*–*V* measurements were carried out using a HP4140B pA meter/DC voltage source through this relation

$$C_{QS} = \frac{I}{\partial V/\partial t} \tag{10.1}$$

where C_{QS} is the quasi-static capacitance, I is the measured current value, and $\partial V/\partial t$ is the gate biasing voltage ramp rate; a ramp rate of $0.1\ \mathrm{V\,s^{-1}}$ was employed. The uni-directional quasi-static and high-frequency (1 MHz) *C*–*V* characteristics are plotted together in Figure 10.6a and b from both p- and n-type Ge substrates, and without the loss of generality, only the post-forming gas anneal data are shown. The *C*–*V* curves on the p-type Ge substrate appear to be more stretched out than the n-type counterparts primarily due to the higher p-type doping concentration ($3 \times 10^{17}\ \mathrm{cm^{-3}}$ for p-type and $2 \times 10^{16}\ \mathrm{cm^{-3}}$ for n-type) rather than an intrinsically higher D_{it} value. After combining both the quasi-static and high-frequency *C*–*V* measurements, D_{it} was computed with

$$D_{it} = \frac{C_{OX}}{q}\left(\frac{C_{QS}/C_{OX}}{1 - C_{QS}/C_{OX}} - \frac{C_{HF}/C_{OX}}{1 - C_{HF}/C_{OX}}\right) \tag{10.2}$$

where q is the electronic charge and C_{HF} is the high-frequency capacitance. Equation (10.2) gives D_{it} over only a limited range of the band gap, typically from the semiconductor flat-band condition to the onset of inversion ($2\psi_B$).

Figure 10.6c and d show extracted D_{it} as a function of Ge surface potential before and after a forming gas anneal at 300°C for 30 min. The surface potential axes (*x*-axes) were labeled with the convention that 0.0 V represents the mid-gap position (E_i), a positive potential samples the upper-half of the band gap towards the conduction band edge (E_C), and a negative value explores the lower-half of the band gap towards the valence band edge (E_V). The plotted surface potential span was in fact less than the Ge band gap as we were not mapping the D_{it} near either the E_C or the E_V edge. Since the surface potential calculations are usually subjected to errors such as doping and interface charge non-uniformities [47], the relative D_{it} distributions within the band gap should be interpreted qualitatively. The as-deposited D_{it} level is around $3 \times 10^{12}\ \mathrm{cm^{-2}eV^{-1}}$ on both p- and n-type Ge substrates with relatively symmetric distribution in both the upper and lower-halves of the band gap. After the forming gas anneal, the D_{it} level dropped by about a factor of 3 to roughly $1 \times 10^{12}\ \mathrm{cm^{-2}eV^{-1}}$ on both substrates.

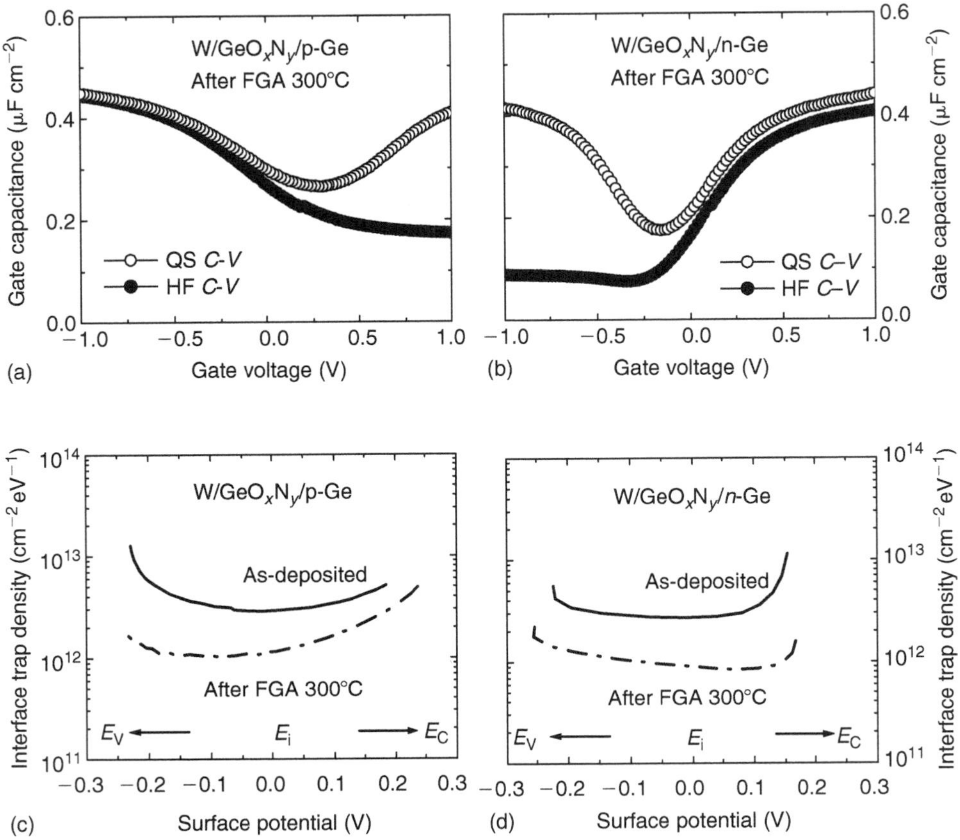

Figure 10.6. Uni-directional quasi-static and high-frequency capacitance–voltage characteristics measured from forming gas annealed W/GeO$_x$N$_y$/Ge capacitors fabricated on (100) oriented (a) p- and (b) n-type Ge substrates. The corresponding interface trap level densities were extracted against the Ge surface potential on (c) p- and (d) n-type Ge substrates before and after the forming gas anneal.

On the n-type Ge sample, the lowest D_{it} value obtained was $8.4 \times 10^{11}\,\text{cm}^{-2}\text{eV}^{-1}$ and the spread of the D_{it} levels was rather even within both halves of the band gap (Figure 10.6d). On the other hand, the post-anneal D_{it} map on the p-type Ge sample indicated a clear asymmetry that has a considerably higher level within the upper-half of the band gap (Figure 10.6c), which has been similarly reported to be composed of acceptor-type states at the hafnia–Ge [50] and hafnia–nitrided Ge [51] interfaces. This important observation may help to explain the asymmetric electron and hole mobility degradation in Ge n- and p-channel enhancement-mode MOSFETs with GeO$_x$N$_y$ gate dielectrics [4,18,19]. In the enhancement-mode MOSFET on-state, the surface inversion would govern the Fermi level to reside within the upper-half of the band gap in p-type Ge for n-channel MOSFET (lower-half of the band gap in n-type Ge for p-channel MOSFET) operation. Therefore, the n-channel carrier (electron) mobility would tend to be degraded more due to coulomb scattering [19] with the filled acceptor states present at a higher D_{it} near the E_C edge (versus those states at a lower D_{it} near the E_V edge in n-Ge for hole scattering in p-channel MOSFET).

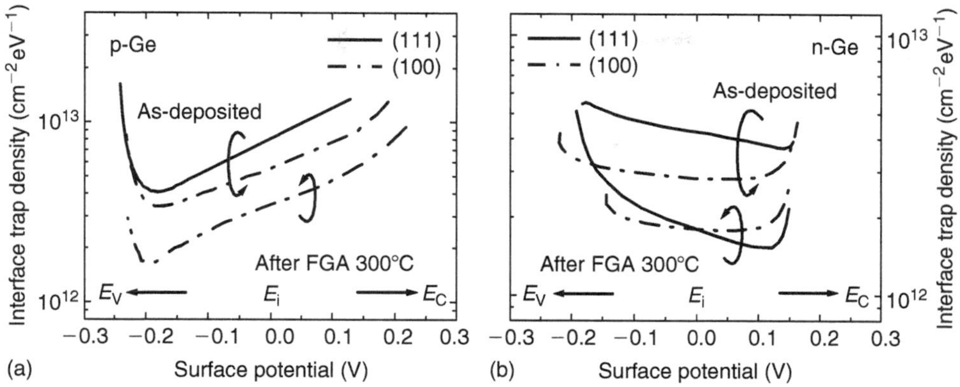

Figure 10.7. Interface trap level densities extracted from W/GeO_xN_y/Ge capacitors fabricated on (a) p- and (b) n-type Ge substrates with different orientation before and after the forming gas anneal.

Utilizing this combined low-high frequency capacitance method, the GeO_xN_y–Ge interface trap level properties were also investigated on the differently oriented substrates discussed in Section 10.2.2. On the same set of samples whose *C–V* and *G–V* characteristics are shown in Figure 10.4, the interface trap level densities were extracted before and after a forming gas anneal at 300°C for 30 min as displayed in Figure 10.7. As similarly observed in Figure 10.6, the D_{it} levels decreased after the forming gas anneal on both Ge substrate types with either (100) or (111) orientation. The post-anneal D_{it} curve from the (111) oriented p-type Ge substrates could not be acquired because the associated dielectric leakage exceeded that required for a reliable quasi-static *C–V* measurement. Generally speaking, the (111) oriented substrates possessed a higher D_{it} level than the (100) samples possibly attributed to the larger number of available bonds per unit area on the Ge surface.

According to Figure 10.7a, both the (100) and (111) oriented p-type Ge samples revealed an asymmetric D_{it} distribution along the band gap. The higher D_{it} values within the upper-half of the band gap would tend to interfere with the operation of n-channel enhancement-mode MOSFET as explained above. Alternatively, the n-type Ge substrates behaved differently with different surface orientation as graphed in Figure 10.7b. After the forming gas anneal, the (100) oriented n-type Ge sample maintained its uniform D_{it} distribution but the (111) sample increased its D_{it} asymmetry otherwise. Comparing the post-anneal D_{it} curves, the (111) surface has respectively a lower and higher D_{it} value within the upper-half and lower-half of the band gap than the (100) surface. After merging all these phenomena, one would recommend the choice of (100) oriented Ge substrates for the fabrication of enhancement-mode complementary MOSFETs from the lower D_{it} perspective. In contrast, the (111) oriented Ge substrates would find attractive applications in building accumulation-mode MOSFETs with either an ultra-thin body [52,53] or non-planar [54] device architecture for better channel length scalability.

As a final note, if some degrees of flexibility and creativity are being allowed in device engineering, the D_{it} asymmetry observed above could be utilized to our advantage. In order to make Ge p-channel MOSFETs on (100) oriented substrate for instance, one could elect either an n-type channel doping for enhancement-mode operation or a p-type channel doping for accumulation-mode operation. Based on the higher D_{it} value sampled from the n-type Ge (Figure 10.6d and 10.7b) than that from the p-type Ge (Figure 10.6c and 10.7a) within the lower-half of the band gap, the latter should thus be selected. Although a fair experimental comparison is still pending, Ge p-channel MOSFETs with SiO_2 dielectric operating in accumulation mode [53] have been reported to deliver a higher hole mobility than the enhancement-mode devices with GeO_xN_y dielectric [4] even though the grown GeO_xN_y should be more compatible with Ge than SiO_2 in principle. On the other hand, since the (111) oriented Ge surface was theoretically shown to carry higher ballistic current in n-channel MOSFETs than the (100) surface [36], having an accumulation-mode n-channel MOSFET on the (111) surface would certainly unite both the benefits of higher carrier mobility and lower D_{it} near the conduction band (Figure 10.7b).

10.2.4. Dielectric leakage behavior

The general gate leakage behavior observed from similar W–GeO_xN_y–Ge capacitors was previously summarized [42] and the typical gate leakage density as a function of voltage bias is shown in Figure 10.8a. These data were captured from a GeO_xN_y sample grown by RTO at 600°C for 5 s followed by RTN at 600°C for 3 min and without PGA on the (100) oriented p-type Ge substrate. Both the gate and substrate injection leakage characteristics were measured using a HP4155A semiconductor parameter analyzer.

Via direct gate voltage bias sweeping, the characteristic leakage density pattern [42] was observed with symmetric leakage level at both positive and negative bias polarities (Figure 10.8a). However, there occasionally exists a leakage current hump around the gate voltage of −0.5 V to 0 V, which could possibly arise from a transient charge trapping effect. In order to verify this proposition, the transient leakage currents were monitored at selected negative and positive voltage biases as illustrated in Figure 10.8b and 10.8c, respectively. After about 1 min of bias holding, the leakage currents appeared to stabilize at their respective gate voltages, an indication that the initially trapped charges are gradually relaxed to restore the normal leakage levels. When these "after bias holding" leakage levels were plotted against those via "direct bias sweeping" (Figure 10.8a), a more reasonable leakage behavior can be comprehended with a monotonic increase in leakages with increasing bias amplitudes from 0 V. Generally speaking, this leakage current hump is often recorded at gate injection biases regardless of substrate dopant type and crystal orientation, which indicates the associated slow traps are located at the dielectric–substrate interface. In Section 10.3.5, the GeO_xN_y leakage levels are further benchmarked against those extracted from various high-permittivity metal oxide dielectric stacks on Ge.

10.2.5. Summary

In this section, we first discussed the synthesizing methodology of the GeO_xN_y by RTN of the intentionally grown Ge oxides. The resulting GeO_xN_y film thickness, refractive index,

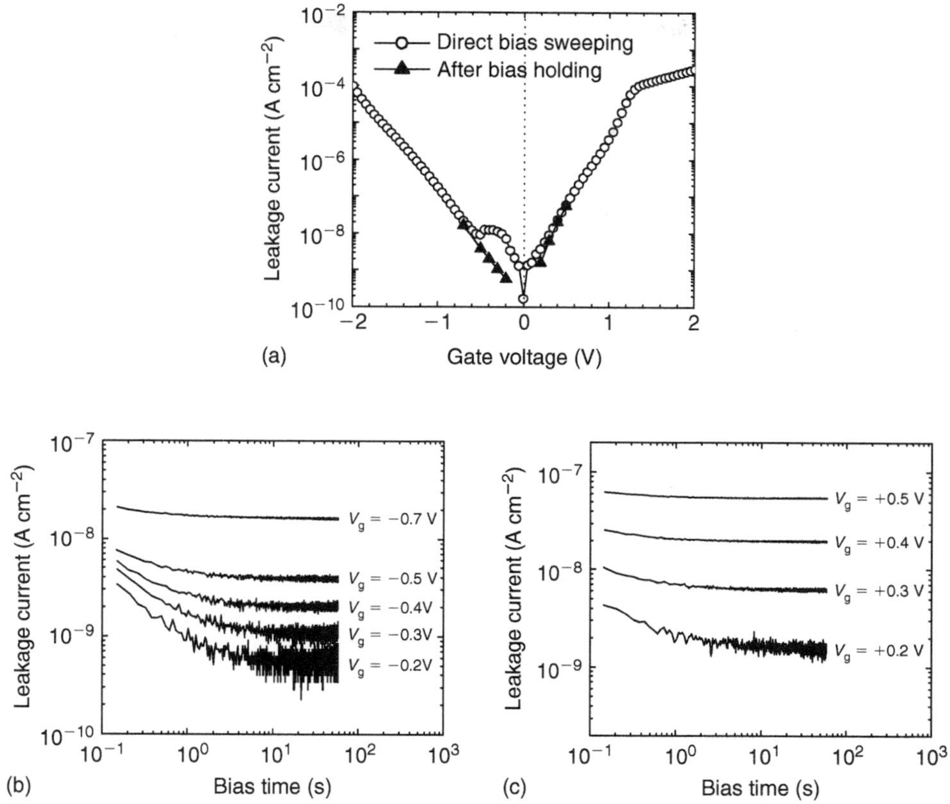

Figure 10.8. (a) The gate leakage density as a function of voltage bias measured from a typical W/GeO_xN_y/Ge capacitor. (b) and (c) are the corresponding transient leakage currents at negative and positive biasing, respectively.

N content and depth profile were examined by optical ellipsometry and X-ray photoemission spectroscopy.

Metal-gated GeO_xN_y MOS capacitors were subsequently fabricated to investigate their basic dielectric characteristics on Ge substrates with different dopant type and crystal orientation. Nitridation of Ge oxides was shown to be useful in lowering the amount of interface charge trapping, however, the degree of nitridation needed to be well optimized to realize high quality Ge MOS capacitors with minimal frequency dispersion and *C–V* hysteresis.

Dielectric–substrate interfaces were also analyzed using the combined low-high frequency capacitance method and the forming gas anneal was found to be effective in reducing the interface trap density. In addition, an asymmetry in the interface trap density distribution within the Ge band gap was revealed that helped to explain the inferior n-channel GeO_xN_y MOSFET performance over their p-channel counterparts. Finally, the general gate leakage behavior from these MOS capacitors was reported with an observed abnormality that could be attributed to a transient charge trapping effect.

10.3. High-permittivity Metal Oxide Dielectrics

Even though the grown GeO_xN_y discussed in Section 10.2 could be scaled down to an EOT of 1.9 nm [42], a sub-1.0 nm EOT solution is still mandated for Ge MOSFETs to advance beyond the 32 nm technology node [1]. Inspired by the research successes to scale down EOT in Si MOS devices with the use of high-permittivity (high-κ) metal oxides [56], the possibility of applying these high-κ materials onto Ge was first demonstrated in 2002 [3,57]. Subsequently, variants of this "high-κ dielectric on Ge" concept have been independently realized [58–67].

Usually, the common high-κ dielectric deposition onto Si either requires or produces a lower-κ Si sub-oxide (SiO_x) interfacial layer between the dielectric film and the substrate [56], which is currently a major bottleneck to scale the EOT thinner than 1.3 nm. In comparison, the thermodynamic instability of GeO_x may permit a true high-κ gate stack on Ge without a performance-limiting lower-κ interfacial layer, and thus overcome the EOT scaling barrier.

In this section, we first identify the high-κ dielectric candidates for Ge MOS device applications according to both the thermodynamic and electrical selection criteria. After which, we present two major high-κ metal oxide deposition techniques: (1) atomic layer deposition (ALD) and (2) ultraviolet ozone oxidation (UVO) of metals. Other contending techniques are also discussed. Various Ge substrate surface preparations were attempted prior to high-κ dielectric deposition to examine the resultant effects on the fabricated metal-gated MOS capacitors. Finally, the leakage current densities measured from various nanoscale Ge MOS dielectrics are compared and the scalability of various high-κ gate stacks on Ge is evaluated against their Si counterparts.

10.3.1. High-κ dielectrics selection criteria

For a high-κ dielectric material to be chosen to integrate onto Ge for MOS device applications, both the thermodynamic and electrical criteria need to be met simultaneously. Various high-κ candidates including metal germanates (MGe_xO_y) and metal oxides (MO_x) have been considered from the perspective of their thermodynamic stability in contact with Ge and several stable MGe_xO_y phases were initially identified from the binary phase diagrams of various metal oxide systems [68]. Owing to their complexity, however, we limited our study to the MO_x high-κ candidates for Si and investigated their application to Ge MOS devices. Based on the thermodynamic analysis using a pseudo-ternary phase diagram [69], the thermodynamic stability of zirconium oxide (ZrO_2) with respect to solid-state reaction with Ge can be predicted as illustrated in Figure 10.9. This thermodynamic calculation was performed at the nominal high-κ dielectric ALD temperature [41]. The existence of a tie line between ZrO_2 and Ge, indicating simultaneous stability of the two phases in contact with one-another, results from the large standard Gibbs free energy of formation of ZrO_2 (-1135 kJ/mol) compared to that of GeO_2. Therefore, solid-state reaction between ZrO_2 and Ge should not occur spontaneously. An interfacial layer between the ZrO_2 dielectric and Ge substrate may be avoided under conditions in which the deposition process does not produce significant Ge oxidation. Likewise, a similar thermal stability of hafnium oxide

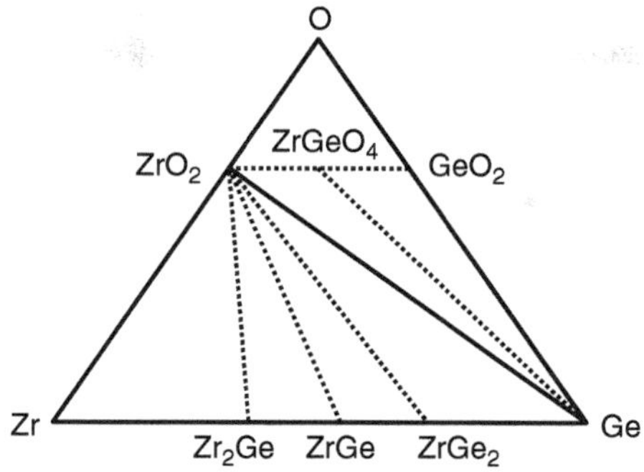

Figure 10.9. Calculated Zr–Ge–O ternary phase diagram at 600 K.

(HfO_2) on Ge was predicted for the Hf–Ge–O ternary system at the same ALD reaction temperature.

On the other hand, the electrical behavior of ZrO_2 and HfO_2 dielectrics on Ge is also of great importance. The κ-values of these two gate dielectrics were extracted from Si MOS capacitors to be around 30 and 20 for ALD ZrO_2 and HfO_2, respectively [69], which should be effective in leakage suppression compared to grown GeO_xN_y at an equivalent EOT. In addition, the conduction band offset (ΔE_C) at the MO_x–Ge interfaces can be computed [70] using the following expression

$$\Delta E_C = \chi_{Ge} - \chi_{MO_x} + (S - 1)(\Phi_{CNL,Ge} - \Phi_{CNL,MO_x}) \tag{10.3}$$

where χ_{Ge} and χ_{MO_x} are the electron affinities of Ge and MO_x, respectively, $\Phi_{CNL,Ge}$ and Φ_{CNL,MO_x} are the charge neutrality levels of Ge and MO_x, respectively, and S is an empirical parameter analogous to the Schottky barrier pinning factor in Bardeen's model of metal-semiconductor junctions. Employing the values reported in the literature [71–73], the theoretical values of ΔE_C are calculated to be 1.63 eV and 1.65 eV for ZrO_2 and HfO_2, respectively, which should be sufficiently large to provide an acceptable barrier for leakage conduction. Merging the thermodynamic and electrical qualifications together, both ZrO_2 and HfO_2 should be applicable for future Ge MOS device applications.

10.3.2. ALD of high-κ dielectrics

ALD deposition of high-κ metal oxide dielectrics was carried out in a cold-wall high vacuum base pressure system [69]. ALD is a particularly attractive method for preparing ultrathin MO_x layers with excellent electrical characteristics and near-perfect film conformality because of the layer-by-layer nature of the deposition kinetics. A typical ALD process for ZrO_2 or HfO_2 is performed at 300°C, using alternating surface-saturating reactions of water (H_2O) and metal tetrachloride (MCl_4). In our reactor, each precursor is pulsed for 2 s and nitrogen purging follows for 30 s and 60 s after each H_2O and MCl_4 pulse, respectively.

The base pressure of the system is $\sim 5 \times 10^{-8}$ Torr and the process pressure is maintained at 0.5 Torr during ALD.

10.3.2.1. ALD of zirconia

Since the ALD reaction on differently prepared substrate surfaces could yield high-κ films with various microstructures and electrical qualities, ALD of ZrO_2 on Ge was investigated on three different surfaces:

(i) Etched for 1 min with HF vapor evaporated from a concentrated (49%) aqueous solution,

(ii) Rinsed in DI water,

(iii) Capped with a native oxide layer.

The first two preparations were intended to produce a non-hydroxylated Ge surface and the last one was designed to provide surface hydroxylation for the ALD reaction. Since the soluble native oxide was examined to be the dominant species on as-received wafers by XPS, DI water rinsing was used to dissolve that. To optimize the dwell time in DI water for minimum resultant surface roughness, atomic force microscopy (AFM) was used to track the root-mean-square roughness (R_{rms}) as highlighted in Figure 10.10 [41].

Starting with the as-received bare Ge substrate, the initial peak in R_{rms} (0–60 s) was due to partial and then complete removal of the soluble native oxide. Inferred from the experience on Si [74–76], the dissociative adsorption of H_2O at room temperature on the Ge surface is believed to preferentially form a GeOH surface intermediate, which could lead to increased oxidation (and R_{rms}) with increasing rinse time (>60 s). The optimal DI water dwell time employed in the subsequent experiments is 1 min.

The starting substrate used for ALD of ZrO_2 was (100) oriented n-type Ge wafers with a net background concentration of $\sim 7 \times 10^{15}\,cm^{-3}$. The three different surface preparations

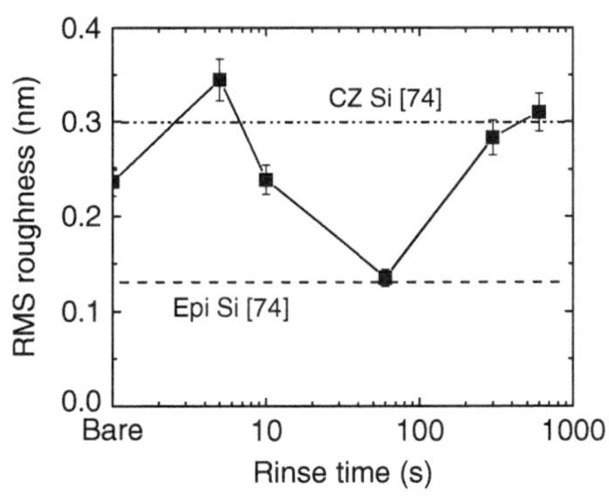

Figure 10.10. Ge surface R_{rms} from AFM as a function of DI water rinse time [41]. The R_{rms} values from both epitaxial and Czochralski Si wafers were also included.

were applied prior to the ALD of ~5.5 nm ZrO_2. Various sizes of 50 nm thick platinum (Pt) gate electrodes were formed by electron beam evaporation through a shadow mask, and subsequently, aluminum (Al) was evaporated on the wafer backside to reduce the sample contact (and series) resistance. Finally, a forming gas anneal was carried out at 400°C for 30 min on the completed Pt/ZrO_2/Ge capacitors.

C–V characteristics were measured on the MOS capacitors using the HP4284A precision LCR meter. On both of the non-hydroxylated (Figure 10.11a and b) and hydroxylated (Figure 10.11c) Ge surfaces, similarly large *C–V* frequency dispersion, bi-directional hysteresis, as well as abnormal inversion behavior could be observed. After correcting for the substrate resistivity induced frequency dispersion [77] in the accumulation capacitance (using the 100 kHz and 800 kHz *C–V* data), EOT values of 2.28, 2.28, and 2.90 nm were computed for the HF vapor etched, DI water rinsed, and native oxides capped samples, respectively. Both of the non-hydroxylated surface passivation schemes result in a smaller

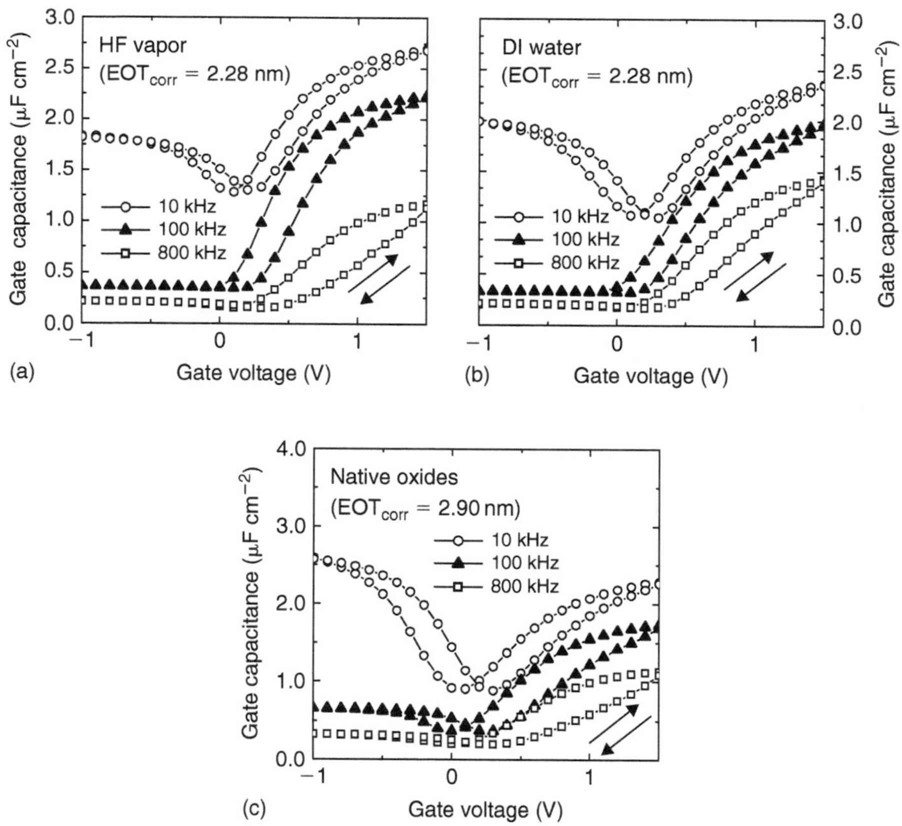

Figure 10.11. Multi-frequency bi-directional capacitance–voltage characteristics measured from forming gas annealed Pt/ZrO_2/Ge capacitors fabricated on (a) HF vapor etched, (b) DI water rinsed, and (c) native oxides capped Ge surfaces. The corrected EOT values were computed from the 100 and 800 kHz traces.

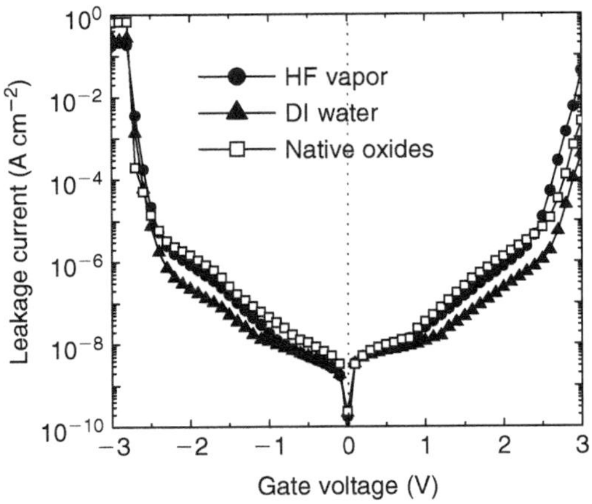

Figure 10.12. Gate leakage–voltage measurements from Pt/ZrO_2/Ge capacitors with different Ge surface preparations.

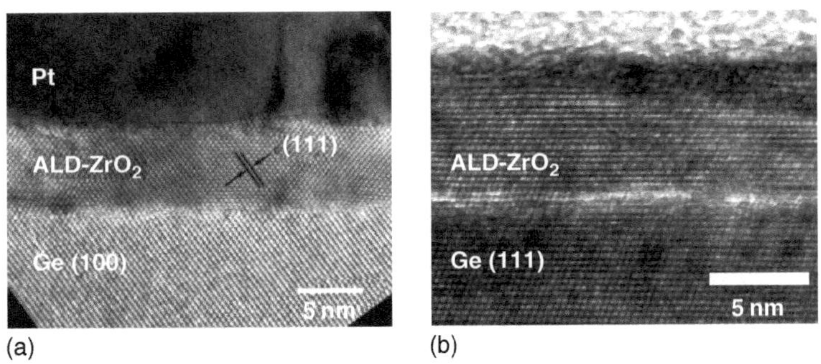

Figure 10.13. Cross-sectional HR-TEM micrographs along the <110> zone-axis of (a) Pt/5.5 nm ZrO_2/Ge(100) and (b) 6.8 nm ZrO_2/Ge(111) (after Kim *et al.* [78]).

EOT owing to their effectiveness at removing the native oxides prior to ALD. Figure 10.12 shows the gate leakage density of these ZrO_2–Ge capacitors as a function of voltage bias for both gate and substrate injections using the Keithley 230 and 6512. Among the Ge capacitors, the leakage measured from the hydroxylated sample (open square symbol) shows a higher leakage even with a larger EOT compared to those measured from the non-hydroxylated samples (solid symbols), attributing to the relatively poor electrical quality GeO_x interfacial layer. Nonetheless, gate leakage current densities were found to be remarkably low and are benchmarked with other dielectric stacks in Section 10.3.5.

The inferior *C–V* characteristics are believed to stem from interfacial defects. For instance, ALD of ZrO_2 films on both Ge(100) and Ge(111) exhibited local epitaxial growth without a distinct interfacial layer [78], as shown in the cross-sectional high-resolution transmission electron micrographs (HR-TEM) from Figure 10.13a and b. This absence of interfacial oxides was also confirmed with sputtered depth profile XPS analysis [69,78].

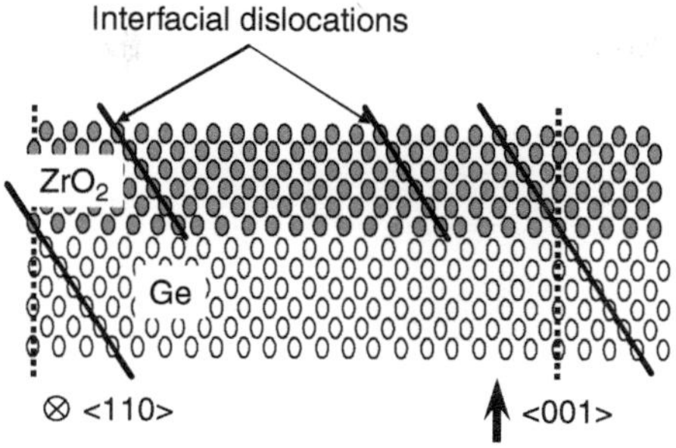

Figure 10.14. Schematic diagram showing the epitaxial relationship and the interfacial dislocations for the sample illustrated in Figure 10.13a (after Kim *et al.* [69]).

The thermodynamic stability for this interfacial layer-free ZrO_2 on Ge substrate is plausible because of the more negative Gibbs free energy of ZrO_2 formation (−1135 kJ/mol at 600 K) compared to that of GeO_2 (−610 kJ/mol at 600 K) as discussed in Section 10.3.1. Even though the reformation of GeO_x on HF-treated Ge surface prior to the ALD process is largely possible (as in the case of ALD of MO_x on Si), it could be easily dissociated and/or dissolved in H_2O during the subsequent ALD steps. Owing to the large lattice mismatch (~10%) between the tetragonal ZrO_2 and Ge, a significant areal density of interfacial dislocations can be seen in the cross-sectional image in Figure 10.13. This phenomenon is schematically illustrated in Figure 10.14 where there exists an extra atomic plane out of every other 10 within the ZrO_2 layer.

The epitaxial relationship between the ZrO_2 film and Ge substrate was further verified using electron diffraction analysis and plan-viewing TEM imaging (Figure 10.15). Indexing of electron diffraction patterns obtained during electron microscopy indicates that the ALD grown ZrO_2 film is quite likely to be in tetragonal phase [69,78]. Due to the large lattice mismatch, local epitaxial growth generated numerous distorted Moiré fringes, as shown in the plan-view images (Figure 10.15a and c). The mosaic spread of the epitaxial film orientation also manifests itself in a distortion of diffraction spots seen in the electron diffraction patterns (Figure 10.15b and d). The (100) Ge//(100) ZrO_2 and [100] Ge//[100] ZrO_2 epitaxial relationship is observed and the existence of an extra ZrO_2 (111) atomic plane per every ten planes (Figure 10.13a) hints that the compressive misfit strain of the ZrO_2 film is almost fully relaxed by these misfit dislocations. On Ge (111) substrate, a similar (111) Ge//(111) ZrO_2 and [111] Ge//[111] ZrO_2 epitaxial correlation was obtained as well.

This 10% lattice mismatch results in a significant areal density of interfacial dislocations ($\sim 7 \times 10^{12}$ cm^{-2}) that may be responsible for the *C–V* frequency dispersion and electron trapping ($\sim 3 \times 10^{12}$ cm^{-2} estimated from the hysteretic bi-directional *C–V* traces). Moreover, excess carrier generation from slow states resulting from interfacial dislocations and/or diffusion of metal impurities into the near-surface Ge substrate may produce the abnormal inversion behavior in the *C–V* characteristics [79]. Insertion of an interfacial passivation and diffusion barrier layer has been found to minimize these problems as discussed in Section 10.3.2.2.

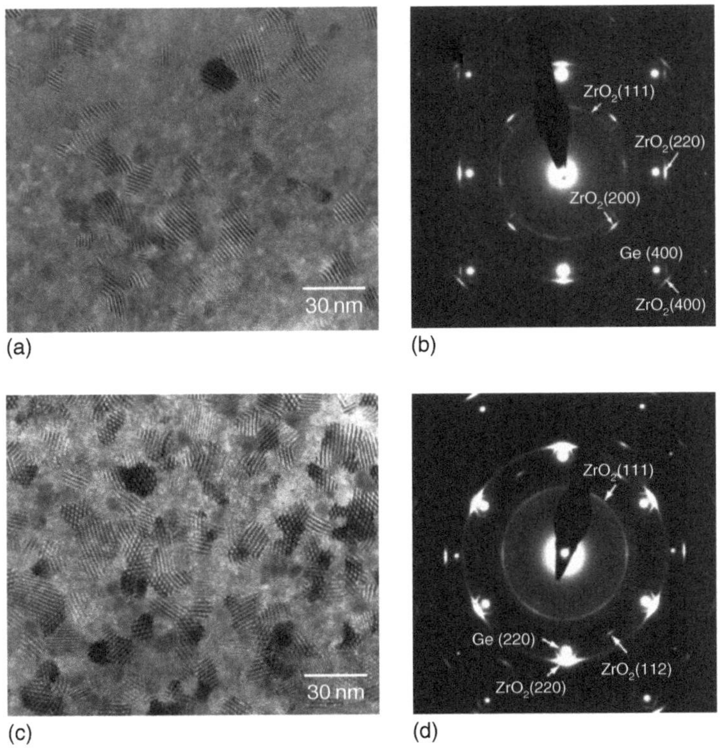

Figure 10.15. (a) Bright-field plan-view TEM image of the 5.5 nm ZrO_2/Ge(100), and (b) selective area diffraction pattern showing the corresponding epitaxial relationship. (c) Bright-field plan-view TEM image of the 6.8 nm ZrO_2/Ge(111), and (d) selective area diffraction pattern showing the corresponding epitaxial relationship (after Kim *et al.* [69,78]).

10.3.2.2. ALD of hafnia

As a leading high-κ dielectric candidate for Si MOS devices, we have also studied the ALD HfO_2 on several Ge surfaces with and without intentional hydroxylation. The starting substrates were (100) oriented p-type Ge wafers with a net background concentration of $\sim 3 \times 10^{17}\,cm^{-3}$. From the ALD ZrO_2 on Ge results above, we learned that the native oxide-capped surface does not yield a good quality dielectric, possibly due to its instability during the ALD reaction. Stimulated by the high quality GeO_xN_y–Ge interface discussed in Section 10.2, we investigated also the possibility of applying a GeO_xN_y interfacial layer prior to the ALD of HfO_2. Pt-gated HfO_2 capacitors were similarly fabricated as previously described on differently prepared Ge surfaces:

(i) CHF (discussed in Section 10.2) to remove both water soluble and insoluble native oxides,

(ii) Rinsed in DI water for 1 min (discussed in Section 10.3.2.1),

(iii) Capped with thick GeO_xN_y grown by furnace oxidation of CHF cleaned Ge at 500°C for 2 min followed by RTN at 600°C for 1 min in NH_3 (discussed in Section 10.2.1),

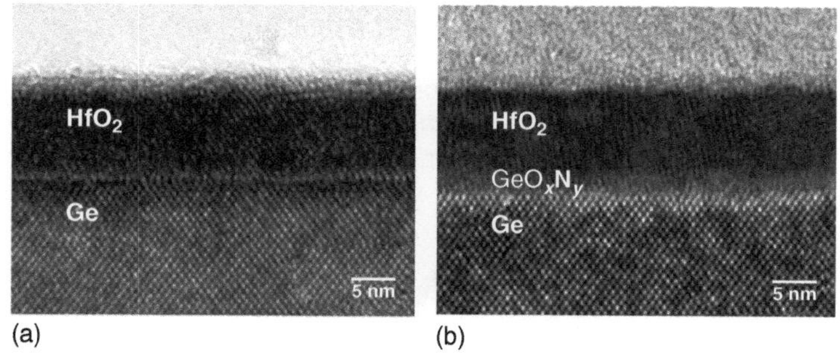

Figure 10.16. Cross-sectional HR-TEM images of ~4.5 nm ALD HfO_2 on (a) CHF cleaned and (b) thin GeO_xN_y capped Ge surfaces.

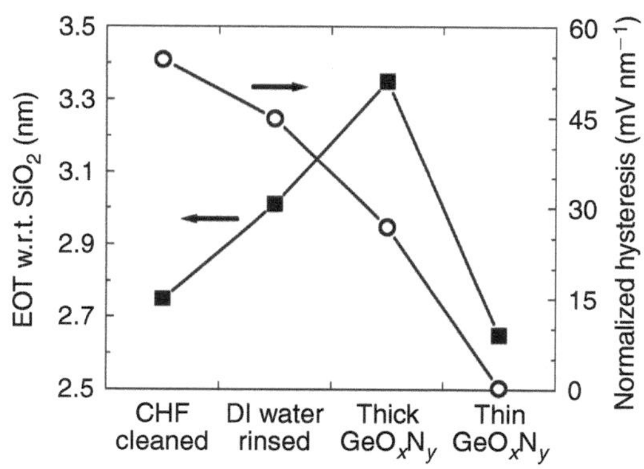

Figure 10.17. EOT values with respect to SiO_2 and normalized bi-directional *C–V* hystereses estimated Pt/HfO_2/n-Ge with various surface preparations prior to ALD.

(iv) Capped with thin GeO_xN_y grown by RTN at 600°C for 1 min in NH_3 on CHF cleaned Ge.

Interfacial microstructures on selected ALD HfO_2 on Ge samples were inspected using TEM (Figure 10.16). The ALD HfO_2/CHF cleaned Ge surface has negligible interfacial oxide (Figure 10.16a) similar to the ALD ZrO_2 experience described in Section 10.3.2.1. From Figure 10.16b, the thin interfacial GeO_xN_y is found to be quite uniform with thickness of ~1.1 nm.

C–V characteristics were extracted from these ALD HfO_2 Ge MOS capacitors using a HP4284A precision LCR meter. Since the HfO_2 film on different samples were identically deposited at the same time with the same number of ALD cycles, the variation in deposition conditions was minimized. After correcting for the substrate resistivity induced accumulation capacitance dispersion, the EOT values and bi-directional hysteresis normalized to the corresponding EOT were excerpted as shown in Figure 10.17.

Among the surface hydroxylated samples, the inclusion of a thicker GeO_xN_y interfacial layer intuitively produces a larger EOT value than a thinner one. On those non-hydroxylated surfaces, the CHF cleaned sample offers a smaller EOT compared to the DI water rinsed

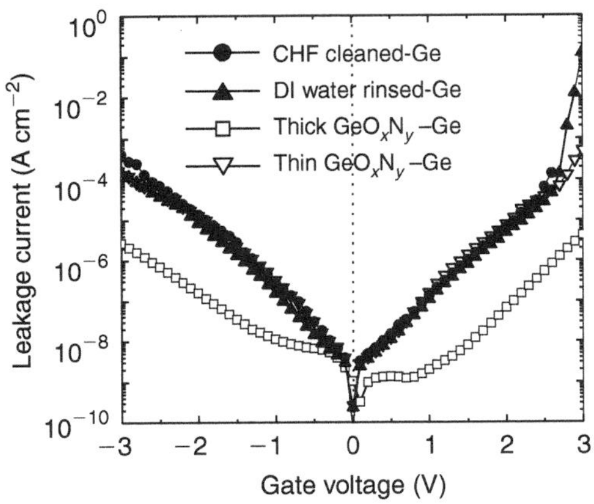

Figure 10.18. Gate leakage-voltage measurements from Pt/4.5 nm HfO_2/Ge capacitors with different Ge surface preparation. Solid and open symbols are used for non-hydroxylated and hydroxylated surfaces, respectively.

sample, indirectly suggesting that the combination of HF and DI water cleaning is more effective to remove GeO_x than DI water alone. On the other hand, the *C–V* hysteresis obtained from different starting surface represents another scenario. The incorporation of a GeO_xN_y interfacial layer delivers a reduction in the amount of charge trapping (and thus hysteresis) versus ALD directly onto non-hydroxylated Ge surfaces. The improvement may be attributed to improved ALD nucleation on stable hydroxylated surfaces over the atomically clean Ge, which allows locally epitaxial growth with a different microstructure [78].

Illustrated in Figure 10.18 is the gate leakage density of these HfO_2/Ge capacitors as a function of voltage bias using Keithley 230 and 6512. With either a gate or substrate injection, the gate leakage for all the samples was similarly low. The sample with a thick GeO_xN_y interfacial layer gave the lowest leakage, which is expected because of its larger EOT. Compared to SiO_2 on Si in the same EOT range [80], the leakage current obtained from the high-κ Ge MOS capacitors demonstrated about 10^3–10^5 times reduction. Combining the accumulation EOT, *C–V* hysteresis, and gate leakage data, we found that the optimal Ge MOS gate dielectric stack, for the processing methods investigated in this work, was composed of ALD HfO_2 on a thin GeO_xN_y interfacial layer to generate acceptable capacitors (Figure 10.19). On both p- and n-type Ge substrates, the kinks that showed up near inversion in lower frequency scans again suggest the presence of slow interface states as observed in GeO_xN_y/Ge capacitors (Section 10.2.2). However, the observed inversion capacitance increase in n-type Ge in lower frequency scans may be due to excess carrier generation over that in p-type Ge. Alternatively, the presence of negative fixed charge within the HfO_2/GeO_xN_y dielectric stack that causes n-type surface depletion (except the capacitor active area) and supplies positive carriers for capacitor channel inversion (depending upon the capacitor area) could also cause the same effect [47].

In order to optimize the hydroxylation conditions for ALD of HfO_2 onto Ge, the Pt/HfO_2/GeO_xN_y/Ge capacitor electrical characteristics with different GeO_xN_y RTN temperature

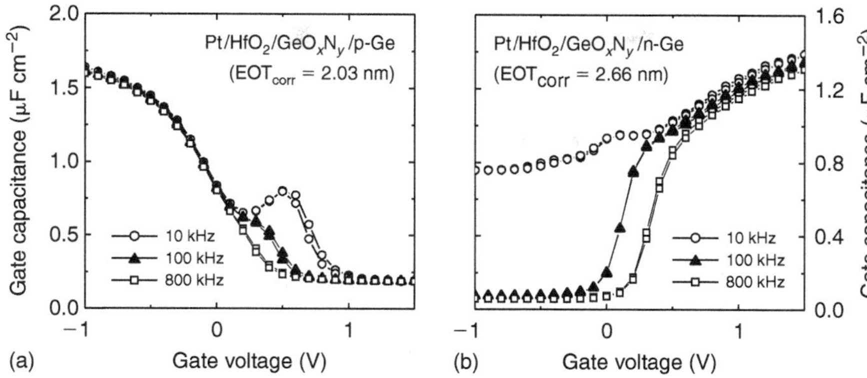

Figure 10.19. Multi-frequency capacitance–voltage characteristics measured from (a) Pt/HfO_2/GeO_xN_y/p-Ge and (b) Pt/HfO_2/GeO_xN_y/n-Ge capacitors.

were examined before and after forming gas anneal. When the *C–V* curves for as-deposited films are compared (Figure 10.20a and b), the 600°C grown GeO_xN_y interfacial layer revealed minimal hysteresis and frequency dispersion compared to the 700°C case. The higher RTN temperature degraded the interfacial quality and triggered an inversion capacitance increase, which are the characteristic phenomena of over-nitridation as observed in Section 10.2.2. This behavior is not believed to result from metal impurity in-diffusion (Section 10.3.2.1) as medium energy ion scattering (MEIS) data [79] suggest that the GeO_xN_y interfacial layer is effective in blocking Hf atoms from diffusing into Ge substrate at the process temperatures investigated.

After a forming gas anneal at 300°C for 30 min, both the 600°C and 700°C GeO_xN_y capacitors showed increased bi-directional hysteresis due to hole trapping (Figure 10.20c and d) owing to a possible de-passivation of the weak Ge–H bonds formed during NH_3 nitridation. Contrary to the typical case of forming gas passivating most of the SiO_2–Si interfacial dangling bonds, the efficacy of forming gas anneals for Ge MOS devices is not obvious. Low gate leakage current densities on these HfO_2/GeO_xN_y/Ge capacitors were observed for both gate and substrate injection, relatively independent of the nitridation condition as shown in Figure 10.20e.

Finally, physical characterization of the GeO_xN_y interfacial layers was carried out to inspect their N content, chemical stability, and the impurity diffusion blocking capability. XPS was employed to estimate the N atomic concentration in GeO_xN_y layers grown on CHF cleaned Ge at various temperatures for 1 min (Figure 10.21). Any surface hydrocarbon layer adsorbed during sample transfer in air was also included in the calculation. The relative ratio of N to the sum of N plus carbon (C) and oxygen (O) increases linearly with the RTN temperature from about 10% to 30%. In addition, the oxidized Ge(3d) peak (Ge^{n+}) shifted toward lower binding energy at higher RTN temperatures, which suggests a gradual modification of the Ge bonding configuration. The chemical stability of the GeO_xN_y layer grown at 600°C was also studied by dipping the sample in either 20:1 HF solution for

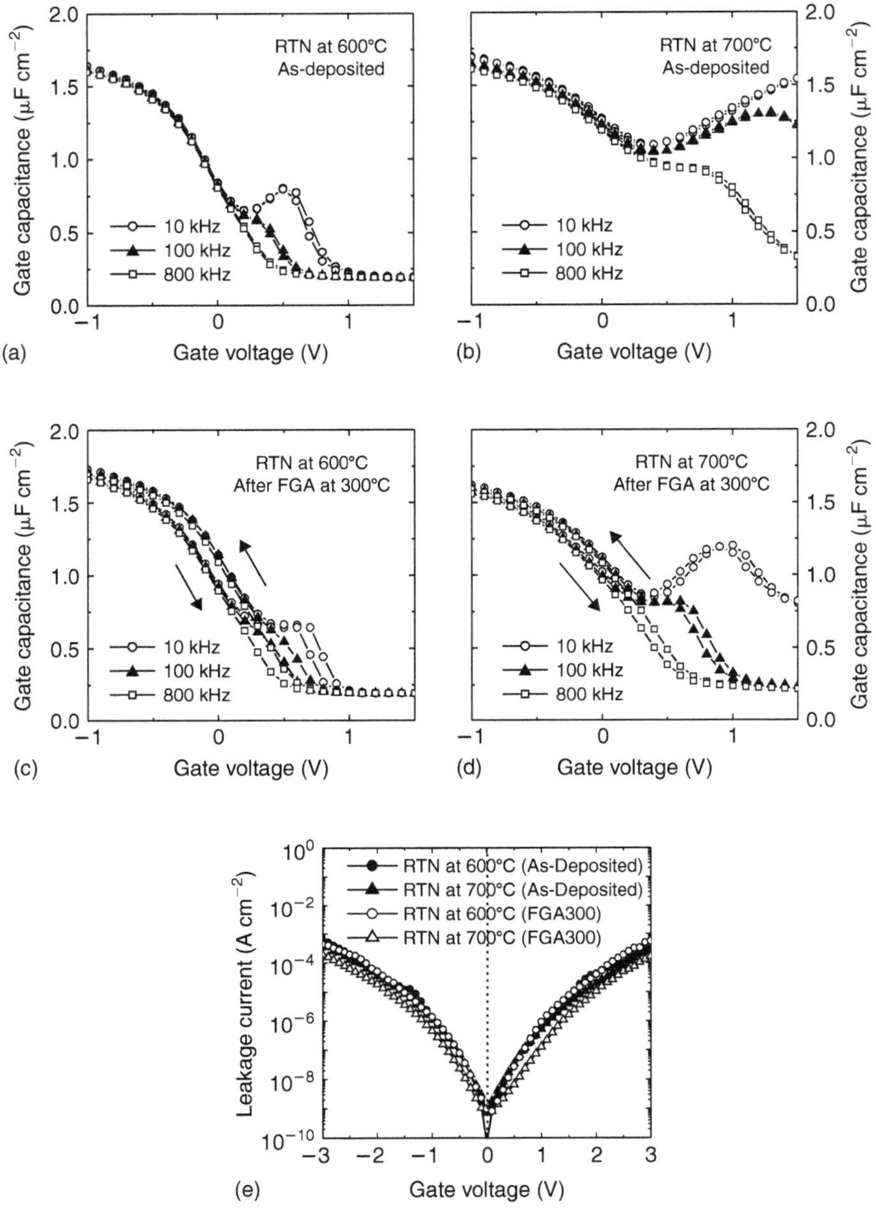

Figure 10.20. (a)–(d) Multi-frequency bi-directional capacitance–voltage characteristics measured from as-deposited and forming gas annealed Pt/HfO_2/GeO_xN_y/Ge capacitors with the GeO_xN_y grown at 600°C and 700°C. (e) The corresponding gate leakage–voltage measurements.

1 min or DI water for 2 min (Figure 10.22). After the N incorporation into GeO_x, the resultant GeO_xN_y layer exhibited a drastic reduction in DI water solubility and an enhanced chemical stability, while this stabilized GeO_xN_y film can easily be etched away in HF solution – the standard SiO_2 etchant.

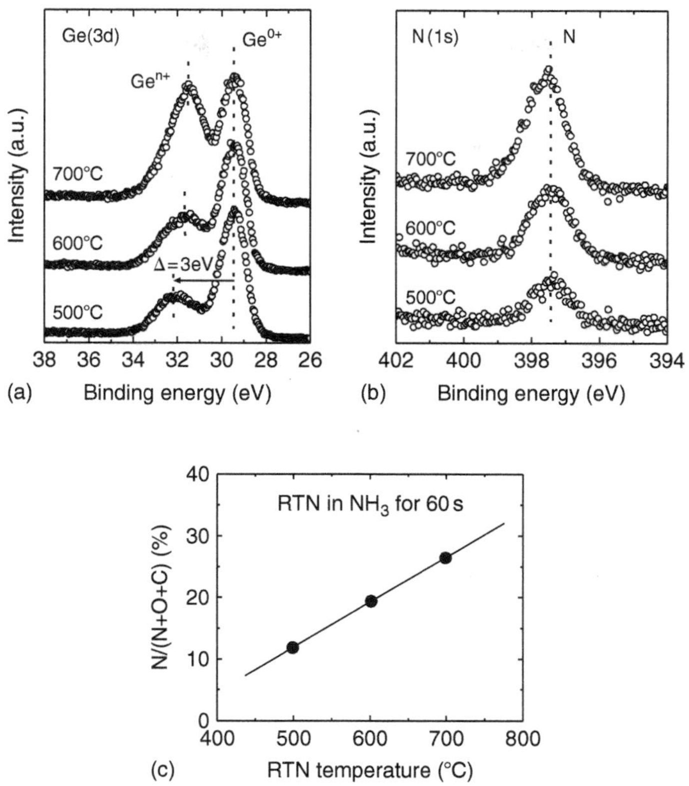

Figure 10.21. Core-level Ge(3d) XPS spectra (a), core-level N(1s) XPS spectra (b), and relative N atomic concentration from the GeO_xN_y films (c) as a function of RTN temperature.

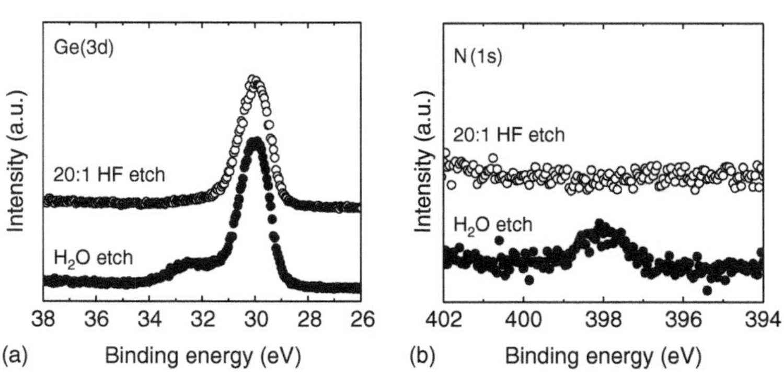

Figure 10.22. Core-level Ge(3d) (a) and N(1s) (b) XPS spectra taken on the GeO_xN_y film grown at 600°C after either a HF or DI water treatment.

Analogous to the experience with silicon oxynitride (SiO_xN_y), the nitrogen incorporation into the GeO_xN_y interfacial layer is expected to block the metal impurities within the metal gate electrode/high-κ dielectric stack from diffusing into Ge substrate. MEIS was chosen for such an investigation, which is a lower energy version of Rutherford backscattering (RBS) in which the energy and angular distributions of scattered ions are used to obtain detailed information about atomic structure and composition within the top 5–20 nm of the sample [81]. About 4 nm of HfO_2 was deposited by ALD on two different Ge (100) surfaces: (i) native oxides capped and (ii) CHF cleaned followed by RTN at 600°C with 1 min soak time in NH_3. MEIS analysis was then performed with a 100 keV proton beam in a double alignment so as to reduce the contributions from the crystalline Ge substrate, allowing deconvolution of spectra into contributions from the HfO_2 layer and Ge-containing underlayer and substrate [79]. The incident ions were directed along the (111) plane and the scattered ions were along the (100) plane with a scattering angle of ~125°. Quantitative depth profiles for different species were extracted with a resolution of <0.5 nm in the near-surface region.

Figure 10.23 depicts the measured MEIS spectra and the ion scattering simulation results from both samples. The overall shapes of both spectra were similar except for a difference in the lower energy edge of the Hf signal as shown in the figure insets, which corresponds to the interfacial layer under the HfO_2 film. On the chemical oxide kept sample, a marked shoulder was observed which is likely due to an intermixing between Hf and GeO_x or large interface roughness even without under-going any post-deposition thermal treatment. In contrast, no significant amount of Hf atoms could be found inside the interfacial GeO_xN_y

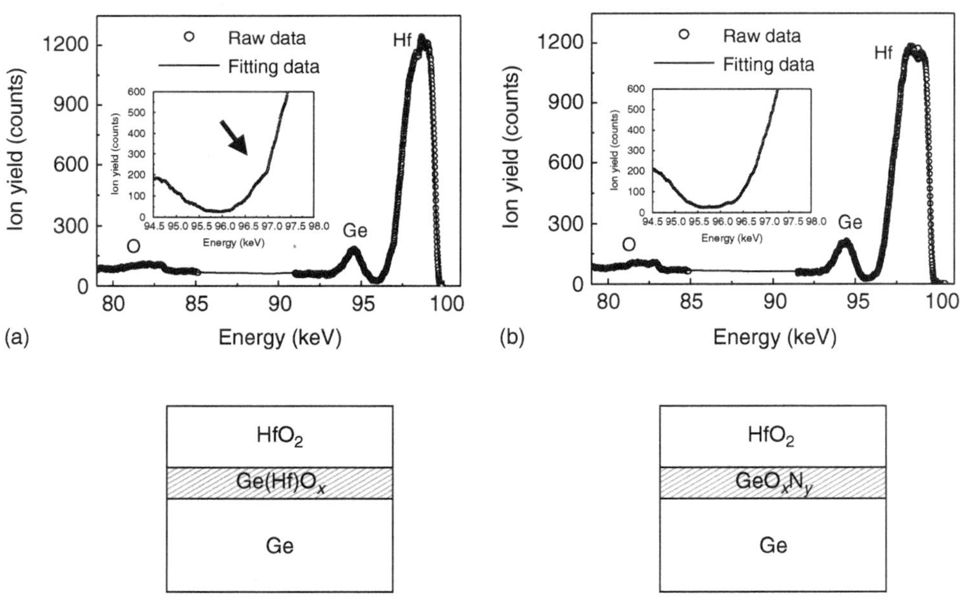

Figure 10.23. MEIS spectra with the simulation curves and the corresponding model structure of (a) HfO_2 on native oxides capped Ge substrate and (b) HfO_2 on CHF cleaned Ge substrate followed by RTN at 600°C for 1 min in NH_3 (after Kim *et al.* [79]).

within the MEIS detection limit (<5 at. %), which is consistent with the expected effectiveness of the GeO_xN_y interfacial layer in blocking Hf out-diffusion from the HfO_2 dielectric into Ge substrates.

Combining both the electrical and physical characterization data, it can be concluded that the insertion of a GeO_xN_y interfacial layer is advantageous for producing high quality ALD high-κ dielectric stacks on Ge for MOS applications, consistent with other reported high-κ deposition results on Ge discussed in Section 10.3.4 [58,59,63–65]. Nevertheless the scalability of the thin GeO_xN_y interfacial layer has to be carefully studied in order to offer a sub-1.0 nm EOT using this approach.

10.3.3. UVO of high-κ dielectrics

As an alternative high-κ dielectric deposition technique, the UVO process offers advantages including very low processing temperatures, low contamination levels, as well as an *in-situ* gate electrode capping capability. In this technique, a thin metal (e.g. Zr or Hf) precursor film is first deposited by sputtering on the wafer surface in ultrahigh vacuum (UHV) followed by an *in-situ* oxidation in atomic oxygen and ozone, which are co-generated by ultraviolet illumination of high-purity (99.999%) oxygen [82]. The typical UVO process proceeds at 600 Torr oxygen partial pressure and low temperature (~50°C) for 60 min. Due to the reactive nature of those metal precursors and instability of surface GeO_x, low-permittivity interfacial layer formation may possibly be avoided using this technique as well.

10.3.3.1. UVO of zirconia

In order to examine the UVO ZrO_2 quality on differently prepared substrate surfaces, MOS capacitors were fabricated on (100) oriented n-type Ge wafers with a net background impurity concentration of $\sim 7 \times 10^{15}\,cm^{-3}$. Three different surface preparations, as described in Section 10.3.2.1, were applied prior to UHV sputtering of ~2 nm Zr film and followed by an *in-situ* UVO at room temperature. About 50 nm of Pt was subsequently deposited by UHV sputtering as the gate electrode layer and circular capacitor structures were defined by optical lithography and argon (Ar) ion sputter etch. No wafer backside Al was applied to these $Pt/ZrO_2/Ge$ capacitors, which were finally subjected to a forming gas anneal at 410°C for 30 min.

C–V characteristics were measured from these UVO ZrO_2 Ge MOS capacitors using a HP4275A multi-frequency LCR meter as shown in Figure 10.24a–c. The corresponding EOT values were computed after correcting for the substrate resistivity induced accumulation capacitance dispersion between the 100 kHz and 1 MHz traces. Very small hystereses, minimal frequency dispersions, as well as sub-1.0 nm EOTs were achieved on the HF vapor etched (Figure 10.24a) and DI water rinsed (Figure 10.24b) samples whose surface GeO_x had been intentionally stripped. Comparing these two treatments, the former offers a smaller EOT which suggests it is more effective in removing GeO_x, a conclusion that is consistent with the ALD ZrO_2 and HfO_2 experience on HF-treated Ge surfaces described in Section 10.3.2.

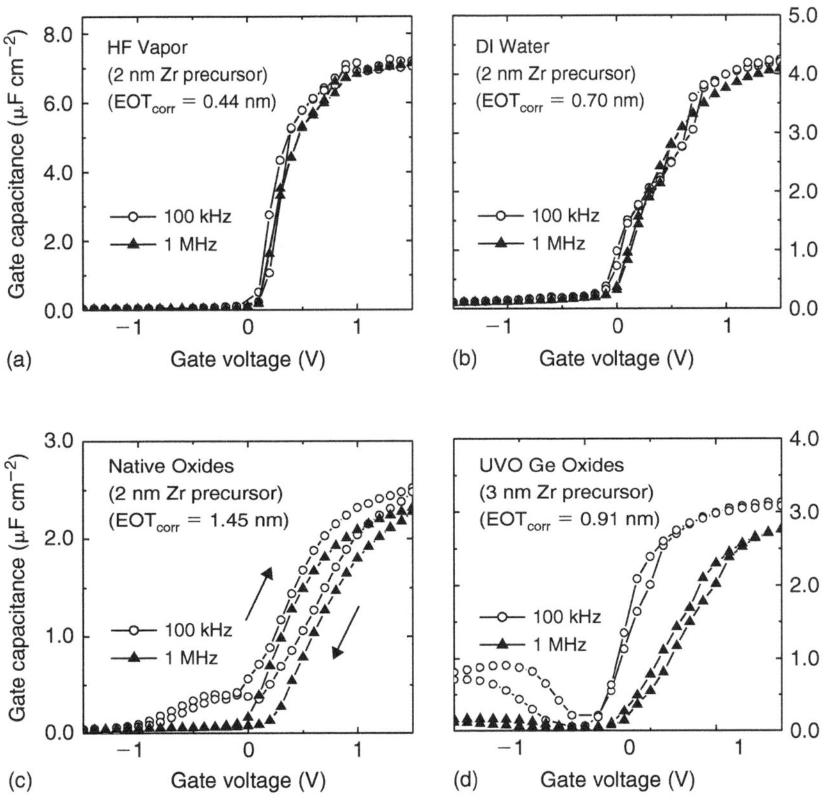

Figure 10.24. Multi-frequency bi-directional capacitance–voltage characteristics measured from forming gas annealed Pt/ZrO_2/Ge capacitors fabricated on (a) HF vapor etched, (b) DI water rinsed, (c) native oxides capped, and (d) UV Ge oxides grown Ge surfaces. The corrected EOT values were computed from the 100 kHz and 1 MHz traces.

On the native oxide-capped sample (Figure 10.24c), a large C–V hysteresis (~0.3 V), substantial frequency dispersion, and relatively large EOT were obtained, which may stem from the inferior ZrO_2–GeO_x–Ge interface quality. To further illustrate the importance of the Zr precursor thickness, an additional Pt/ZrO_2/Ge sample was fabricated, whose Ge surface had been intentionally UV oxidized followed by sputtering a thicker (~3 nm) Zr film and its UVO oxidation. Although the UVO-grown Ge oxides should be thicker than native GeO_x, the thicker Zr precursor produced a thinner EOT value, suggesting that reaction of the thick Zr precursor layer with the underlying oxide produces a substantially different high-κ film and interfacial layer structure (Figure 10.24d) than does the thinner precursor layer. The frequency dispersion observed, which may be attributed to oxygen deficiency during the UVO oxidation of this thicker Zr precursor layer [83], can be removed by lengthening the UVO process duration.

Illustrated in Figure 10.25 is the gate leakage density of these ZrO_2/Ge capacitors as a function of voltage bias for both gate and substrate injections using the HP4155A. The native

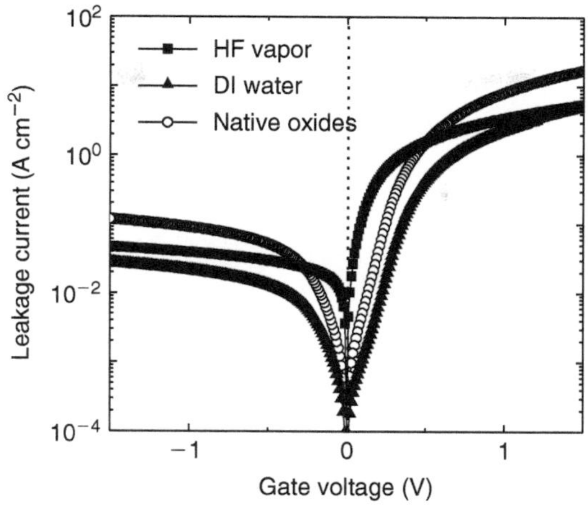

Figure 10.25. Gate leakage–voltage measurements from Pt/ZrO_2/Ge capacitors with different Ge surface preparation. Open and solid symbols are used for native oxides capped and stripped surfaces, respectively.

oxide capped sample (open symbol) gives the highest leakage even with the largest EOT that implies the detrimental effect with the presence of interfacial GeO_x layer. The leakage measured from the native oxides stripped samples (solid symbols) however scales with their individual EOT level.

From the cross-sectional HR-TEM image of the HF vapor surface etched sample, the physical ZrO_2 thickness is roughly 3.6 nm with an atomically flat ZrO_2–Ge interface (Figure 10.26a). Although these HR-TEM images are relatively insensitive to local composition variations, the uniform image contrast between the Ge substrate and Pt electrode suggests that the high-κ dielectric stack is free of any significant interfacial GeO_x layer between the ZrO_2 film and Ge substrate. Compared with previous observations of well-defined polycrystallinity in ZrO_2 on SiO_2 [82,84], the ZrO_2 film grown directly on Ge has an amorphous or, at most, partially microcrystalline nature. Taking the κ value to be ~25 from the reported bulk value for the monoclinic phase of ZrO_2 [56], the calculated physical EOT is 0.56 nm. Combining both the electrical (Figure 10.24a) and physical extractions together with the anticipated quantum mechanical capacitance correction, the representative EOT of the HF vapor etched sample is about 0.5–0.7 nm. Similar extraction on the DI water rinsed sample reveals a thicker EOT of 0.6–0.8 nm attributed to a possible non-uniform removal of GeO_x prior to the UVO process; this is also the primary cause of a rougher ZrO_2–Ge interface (Figure 10.26b) even though this dielectric stack is interfacial layer free as well.

10.3.3.2. Zirconia–germanium interface photoemission spectroscopy

While the sub-1.0 nm EOT ZrO_2 on Ge gate stack revealed no apparent interfacial layer in between as illustrated in Figure 10.26, the conventional TEM microscopy is not well suited for identifying an interfacial GeO_x layer in this material system due to the closeness in

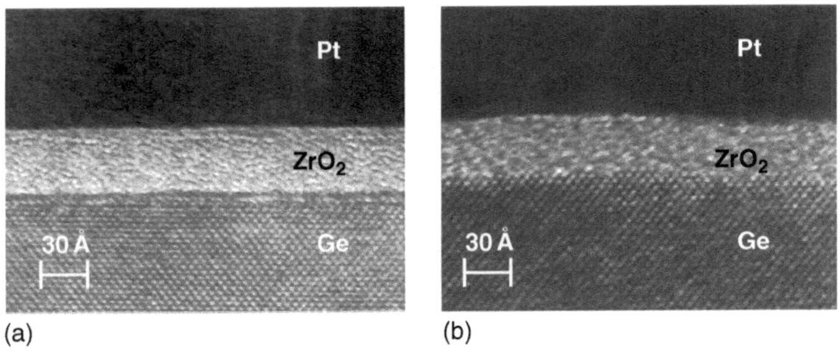

Figure 10.26. Cross-sectional HR-TEM images of Pt/ZrO_2/Ge capacitors with different starting Ge surfaces: (a) HF vapor etched and (b) DI water rinsed.

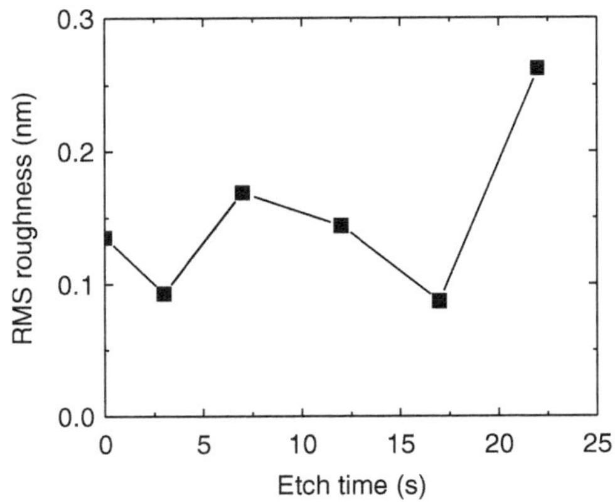

Figure 10.27. ZrO_2/Ge sample surface root-mean-square roughness measured by AFM as a function of the etch time in 100:1 HF solution.

atomic number of Zr and Ge resulting in poor image contrast. In order to understand whether it is technologically feasible to achieve an interfacial layer-free high-κ film growth on Ge, an alternative physical characterization should be considered. The synchrotron radiation photoemission spectroscopy (SR-PES) was used due to its high surface sensitivity and fine energy resolution. Upon layer-by-layer removal of the ZrO_2 film on Ge, the compositional and chemical evolution was monitored to determine the existence of any interfacial GeO_x layer using SR-PES. In addition, the ZrO_2–Ge interface energy band alignment was determined by monitoring offsets in the valence band spectra.

The ZrO_2 film was etched in a layer-by-layer fashion with numerous SR-PES spectra taken between these etches. These layer-by-layer etches were carried out inside an Ar purged glove bag attached to the analytical chamber load-lock using 100:1 HF solution. The ZrO_2 film etch rate was about 0.2–0.3 nm s^{-1}. In order to guarantee the wet etched ZrO_2 surfaces were smooth and pinhole free after each HF etch, AFM was used to monitor surface morphology and extract R_{rms} as a function of etch time as illustrated in Figure 10.27. The R_{rms} level was maintained around 0.15 nm (<half of a monolayer) roughly before the entire ZrO_2 layer was removed, beyond which the roughness increased owing to the HF induced

Ge surface roughening. These results indicated that this wet etching technique could provide pinhole-free surfaces throughout the layer-by-layer removal.

The synchrotron radiation source is provided by the Stanford Positron Electron Asymmetric Ring (SPEAR) located at the Stanford Synchrotron Radiation Laboratory (SSRL). The SR-PES experiments were performed at beam line 8–1 of SSRL with photon monochromator energy range of 20–180 eV and photon flux spot size of 2 mm $\times$ 0.5 mm at the sample position. During data acquisition, the analytical chamber base pressure was kept at $\sim 5 \times 10^{-11}$ Torr. The photoemission spectra were measured with a PHI model 10–360 hemispheric capacitor electron energy analyzer with the Omni Focus III small area lens mounted on the chamber with an angle of 54° with respect to the incoming photon beam direction. The analyzer has a multi-channel detector that allows a detectable kinetic energy range of 0–1000 eV with energy resolution of 0.05 eV. The photoelectron take-off angle is set normal to the sample surface to maximize the depth information.

In order to maximize the SR-PES sensitivity from the sample, it is necessary to select Ge core-level peaks carefully with regards to the beam line capability. Since the available photon energy range of beam line 8–1 allows us to observe both the Ge(3d) (with binding energy of 29.2–29.8 eV) and Ge(3p) (with binding energy of 120.8–124.9 eV) core levels, their corresponding sub-shell photoelectron cross-sections should be compared. Within the photon energy range of interest, the Ge(3d) peaks possess at least one and a half order of magnitude larger photoelectron cross-sections than the Ge(3p) peaks [85] and therefore the Ge(3d) peaks were chosen to be monitored. It is worth noting that the Zr(4p) core-level peaks also have a similar binding energy (27.1–28.5 eV), which may overlap with the Ge(3d) spectra. Fortunately, the photoelectron cross-sections of the Zr(4p) signals are at least an order of magnitude lower than those of the Ge(3d) signals; and careful peak fittings allow clear distinction between the two for our purpose.

10.3.3.2.1. Interfacial Layer identification and quantification

The first objective of this work was to trace the compositional and chemical evolutions within the ZrO_2–Ge stack during the layer-by-layer wet etching of the ZrO_2 film. Due to the high surface sensitivity of SR-PES technique, the captured core-level spectra could be utilized to discover the existence of any interfacial GeO_x layer inside the stack. The 100 eV input photon energy was selected to seize both core-level and valence band spectra simultaneously with the same system setting. Since the etch duration (which was on the order of seconds) was difficult to control, the spectra from two samples were interleaved together without compromising the general trend as illustrated in Figure 10.28a. Qualitatively, no Ge(3d) core-level peak appears before the sample had been etched for ~11 s, which corresponds to the removal of ~2.8 nm of ZrO_2. Beyond this point, the elemental Ge signals begin to appear since the inelastic mean-free-paths (IMFPs) [86] of Ge(3d) core-level electrons are 0.4–0.8 nm in this kinetic energy range, which are long enough to penetrate either the ZrO_2 or GeO_x overlayer.

From the 11 s spectrum, there exist features with binding energies (30–33 eV) higher than the Ge^{0+} peak, which could plausibly originate from either ZrO_2 or GeO_x. In an attempt to determine the source of the peak, the thicker ZrO_2 spectra (from the 3 and 6 s etched samples) were scaled and zoomed together with the 11 s spectrum, as plotted in Figure 10.28b.

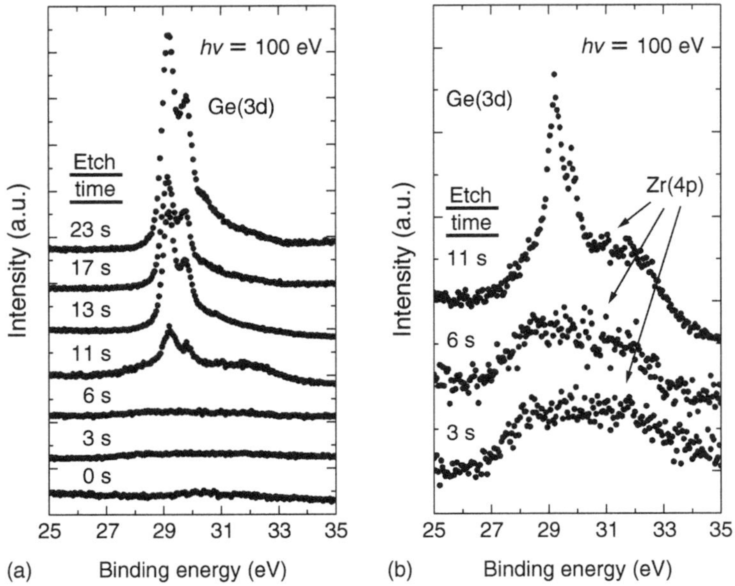

Figure 10.28. (a) Evolution of the Ge(3d) core-level SR-PES spectra as a function of the ZrO_2–Ge stack wet etching time taken at a photon energy of 100 eV. (b) Scaled and zoomed spectra after 3, 6, and 11 s of etch disclosing the embedded Zr(4p) core-level signals of similar binding energy.

The 0 s spectrum was not considered owing to possible surface contaminations as previously suggested. Clearly, most of the plateau within the 11 s spectrum that has a binding energy range of 30–33 eV stems from the Zr(4p) core-level signal, yet the possibility of GeO_x could not be completely neglected.

To identify any interfacial GeO_x layer and quantify its thickness if necessary, the ensemble of core-level SR-PES spectra showing the evolution of the etching sequence were first normalized to the photon beam current (I_0) to remove any real-time fluctuation in the synchrotron radiation flux. Errors in such a computation exercise were further minimized by performing quantification only on spectra from the same sample (not both). Given our primary interest is the discovery of any oxidized Ge features during the evolution, the overlapping ZrO_2 signals were identified upfront and subsequently subtracted through an iterative peak fitting procedure. In brief, the Zr(4p) peaks (binding energies) were first located by deconvolving the spectra from thicker ZrO_2. By incorporating these ZrO_2 signals, extensive peak fittings [87] were then carried out on the remaining spectra by assigning the Ge(3d) signals with every possible oxidation states. After all the Ge(3d) and Zr(4p) peak (binding energy) positions had been defined, they were re-applied to the ensemble of evolution spectra to further refine the deconvolutions and fitted areas. Generally speaking, rigorous fitting requirements were imposed by fixing the core-level doublet peak splitting energies and branching ratios [88], the Gaussian and Lorentzian line widths, as well as the peak energy positions [89], while freeing only the peak areas to obtain the best possible fits. As an example, Figure 10.29 depicts one such fitting result of the spectra with 17 s etch with both elemental (Ge^{0+}) and oxidized (Ge^{n+} where $1 \leq n \leq 4$) peaks assigned.

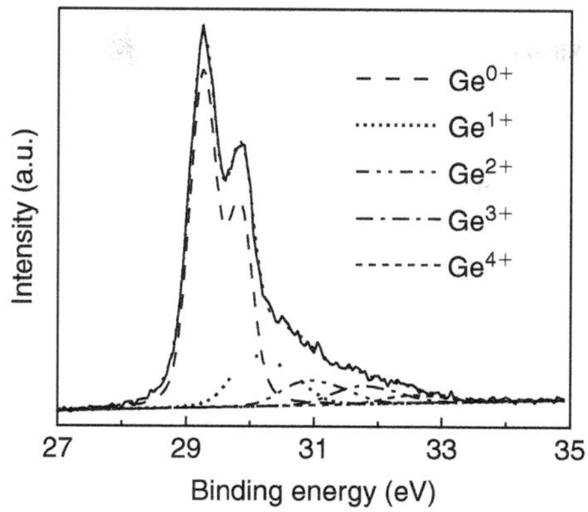

Figure 10.29. Peak-fitted SR-PES spectra from the ZrO_2–Ge stack after ~17 s of etch with both elemental and oxidized Ge peaks assigned.

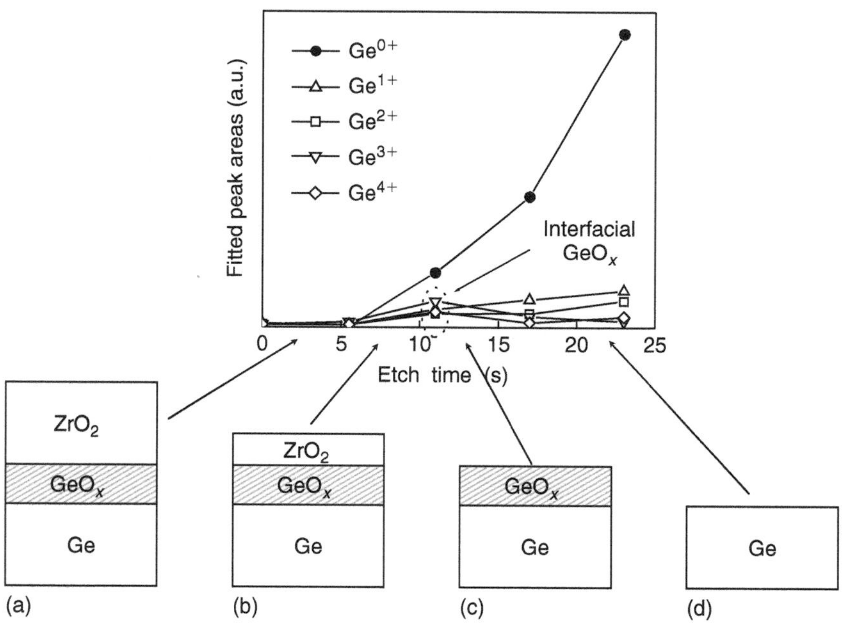

Figure 10.30. Extracted core-level Ge(3d) fitted peak areas as a function of etch time. Cross-sectional schematics of the stack are also included to illustrate the layer-by-layer etching progress.

The areas of the fitted Ge(3d) core-level peaks as a function of etch time are summarized in Figure 10.30. To illustrate the etching progress, cross-sectional schematics of the stack are also included. First of all, a sanity check on the overall fitting validity was done by monitoring the substrate (Ge^{0+}) peak evolution. The fitted peak areas increase with etch time, which is consistent with the theoretical prediction that elemental signals from a Ge substrate would be less attenuated with a thinner overlayer on top (e.g. ZrO_2 and/or GeO_x) upon etching. Next, the evolution of the oxidized Ge peaks was studied. Within the first few seconds of etching (Figure 10.30a), all oxidized signals were absent as any possible interfacial GeO_x layer would be buried by the relatively thick ZrO_2 film above. When the ZrO_2 film got sufficiently thin (Figure 10.30b), the more oxidized Ge signals (Ge^{n+} where $n \geq 2$) began to rise and reach a maximum just when the entire ZrO_2 film was removed, showing clear evidence of an interfacial GeO_x layer within the ZrO_2–Ge stack. Further etches would start to consume the GeO_x layer (Figure 10.30c) and cause the more oxidized Ge signal intensity to fall back down. Once all GeO_x were removed (Figure 10.30d), the oxidized Ge signals increased slightly again with the least oxidized peak (Ge^{1+}) dominating. This re-appearance of oxidized Ge signature may be due to Ge surface re-oxidation right after a complete GeO_x removal even inside the Ar purged glove bag and formation of sub-stoichiometric surface oxides. Finally, the questionable continued increase of the Ge^{0+} intensity even after complete GeO_x removal (Figure 10.30d) could be attributed to the surface roughening of Ge surface as observed in Figure 10.27.

Next, extraction of the interfacial GeO_x layer thickness was done by first computing the escape depth of photoelectrons from the Ge substrate as illustrated in Figure 10.31. In doing so, the extraction time points throughout the evolution were judiciously selected to simplify the calculation process: (i) right around the complete ZrO_2 removal, and (ii) right around the complete GeO_x removal. Thus, the fitted peak area right after ZrO_2 removal, $N_{Ge(i)}$, could be expressed with

$$N_{Ge(i)} \propto e^{\frac{-s}{\lambda_{GeO_x}}} \int_s^{\infty} e^{\frac{-z}{\lambda_{Ge}}} dz = \lambda_{Ge} e^{\frac{-s}{\lambda_{GeO_x}}} e^{\frac{-s}{\lambda_{Ge}}} \quad (10.4)$$

and the fitted peak area right after GeO_x removal, $N_{Ge(ii)}$, could be related using

$$N_{Ge(ii)} \propto \int_0^{\infty} e^{\frac{-z}{\lambda_{Ge}}} dz = \lambda_{Ge} \quad (10.5)$$

where s is the GeO_x thickness, and λ_{GeO_x} and λ_{Ge} are the electron IMFPs in GeO_x and Ge, respectively. By substituting Equation (10.5) into Equation (10.4), the GeO_x thickness

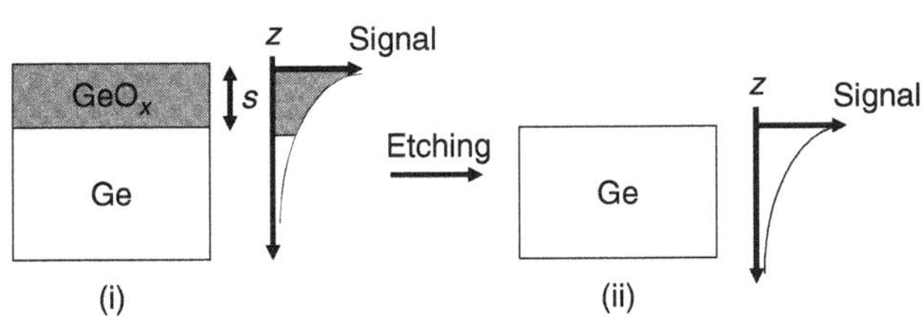

Figure 10.31. Illustrations of escape depth calculation of the photoelectrons from Ge substrate at two different time points: (i) right around the complete ZrO_2 removal, and (ii) right around the complete GeO_x removal.

could thus be extracted using

$$s = \frac{\lambda_{Ge}\lambda_{GeO_x}}{\lambda_{Ge} + \lambda_{GeO_x}} \ln\left(\frac{N_{Ge(ii)}}{N_{Ge(i)}}\right) \tag{10.6}$$

The λ_{Ge} is estimated to be about 0.44 nm and λ_{GeO_x} could range between 0.4–0.8 nm at our kinetic energy range [86]. Taking into account any potential errors in determination of extraction time point and uncertainty in λ_{GeO_x}, the interfacial GeO_x thickness is calculated to be 0.19–0.36 nm. Since this is less than a complete monolayer thickness of GeO_x, the oxidized Ge signature was therefore attributed to a possible Zr–O–Ge interfacial bonding configuration. Besides, the interfacial GeO_x thickness as determined by SR-PES is comparable to those obtained by an angle-resolved XPS study on similar gate stacks [90].

10.3.3.2.2. Interfacial energy band alignment

In addition to the possession of an ultra-thin interfacial layer, having sufficiently large barrier heights for electron and hole injection would be considered an equally important high-κ dielectric qualification criterion to lower gate leakage for better scalability. Among other techniques, photoemission is one of the most reliable methods to study these systems. In particular, offsets in the valence band spectra could be monitored to map out the interface energy band alignment for many high-κ on Si and Ge stacks [91]-[93], and its application to our ZrO_2–Ge interface is the focus of this section.

The basic principle is to first locate the E_V (binding energy position) from those thicker ZrO_2 photoemission spectra throughout the evolution as shown in Figure 10.32a and then compare with the E_V from a clean bulk Ge. For an accurate valence band offset measurement,

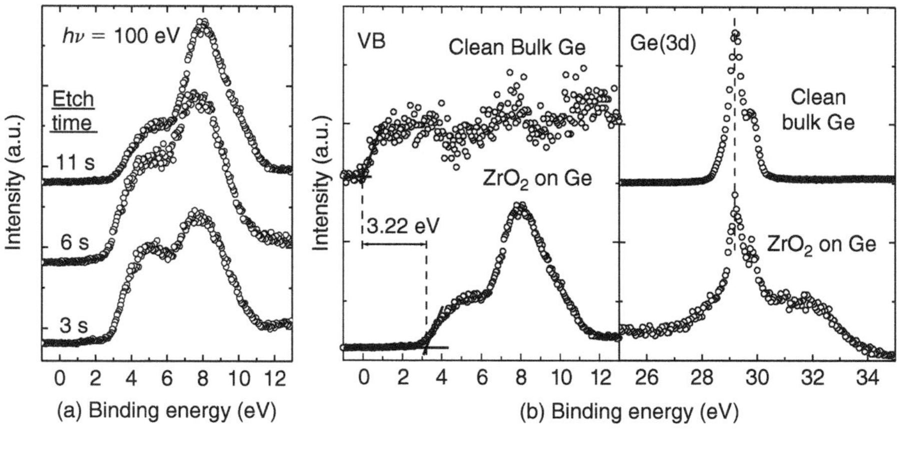

Figure 10.32. (a) Valence band photoemission spectra from thicker ZrO_2 captured during the etching evolution. (b) Valence band spectra (left) from a clean bulk Ge sample and the ZrO_2–Ge stack after 11 s of etch with their corresponding Ge(3d) core-level peaks aligned (right).

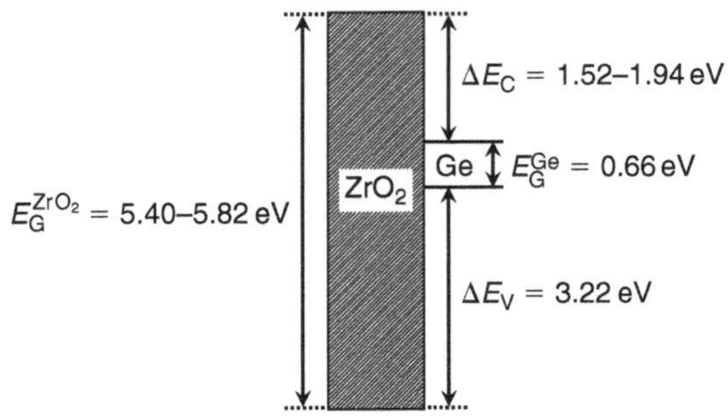

Figure 10.33. Energy band diagram of the ZrO_2–Ge structure inferred from the SR-PES measurements.

any energy shift induced by sample surface charging during photoemission should be corrected by aligning to the Ge(3d) core-level peaks. Since a definitive Ge^{0+} signal had only been discerned from the 11 s spectrum in Figure 10.28b, this spectrum was chosen for ZrO_2 valence band offset estimation as detailed in Figure 10.32b. The clean bulk Ge spectra were prepared by *in-situ* heating of a bare Ge substrate to 700°C for ~25 min to desorb any surface native oxides [94] after pumping down the inside of the analytical chamber. The E_V was determined to be the point of intersection of a "best" straight line fit to represent the decay in valence band spectrum and the baseline [91–93]. After aligning their corresponding Ge^{0+} core-level peaks, the valence band offset (ΔE_V) between bulk Ge and ZrO_2 was found to be 3.22 eV as highlighted in Figure 10.32b, a value that is comparable to the 3.36 eV obtained by X-ray photoemission spectroscopy [93].

Before the entire ZrO_2–Ge interface energy band alignment could be determined, the energy band gap of ZrO_2 needed to be researched in the literature. From those reports on both amorphous and polycrystalline ZrO_2 that were deposited using various methods on either Si or Ge, the energy band gap of ZrO_2 ($E_G^{ZrO_2}$) was experimentally measured to be 5.40–5.82 eV [91,93,95]. Together with the interface data from the last section, the energy band diagram of the ZrO_2–Ge structure was constructed as displayed in Figure 10.33. Taking the Ge energy band gap (E_G^{Ge}) to be 0.66 eV at 300 K [71], the ΔE_C is deduced to be 1.52–1.94 eV, which is in good agreement with the theoretical values calculated in Section 10.3.1.

10.3.3.3. UVO of hafnia

MOS capacitors were fabricated on a (100) oriented n-type Ge wafer with a net background concentration of $\sim 7 \times 10^{15}\,cm^{-3}$. UVO HfO_2 was prepared on a DI water rinsed Ge surface. A nominal thickness of 2 nm of Hf precursor was deposited by UHV sputtering onto the rinsed surface followed by an *in-situ* UVO oxidation at room temperature. Finally, the $Pt/HfO_2/Ge$ capacitors were made as described in Section 10.3.3.1 and subjected to a forming gas anneal at 300°C for 30 min.

C–V characteristics were measured from these UVO HfO_2 Ge MOS capacitors using HP4275A multi-frequency LCR meter with minimal bi-directional hysteresis and frequency dispersion as shown in Figure 10.34a. The raw data show non-saturation of the

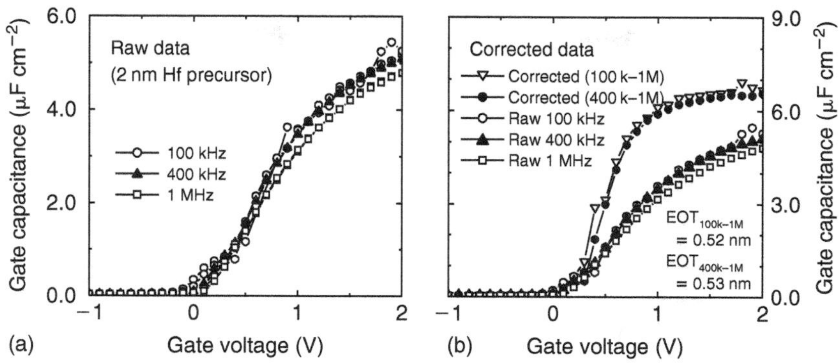

Figure 10.34. (a) Multi-frequency bi-directional capacitance–voltage characteristics measured from forming gas annealed Pt/HfO_2/Ge capacitors fabricated on DI water rinsed Ge surface. (b) The substrate resistivity corrected multi-frequency uni-directional capacitance–voltage characteristics with the respective EOT values computed.

accumulation capacitance even at high enough gate voltage. After correcting for the high substrate resistivity with respect to thin MOS dielectrics using the two-frequency method [77], well behaved *C–V* characteristics with higher capacitance could be discerned, as depicted in Figure 10.34b. The EOT values computed from either pair of frequencies coincided very well at a number below 1.0 nm. Recently, another HfO_2 gate stack on Ge with sub-1.0 nm EOT had also been demonstrated by atomic oxygen beam deposition [64].

In good correlation with the thermodynamic prediction (Section 10.3.1) and physical characterizations [70,90], the minimal *C–V* hysteresis and sub-1.0 nm EOTs measured on these UVO ZrO_2 and HfO_2 Ge capacitors unambiguously demonstrated the feasibility of low-permittivity interfacial layer-free high-κ dielectric growth on Ge substrate. Likewise, substantial improvements in electrical performance can be achieved by complete elimination of the poor quality and readily removed GeO_x interfacial layer, followed by direct high-κ growth on Ge. The gate leakage current densities extracted from these UVO Ge capacitors are benchmarked in Section 10.3.5.

10.3.4. Other high-κ deposition techniques

In addition to the two techniques discussed in Sections 10.3.2.2 (ALD) and 10.3.3.3 (UVO), metal-organic chemical vapor deposition (MOCVD) [58,59,61,62,65], physical vapor deposition (PVD) [60], [67], and atomic oxygen beam deposition [63,64] are the other contending high-κ deposition techniques on Ge.

10.3.4.1. Metal-organic chemical vapor deposition of hafnia

The starting substrates were (100) oriented n-type Ge wafers with a net background concentration of $\sim 5 \times 10^{16}$ cm^{-3}. The substrates were first dipped in ammonium hydroxide (NH_4OH) solution to remove native oxide, then in H_2O_2 to form chemical oxide, and again

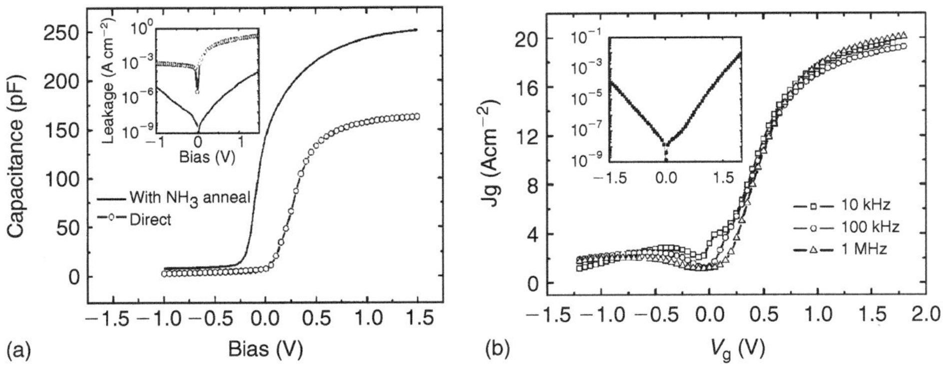

Figure 10.35. Capacitance–voltage characteristics measured from forming gas annealed TaN/HfO_2/Ge capacitors fabricated on (a) bare and NH_3 annealed, and (b) SiH_4 passivated Ge surfaces (after Wu *et al.* [59,62]).

in NH_4OH for chemical oxide removal [59]. A typical MOCVD process uses Hf-tert-butoxide ($C_{16}H_{36}HfO_4$) as the metal-organic precursor in an $N_2 + O_2$ ambient at 400°C with a base pressure of 3×10^{-3} Torr. A vacuum load-lock is used to store and transfer wafers. A post-deposition anneal (PDA) is then performed in an RTP system in N_2 ambient under 760 Torr at 600°C for 30 s. After that, tantalum nitride (TaN) is sputtered at room temperature followed by lithography and dry etch to define the MOS capacitor gate electrodes. The final step is a forming gas anneal at 300°C.

Compared to the direct MOCVD HfO_2 on Ge, an additional surface annealing step in NH_3 prior to the HfO_2 deposition could result in reduction in both gate leakage and EOT [59], which is consistent with the ALD HfO_2 deposition experience on Ge as discussed in Section 10.3.2.2. It was possible to achieve an EOT value of 1.05 nm (Figure 10.35a). As an alternative surface passivation, silane (SiH_4) was attempted prior to MOCVD HfO_2, which was shown to simultaneously prevent the surface GeO_x formation and suppress Ge out-diffusion after the HfO_2 deposition [62]. An EOT value of 1.35 nm was thus achieved with low gate leakage (Figure 10.35b).

10.3.4.2. PVD of zirconia and hafnia

As one of the simplest high-κ deposition techniques on Si, PVD of ZrO_2 and HfO_2 were recently demonstrated on Ge [67]. The starting substrates were (100) oriented n-type Ge wafers with a net background concentration of $\sim 9 \times 10^{15}$ cm^{-3}. The substrates were cleaned with 1% HF solution followed by rinsing in DI water. Either a ZrO_2 or HfO_2 layer was deposited by sputtering with an oxide target at room temperature. Pt gate electrodes were formed by electron beam evaporation through a shadow mask. Finally, a PDA in N_2 was carried out at 500°C for 30 min on the completed Pt/ZrO_2/Ge capacitors.

Direct comparison of ZrO_2 and HfO_2 gate dielectrics on Ge substrates was performed in terms of obtaining ultra-thin gate stacks with low gate leakage current. Single-layer gate dielectric of PVD ZrO_2 on Ge could be obtained with a low EOT value of 1.2 nm even after a PDA at

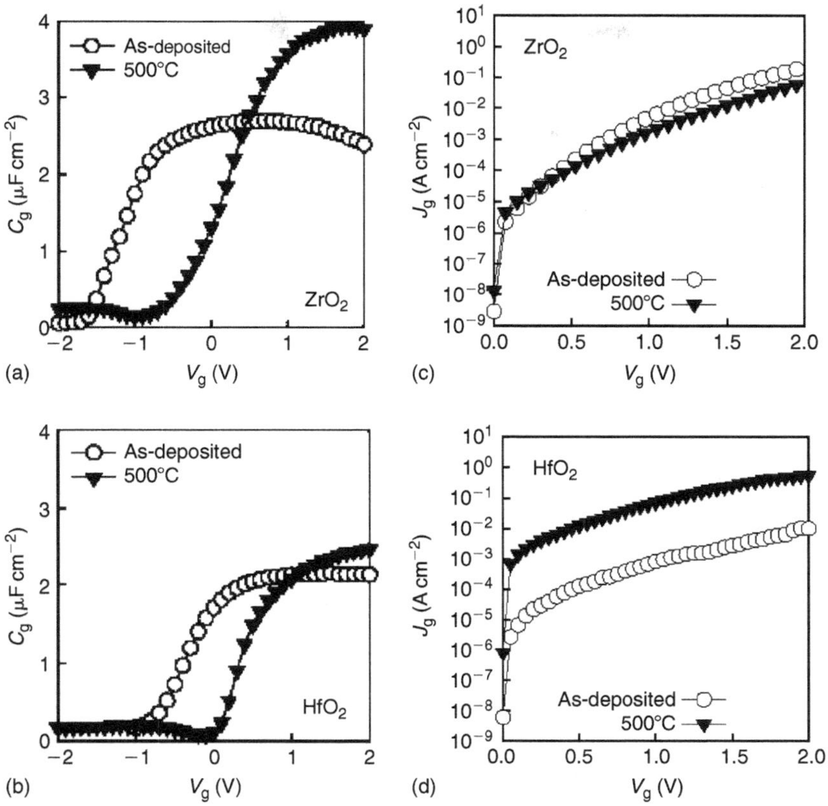

Figure 10.36. Capacitance–voltage characteristics of (a) Pt/ZrO_2/Ge and (b) Pt/HfO_2/Ge samples before (open circles) and after 500°C N_2 anneal (solid triangles). Gate leakage–voltage characteristics of (c) Pt/ZrO_2/Ge and (d) Pt/HfO_2/Ge samples before (open circles) and after 500°C N_2 anneal (solid triangles) (after Kamata *et al.* [67]).

temperature as high as 500°C (Figure 10.36a). The Ge-rich interfacial layer, which was formed during PVD, diffused into the ZrO_2 layer after PDA and resulted in a single high-κ layer. On the other hand, the interfacial layer remained intact after PVD in the case of HfO_2 on Ge, resulting in a relatively large EOT value of 1.6 nm (Figure 10.36b). Moreover, a Ge-rich oxide layer was formed on top of HfO_2, presumably due to germanium monoxide (GeO) desorption from the HfO_2–Ge interface [67]. No leakage current increase was observed for the ZrO_2 stack (Figure 10.36c), whereas a large increase occurred in the case of HfO_2 (Figure 10.36d).

10.3.4.3. Atomic oxygen beam deposition of hafnia

Being an UHV technique, atomic oxygen beam deposition offers alternative ways of cleaning the surface such as *in-situ* oxide thermal desorption, which cannot be easily implemented in CVD-based methodologies [64]. The starting substrates were (100) oriented p-type Ge wafers with a net background concentration of $\sim 1 \times 10^{15}\,cm^{-3}$. The native oxide was desorbed *in-situ* under UHV conditions by heating the substrate to 360°C for 15 min until a (2×1) reconstruction appears in the reflection high electron energy diffraction (RHEED) pattern, indicative of a clean Ge (100) surface. Subsequently, at a substrate temperature of 225°C, the surface was exposed to atomic O and N beams generated by a radio-frequency (RF) plasma source with the simultaneous evaporation of Ge at a very low rate, aiming to form an ultra-thin GeO_xN_y starting layer. Finally, thin HfO_2 films were deposited at several temperatures between 60°C and 310°C spanning a range of thickness between 3 and 8 nm. Hf was electron beam evaporated under atomic O beams generated by the RF plasma source. 30 nm thick Pt gate electrodes were formed by electron beam evaporation through a shadow mask. The backside ohmic contact was made using eutectic In–Ga alloy.

The atomic oxygen beam deposited HfO_2 formed atomically sharp interfaces with Ge and behaved as an excellent dielectric with $\kappa \sim 25$. Very low EOT of 0.65–0.85 nm (Figure 10.37a–d) with low gate leakage was achieved (Figure 10.37e). Strong frequency dispersion of the inversion capacitance and "low-frequency" behavior of the high-frequency *C–V* curves was observed. This dispersion is attributed to a combined effect of a high generation rate of minority carriers due to impurity traps and high intrinsic carrier concentration in Ge, which result in a short minority carrier response time [48].

10.3.5. Nanoscale dielectrics leakage and scalability

From both the dielectric leakage and scalability perspectives, it is important to objectively assess the nanoscale Ge MOS gate dielectrics discussed in this chapter and those published

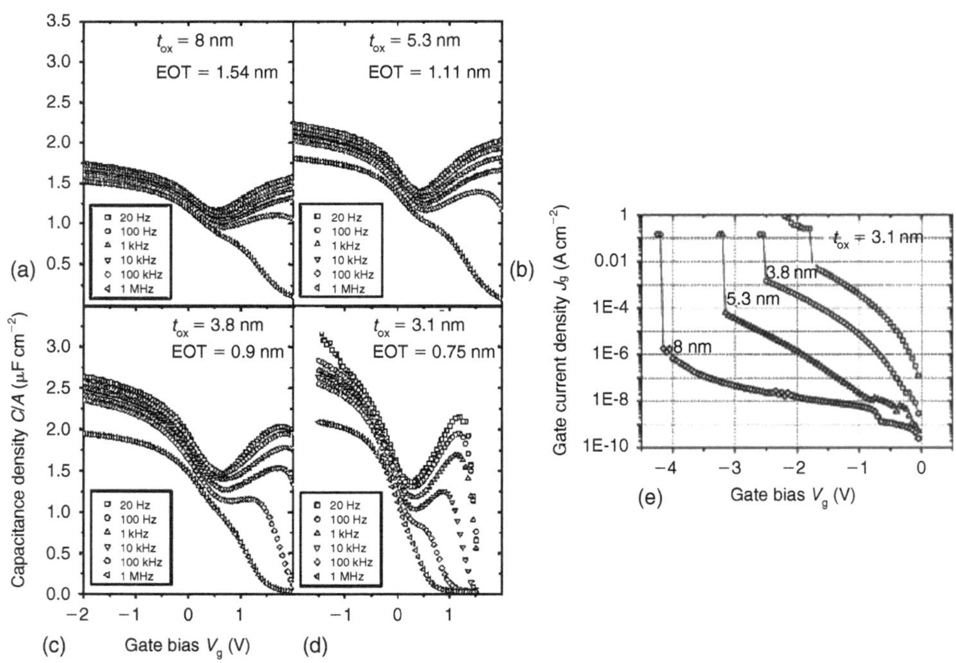

Figure 10.37. (a)–(d) Multi-frequency capacitance–voltage characteristics measured from $Pt/HfO_2/Ge$ capacitors with the respective physical thickness (t_{ox}) and EOT shown. (e) The corresponding gate leakage–voltage characteristics (after Dimoulas *et al.* [64]).

in the literature [58,59,63–65,67,96]. These dielectric stacks can be classified into five categories:

(i) germanium oxynitrides (GeO_xN_y),

(ii) alumina (Al_2O_3),

(iii) hafnia with an intentionally prepared interfacial layer (HfO_2 (with IL)),

(iv) hafnia without an intentionally prepared interfacial layer (HfO_2 (no IL)) and

(v) zirconia without an intentionally prepared interfacial layer (ZrO_2 (no IL)),

Reasonable *C–V* characteristics had been first obtained on the MOS capacitors before their corresponding dielectric leakage data were further analyzed. As one of the most accepted benchmarking strategies, we recorded their leakage density at a voltage bias of 1 V into the accumulation from V_{FB} – in order to achieve a similar carrier concentration for similar dielectric thickness and substrate doping [71]. In other words, the MOS dielectric leakage was taken at either 1 V above V_{FB} on n-type substrate or 1 V below V_{FB} on p-type substrate, and then plotted against their respective EOT values in Figure 10.38.

Although the differences in metal gate electrode work function and substrate doping polarity may inevitably introduce errors in this benchmarking effort, a distinct trend can clearly be observed in each of the categories aforementioned. Among them, GeO_xN_y possesses the highest leakage followed by Al_2O_3, HfO_2, and ZrO_2, a sequence that corresponds well with their individual κ-values. Even though small-EOT HfO_2 stacks can be attained [67] without an intentionally prepared interfacial layer, their electrical qualities are usually not optimal and they exhibit relatively high leakage current densities. As a result, the strategy of inserting an intentionally prepared interfacial layer has been widely adopted in many HfO_2 gate

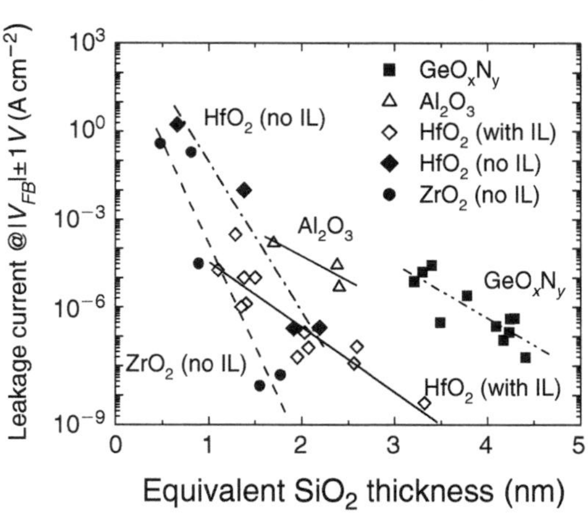

Figure 10.38. Benchmarking the gate leakage levels from various nanoscale Ge MOS dielectrics. The EOT values were lifted from the 100 kHz *C–V* data.

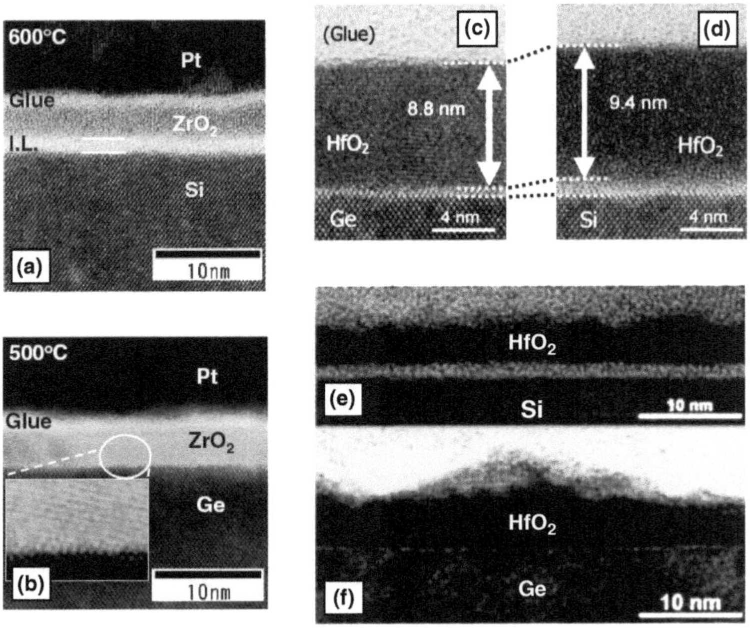

Figure 10.39. Cross-sectional HR-TEM images of (a) PVD ZrO_2 on Si, (b) PVD ZrO_2 on Ge, (c) PVD HfO_2 on Ge, (d) PVD HfO_2 on Si, (e) MOCVD HfO_2 on Si, and (f) MOCVD HfO_2 on Ge (after Kamata *et al.* [67], Kita *et al.* [60], and Van Elshocht *et al.* [61]).

stacks on Ge [58,59,63–65] that promises a better electrical performance. However, the added EOT from the interfacial layer might limit the ultimate scalability of HfO_2 gate stacks on Ge.

On the other hand, the sub-1.0 nm EOT ZrO_2 gate stacks on Ge without an intentionally prepared interfacial layer [57,67] do reveal promising *C–V* characteristics and lower leakage current densities than the HfO_2 counterparts. These marked differences could primarily be attributed to the higher κ-value of ZrO_2. On those interfacial layer-free local epitaxial ZrO_2 gate stacks on Ge discussed in Section 10.3.2.1, remarkably low gate leakage was evident even in the presence of high interface defect densities and sub-grain boundaries (Figure 10.12). These findings suggest that it may be beneficial to employ crystalline MO_x dielectrics with a closer lattice match to Ge in order to avoid interfacial misfit dislocation formation.

Compared to many high-κ gate stacks on Si, those on Ge are in fact exhibiting a better scalability due to their larger dielectric–substrate conduction band offset [70,73] and thinner low-κ interfacial layers. In addition to the ultra-thin interfacial layer within the ZrO_2 gate stack reported earlier [70,90], much thinner interfacial layers have been achieved against their Si counterparts (Figure 10.39) for direct PVD ZrO_2 [67], reactive PVD HfO_2 [60], as well as MOCVD HfO_2 [61] on Ge.

10.3.6. Summary

In this section, we first compared the contending high-κ dielectric candidates to be applied to Ge MOS device applications based on both the thermodynamic and electrical properties selection requirements. ZrO_2 and HfO_2 are promising owing to their thermodynamic stability in direct contact with Ge and the inherent large conduction band offsets between the dielectric and the substrate. Prior to their deposition onto Ge, various surface preparation strategies were employed to investigate the resultant effects on the fabricated metal-gate MOS capacitors.

ALD of metal oxides and UVO of metal thin films were the two major high-κ deposition techniques investigated together with others. Although interfacial layer-free epitaxial growth of ALD high-κ could be obtained, a more stable interfacial GeO_xN_y layer is necessary to produce high quality Ge MOS capacitors with ALD high-κ dielectrics. This may compromise future dimensional scalability of Ge devices with ALD metal oxide. Next, viable interfacial layer-free UVO high-κ growth on Ge was studied due to the reactive nature of metal precursors and the relative instability of GeO_x surface passivation. Experimental results suggested that ZrO_2 and HfO_2 gate stacks on Ge with minimal C–V hysteresis and sub-1.0 nm EOT can indeed be achieved with a negligible interfacial layer contribution to the capacitance. In addition, considerable improvements in electrical performance were readily achieved by eliminating the poor-quality interfacial layer followed by direct high-κ growth on Ge.

When the leakage current densities of these nanoscale Ge MOS dielectrics were benchmarked together, an unambiguous trend can be observed in which the higher κ gate insulator gives lower leakage. Compared to similar high-κ gate stacks on Si, these high-κ gate stacks on Ge appear to have better scalability due to their larger conduction band offsets and the relative ease with which thinner low-permittivity interfacial layers can be produced. Besides our analysis, these advantages have also been reported by researchers worldwide, consolidating the future promise of "high-κ dielectric on Ge" technology.

10.4. Shallow Junctions in Germanium

For advanced MOSFET applications, shallow source and drain junctions with high dopant activation level are mandated to simultaneously suppress short-channel effects and minimize parasitic resistances. In the past decades, tremendous theoretical and technological progresses have been made in both understanding and making shallow junctions in Si. From the experience on Si, we can identify two determining factors in forming shallow junctions with low sheet resistance: low dopant diffusivity and high dopant solid solubility.

For Ge to be a successful candidate for transport-enhanced FET channel material specified in the ITRS [1], similar technology should exist to allow fabrication of heavily doped shallow junctions in Ge. As a historically important semiconductor material, many basic physical properties of Ge including impurity diffusivity and solid solubility were archived during the mid-1950s.

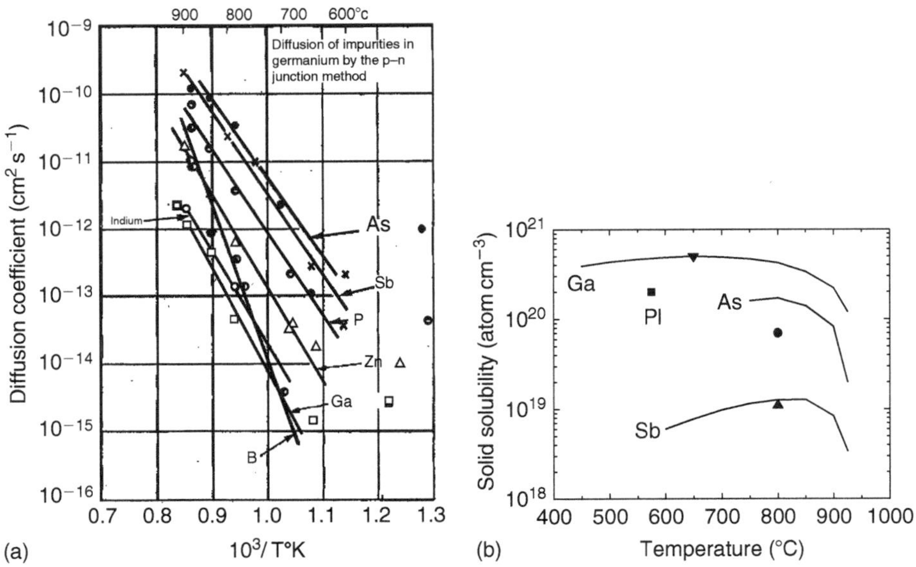

Figure 10.40. (a) Diffusivity of various impurities in Ge (After Dunlap [97].) (b) Solid solubility values for various dopants in germanium. The Ga, As, and Sb data (solid lines) were taken in 1960 [99] and the scattered solubilities (Ga (▼), P (■), As (●), and Sb (▲)) were summarized in 1990 [100].

Diffusivities of various impurities in Ge are shown in Figure 10.40a, which were obtained in 1954 by the observation of the p–n junctions formed as the diffusing impurities penetrate the specimen [97]. When these values are compared to those of the same impurities in Si [98], a few orders of magnitude higher diffusivities could be found in Ge. In addition, the common p-type dopant (boron (B)) diffuses much slower than n-type dopants (phosphorus (P), arsenic (As), and antimony (Sb)), a phenomenon that is exactly opposite to the much faster p-type dopant diffusion than n-type dopants diffusion in Si. On the other hand, a systematic investigation [99] on the solid solubility of various impurities in Ge was performed in 1960; however, the solubility of common dopants like B and P in Ge was not available. These solubilities are plotted in Figure 10.40b together with some scattered data that were documented in a later report [100]. Compared to the data from the same dopants in Si [99], about an order of magnitude lower solid solubility is discerned for n-type dopants (P, As, and Sb) in Ge with the exception of Ga (a p-type dopant) showing an order of magnitude higher solubility in Ge.

From the brief review of both dopant diffusivity and solid solubility, one could imagine the formation of heavily doped shallow p^+ junctions in Ge to be much less challenging than the n^+ junction fabrication. Even though the higher p-type dopant diffusivity and solid solubility in Ge may compensate with each other and lead to comparable p^+ junctions as in Si, the higher n-type dopant diffusivity and lower solid solubility in Ge (over Si) would literally double the hindrance in shallow n^+ junction construction. Nevertheless, there would still be a slight chance for the resultant n^+ layer to be less resistive primarily owing to the higher intrinsic carrier mobility in Ge [101].

In order to obtain decent junctions for radiation detector and MOSFET applications, various Ge doping experiments were carried out during the last three decades. For p-type dopant incorporation, implantation of $^{11}B^{+}$ ions was studied [102–105] at a dose range of 5×10^{11} to $6 \times 10^{17}\,cm^{-2}$ and energy spanning from 20 to 100 keV; these implants were then anneal activated at temperatures of 200–850°C for 30–60 min. Also, implantation of 5×10^{11} to $1 \times 10^{14}\,cm^{-2}$ of boron difluoride ions ($^{49}BF_2^{+}$) at 100 keV was employed [104] followed by thermal anneals at temperatures as high as 550°C for 30 min. All these experiments, however, utilized a conventional furnace for activation anneals which may have limited applicability for making shallow junctions.

For n-type doping in Ge, implantations of $^{31}P^{+}$,$^{75}As^{+}$, and $^{121}Sb^{+}$ ions were examined with a dose and energy range of 1×10^{12} to $1 \times 10^{15}\,cm^{-2}$ and 20–700 keV, respectively [106] – [108], and were anneal activated at 200–650°C for 20–80 min. Nonetheless, either a too low dose or too high implant energy was used in any single study together with furnace anneals. As implant activation was further inspected with both furnace anneal and rapid thermal anneal (RTA) at temperatures 430–650°C [109]. Although, the more practical implant energy (50 keV) was used for our purpose, the implant dose range (8×10^{12} to $1 \times 10^{14}\,cm^{-2}$) and resultant level of activation ($<10^{19}\,cm^{-3}$) were low for futuristic device applications.

In this section, the activation of various p-type and n-type dopants in Ge are thoroughly studied. Two techniques employed in dopant incorporation are ion implantation and solid source diffusion (SSD). The incorporated dopants were activated mostly using RTA to minimize their diffusion and occasionally with furnace anneal to update the literature. In addition, the related activation dependences on common n-type dopants are investigated. Through the examination of thermal stability on the activated dopants, their deactivation is observed for the first time. Lastly, a possible mechanism for such a deactivation process is proposed.

10.4.1. Ion implantation doping

Ion implantation has been the dominating doping technique for Si ICs for the past two decades or so. In this process, dopant ions are accelerated to hundreds or thousands of volts of energy and smashed into a perfect Si or Ge lattice, creating a cascade of damage that may displace a thousand Si atoms for each implanted ion [98]. It is a very precise method to introduce a specific amount of dopant atoms into crystal lattices like Si or Ge. The electronic charge on the ions is counted by collection in a Faraday cup, and doses ranging from 10^{12} to 10^{16} cm^{-2} are routinely introduced during the fabrication of a MOS device.

10.4.1.1. p-type junction activation with furnace anneal

The starting substrates used to fabricate p^{+} junctions were (100) oriented n-type Ge wafers with a net background concentration of $\sim 7 \times 10^{15}\,cm^{-3}$ at room temperature. No special Ge surface preparation or cleaning was performed before ion implantation. $^{49}BF_2^{+}$ ions were implanted with dose and energy of $4 \times 10^{15}\,cm^{-2}$ and 35 keV, respectively, at room temperature. These implantation conditions were calculated to produce junctions with

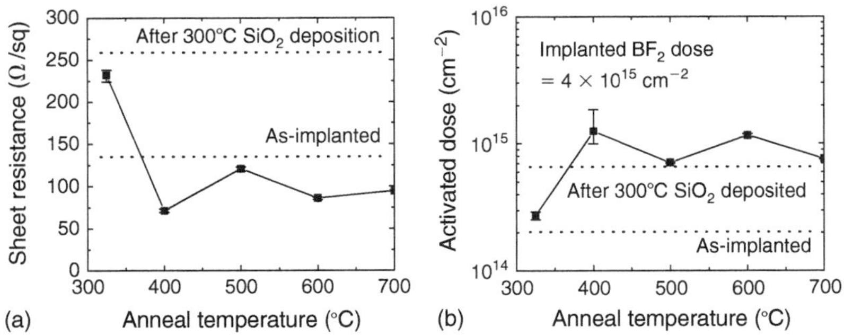

Figure 10.41. (a) Sheet resistance and (b) activated dose measured from surface implanted p^+ layers on n-type Ge substrate as a function of furnace anneal temperature. These Hall Effect measurements were carried out at room temperature after the capping SiO_2 removal. The values extracted from the as-implanted sample and sample right after SiO_2 deposition are also included.

depth of about 100 nm and hence the ion beam was also tilted by 7° to the substrate surface normal to minimize channeling effect. These implanted samples were first degreased in PRS1000 (positive resist stripper) and then loaded into a hot-wall low-pressure chemical vapor deposition (LPCVD) furnace to deposit roughly 30–50 nm of SiO_2 at 300°C. Activation anneals were carried out in a conventional furnace at 325–700°C for 30 min in N_2 gas, a choice over RTA that was primarily picked to monitor the dopant activation levels rather than to obtain shallow junctions. The surface capping SiO_2 was removed in 50:1 HF solution prior to the subsequent electrical measurements.

Hall Effect measurement can be utilized to directly reveal the resistivity of the material, the dominant carrier type, and thus the electrically active doping concentration. On these BF_2 implanted n-type Ge samples, room temperature Hall Effect measurements were applied to extract both sheet resistance (ρ_s) and activated dose from the surface p^+ layers. In Figure 10.41a, the low as-implanted ρ_s is probably originated from implant damage-related acceptor centers [104] as well as the as-implanted substitutionally active p-type doping by B implantation, which has been re-confirmed very recently [110]. After the SiO_2 capping at 300°C, some of the implant damages began to be repaired but the boron had not been substitutionally active and led to an increase in ρ_s. A furnace anneal at 325°C after SiO_2 capping (at 300°C) only slightly lowered the ρ_s; however, a more dramatic ρ_s reduction could be observed if the anneal was executed at 400°C. This could be explained by both the effectiveness of the 400°C anneal to completely recrystallize Ge surface layers (amorphized by ion implantation with dose of less than $6 \times 10^{15} cm^{-2}$ [102]) giving rise to high carrier mobility, and the substitutional B incorporations into Ge crystal lattice providing sufficient carriers for current conduction. When the anneal temperature was further raised to 500–700°C, no significant change in ρ_s was perceived, hypothesizing that once the B atoms have substitutionally incorporated into the Ge lattice, they do not tend to redistribute and diffuse.

Similarly, the activated dose illustrated in Figure 10.41b follows the same activation temperature dependence. For instance, the activated dose increases with the anneal temperature rise from 325°C to 400°C reinforcing the idea of increased substitutional B incorporations

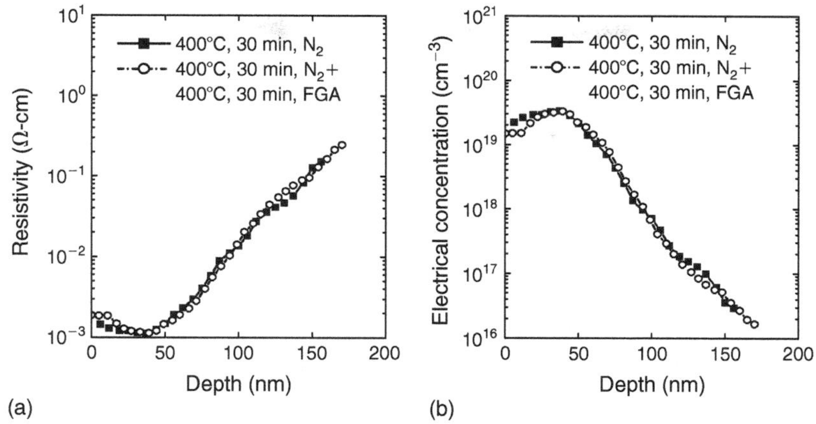

Figure 10.42. (a) Resistivity and (b) electrical concentration measured on two BF_2 implanted Ge samples with one annealed at 400°C for 30 min in N_2 and the other first annealed at 400°C for 30 min in N_2 and then in forming gas. The SPR measurements were carried out at room temperature after the capping SiO_2 removal.

aforementioned; but none of those higher temperature anneals improve the activated dose any further. As these Hall Effect measurements indicate, one can effectively activate p-type dopant in Ge at a temperature as low as 400°C with about 31% of the total implanted dose activated.

Next, spreading resistance probe (SRP) analysis was chosen to measure the electrical concentration depth profiles and study dopant diffusion. In brief, SRP analysis provides an electrical measurement of the active concentration using a pair of fine metal probes to step down the beveled surface of a sample. The raw data captured are first compared against calibration standards to generate resistivity plots, which are then computed together with published carrier mobility values to obtain the carrier concentration as a function of sample depth. Extra caution should be exercised in analyzing Ge samples including the use of softer probes owing to the softer substrate and the employment of Ge mobility numbers for data conversion.

Figure 10.42 depicts the SRP data on the 400°C furnace annealed sample discussed in Figure 10.41. To study the dopant diffusion behavior, part of the sample was further annealed at 400°C for 30 min in forming gas. The resistivity plots in Figure 10.42a were used to calculate the electrically active dopant concentrations in Figure 10.42b, where both of them indicate a negligible amount of dopant redistribution or diffusion after the additional forming gas anneal. This phenomenon is coherent with the extremely low B diffusivity at 400°C shown in Figure 10.40a.

Additionally, the peak and surface electrical concentrations are $3.3 \times 10^{19}\,cm^{-3}$ and $\sim 2 \times 10^{19}\,cm^{-3}$, respectively exhibiting decent p^+–n junctions in Ge (Figure 10.42b). However, the integrated electrically active dose from SRP ($1.6 \times 10^{14}\,cm^{-2}$) is about an order of magnitude lower than the activated Hall dose extracted previously. This big discrepancy is possibly originated from the lack of an experimentally calibrated standard sample for both SRP analysis and Hall correction factor adjustment. Nevertheless, by using conventional ion implantation doping followed by RTA, a shallow p^+ junction in Ge with surface concentration exceeding $10^{20}\,cm^{-3}$ is readily demonstrated as will be discussed together with n-type junction activation in Section 10.4.1.2.

10.4.1.2. Complementary junction activation with rapid thermal anneal

Highly activated p^+ and n^+ junctions are the utmost important components in building complementary MOS (CMOS) logic, which is also the major focus of this sub-section. The starting substrates used to fabricate complementary junctions were (100) oriented n-type and p-type Ge wafers with a net background concentration at room temperature of $\sim 6 \times 10^{15}$ and $\sim 2 \times 10^{14}\,cm^{-3}$, respectively. No special Ge surface preparation or cleaning was performed before ion implantation. Room temperature implantations of various common ionic species at a fixed dose of $4 \times 10^{15}\,cm^{-2}$ were carried out at energies [111] corresponding to a similar projected range (R_p). $^{49}BF_2^+$ ions with 20 keV were used to form p^+ junctions in n-type Ge, while 18 keV of $^{31}P^+$, 30 keV of $^{75}As^+$, and 45 keV of $^{121}Sb^+$ ions were employed to make n^+ junctions in p-type Ge. The ion beam was tilted by 7° to the substrate surface normal to minimize channeling effects. These implanted samples were first degreased in PRS1000 and then deposited with about 30–50 nm of SiO_2 at 300°C in an LPCVD furnace. RTA was then performed at temperatures of 600–850°C with 1–300 s soak time in N_2 ambient to activate dopants and remove the implantation damage while minimizing non-equilibrium diffusion effects such as transient enhanced diffusion (TED). Without removing the capping SiO_2, SRP analysis was applied to track the electrically active dopant depth profiles.

As illustrated in Figure 10.43, the ion implanted B can be activated to produce shallow p^+ junctions with no substantial B diffusion for the RTA thermal budgets of interest. Surface electrical concentrations on the order of $1 \times 10^{20}\,cm^{-3}$ could be obtained with about 2.9–5.9% of the implanted dose activated, which might also reflect its solid solubility level at these temperatures. On the other hand, RTA of ion implanted n-type dopants causes significant diffusions and delivers not as high surface concentrations. From all n-type doping profiles, the peak concentrations were never shown to occur at the surface which could be attributed to dopant out-diffusion from the Ge substrate and segregation into the SiO_2–Ge interface. Among the three n-type dopant candidates, P apparently gives the highest peak and surface concentration of $\sim 8.5 \times 10^{19}\,cm^{-3}$ and $\sim 2 \times 10^{19}\,cm^{-3}$ respectively after an RTA at 650°C for 10 s. In addition, all profiles exhibit an abrupt edge near the metallurgical n^+–p junctions except for P that has tails extending into bulk Ge (Figure 10.43), as similarly reported before [106]. These tails are related to defects at the original implant-induced amorphous–crystalline interface referred to as end-of-range (EOR) defects. They occur because there is a large amount of damage which is just below the threshold of amorphization beyond the amorphous–crystalline interface, where the maximum possible amount of damage can exist in the crystal without itself being amorphous. After all, these EOR defects could only be annealed out with larger thermal budgets as will be elaborated next.

As demonstrates a slightly lower activation level than P but this is still an order of magnitude higher than that of Sb. The SRP peak concentration of these three elements in Ge when compared to the solid solubility shown in Figure 10.40b, obeys the same order ($P > As > Sb$). Figure 10.44 plots the fractional activation of the three n-type dopant

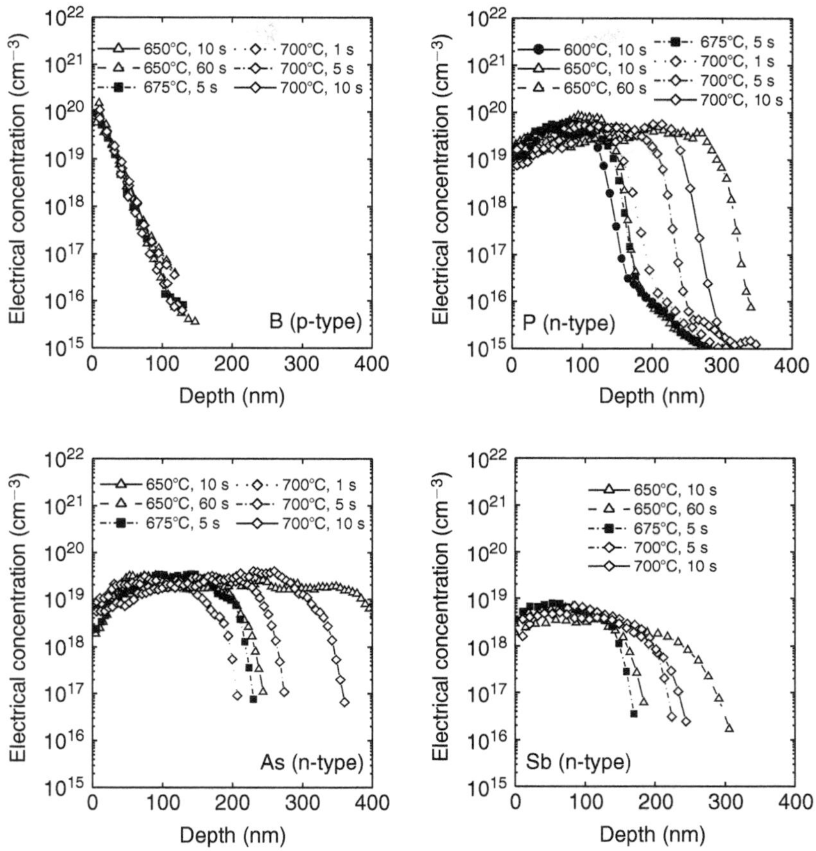

Figure 10.43. Electrical concentration measured on BF_2, P, As, and Sb implanted Ge samples after various RTA. The SRP measurements were carried out at room temperature with the capping SiO_2 on top.

implants as a function of anneal temperature for two RTA soak times. Regardless of the low Sb fractional activation, the experimental peak concentration matches the solid solubility, hinting any lower resistivity n^+ layer would not be feasible owing to a fundamental limit. Contrarily, the experimental P and As peak concentrations are a few times lower than their respective solid solubilities while their fractional activations are far from unity, leaving room for further improvement. A probable explanation of such a discrepancy is the ability for Sb to completely amorphize Ge but only partial amorphization is possible with either P or As at the same implant dose. Even though the dose of $4 \times 10^{15} cm^{-2}$ in our experiment is above the usual amorphization threshold ($\sim 1 \times 10^{15} cm^{-2}$) in Si, the larger lattice constant in Ge may require an even higher threshold for some lighter ions. From the experience on Si, it is much more difficult to anneal a sample that has only been partially damaged at levels below the amorphization threshold [112]. This difficult regime is where secondary damage forms that has a more complex annealing behavior. Usually when this happens, a gradual increase in fractional activation with anneal temperature would be discerned (like from medium dose P implants into Si [112]); now, similar behaviors are also found in P and As implants into Ge (Figure 10.44). In principle, a higher activation level could thus be

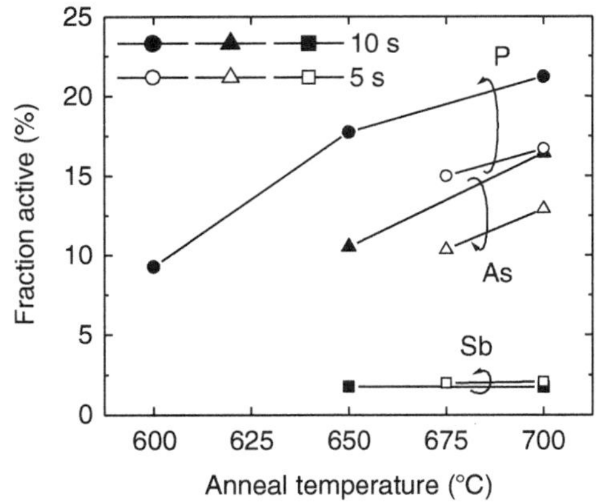

Figure 10.44. Fractional activation of various n-type dopant implants as a function of anneal temperature for two isochronal anneals (solid symbols for 10 s and open symbols for 5 s).

achieved at higher anneal temperature. However, the accelerated dopant diffusion at these temperatures compromises shallow junction formation, which instead mandates a more careful selection of implant doses.

In all commercial CMOS IC fabrications, both p^+ and n^+ junctions are simultaneously activated with the same thermal anneal budget. Therefore, we should also compile with this requirement for Ge CMOS applications by selecting the most symmetric pair of junctions from the above complementary junctions activation experiment. Besides using SRP to monitor electrical depth profiles, secondary ion mass spectroscopy (SIMS) is also employed for chemical concentration profiling.

Figure 10.45 shows the depth profiles of B and P both as-implanted and after an RTA at 650°C for 10 s. Only SIMS data is shown for the as-implanted samples and both SIMS and SRP results are presented for the post-RTA case. Negligible B diffusion is again confirmed the overlapping of the as-implanted and post-RTA SIMS profiles. Conversely, P diffuses substantially even after a very low thermal budget anneal. Electrically, the resultant p^+–n junction has a depth of ~150 nm and a peak active concentration of $\sim 1.5 \times 10^{20}\,\mathrm{cm}^{-3}$ (with a corresponding sheet resistance of 210 Ω cm/square while those from the n^+–p junction are ~280 nm and $\sim 8.5 \times 10^{19}\,\mathrm{cm}^{-3}$ (sheet resistance of 42 Ω cm/square), respectively. It is worth noting that the electrical active concentration of P in Ge after RTA from SRP shows a higher value than the chemical concentration from SIMS. This apparent contradiction could be attributed to the inaccuracy in raw data conversion, due to the lack of some well-calibrated standards in Ge. To summarize, symmetrically high levels of electrical activation of both p-type and n-type dopants in Ge are plausible and have been demonstrated at concentrations directly applicable to advanced CMOS devices.

10.4.1.3. n-type junction activation dependences

The starting substrates used to fabricate n^+ junctions were (100) oriented p-type Ge wafers with a net background concentration of $\sim 1.5 \times 10^{15}\,\mathrm{cm}^{-3}$ at room temperature. No special

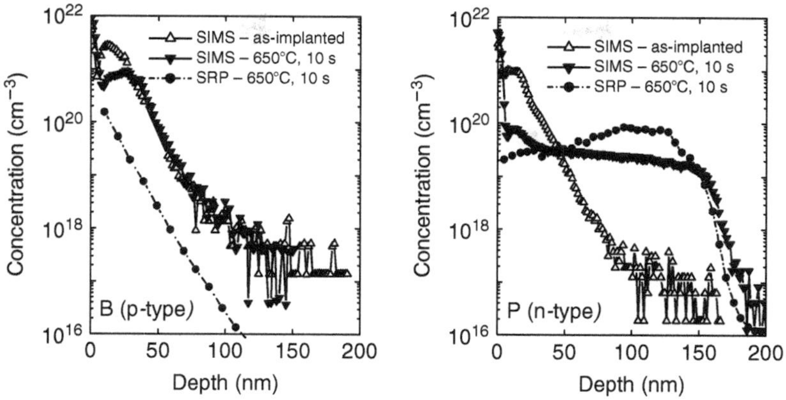

Figure 10.45. SIMS and SRP depth profiles of B and P before and after an RTA at 650°C for 10 s. BF_2 and P were implanted at 20 and 18 keV, respectively, at a dose of 4×10^{15} cm^{-2}. Both SIMS and SRP measurements were carried out with the capping SiO_2 on top.

Ge surface preparation or cleaning was performed before ion implantation. Either $^{31}P^+$ or $^{75}As^+$ ions were implanted at the respective energy of 32 keV and 64.5 keV to match R_p. The input dose was ranged from 4×10^{14} cm^{-2} to 2×10^{15} cm^{-2} and the ion beams were tilted by 7° to the substrate surface normal to minimize channeling effects. These implanted samples were first degreased in PRS1000 and deposited with about 50 nm of SiO_2 at 300°C in an LPCVD furnace. Isochronal RTA was then performed in N_2 ambient at temperatures of 500–700°C to activate dopants and remove the implantation damage. The chemical and electrically active concentration profiles were measured with SIMS and SRP. These SIMS concentration-depth profiles were carried out using the Cameca™ 6F instrument under cesium primary ion (Cs^+) bombardment followed by two different secondary ion detection approaches. Positive secondary ion detection was employed on samples with capping SiO_2 for better depth resolution and negative mode was used after stripping SiO_2 for better detection sensitivity. During the spectra acquisition, the SiO_2 capped samples were sputtered at an estimated rate of ~0.34 nm s^{-1} in SiO_2 and ~0.7 nm s^{-1} in Ge, while that for the SiO_2 stripped samples was ~0.9 nm s^{-1}.

Figure 10.46 shows the SIMS and SRP depth profiles of P and As in Ge after isochronal (10 s) RTA at varying temperatures of 500–700°C. Since the thermal activation trends monitored on samples with different dose were similar for the individual species, only the data from an implanted dose of 8×10^{14} cm^{-2} are illustrated without losing generality. For the P implants into Ge, only the RTA at 500°C could retain a shallow junction with high surface chemical concentration whereas any higher temperatures degraded both merits (Figure 10.46a). In addition, the characteristic "box-shape" concentration-dependent diffusions were observed after RTA at 600°C and above, a phenomenon that was also apparent in the accompanying electrical concentration profiles shown in Figure 10.46b. Nonetheless, the maximum achievable electrical concentration saturated at around 5×10^{19} cm^{-2} even after the lowest temperature RTA with minimum diffusion. When these data are compared with the previous experimentations [113], this saturation in active concentration could arguably

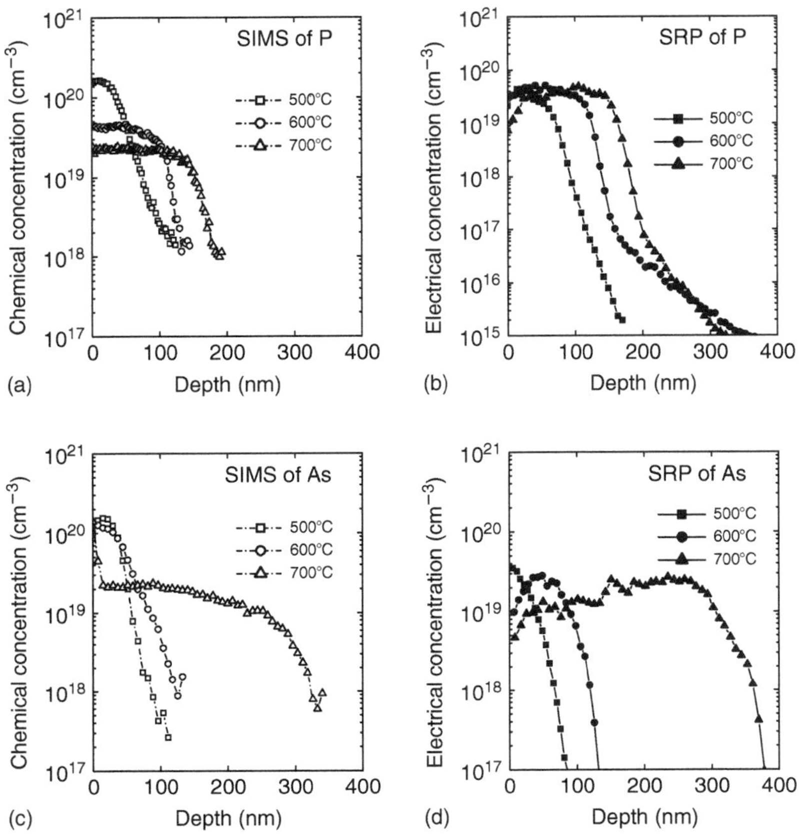

Figure 10.46. SIMS and SRP depth profiles of P and As in Ge after isochronal (10 s) RTA at varying temperatures of 500–700°C. $^{31}P^{+}$ and $^{75}As^{+}$ ions were, respectively, implanted at 32 and 64.5 keV at a dose of $8 \times 10^{14} cm^{-2}$.

be attributed to the substitutional solid-solubility limitation (SSL). It is noteworthy that this SSL estimated directly from the electrically active dopant concentration could perhaps be lower but correlated to the commonly quoted impurity solid solubility in Ge [99], which also includes the electrically inactive concentration. Regarding the As implants into Ge, an analogous thermal budget dependence on activation anneals could be identified except with a slight difference in demarcating the aforementioned regimes. In other words, shallow As junctions with high surface concentration were obtained after RTA at a temperature as high as 600°C, and above which fast As diffusion began to kick-in (Figure 10.46c and d). The highest As electrical concentration was $\sim 3.5 \times 10^{19} cm^{-2}$.

Viewing from a defect annealing perspective, we can gain some insights on the factors governing the above demarcation process. In order to eliminate the primary damage created by the implant and restore the Ge lattice to its perfect crystalline state with dopants sitting on substitutional sites, only a relatively low temperature anneal would be necessary, whereby the associated diffusion should also be minimal. After which, a higher thermal budget is required to annihilate the residual interstitial-type defects originated from the extra atom introduction into the lattice. During this phase, the interstitials could contribute an extrinsic mechanism in addition to intrinsic diffusions that is responsible for the concentration-dependent box-shape profiles. Moreover, the higher demarcation temperature (between shallow and box-shape junctions) for As implants could be explained by the larger amount of primary damage they had generated as a heavier species than P, even though the same implanted dosage was employed.

As another clear distinction between these two dopants species, the As electrical profiles indicated relatively abrupt metallurgical n^+–p junctions (Figure 10.46d) but there existed EOR tails in some P profiles (after RTA at 600°C and 700°C) (Figure 10.46b). Although in principle these tails should also appear in the 500°C profile given the same amount of dose (at the same energy) was implanted, the much slower P diffusion at the reduced temperature prevented its exposure. On increasing annealing temperatures, some of the EOR defects began to disappear or possibly condense into fewer but bigger loops like in Si [114], giving rise to the steeper tail in the 700°C profile. For a complete removal of these defects, an even higher thermal budget anneal with longer time and/or higher temperature would be mandatory. Nevertheless, such an anneal could in turn lower the surface dopant concentration and deteriorate the contact resistance due to an excessive diffusion or solid-solubility reduction at elevated temperatures [99].

Besides, it is worth mentioning that the presence of these EOR defects within the n^+–p diode depletion region may cause significant junction leakage and disturb normal device operations. This situation could however be rectified by engineering these defects to stay inside the n^+ regions. According to a previous report [19], the minimum thermal budget to achieve good rectifying n^+–p diodes is 500°C for 5 min using $4 \times 10^{15}\,\text{cm}^{-2}$ of implanted P. In fact, it is perfectly legitimate to assume also the existence of a comparable amount of EOR defects for As implants into Ge; the fast As diffusion (e.g. after RTA at 700°C in Figure 10.46d) had conversely pushed the metallurgical n^+–p junction deeper than the EOR location and thus shielded the defects.

With the aim of understanding the implantation dosage dependence in shallow junction formation, low thermal budget activations are investigated to avoid any fast extrinsic diffusion discussed earlier. Figure 10.47 plots the SIMS and SRP depth profiles of P and As in Ge after isochronal (10 s) RTA at 500°C with varying implanted dose from 4×10^{14} to $2 \times 10^{15}\,\text{cm}^{-2}$. Generally speaking, slightly deeper junctions with higher chemical concentration were obtained with heavier doses (Figure 10.47a and c). The P electrical concentrations (Figure 10.47b) suggested otherwise a rather constant peak value, probably owing to SSL. On the other hand, the As junctions of a similar depth (Figure 10.47d) revealed a drop in electrical concentration with the heaviest dose ($2 \times 10^{15}\,\text{cm}^{-2}$), which could merely be attributed to a dopant deactivation since a chemical dose loss due to surface segregation was unlikely from Figure 10.47b. Given a significant As diffusion was absent, these As in Ge may deactivate through the formation of small inactive clusters as observed in Si [115].

Among the figures of merit in evaluating shallow junctions, the junction depth and sheet resistance are the most important. By matching the implantation R_p for both species in our

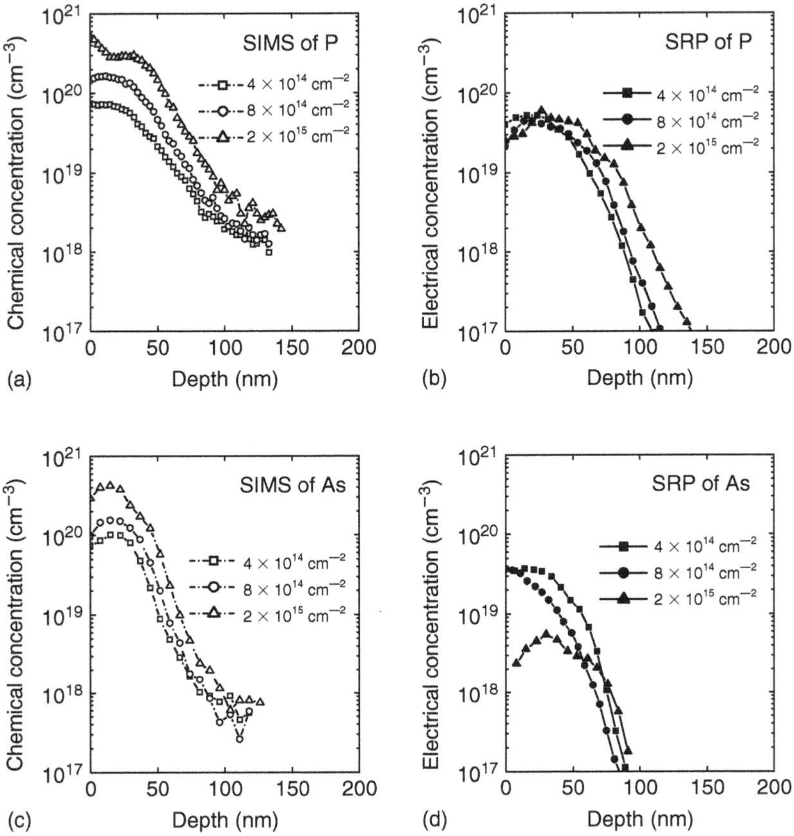

Figure 10.47. SIMS and SRP depth profiles of P and As in Ge after isochronal (10 s) RTA at 500°C with varying implanted doses from 4×10^{14} to $2 \times 10^{15}\text{cm}^{-2}$.

case, whose longitudinal straggle is smaller for As (~15.8 nm) than for P (~21.5 nm), shallower and more abrupt As junctions were obtained over those with P (Figure 10.47b and d). From these shallow n^+ junctions, electrically active fraction of the implanted dose and the corresponding sheet resistance were extracted in Figure 10.48. The active fractions of P in Ge decreased with increasing implanted dose while the sheet resistance stayed about the same; which is a strong evidence to prove that any excess dose beyond the point of SSL would not improve the junction sheet resistance but rather worsen the crystal damage due to the surplus implants. Compared to the P data, As junctions indicated a similar trend with lower active fractions and higher sheet resistances owing to their shallower depths and lower dopant SSL. The only exception to the trends arose from the heaviest dose sample that suffered from dopant deactivation.

In conclusion, several dependences associated with the activation of common n-type dopants in Ge during shallow junction formation have been examined. As a rule of thumb, low thermal budget anneals are preferred as they should be sufficient to annihilate the primary implantation damage while minimizing any fast extrinsic diffusion. Any attempts to

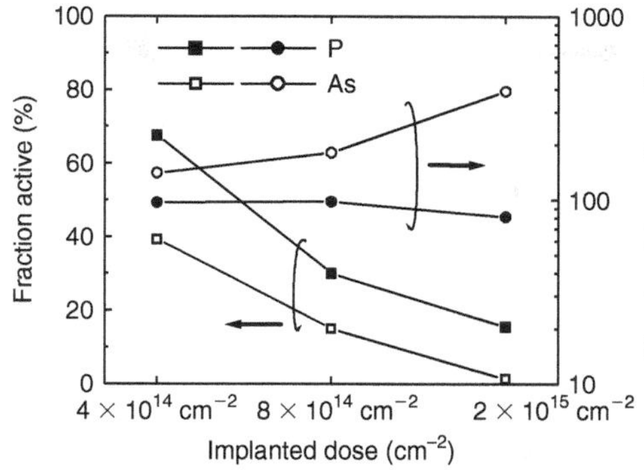

Figure 10.48. Electrically active fractions of the implanted dose and the corresponding sheet resistances extracted from the shallow n^+ junctions in Ge after RTA at 500°C.

over-dose the junctions beyond the point of SSL could only cause more physical damages but not reduction in resistance. For those slightly deeper junctions with EOR defects near the vicinity of the metallurgical n^+–p depletion region, higher thermal budgets would be necessary to anneal these defects at the expense of deepening the junctions. All these findings could help to provide explanations to many unsatisfactory Ge n-channel MOSFET demonstrations.

10.4.2. SSD doping

Since the ion implantation doping results in fast diffusion in Ge especially for n-type dopants, its efficacy to form shallow junctions might eventually be limited. As an alternative doping technology, SSD doping is free from problems such as channeling effects, TED, and extended defect formation, even at very low implant energies [116]. In addition to these advantages, shallow junctions with low sheet resistance have been achieved in silicon using SSD [116,117]. In an attempt to obtain shallow junctions in Ge via diffusion from doped SiO_2, we have studied the SSD doping in Ge. RTP was utilized to maximize the activation level of the out-diffused dopants and minimize their redistribution.

10.4.2.1. n-type junction activation and diffusion

The starting substrates used to fabricate n^+ junctions were (100) oriented p-type Ge wafers with a net background concentration of $\sim 3 \times 10^{17}\,cm^{-3}$ at room temperature. The surfaces were first cleaned by CHF (discussed in Section 10.2), blown dry with N_2 and immediately loaded into a LPCVD furnace to deposit about 200 nm of 8 wt.% phosphosilicate glass (PSG) as the solid dopant source from a mixture of SiH_4 and phosphine (PH_3) gases at 400°C. This choice of PSG concentration was intended to maximize the P amount while preventing the film from becoming hygroscopic. The SSD doping was subsequently carried out in a commercial RTP system at 600–900°C with 5–300 s soak time in N_2 ambient. The RTP ramp rate was 100°C s^{-1}. Without removing the capping SiO_2, SRP analysis was applied to track the electrically active dopant depth profiles.

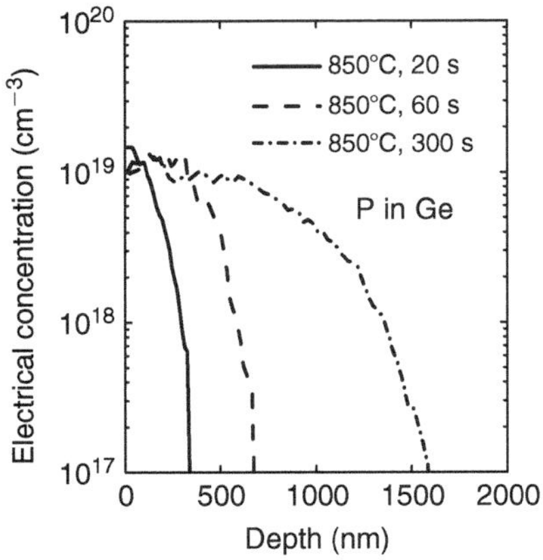

Figure 10.49. SRP depth profiles of the out-diffused P into a p-type Ge substrate as a function of RTP soak time at 850°C in N_2 ambient.

Figure 10.49 shows the electrically active depth profiles of the out-diffused P into p-type substrate as a function of the RTP soak time at 850°C in N_2 ambient. These resultant n^+–p junctions always showed the peak concentration at the surface of about $1 \times 10^{19}\,cm^{-3}$. This highest achievable peak concentration depends on several factors including the P concentration and diffusivity within the PSG layer, the P segregation at the PSG-Ge interface, as well as the P solid solubility in Ge. For the 8 wt.% PSG that we used in our experiment, no appreciable out-diffusion was observed below 800°C, which could be attributed to the low diffusivity of P within the PSG layer at such low temperatures. In order to simultaneously lower the out-diffusion temperature and raise the surface peak concentration, a solid source with either higher dopant concentration or diffusivity could be employed.

Before the extraction of intrinsic dopant diffusivity, a constant-source diffusion behavior has to be verified by examining the out-diffused junction profiles. The junctions formed by out-diffusion at 800°C did not produce the same surface peak concentration with different RTP soak times which represents non-ideal constant-source diffusion. A true constant-source diffusion regime instead became apparent after SSD at 850°C and 900°C. By plotting the resultant junction depth versus the square root of diffusion time, a linear relationship could be observed as illustrated in Figure 10.50(a).

The diffusion coefficient, D, could then be estimated from the complementary error-function for constant-source diffusion

$$x_j = 2\sqrt{Dt} \times \left[\mathrm{erfc}^{-1}\left(\frac{N_B}{N_0}\right)\right] \tag{10.6}$$

where x_j is the junction depth, t is the RTP soak time, N_B is the p-type Ge substrate concentration, and N_0 is the n^+ Ge surface peak concentration. Assuming most of the commonly

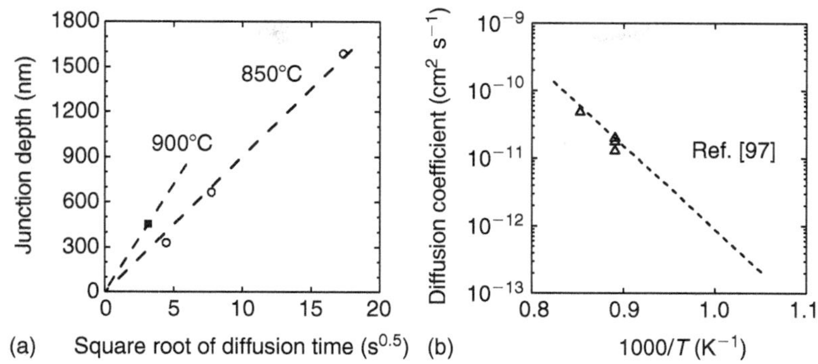

Figure 10.50. (a) Metallurgical junction depths versus the square root of RTP soak time extracted from the out-diffused junctions at 850°C and 900°C, and (b) Arrhenius plot of the extracted intrinsic diffusivity of P in Ge together with the reference data from Ref. [97].

known extrinsic diffusion mechanisms were absent during the SSD process, the diffusion coefficients extracted from Equation 10.6 and Figure 10.50a should represent the intrinsic diffusivities of P in Ge.

When these intrinsic diffusivities are plotted as a function of temperature as shown in Figure 10.50b, they follow the Arrhenius relationship

$$D = D_0 \exp\left(\frac{-H}{kT}\right) \tag{10.7}$$

where D_0 is the diffusivity pre-factor, H the activation enthalpy of diffusion, k is the Boltzmann's constant, and T is the absolute temperature. This equation has been discussed in detail in Chapter 3. Also included in Figure 10.50b are the P diffusion coefficients in Ge from Ref. [97], which indicates a very good match with the experimental data in the present study; the extracted E_A is 2.45 eV and D_0 is 2.05 $cm^2 s^{-1}$ for such an intrinsic SSD.

In order to compare the intrinsic diffusivity of P in both Ge and Si and thus to find out which material would ultimately permit shallower junctions, their temperature dependences were plotted together in Figure 10.51a. However, because of their large difference in melting point, a more objective assessment could only be made by normalizing the same intrinsic diffusivity dependence with respect to their individual melting point (Figure 10.51b). Still, P in Ge is found to be more diffusive than in Si. Together with its lower solid solubility (Section 10.4), the shallow n^+ junction formation in Ge would be relatively more challenging.

To sum up, shallow n^+ junctions in Ge have been demonstrated through controlled SSD doping with low thermal budget RTP. In addition to the insulating solid dopant source aforementioned, the use of *in-situ* p-type and n-type doped silicon germanium (Si_xGe_{1-x}) alloys as solid sources have also been attempted. The initial results suggest their suitability by enabling p^+–n and n^+–p diodes with ideality factor close to unity, and such idea should deserve more research effort to show further promises.

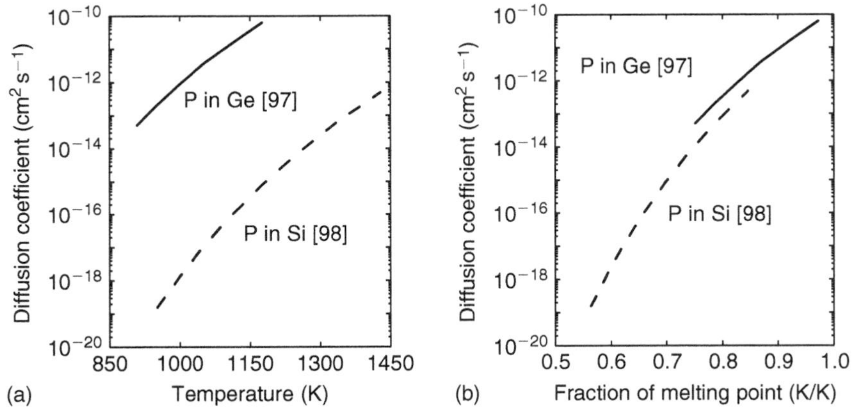

Figure 10.51. (a) Temperature dependence of the intrinsic diffusion coefficient of P in Ge [97] and Si [98] and (b) the same intrinsic diffusivity dependence normalized to the corresponding crystal melting point in K.

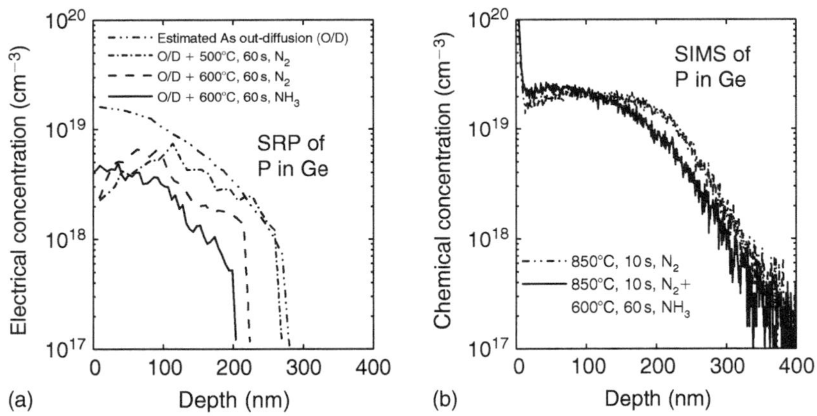

Figure 10.52. (a) SRP depth profiles of the junction first formed by out-diffusion from PSG at 850°C for 10 s (estimated) in N_2 and those subsequently annealed with different thermal budgets and in different ambient. (b) SIMS depth profiles of the junction first formed by out-diffusion from PSG at 850°C for 10 s in N_2 and that subsequently annealed at 600°C for 60 s in NH_3 ambient.

10.4.2.2. Dopant deactivation within activated junctions

Being able to maintain the high level of dopant activation during the subsequent thermal steps in a MOS transistor fabrication process is crucial to guarantee good device performance. Working towards this goal, we have studied the activation level of the out-diffused dopants after various thermal anneals. On the same n^+–p junction out-diffused from PSG at 850°C for 10 s, RTA were individually carried out at either 500°C in N_2, 600°C in N_2, or 600°C in NH_3 [118]. Figure 10.52a depicts their measured electrically active depth profiles together with the estimated as-out-diffused junction profile.

After either one of these subsequent thermal anneals, the electrically activated n-type dopants were distinctly observed to be deactivated for the first time. Comparing the thermal anneals in N_2 ambient, the 600°C treatment caused more severe deactivation then the 500°C case. When the treatments were carried out at the same 600°C, the anneal in NH_3 deactivated dopants more than in N_2. On the other hand, the surface dopant concentration dropped after anneals in N_2 but not in NH_3. This phenomenon indicates that the PSG–Ge interfacial NH_3 nitridation might help to suppress surface deactivation or to minimize surface dose loss, possibly due to a similar nitridation-retarded surface P diffusion observed in Si [119].

In order to propose a plausible deactivation mechanism, we examined the change in P chemical concentration profiles with SIMS on these junctions before and after the deactivation anneal at 600°C in NH_3 (Figure 10.52b). Prior to the SIMS analysis on each sample, the PSG layer on Ge surface was removed in 50:1 HF to avoid any interference between the out-diffused dopants inside Ge, and those within the PSG layer which were present in a very high concentration. This PSG stripping was not necessary in previous SRP measurements because the dopants within the PSG layer are not electrically active. After the deactivation anneal, no significant chemical dopant redistribution was observed. The slight reduction in junction depth could either due to dopant diffusion and segregation at the PSG-Ge interface (which was unfortunately not captured owing to PSG removal) or merely be an error in SIMS craters height measurement for depth calibration. When these chemical SIMS profiles are compared to the electrically active SRP profiles in Figure 10.52a, we notice that only the electrical activation level dropped after the 600°C anneal in NH_3 while the chemical concentration stayed about the same. Combining these experimental facts together, we suggest that activated P in Ge deactivates through the formation of small inactive dopant-defect clusters [115], which also indirectly indicates the existence of an appreciable amount of point defects even in a crystalline Ge substrate. On the contrary, deactivation through precipitation is less likely as it usually occurs only after significant dopant diffusion, an observation that is missing from the SIMS data after such a short anneal for 60 s. More systematic studies into these mechanisms would be timely and essential as such a deactivation behavior would not only increase the MOS transistor source and drain sheet resistance, but more importantly degrade the contact resistivity which is currently the major bottleneck in the development of advanced transistors.

10.4.3. Metal germanide contacts

Last but not the least, low-resistance contact technology to Ge has to be developed in order to fully exploit the superior transport properties of Ge MOSFETs. As an obvious extension of metal silicide contact technology to Si MOSFETs, metal germanide contacts have recently attracted some attentions [120,121]. Nickel germanide (NiGe) was initially chosen to be studied owing to the wealth of knowledge on nickel silicide properties and its formation.

The starting substrates used to form NiGe were (100) oriented p-type Ge wafers having a resistivity of 0.05 Ω-cm at room temperature. The surfaces were first cleaned in dilute HF solution and immediately loaded into an electron beam evaporation system operating at a base pressure of 1×10^{-6} Torr. About 20 nm thick of Ni was then deposited onto the Ge

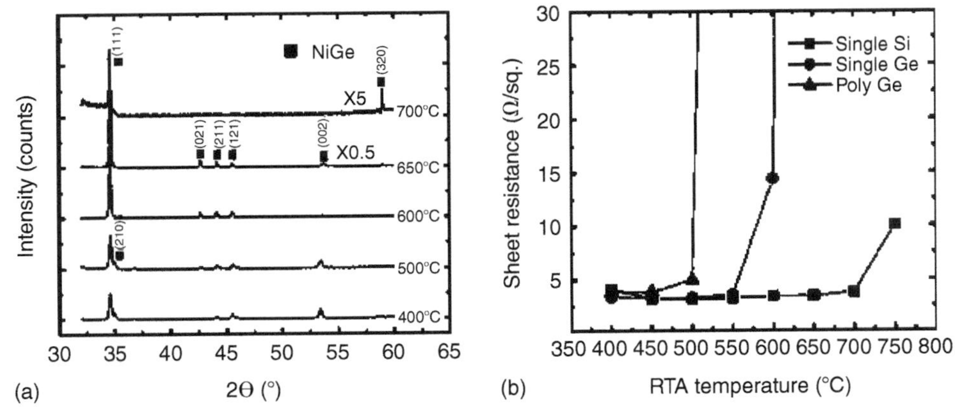

Figure 10.53. (a) XRD spectra of NiGe formed on Ge at different RTA temperatures. (b) The sheet resistance of NiGe film plotted as a function of the RTA temperature. The sheet resistances of NiSi and NiGe formed on polycrystalline Ge are also included for comparison (after Hsu *et al.* [120]).

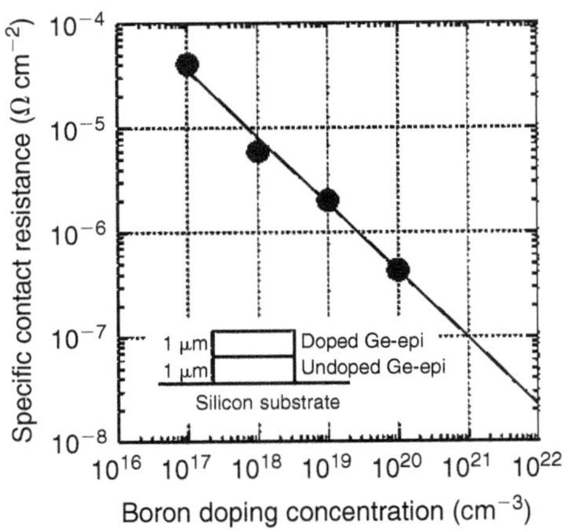

Figure 10.54. Specific contact resistance of NiGe films to *in-situ* B doped epitaxial Ge. The inset shows a schematic cross-section of the etched mesa structure used for the TLM measurements (after Spann *et al.* [121]).

surfaces at a rate of ~0.1 nm s^{-1}. Nickel germanidation reactions were performed through RTA at various temperatures for 30 s in N_2 ambient.

Figure 10.53a presents the X-ray diffraction (XRD) spectra of NiGe formed at different temperatures [120]. All of the peaks observed in the temperature range of 400–700°C revealed the orthorhombic NiGe phase. The relative intensity of the NiGe peak in the (111) orientation increased and the full width at half maximum narrowed with increasing temperature, which indicated that (111) orientation is preferred for NiGe grains during high-temperature annealing. Figure 10.53b plots the sheet resistance of the NiGe film as a function of the RTA temperature. These NiGe films were thermally stable up to 550°C, and beyond which, agglomeration occurred that resulted in dramatic increase in its sheet resistance.

The specific contact resistance (ρ_c) of NiGe on heavily doped Ge was determined using standard transmission line measurements (TLM) [121]. The samples consisted of 2 μm of epitaxially grown Ge on high-resistivity Si substrates. The top 1 μm of Ge was *in-situ* doped with B to various concentrations measured by SIMS. Device isolation was achieved by mesa etching the TLM structures down to the Si substrate. NiGe contacts were formed from 50 nm of Ni annealed at 400°C and the ρ_c was therefore extracted (Figure 10.54). The results suggested that B doping in excess of $10^{22}\,cm^{-3}$ will be required to deliver a ρ_c of $10^{-8}\,\Omega\text{-cm}^2$ if NiGe-based contacts to Ge MOSFETs are to be used to augment Si in the future technology nodes [1].

10.4.4. Summary

In this section, we have investigated two techniques to form shallow junctions for Ge MOSFET contact applications including ion implantation and SSD doping. The incorporated dopants were activated using furnace anneal and RTA. The activation and diffusion of various p-type and n-type dopants in Ge were studied using Hall Effect measurement, SRP, and SIMS.

Ion implanted p-type dopants in Ge can be effectively activated with negligible diffusion at temperature as low as 400°C using furnace anneal. However, the higher temperature required to activates ion implanted n-type dopants would inherently cause fast extrinsic diffusion. RTA was then employed to circumvent the problem and led to the demonstration of symmetrically high levels of both p-type and n-type dopant activation at concentrations directly applicable to advanced CMOS devices. In addition, RTP was utilized together with SSD doping to obtain shallow n^+–p junctions in Ge. By examining the thermal stability of the out-diffused dopants, we observed P deactivation in Ge for the first time, possibly through the formation of small inactive dopant-defect clusters.

10.5. General Conclusion

In the past few years, there has been a tremendous progress made in advancing the Ge MOS technology. In particular, nanoscale gate dielectric and shallow junction demonstrations are the mostly researched yet promising areas.

Ge oxynitride and high-κ dielectrics are thoroughly analyzed for Ge MOS dielectric applications. Their syntheses, physical properties, and electrical characteristics are discussed. Even though the Ge oxynitride is relatively stable and scalable down to an equivalent oxide thickness of 1.9 nm, a sub-1.0 nm solution should still be pursued. Insertion of an interfacial layer prior to CVD-based high-κ depositions has been shown to produce high quality Ge MOS capacitors yet with compromised scalability. Sub-1.0 nm Ge gate stacks have otherwise been demonstrated using PVD approaches. Generally speaking, the high-κ gate stacks on Ge are shown to be more scalable than their Si counterparts.

On the other hand, low-resistance shallow p-type junctions in Ge are readily demonstrated, however the n-type version is fundamentally challenging owing to their low dopant solid solubility and fast diffusion. Several activation dependences of common n-type dopants have been examined, which leads to the conclusion that over-dosing the junctions would only cause more physical damages but not resistance reduction. Moreover, the deactivation of solid source diffused dopants indirectly indicates the existence of an appreciable amount of point defects even in a crystalline Ge substrate. In order to bring the overall Ge MOS technology to the next level, several breakthroughs have to be made including dielectric–substrate interface state density reduction, n-dopant solubility enhancement and diffusion suppression, and others.

References

[1] *The International Technology Roadmap for Semiconductors*, Semiconductor Industry Association, 2004 Update. (http://public.itrs.net/)

[2] M.L. Lee, C.W. Leitz, Z. Cheng, A.J. Pitera, T. Langdo, M.T. Currie, G. Taraschi, E.A. Fitzgerald and D.A. Antoniadis, *Appl. Phys. Lett.* **79** (2001) 3344.

[3] C.O. Chui, H. Kim, D. Chi, B.B. Triplett, P.C. McIntyre and K.C. Saraswat, *IEEE IEDM Tech. Dig.* (2002) 437.

[4] H. Shang, H. Okorn-Schmidt, K.K. Chan, M. Copel, J.A. Ott, P.M. Kozlowski, S.E. Steen, S.A. Cordes, H.-S.P. Wong, E.C. Jones and W.E. Haensch, *IEEE IEDM Tech. Dig.* (2002) 441.

[5] M. Lundstrom, *IEEE Electron Dev. Lett.* **18** (1997) 361.

[6] J. Bardeen and W.H. Brattain, *Phys. Rev.* **74** (1948) 230.

[7] O.J. Gregory, E.E. Crisman, L. Pruitt, D.J. Hymes and J.J. Rosenberg, *MRS Symp. Proc.* **76** (1987) 307.

[8] M.D. Jack, J.Y.M. Lee and H. Lefevre, *J. Electron. Mater.* **10** (1981) 571.

[9] E.E. Crisman, J.I. Lee, P.J. Stiles and O.J. Gregory, *Electron. Lett.* **23** (1987) 8.

[10] Y. Wang, Y.Z. Hu and E.A. Irene, *J. Vac. Sci. Technol. A* **12** (1994) 1309.

[11] V. Craciun, I.W. Boyd, B. Hutton and D. Williams, *Appl. Phys. Lett.* 75 (1999) 1261.

[12] R.S. Johnson, H. Niimi and G. Lucovsky, *J. Vac. Sci. Technol. A* **18** (2000) 1230.

[13] D.J. Hymes and J.J. Rosenberg, *J. Electrochem. Soc.* **135** (1988) 961.

[14] J.J. Rosenberg and S.C. Martin, *IEEE Electron Dev. Lett.* **9** (1988) 639.

[15] S.C. Martin, L.M. Hitt and J.J. Rosenberg, *IEEE Electron Dev. Lett.* **10** (1988) 325.

[16] T.N. Jackson, C.M. Ransom and J.F. DeGelormo, *IEEE Electron Dev. Lett.* **12** (1991) 605.

[17] C.M. Ransom, T.N. Jackson and J.F. DeGelormo, *IEEE Trans. Electron Dev.* **38** (1991) 2695.

[18] H. Shang, H. Okorn-Schimdt, J. Ott, P. Kozlowski, S. Steen, E.C. Jones, H.-S.P. Wong and W. Haesch, *IEEE Electron Dev. Lett.* **24** (2003) 242.

[19] H. Shang, K.-L. Lee, P. Kozlowski, C. D'Emic, I. Babich, E. Sikorski, M. Ieong, H.-S.P. Wong, K. Guarini and W. Haensch, *IEEE Electron Dev. Lett.* **25** (2004) 135.

[20] Z. Sun and C. Liu, *Semicond. Sci. Technol.* **8** (1993) 1779.

[21] L.L. Chang and H.N. Yu, *Proc. IEEE* **53** (1995) 316.

[22] T.O. Sedgwick, *J. Appl. Phys.* **39** (1968) 5066.

[23] T. Yashiro, *Jpn. J. Appl. Phys.* **9** (1970) 740.

[24] K.L. Wang and P.V. Gray, *IEEE Trans. Electron Dev.* **22** (1975) 353.

[25] K.L. Wang and P.V. Gray, *J. Electrochem. Soc.* **123** (1976) 1392.

[26] A.V. Rzhanov and I.G. Neizvestny, *Thin Solid Film.* **58** (1979) 37.

[27] G.G. Fountain, S.V. Hattangady, R.A. Rudder, R.J. Markunas, G. Lucovsky, S.S. Kim and D.V. Tsu, *J. Vac. Sci. Technol. A* **7** (1989) 576.

[28] M. Randolph and L.G. Meiners, *J. Electrochem. Soc.* **136** (1989) 2699.

[29] D.G. Leupp and G.M. Oleszek, *Symp. on Low Temp. Electron. Dev. Operat. Tech. Dig.* (1991) 141.

[30] Y. Wang, Y.Z. Hu and E.A. Irene, *J. Vac. Sci. Technol. B* **14** (1996) 1687.

[31] G.G. Fountain, R.A. Rudder, S.V. Hattangady, D.J. Vitkavage, R.J. Markunas and J.B. Posthill, *Electron. Lett.* **24** (1988) 1010.

[32] D.J. Vitkavage, G.G. Fountain, R.A. Rudder, S.V. Hattangady and R.J. Markunas, *Appl. Phys. Lett.* **53** (1988) 692.

[33] S.V. Hattangady, G.G. Fountain, R.A. Rudder, M.J. Mantini, D.J. Vitkavage and R.J. Markunas, *Appl. Phys. Lett.* **57** (1990) 581.

[34] S.V. Hattangady, M.J. Mantini, G.G. Fountain, R.A. Rudder and R.J. Markunas, *J. Appl. Phys.* **71** (1992) 3842.

[35] D. Reinking, M. Kammler, N. Hoffmann, M. Horn-von Hoegen and K.R. Hofmann, *Electron. Lett.* **35** (1999) 503.

[36] T. Yashiro, *Jpn. J. Appl. Phys.* **10** (1971) 1691.

[37] T.O. Sedgwick and J.A. Aboaf, *IEEE Trans. Electron Dev.* **15** (1968) 1015.

[38] S. Iwauchi and T. Tanaka, *Jpn. J. Appl. Phys.* **10** (1971) 260.

[39] Z. Benamara and B. Gruzza, *Mater. Chem. Phys.* **39** (1994) 85.

[40] R.P.H. Chang and A.T. Fiory, *Appl. Phys. Lett.* **49** (1986) 1534.

[41] C.O. Chui, H. Kim, P.C. McIntyre and K.C. Saraswat, *IEEE Electron Dev. Lett.* **25** (2004) 274.

[42] C.O. Chui, F. Ito and K.C. Saraswat, *IEEE Electron Dev. Lett.* **25** (2004) 613.

[43] A. Uchiyama, H. Fukuda, T. Hayashi, T. Iwabuchi and S. Ohno, *IEEE IEDM Tech. Dig.* (1990) 425.

[44] M.M. Moslehi, C.J. Han, K.C. Saraswat, C. R. Helms and S. Shatas, *J. Electrochem. Soc.* **132** (1985) 2189.

[45] E.E. Crisman, Y.M. Ercil, J.J. Loferski and P.J. Stiles, *J. Electrochem. Soc.* **129** (1982) 1845.

[46] L.M. Terman, *Solid-State Electr.* **5** (1962) 285.

[47] E.H. Nicollian and J.R. Brews, *MOS (Metal Oxide Semiconductor) Physics and Technology*, Wiley, New York (2003).

[48] A. Dimoulas, G. Vellianitis, G. Mavrou, E.K. Evangelou and A. Sotiropoulos, *Appl. Phys. Lett.* **86** (2005) 223507.

[49] R. Castagné and A. Vapaille, *Surf. Sci.* **28** (1971) 157.

[50] V.V. Afanas'ev, Y.G. Fedorenko and A. Stesmans, *Appl. Phys. Lett.* **87** (2005) 032107.

[51] M. Houssa, T. Conard, J. Van Steenbergen, G. Nicholas, G. Mavrou, Y. Panayiotatos, A. Dimoulas, M. Meuris, M. Caymax and M.M. Heyns, *Proc. of Electrochem. Soc.* (2005).

[52] O. Faynot, S. Cristoloveanu, A.J. Auberton-Hervé and C. Raynaud, *IEEE Trans. Electron Dev.* **42** (1995) 713.

[53] T. Maeda, K. Ikeda, S. Nakaharai, T. Tezuka, N. Sugiyama, Y. Moriyama and S. Takagi, *IEEE Electron Dev. Lett.* **26** (2005) 102.

[54] J.W. Park, W. Xiong and J.P. Colinge, *Proc. IEEE Int. SOI. Conf.* (2003) 65.

[55] S. Takagi, *VLSI Symp. Tech. Dig.* (2003) 115.

[56] G.D. Wilk, R.M. Wallace and J.M. Anthony, *J. Appl. Phys.* **89** (2001) 5243.

[57] C.O. Chui, S. Ramanathan, B.B. Triplett, P.C. McIntyre and K.C. Saraswat, *IEEE Electron Dev. Lett.* **23** (2002) 473.

[58] W.P. Bai, N. Lu, J. Liu, A. Ramirez, D.L. Kwong, D. Wristers, A. Ritenour, L. Lee and D. Antoniadis, *VLSI Symp. Tech. Dig.* (2003) 121.

[59] N. Wu, Q. Zhang, C. Zhu, C.C. Yeo, S.J. Whang, D.S.H. Chan, M.F. Li, B.J. Cho, A. Chin, D.-L. Kwong, A.Y. Du, C.H. Tung and N. Balasubramanian, *Appl. Phys. Lett.* **84** (2004) 3741.

[60] K. Kita, K. Kyuno and A. Toriumi, *Appl. Phys. Lett.* **85** (2004) 52.

[61] S. Van Elshocht, B. Brijs, M. Caymax, T. Conard, T. Chiarella, S. De Gendt, B. De Jaeger, S. Kubicek, M. Meuris, B. Onsia, O. Richard, I. Teerlinck, J. Van Steenbergen, C. Zhao and M. Heyns, *Appl. Phys. Lett.* **85** (2004) 3824.

[62] N. Wu, Q. Zhang, C. Zhu, D.S. H. Chan, M. Li, N. Balasubramanian, A. Chin and D.-L. Kwong, *Appl. Phys. Lett.* **85** (2004) 4127.

[63] J.J.-H. Chen, N.A. Bojarezuk Jr., H. Shang, M. Copel, J. Hannon, J. Karasinski, E. Preisler, S.K. Banerjee and S. Guha, *IEEE Trans. Electron Dev.* **51** (2004) 1441.

[64] A. Dimoulas, G. Mavrou, G. Vellianitis, E. Evangelou, N. Boukos, M. Houssa and M. Caymax, *Appl. Phys. Lett.* **86** (2005) 032908.

[65] F. Gao, S.J. Lee, J.S. Pan, L.J. Tang and D.-L. Kwong, *Appl. Phys. Lett.* **86** (2005) 113501.

[66] A. Delabie, R.L. Puurunen, B. Brijs, M. Caymax, T. Conard, B. Onsia, O. Richard, W. Vandervorst, C. Zhao, M.M. Heyns, M. Meuris, M.M. Viitanen, H.H. Brongersma, M. de Ridder, L.V. Goncharova, E. Garfunkel, T. Gustafsson and W. Tsai, *J. Appl. Phys.* **97** (2005) 064104.

[67] Y. Kamata, Y. Kamimuta, T. Ino and A. Nishiyama, *Jpn. J. Appl. Phys.* **44** (2005) 2323.

[68] E.M. Levin, H.F. McMurdie, H.P. Hallm, M.K. Reser and H. Insley, *Phase Diagrams for Ceramists,* American Ceramic Society, Columbus, Ohio (1956).

[69] H. Kim, PhD dissertation, Stanford University (2004).

[70] C.O. Chui, D.-I. Lee, A.A. Singh, P.A. Pianetta and K.C. Saraswat, *J. Appl. Phys.* **97** (2005) 113518.

[71] S.M. Sze, *Physics of Semiconductor Devices,* 2nd edition, Wiley, New York (1981).

[72] J. Tersoff, *Phys. Rev. B* **30** (1984) 4874.

[73] J. Robertson, *J. Vac. Sci. Technol. B* **18** (2000) 1785.

[74] B.B. Triplett, *Silicon Symposium VII*, Electrochemical Society (1994).

[75] S. Rajagopalan, U. Mitra, S. Pan, K. Gupta, C.M. Lin, G. Sery, S. Mittal, K. Hasserjian, W.J. Lo and G. Neubauer, *IEEE IRPS Proc.* (1993) 28.

[76] A.T.S. Wee, C.H.A. Huan, P.S.P. Thong and K.L. Tan, *Corros. Sci.* **36** (1994) 9.

[77] K.J. Yang and C. Hu, *IEEE Trans. Electron Dev.* **46** (1999) 1500.

[78] H. Kim, C.O. Chui, K.C. Saraswat and P.C. McIntyre, *Appl. Phys. Lett.* **83** (2003) 2647.

[79] H. Kim, P.C. McIntyre, C.O. Chui, K.C. Saraswat and M.-H. Cho, *Appl. Phys. Lett.* **85** (2004) 2902.

[80] S.-H. Lo, D.A. Buchanan, Y. Taur and W. Wang, *IEEE Electron Dev. Lett.* **18** (1997) 209.

[81] A.C. Diebold, D. Venables, Y. Chabal, D. Muller, M. Weldon and E. Garfunkel, *Mater. Sci. Semicond. Process* **2** (1999) 103.

[82] S. Ramanathan, G.D. Wilk, D.A. Muller, C.-M. Park and P.C. McIntyre, *Appl. Phys. Lett.* **79** (2001) 2621.

[83] S. Ramanathan, D.A. Muller, G.D. Wilk, C.M. Park, and P.C. McIntyre, *Appl. Phys. Lett.* **79** (2001) 3311.

[84] C.M. Perkins, B.B. Triplett, P.C. McIntyre, K.C. Saraswat, S. Haukka and M. Tuominen, *Appl. Phys. Lett.* **78** (2001) 2357.

[85] J.J. Yeh and I. Lindau, *At. Data Nucl. Data Tables* **32** (1985) 1.

[86] NIST Electron Inelastic-Mean-Free-Path Database 71 (Ver. 1.1), National Institute of Standards and Technology (2000).

[87] A. Herrera-Gómez, F.S. Aguirre-Tostado, Y. Sun, P. Pianetta, Z. Yu, D. Marshall, R. Droopad and W.E. Spicer, *J. Appl. Phys.* **90** (2001) 6070.

[88] X. Yang, R. Cao, J. Terry and P. Pianetta, *J. Vac. Sci. Technol. B* **10** (1992) 2013.

[89] D. Schmeisser, R.D. Schnell, A. Bogen, F.J. Himpsel, D. Rieger, G. Landgren and J.F. Morar, *Surf. Sci.* **172** (1986) 455.

[90] D. Chi, C.O. Chui, K.C. Saraswat, B.B. Triplett and P.C. McIntyre, *J. Appl. Phys.* **96** (2004) 813.

[91] S. Miyazaki, *J. Vac. Sci. Technol. B* **19** (2001) 2212.

[92] S. Sayan, E. Garfunkel and S. Suzer, *Appl. Phys. Lett.* **80** (2002) 2135.

[93] S.J. Wang, A.C.H. Huan, Y.L. Foo, J.W. Chai, J.S. Pan, Q. Li, Y.F. Dong, Y.P. Feng, and C.K. Ong, *Appl. Phys. Lett.* **85** (2004) 4418.

[94] K. Prabhakaran, F. Maeda, Y. Watanabe, and T. Ogino, *Appl. Phys. Lett.* **76** (2000) 2244.

[95] M. Houssa, M. Tuominen, M. Naili, V. Afanas'ev, A. Stesmans, S. Haukka and M.M. Heyns, *J. Appl. Phys.* **87** (2000) 8615.

[96] C.H. Huang, M.Y. Yang, A. Chin, W.J. Chen, C.X. Zhu, B.J. Cho, M.-F. Li, D.L. Kwong, *VLSI Symp. Tech. Dig.* (2003) 119.

[97] W.C. Dunlap, *Phys. Rev.* **94** (1954) 1531.

[98] J.D. Plummer, M.D. Deal, and P.B. Griffin, *Silicon VLSI Technology*, Prentice Hall, New Jersey (2000).

[99] F.A. Trumbore, *Bell Syst. Tech. J.* **39** (1960) 205.

[100] B.L. Sharma, *Defect Diffus.* Forum **70–71** (1990) 1.

[101] S.M. Sze and J.C. Irvin, *Solid-State Electron.* **11** (1968) 599.

[102] V.M. Gusev, M.I. Guseva, E.S. Ionova, A.N. Mansurova and C.V. Starinin, *Phys. Status Solidi A* **21** (1974) 413.

[103] M. Metzger, Z. Zhang, B. Schmiedt and H. Ryssel, *Proc. 4th International Conference on Ion Implantation: Equipment and Techniques* (1982) 458.

[104] K.S. Jones and E.E. Haller, *J. Appl. Phys.* **61** (1987) 2469.

[105] S. Uppal, A.F.W. Willoughby, J.M. Bonar, A.G.R. Evans, N.E.B. Cowern, R. Morris and M.G. Dowsett, *J. Appl. Phys.* **90** (2001) 4293.

[106] A. Axmann, M. Schulz and C.R. Fritzsche, *Appl. Phys.* **12** (1977) 173.

[107] K. Benourhazi and J.P. Ponpon, *Nucl. Instrum. Methods Phys. Res. B* **71** (1992) 406.

[108] T. Ahlgren, J. Likonen, S. Lehto, E. Vainonen-Ahlgren and J. Keinonen, *AIP Conf. Proc.* **576** (2001) 887.

[109] S.V. Hattangady, G.G. Fountain, E.H. Nicollian and R.J. Markunas, *J. Appl. Phys.* **63** (1988) 68.

[110] Y.S. Suh, M.S. Carroll, R.A. Levy, G. Bisognin, D. De Salvador, M.A. Sahiner and C.A. King, *IEEE Trans. Electron Dev.* **52** (2005) 2416.

[111] B. Smith, *Ion Implantation Range Data for Si and Ge Device Technologies*, Harwell, Research Studies, Oregon (1977).

[112] B.L. Crowder and F.F. Morehead, *Appl. Phys. Lett.* **14** (1969) 313.

[113] C.O. Chui, K. Gopalakrishnan, P.B. Griffin, J.D. Plummer and K.C. Saraswat, *Appl. Phys. Lett.* **83** (2003) 3275.

[114] G.Z. Pan, K.N. Tu and A. Prussin, *J. Appl. Phys.* **81** (1997) 78.

[115] Y. Takamura, P.B. Griffin and J.D. Plummer, *J. Appl. Phys.* **92** (2002) 235.

[116] J. Schmitz, M. van Gestel, P.A. Stolk, Y.V. Ponomarev, F. Roozeboom, J.G.M. van Berkum, P.C. Zalm and P.H. Woerlee, *IEEE IEDM Tech. Dig.* (1998) 1009.

[117] J.M. Hergenrother, S.-H. Oh, T. Nigam, D. Monroe, F.P. Klemens, and A. Kornblit, *Solid-State Electr.* **46** (2002) 939.

[118] C.O. Chui, H. Kim, P.C. McIntyre and K.C. Saraswat, *IEEE IEDM Tech. Dig.* (2003) 437.

[119] P.M. Fahey, P.B. Griffin, and J.D. Plummer, *Rev. Mod. Phys.* **61** (1989) 289.

[120] S.-L. Hsu, C.-H. Chien, M.-J. Yang, R.-H. Huang, C.-C. Leu, S.-W. Shen and T.-H. Yang, *Appl. Phys. Lett.* **86** (2005) 251906.

[121] J.Y. Spann, R.A. Anderson, T.J. Thornton, G. Harris S.G. Thomas, C. Tracy, *IEEE Electron Dev. Lett.* **26** (2005) 151.

CHAPTER 11

Advanced Germanium MOS Devices

Chi On Chui and Krishna C. Saraswat

11.1. Introduction

This Chapter provides an overview on numerous advanced germanium metal-oxide-semiconductor (MOS) devices with a special focus on field-effect transistor (FET) demonstrations reported since the year 2002. Section 11.2 discusses the urgency and justification to consider a high mobility channel for MOSFET applications and the candidate materials. In Section 11.3, the p- and n-channel MOSFETs fabricated on relaxed bulk germanium wafers with different kinds of gate dielectric will be analyzed. Section 11.4 presents the experimental results on germanium MOSFETs with a strained epitaxial channel demonstrated in either a surface or buried channel architecture. The next three sections review the characteristics and advantages of three new types of germanium MOSFETs – germanium-on-insulator (GOI) MOSFETs, Schottky source-drain germanium MOSFETs, and germanium nanowire MOSFETs. Finally some general conclusions are drawn in Section 11.8, which also projects the challenges and opportunities for the next generation research and development on Ge MOS devices.

11.2. The Quest for High Mobility MOSFET Channel

For over three decades, there has been a quadrupling of transistor density and a doubling of electrical performance every 2–3 years. Silicon transistor technology, in particular complementary-metal-oxide-semiconductor (CMOS) has played a pivotal role in this. It is believed that continued scaling will take the industry down to the 32 nm technology node, at the limit of the "long-term" range of the International Technology Roadmap for Semiconductors (ITRS) [1].

Germanium-Based Technologies: From Materials to Devices. C. Claeys and E. Simoen (Editors)

However, it is also well accepted that this long-term range of the 65–32 nm nodes remains solidly in the "no-known solution" category. The difficulty in scaling the conventional MOSFET makes it prudent to search for alternative device architectures. This will require new structural, material and fabrication technology solutions that are generally compatible with current and forecasted installed Si manufacturing.

11.2.1. Challenges to scaling conventional CMOS

It is well accepted that below the 65 nm node although the conventional bulk CMOS can be scaled, however, without appreciable performance gains. There are several technical issues that make proper device scaling increasingly difficult, as shown schematically in Figure 11.1. The drive current is far from that required at the power supply voltage of the sub-65 nm nodes. Suppressing the short channel effect (SCE) and the subthreshold leakage current requires tight control of intricate vertical and lateral channel doping profiles involving heavy doping, very shallow source-drain junctions, and ultrathin gate dielectrics. Some of these approaches directly conflict with the goal of obtaining high carrier mobility, steep subthreshold slope, low series resistance, high drive current, and low leakage current at low operating voltage.

In a long channel bulk MOSFET the physics of transistor operation can be partitioned into two independent portions, i.e. gate-controlled charge formation in the channel, and drain-controlled charge transport. The threshold voltage (V_T), at which the device turns on, is dependent only on the gate voltage and is independent of the drain voltage. Application of the gate voltage lowers a potential barrier near the source and allows electrons to flow from source to drain. Under normal transistor operation fundamental thermodynamics constrains

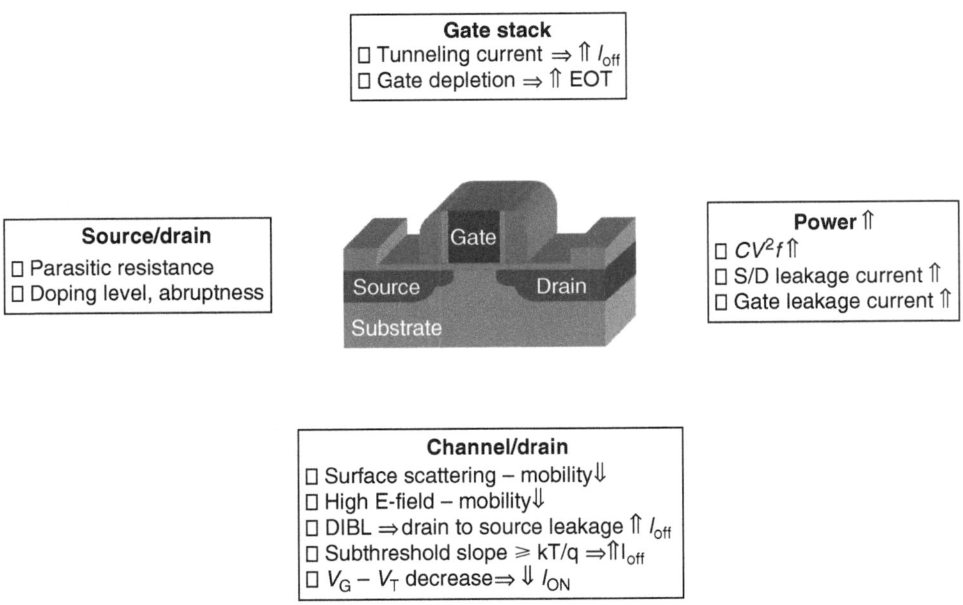

Figure 11.1. Physical limits in scaling the conventional bulk Si MOSFET.

the subthreshold swing to be greater than 60 mV/decade at room temperature. Degraded two-dimensional (2-D) electrostatics at short gate lengths worsen (increase) this value – leading to higher off-state leakage current for the same V_T.

In reality, the potential barrier at the source is controlled by the gate as well as the drain through their respective capacitive coupling to that point. As the gate length is reduced, the drain influence becomes stronger. As a result, it becomes harder for the gate to control the source barrier and turn off the channel. The 2-D effects are manifested in various ways:

(i) V_T reduction with shrinking gate length (V_T roll-off),

(ii) V_T reduction with increasing drain voltage (drain induced barrier lowering – DIBL),

(iii) Degradation in subthreshold swing.

Collectively, these phenomena known as SCE tend to increase the off-state static leakage power. Thus far, device designers have tried to suppress SCE in short gate length devices by a number of methods:

(a) Gate oxide thickness reduction to improve the gate control over the channel,

(b) Source–drain junction depth lowering (especially near the gate edge, where the source–drain regions are called "extensions") to reduce the drain coupling to the source barrier,

(c) Channel doping increase to terminate the electric field lines which originate from the drain and propagate towards the source. In modern bulk MOSFETs, the channel doping is tailored to have complicated vertical and lateral profiles so as to minimize the impact of gate length variations on the SCE.

Each of these approaches comes at a cost which either degrades transistor performance (speed) or introduces a new static leakage mechanism:

(a) As the gate oxide gets very thin, quantum mechanical tunneling allows a leakage gate current to flow. In the direct-tunneling regime, encountered for oxides thinner than about 3 nm, the gate leakage current increases dramatically ($\sim 3\times$ for every 0.1 nm of thickness reduction). The gate leakage can increase standby power as well as compromise proper logic gate operation [2]. Many people have proposed replacing the silicon dioxide (SiO_2) with higher permittivity (high-κ) gate dielectrics [3] such as zirconia (ZrO_2) or hafnia (HfO_2). These enable high gate capacitance with physically thick insulators through which tunneling is low. However, the introduction of such new materials without the accompanying degradation of mobility and reliability is very challenging and remains an area of intensive ongoing research.

(b) As the source–drain junction depths get shallow, their doping must be increased so as to keep the sheet resistance constant. Solid solubility of dopants puts an upper limit ($\sim 10^{20}\,cm^{-3}$) on the doping density. Therefore, further reduction in junction depth causes an increase in the series resistance encountered in accessing. This degrades the overall transistor performance. Also, from a technological point of view, it becomes difficult to form ultra-shallow junctions that remain abrupt after the annealing steps needed to activate the dopants and achieve low resistivity [4].

(c) As the doping density in the channel is increased for SCE suppression, the carrier mobility is degraded due to increased scattering from the ionized dopant atoms. Besides, the subthreshold swing gets worse due to higher depletion capacitance that "steals" away part of the gate voltage from the surface potential. For very high channel doping near the source–drain extensions, another component of static leakage, band-to-band tunneling, becomes important. Finally, as the channel volume reduces in extremely scaled transistors, the random placement of discrete dopant atoms causes stochastic inter-device variations [5].

As a result of these (and other) problems, it is becoming clear that new materials and/or structures will be needed to supplement or even replace the conventional bulk MOSFET in future technology generations.

The enhanced speed and complexity of integrated circuit (IC) chips has been accompanied by an increase in power dissipation [6]. Figure 11.2 depicts the evolution of power density as the gate length is scaled. The active power arises due to the dissipative switching of charge between the transistor gates and supply/ground terminals during logic operations. The subthreshold power, also known as static or standby power, is dissipated even in the absence of any switching operation. It arises due to the fact that the MOS transistor is not a perfect switch – there is some leakage current that flows through it in the off-state.

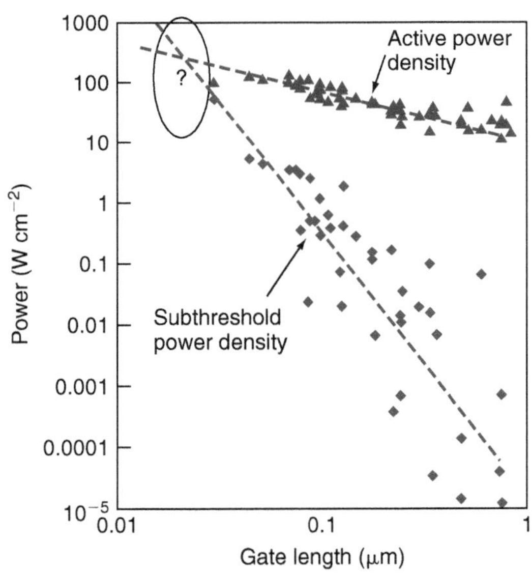

Figure 11.2. Active and subthreshold (standby) power density trends plotted from industry data. The extrapolations indicate a cross-over below 20 nm gate length. As devices scale towards that point, it is questionable if the traditional approaches and reasons for scaling will still be valid (after Nowak *et al.* [6]).

While the active power density has steadily increased with gate length scaling, the static power density has grown at a much faster rate [6]. The latter was a relatively insignificant component of power just a few generations back, but it is now comparable in magnitude to the active power. Management and suppression of static power is one of the major challenges to continued gate length reduction for higher switching speed. Traditional MOSFET scaling has begun to face impediments of both a fundamental as well as practical nature. It is now widely accepted that novel (i.e. non-classical) transistors will be needed to prolong device scaling with commensurate improvements in performance. The double-gate (DG) FET in conjunction with a high mobility channel and high-κ gate dielectric is a promising device structure that can potentially replace conventional transistors in future technology generations.

11.2.2. High mobility channel justification and selection

The saturation of bulk Si MOSFET drive current (I_{Dsat}) upon dimension shrinkage is limiting the prospect of future scaling. To understand this saturation phenomenon, numerous theoretical and experimental analyses were carried out [7–9]. First of all, the I_{Dsat} (and transconductance) in very short-channel MOSFETs is believed to be limited by carrier injection from the source into the channel [7]. In other words, the source injection velocity (v_{src}) saturates during scaling and its limit is set by the thermal injection velocity (v_{inj}) [8]. Also, the carrier density at the top of the source to channel barrier is fixed by MOS electrostatics and the scattering in a short region near the beginning of the channel limits the I_{Dsat}. In deeply scaled MOSFETs, v_{src} was experimentally shown to be at most 40% of v_{inj} [9].

The lower effective mass and lower valley degeneracy of high mobility materials like Ge [10] could alleviate the problem by providing a higher v_{inj}, which translates into higher drive current and smaller gate delay:

$$I_{Dsat} = W \times Q_{inv} \times v_{inj} \tag{11.1}$$

$$\frac{C_{LOAD} V_{DD}}{I_{Dsat}} = \frac{L_{gate} \times V_{DD}}{(V_{DD} - V_T) \times v_{inj}} \tag{11.2}$$

where W is the channel width, Q_{inv} is the inversion charge, C_{LOAD} is the load capacitance, V_{DD} is the supply voltage, L_{gate} is the gate length.

Table 11.1 shows properties of several high mobility materials which are possible candidates for the channel of the nanoscale MOSFETs. Due to their small Γ-valley electron mass, III–V materials like gallium arsenide (GaAs), indium arsenide (InAs), and indium antimonide (InSb) are being investigated as high mobility channel materials for high performance n-channel MOSFETs [11]. Under ballistic conditions, the main advantage of a semiconductor with a small transport mass is its high v_{inj}. However, materials also have a

Table 11.1. Properties of high mobility semiconductors.

Property \ Material	Si	Ge	GaAs	InAs	InP	InSb
Electron mobility (cm^2/V-s)	1400	3900	8500	40,000	5,400	77,000
Hole mobility (cm^2/V-s)	450	1900	400	500	200	850
Bandgap (eV)	1.12	0.66	1.42	0.35	1.34	0.17
Lattice constant (Å)	5.431	5.658	5.653	6.058	5.869	6.749
Dielectric constant	11.7	16.2	12.9	15.2	12.5	16.8

very low density of states in the Γ-valley, which tends to greatly reduce the inversion charge and hence reduce drive current. Further, the very high mobility III–V materials like InAs and InSb have a much smaller direct band gap which gives rise to high band-to-band tunneling (BTBT) leakage. Materials like InAs and InSb have a high dielectric constant and hence are more prone to SCE. All the III–V materials have a severe problem in surface passivation and hence fabrication of a MOSFET is problematic. High performance high electron mobility transistors (HEMTs) and modulation-doped FETs (MODFETs) have been demonstrated in these materials but suffer from high gate leakage and hence are not too useful for conventional logic applications.

Belonging to the same group in the Periodic Table as Si, Ge offers several attractive physical properties over Si. In Ge, the lower electron transverse and light hole (and heavy) effective masses are primarily responsible, respectively, for the higher electron and hole drift mobility. This property is the most advantageous over Si for deeply scaled MOSFET applications as previously discussed regardless of the higher Si saturation velocity. The more symmetric electron and hole mobility in Ge would not only reduce the real estate of p-channel MOSFETs, but would permit more CMOS logic gates as well. Moreover, its smaller bandgap is more compliant with the supply voltage scaling as specified in ITRS [1]; at the same time, this also broadens the optical absorption spectrum to cover telecommunication wavelengths (1.3 and 1.55 μm) allowing optoelectronic integration [12] to enhance CMOS functionality. Furthermore, its lower melting point reflects a possibility to fabricate Ge MOSFETs with much lower thermal budget processes while relaxing some stringent thermal stability requirements in integrating novel materials like metal gate electrode and high-κ gate dielectric into advanced transistors.

11.3. Relaxed Bulk Channel Germanium MOSFETs

The first surface channel bulk Ge n-channel MOSFET demonstration in the literature could be dated back to 1965 with pyrolytically decomposed SiO_2 gate dielectric and antimony (Sb) out-diffused junctions [13]. In contrast, the first surface channel bulk Ge p-channel MOSFET was not realized until 1975 using chemical vapor deposited (CVD) SiO_2 gate dielectric and boron (B) ion implanted source and drain [14]. Subsequently, several Ge n-channel MOSFETs [15–20] and p-channel MOSFETs [18–24] were fabricated using various MOS gate dielectric and junction formation technologies. Even so, none of these technologies would be applicable for deeply scaled MOSFETs. In this section, we summarize the latest developments on surface channel bulk Ge MOSFETs with advanced gate stacks in the recent years.

11.3.1. P-channel MOSFETs

Since the common p-type dopants in Ge could be activated at temperatures as low as 400–500°C as discussed in Chapter 10, gate stacks with relatively lower thermal stability could be employed in the conventional self-aligned gate-first MOSFET fabrication process.

11.3.1.1. Germanium oxynitride gate dielectric

One of the gate stacks adopted on surface channel bulk Ge p-channel MOSFETs was aluminum (Al) gate electrode with SiO_2 on Ge oxynitride (GeO_xN_y) gate dielectric [25]. The starting substrates were (100) oriented n-type Ge wafers with a resistivity of ~0.3 Ω-cm. The surfaces were first wet cleaned and dried with nitrogen (N_2) prior to the formation of the SiO_2/Ge_2ON_2 gate dielectric layers. About ~3 nm of low temperature SiO_2 (LTO) was deposited on ~6 nm of Ge_2ON_2, which was formed by nitridation of a thermally grown Ge oxide. After the Al gate electrodes had been defined, B ions were implanted to form the self-aligned source–drain regions. After dopant activation at 400°C in N_2 ambient, titanium (Ti)/Al ohmic contacts were formed through the opened contact windows. The fabricated devices had a self-isolated ring transistor structure.

From the Ge p-channel MOSFETs with an $Al/SiO_2/Ge_2ON_2$ gate stack, an equivalent SiO_2 thickness (EOT) of ~8 nm was measured using the split capacitance–voltage (C–V) technique that showed very low gate leakage. Figure 11.3a shows the transfer characteristics at low drain bias, which revealed a subthreshold slope as low as 82 mV dec^{-1} (close to the ideal value of 69 mV dec^{-1}). The dielectric–substrate interface trap level density (D_{it}) was then extracted as $\sim 5 \times 10^{11} cm^{-2} eV^{-1}$. The effective mobility was also extracted versus the effective electric field (F_{eff}) from the Ge p-channel MOSFETs as shown in Figure 11.3b. At relatively high F_{eff}, the effective mobility extracted from the Ge device showed ~40% mobility enhancement over the Si control and a similar mobility enhancement over the Si universal hole mobility.

11.3.1.2. Zirconium-based gate dielectric

In order to allow much better EOT scalability, a platinum (Pt) gate electrode on ZrO_2 gate dielectric stack was proposed and integrated onto the Ge channel MOSFETs [26]. The starting substrates were (100) oriented n-type Ge wafers with a net background concentration of $\sim 10^{16} cm^{-3}$. The surfaces were either rinsed in deionized (DI) water or hydrofluoric acid (HF) vapor in an attempt to remove native oxides, followed by 3.5–5.0 nm ZrO_2 deposition using the ultraviolet ozone oxidation (UVO) technique described in Chapter 10. After the Pt gate electrodes had been defined, boron difluoride ions (BF_2^+) were implanted to form the self-aligned source–drain regions. After dopant activation at 400°C in N_2 ambient, Ti/Al ohmic contacts were formed through the opened contact windows. Finally, all the samples

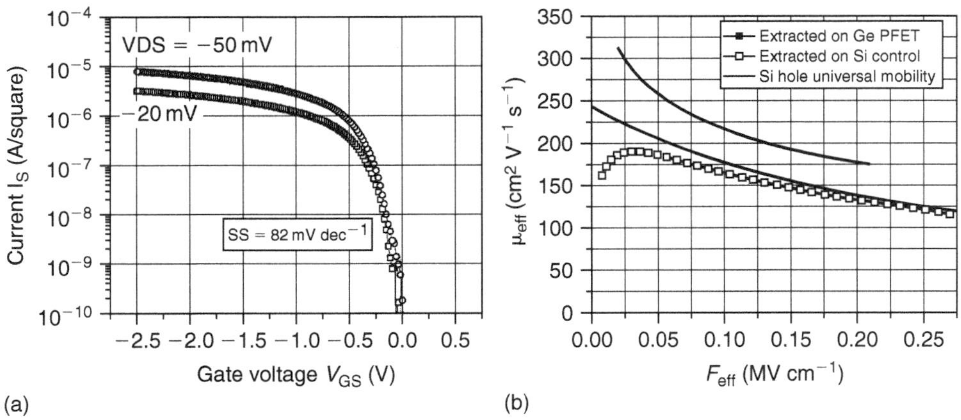

Figure 11.3. (a) Transfer characteristics and (b) effective mobility extracted on Ge p-channel MOSFETs with an Al/SiO_2/Ge_2ON_2 gate stack. Effective mobility extracted on a Si control device is also included for comparison (after Shang *et al.* [25]).

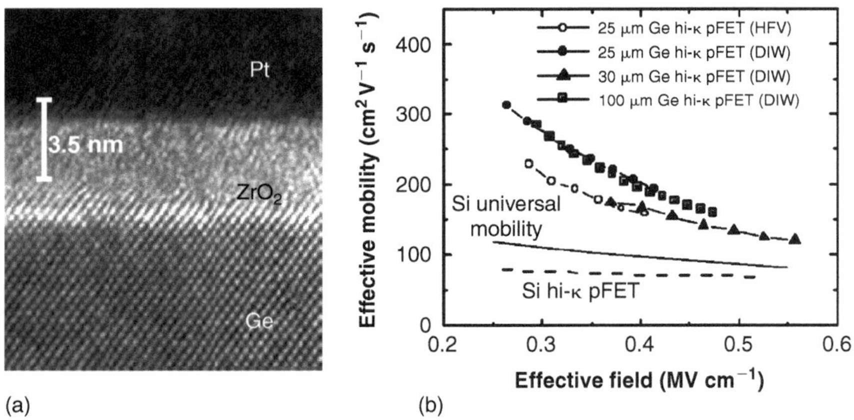

Figure 11.4. (a) Cross-sectional HR-TEM micrograph taken from a fully processed Pt/ZrO_2/Ge p-channel MOSFET with DI water rinsed Ge surface. (b) Effective hole mobility extracted from Pt/ZrO_2/Ge p-channel MOSFETs with either DI water rinsed or HF vapor etched Ge surfaces. Mobility from the Si universal model [27] and a high-κ/Si p-channel MOSFET [28] are also included.

were subjected to forming gas anneal at 300°C. The fabricated devices had a self-isolated ring transistor structure.

From the cross-sectional high-resolution transmission electron micrographs (HR-TEM) shown in Figure 11.4a, no apparent degradation of the metal gate electrode and high-κ gate dielectric stack could be observed after the entire MOSFET fabrication process. The EOT achieved was on the order of 1.0–1.5 nm and respectable on-state drive currents were obtained accompanied with a parasitic substrate bipolar leakage in the off-state. Since the effective

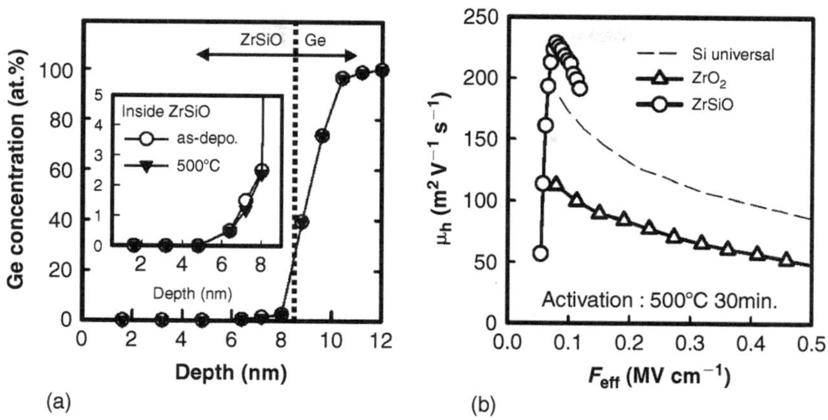

Figure 11.5. (a) HR-RBS depth profile of Ge atoms in the ZrSiO/Ge stacked structure before and after a 500°C anneal. (b) Effective hole mobility extracted from Mo/ZrSiO/Ge p-channel MOSFETs (after Kamata *et al.* [29]).

hole mobility extracted from a limited number of devices exhibited only a slight distribution (Figure 11.4b), no preferential Ge surface cleaning scheme could be identified. In general, the Pt/ZrO_2/Ge p-channel MOSFETs revealed ~60–100% enhancement in hole mobility over the Si universal mobility model [27] and about 3 times higher mobility than that of high-κ/Si p-channel MOSFETs [28], both at low E_{eff}.

Very recently, Zr silicate (ZrSiO) was also integrated onto the Ge p-channel MOSFETs [29] to completely suppress any substrate Ge diffusion into the high-κ gate dielectric layer. On cyclic dilute HF and DI water rinsed (100) oriented n-type Ge substrates with a resistivity of ~0.05 Ω-cm, ZrSiO and ZrO_2 films were deposited by sputtering. After the molybdenum gate electrodes had been defined, BF_2^+ ions were implanted and activated from 400°C to 600°C in N_2. The final devices were subjected to forming gas anneal at 350°C.

As revealed in the high-resolution Rutherford backscattering (HR-RBS) depth profile (Figure 11.5a), no detectable Ge out-diffusion occurred after a 500°C anneal in the ZrSiO/Ge stack. In the same study, substantial out-diffusion was otherwise observed in the ZrO_2/Ge stack. The fabricated Mo/ZrSiO/Ge p-channel MOSFETs exhibited well-behaved output and transfer characteristics with a subthreshold slope of 99 mV dec^{-1}. As extracted in Figure 11.5b, the Mo/ZrSiO/Ge transistor exhibited ~23% and ~100% enhancement in effective hole mobility respectively over the Si universal mobility model and the co-processed Mo/ZrO_2/Ge device at low F_{eff} of 0.1 MV cm^{-1}.

11.3.1.3. Hafnia gate dielectric

As the most extensively studied high-κ gate dielectric for Si MOSFETs, HfO_2 gate dielectric [30–32] were also investigated for Ge MOSFET applications. Among various surface passivations, the plasma phosphine (PH_3) treatment has produced the lowest EOT [32].

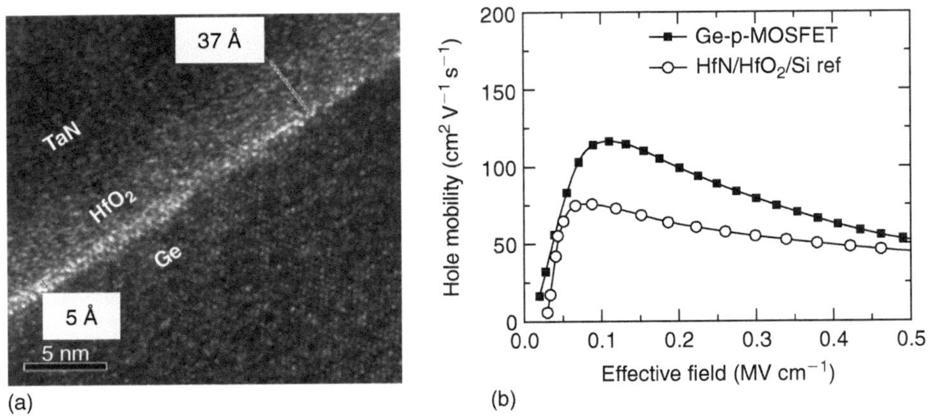

Figure 11.6. (a) Cross-sectional HR-TEM micrograph and (b) effective hole mobility extracted from a fully processed TaN/HfO_2/Ge p-channel MOSFET with plasma PH_3 treated Ge surface (after Whang *et al.* [32]).

The starting substrates were n-type Ge wafers with a resistivity of ~0.5 Ω-cm. The surfaces were first cleaned with dilute HF followed by plasma PH_3 treatment at 400°C. About 3.7 nm of HfO_2 was deposited by both metal-organic chemical vapor deposition (MOCVD) and atomic layer deposition (ALD). After the tantalum nitride (TaN) gate electrodes had been defined, B ions were implanted and activated at 400°C to form the self-aligned source–drain regions.

An EOT of 0.75 nm was extracted from standalone TaN/HfO_2/Ge capacitors with very low leakage. As shown in the cross-sectional HR-TEM in Figure 11.6a, only a very thin (~0.5 nm) interfacial layer was incurred after the plasma PH_3 treatment. Good thermal and electrical stability up to below 600°C had been demonstrated. The calculated dielectric–substrate interface D_{it} was $\sim 8 \times 10^{11}\,cm^{-2}eV^{-1}$ that could be further improved. The fabricated TaN/HfO_2/Ge MOSFETs exhibited well-behaved output and transfer characteristics with a subthreshold slope of 97 mV dec^{-1}. Finally, about 1.6 times increase in the peak hole mobility (Figure 11.6b) was achieved compared to the reported hafnium nitride (HfN)/ HfO_2/Si devices [33], yet it is still below the Si universal mobility.

11.3.2. n-channel MOSFETs

Using similar gate dielectrics as introduced in the last sub-section, several Ge n-channel MOSFETs have also been reported [32,34,35]. Since most of the common n-type dopants in Ge need to be activated in the temperature range of 500–600°C, alternative metal gate electrodes with higher thermal stability would be mandatory. For instance, the Al gate electrode was substituted with tungsten (W) on the SiO_2/Ge_2ON_2 gate dielectric layer [34]. Using an optimized n-type dopant activation thermal budget at 500°C, the fabricated W/ SiO_2/Ge_2ON_2/Ge n-channel MOSFETs demonstrated a minimized off-state leakage, on-to-off current ratio of $\sim 10^4$, and a subthreshold slope of 150 mV dec^{-1}. The extracted electron

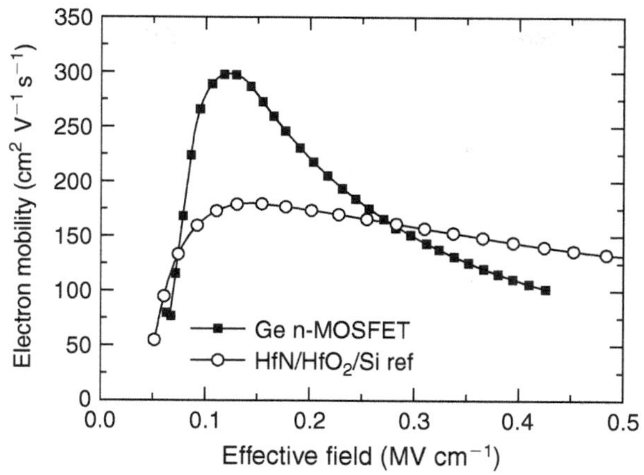

Figure 11.7. Effective electron mobility extracted from a fully processed TaN/HfO_2/Ge n-channel MOSFET with plasma PH_3 treated Ge surface (after Whang *et al.* [32]).

mobility was, however, much lower than the expected ~100 cm^2/V-s at E_{eff} of 0.25 MV/cm, possibly due to Coulomb scattering from the high fixed charge inside the gate dielectric and high source–drain resistance.

Using the same gate stack and surface passivation strategy described in Section 11.3.1.3, TaN/HfO_2/Ge n-channel MOSFETs on plasma PH_3 treated Ge were also fabricated [32]. The implanted phosphorous (P) ions were activated at 500°C, which was shown to guarantee an at least three orders of magnitude difference between the n^+–p diode forward and reverse currents. Good device integrity was demonstrated with a subthreshold slope of 89 mV dec^{-1}. Even though the extracted peak electron mobility (Figure 11.7) was 1.8 times higher than for the HfN/HfO_2/Si devices [33], the enhancement diminished quickly at higher F_{eff}. Besides, the overall effective electron mobility was substantially lower than the Si universal model.

In order not to challenge the gate stack thermal stability and incur implantation damages to the source–drain regions, a simple and novel self-aligned gate-last process was proposed and developed to fabricate Ge n-channel MOSFETs with metal gate electrode and high-κ gate dielectric [35]. The key idea of this process is to selectively place a solid dopant source layer above the MOSFET source–drain regions but not right above the channel. Instead of the conventional gate self-alignment by ion implantation, a source–drain self-registration is obtained by dopant out-diffusion from the solid sources.

The starting substrates were (100) oriented lightly doped p-type Ge wafers. Field isolation composed of an LTO/GeO_xN_y layer stack was first formed followed by the active area patterning. A p-well implant was then applied followed by a blanket phospho-silicate glass (PSG) deposition as the solid dopant source. The PSG layer right above the channel was removed before the source–drain formation by solid source diffusion. On the channel surface, rapid thermal nitridation (RTN) was then performed at 600°C in ammonia (NH_3)

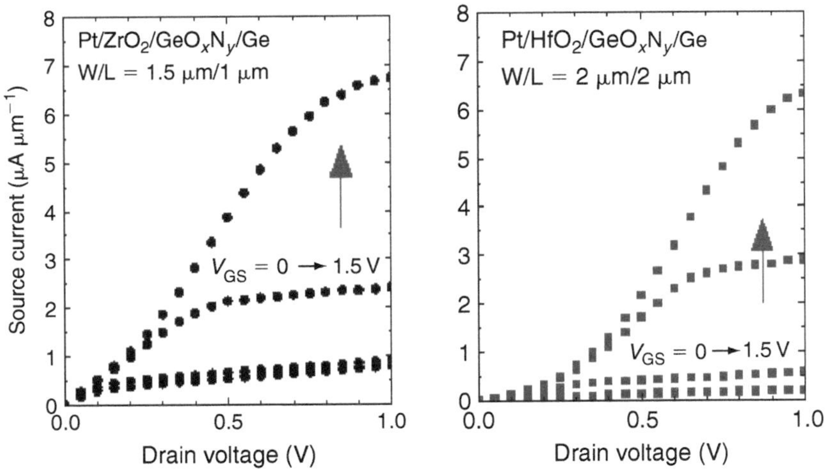

Figure 11.8. $Pt/ZrO_2/GeO_xN_y/Ge$ (left) and $Pt/HfO_2/GeO_xN_y/Ge$ (right) n-channel MOSFET output characteristics measured from 1 to 2 μm channel length device.

ambient for 1 min followed by ALD of ~3 nm ZrO_2 and HfO_2 gate dielectrics. The self-aligned gate-last process was finished with Pt gate electrode definition, back-end-of-line isolation, and standard metallization.

On the fabricated $Pt/ZrO_2/GeO_xN_y/Ge$ and $Pt/HfO_2/GeO_xN_y/Ge$ n-channel MOSFETs, reasonable p-well surface inversions were obtained from capacitance–voltage measurements, which also showed an EOT of ~2.2 nm. Decent rectifying diode characteristics were also confirmed on the MOSFET source–drain to substrate junctions made by solid-source diffusion. Although functional n-channel MOSFETs could be demonstrated in the current experiment (Figure 11.8), the device on-state drive currents were surprisingly low. The primary reason for the low performance was the loss of ohmicity due to dopant deactivation during the high temperature RTN step, which thermal budget should be lowered in any future MOSFET experimentations.

In summary, the achievable Ge n-channel MOSFET performance from recent reports has been disappointing. Although the root causes for the low performance remain controversial, the high dielectric–substrate interface D_{it} and un-optimized source–drain junction formation [36] are believed to play an important role. If both issues will later be proven to be fundamental, alternative MOSFET structures with buried channels (Section 11.4.2) and/or Schottky source–drains (Section 11.6) might remedy the situation.

11.4. Strained Epitaxial Channel Germanium MOSFETs

As a widely adopted technique in manufacturing the state-of-the-art Si MOSFETs, strain engineering has also been applied to boost the performance of Ge channel MOSFETs. Through the pseudomorphic growth of a Ge layer above the SiGe or Si substrate, biaxial compressive strain is induced along the Ge channel. According to a theoretical calculation

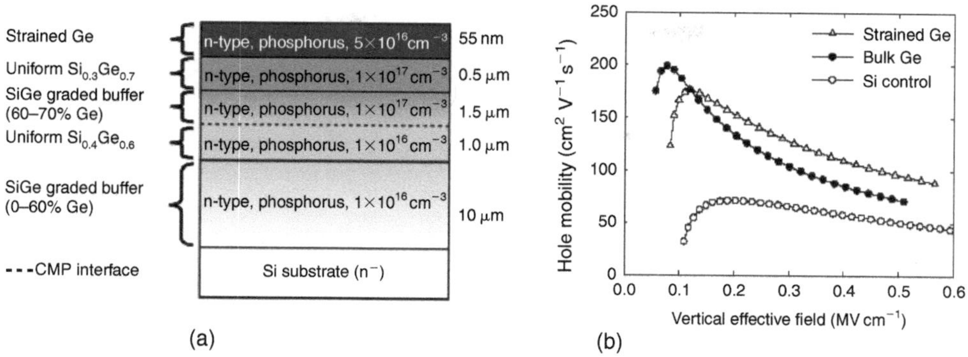

Figure 11.9. (a) Schematic of Ge/SiGe heterostructure grown using UHVCVD. (b) Effective hole mobility extracted from TaN/HfO_2/Ge strained surface p-channel MOSFETs (after Ritenour *et al.* [39]).

[37], the applied compressive strain would cause the degeneracy (heavy-hole and light-hole sub-bands) of the valence bands at the Γ-valley to split. Combining the reduction of the conductivity mass and inter-valley scattering, very high-hole mobility p-channel MOSFETs have been realized with a compressively strained pure Ge channel [38].

11.4.1. Surface strained epitaxial channel

In order to effectively suppress SCE for better gate length scalability, the strained Ge surface p-channel MOSFETs were fabricated using a TaN/HfO_2 gate stack [39]. The (100) oriented n-type strained Ge layers, with a net background concentration of $\sim 5 \times 10^{16}\,cm^{-3}$, were pseudomorphically grown on $Si_{0.3}Ge_{0.7}$ virtual substrates using ultrahigh vacuum CVD (UHVCVD). The details of the virtual substrate are shown in Figure 11.9a.

Prior to the 5 nm MOCVD HfO_2 gate dielectric deposition, the strained Ge surfaces were first cleaned with dilute HF followed by RTN at 500°C in NH_3 ambient for 2 min. After the TaN gate electrodes had been defined, BF_2^+ ions were implanted and activated at 450°C to form the self-aligned source–drain regions. While the compressive strain was partially relaxed by misfit dislocations at the hetero-interface, the strained Ge maintained approximately 0.8% elastic strain after the entire fabrication process. Well-tempered MOSFET characteristics could be obtained with a subthreshold slope of $90\,mV\,dec^{-1}$, on-to-off current ratio of $\sim 10^4$, and EOT of 1.6 nm. Compared to the Si control devices, the strained Ge devices showed an enhancement of about 2 times in the extracted effective hole mobility (Figure 11.9b), which solidly demonstrated the promise of this channel straining technique on Ge MOSFETs. An even larger enhancement could be expected by incorporating higher levels of strain and perfecting the dielectric–substrate interface.

11.4.2. Buried strained epitaxial channel

In the absence of a reasonable dielectric–substrate interface solution, a thin Si cap was pseudomorphically grown above the strained Ge layer to implement the buried channel MOSFET architecture. All recent strained Ge buried channel MOSFETs were grown either

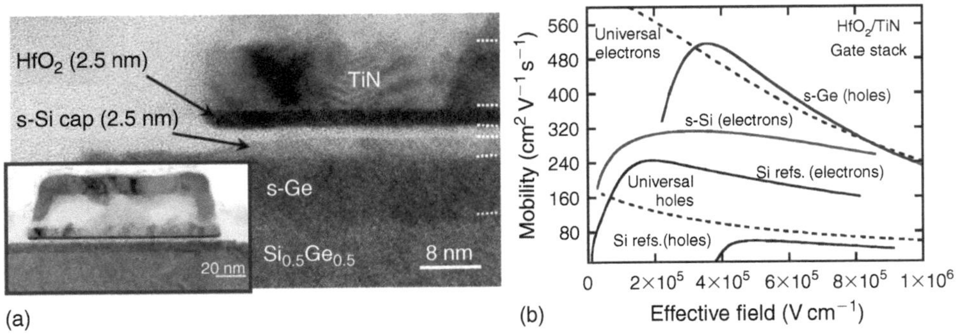

Figure 11.10. (a) Cross-sectional TEM of a 0.2 μm gate length TiN/HfO_2/Si/Ge strained buried p-channel MOSFET. (b) Effective mobility extracted from strained Ge buried p-channel MOSFETs and strained Si surface n-channel MOSFETs (after Weber *et al.* [43]).

on a SiGe [40–43] or Si [44,45] substrate. About 6–8 times hole mobility enhancement had only been reached with either a thicker Si cap [40] or thick SiO_2 gate dielectric [40,41], which reduced to about 3 times with a thin cap and/or thin SiO_2 [42,44]. By incorporating metal gate electrode and high-κ gate dielectric onto a thin Si capped strained Ge quantum-well channel, a very substantial hole mobility improvement could be realized [43]. On the (100) oriented $Si_{0.5}Ge_{0.5}$ buffers, strained Ge epitaxial channels (5.4–8.5 nm) were grown at 350°C by reduced pressure CVD followed by the strained Si cap (2–5 nm) growth at 650°C. After the ALD of 3 nm HfO_2 gate dielectric, polysilicon/titanium nitride (TiN) gate electrode definition, source–drain activation (at 600°C), and nickel (Ni) salicidation were sequentially carried out as shown in Figure 11.10a.

In spite of the presence of a 2.1 nm Si cap, an inversion EOT of 2.1 nm could be achieved with the TiN/HfO_2/Si/Ge stack with very low leakage. Compared to the Si reference devices, the strained Ge buried p-channel MOSFETs exhibited a 6-fold enhancement in on-state drive current. When the linear transconductance (G_M) was normalized to the inversion capacitance, these Ge MOSFETs showed a 9 × improvement at low F_{eff}. Besides, the G_M measured at high F_{eff} was the same from either a 3 or 5 nm Si cap sample, which is indicative of no parasitic conduction along the Si cap owing to the large valence band offset. The effective hole mobility extracted at high F_{eff} (Figure 11.10b) was one of the best ever reported, demonstrating the high quality of the channel stack that did not suffer from any significant strain relaxation after the entire sub-700°C process. Scientifically, such spectacular improvements are believed to be originated from the strain induced hole conductivity effective mass reduction and carrier confinement within the Ge quantum-well for hole transportation.

On the other hand, there has been only one demonstration on strained Ge buried n-channel MOSFETs with metal gate electrode and high-κ gate dielectric [45]. Using the UHVCVD technique, a 44 nm Si buffer was first epitaxially grown on a (100) oriented p-type Si wafer. The strained Ge layer (3.4 nm) was then pseudomorphically grown on top followed by the strained Si cap (0.7–2.3 nm) growth (Figure 11.11a). Hafnium aluminate (HfAlO) gate dielectric was deposited at 450°C and underwent a post deposition anneal at 450°C in N_2 ambient for 1 min. After the TaN gate electrodes were defined, the source–drain junctions

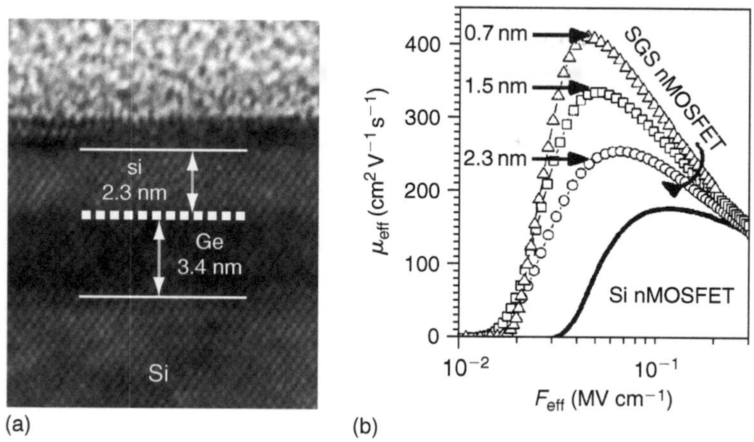

Figure 11.11. (a) Cross-sectional HR-TEM micrograph of a pseudomorphically grown strained Si cap on strained Ge layer on Si substrate. (b) Effective electron mobility extracted from fully processed TaN/HfAlO/Si/Ge strained buried n-channel MOSFETs with varying Si cap thickness (after Yeo *et al.* [45]).

were subsequently formed using a method of amorphization and solid-phase epitaxy re-growth (SPER) at low temperature. Arsenic ions were implanted and then annealed at 550°C in N_2 ambient for 10 min for the SPER. Finally, standard Al contact metallization and forming gas anneal were carried out.

Reasonably well-behaved TaN/HfAlO/Si/Ge MOSFET characteristics could be obtained with a subthreshold slope of 95–110 mV dec^{-1} and EOT of 2.8–3.1 nm. Compared to the bulk Si control, the strained Ge buried n-channel MOSFETs exhibited 19–48% and 40–120% improvement respectively in on-state drive current and peak electron mobility (Figure 11.11b). This is a remarkable yet inspiring result as the biaxial compressive strain degrades the in-plane electron mobility in Ge substantially [37]. Moreover, the carriers (electrons) traveling along the strained Ge n-channel would not be fundamentally confined due to the absence of a quantum-well for electrons. One highly conceivable explanation for the mobility enhancement is thus the majority electron conduction along the thin Si cap, which is likely to possess tensile strain and hence raises mobility. It was shown, however, that the electron mobility kept increasing with decreasing Si cap thickness down to 0.7 nm, a phenomenon that suggested most electrons did travel along the strained Ge n-channel and lead to the high mobility. By putting all the facts together, one can easily visualize the advantage of the buried channel architecture in decoupling the channel carriers from most scattering events with the dielectric interface to resume the intrinsic mobility.

11.5. Germanium-on-Insulator MOSFETs

Owing to the small bandgap in Ge, the high MOSFET off-state (junction) leakage significantly challenges the future device scaling. The GOI MOSFET architecture has therefore been proposed to overcome this issue [46]. Using a thin body GOI channel, a much improved MOSFET electrostatic, reduced junction leakage, and reduced parasitic capacitance could be guaranteed to boost the performance of the CMOS circuits. Particularly, if an ultra-thin GOI layer with very low doping concentration is employed for fully-depleted MOSFET operation, an even higher on-state drive current would be expected thanks to a reduced Coulomb scattering with the mostly eliminated channel depletion charges. In addition, higher electron mobility could be obtained by decreasing the GOI thickness to less than 3 nm [10] as a result of an increased occupancy of the lowest energy sub-band up to 100%, which possesses the lowest effective mass. Taking into account the self-heating effect commonly encountered in thin body devices, the GOI MOSFETs have also been shown to provide at least 30% performance advantage over similar silicon-on-insulator (SOI) counterparts [47] even though the bulk Ge thermal conductivity is only 40% of that in bulk Si.

As of today, there have been four major techniques, as discussed in Chapter 1, to prepare GOI substrates for MOSFET demonstrations:

(a) Ge wafer bonding and etch-back [48,49]. The donor Ge wafer was first implanted with hydrogen ions followed by SiO_2 capping on both the donor Ge and host Si wafers. Oxygen (O_2) plasma enhanced SiO_2–SiO_2 bonding and "smart cut" donor wafer release were subsequently carried out. After an extended annealing and slight polishing, the GOI layer was further thinned down to about 1.6 μm.

(b) Ge solid-phase heteroepitaxy on epitaxial oxide [50]. Lattice matched insulating lanthanum yttria layer was first grown on hydrogen passivated Si wafers with (111) surface orientation in a molecular beam epitaxial (MBE) chamber. An amorphous and uniform Ge layer was then deposited onto the oxide layer and transformed to a 4 nm epitaxial layer at ~450°C.

(c) Ge liquid-phase heteroepitaxial re-growth [51]. A Ge film was first CVD deposited on silicon nitride or SiO_2 capped Si wafers, with seeding windows in the cap layer opened. The Ge film was then patterned into narrow fins and encapsulated by an LTO layer that formed many self-aligned micro-crucibles during the Ge rapid melt growth. Rapid thermal anneal (RTA) was applied to melt the Ge at 940°C. Upon cooling, the heteroepitaxy re-growth started from the Ge–Si interface and propagated along the Ge fins. The resultant GOI fin thickness could easily be controlled by the initial CVD step.

(d) Ge condensation by SiGe oxidation [46,52]. A $Si_{0.85}Ge_{0.15}$ layer was first epitaxially grown by UHVCVD on commercially available (100) oriented bonded SOI wafers. Thermal oxidation in O_2 ambient was then carried out at temperatures below the melting point of $Si_{0.85}Ge_{0.15}$. Since the Si atoms had been preferentially oxidized, the Ge fraction kept increasing within the remaining SiGe layer. An optional SiO_2 removal could be inserted that should be followed by another oxidation step. When almost all Si atoms got oxidized, the condensation process would be complete. A GOI thickness as thin as 7 nm has been demonstrated using this technique.

Among the four techniques, Ge condensation has been the most promising and generated the best devices. In addition, it carries an extra advantage of localized GOI layer formation

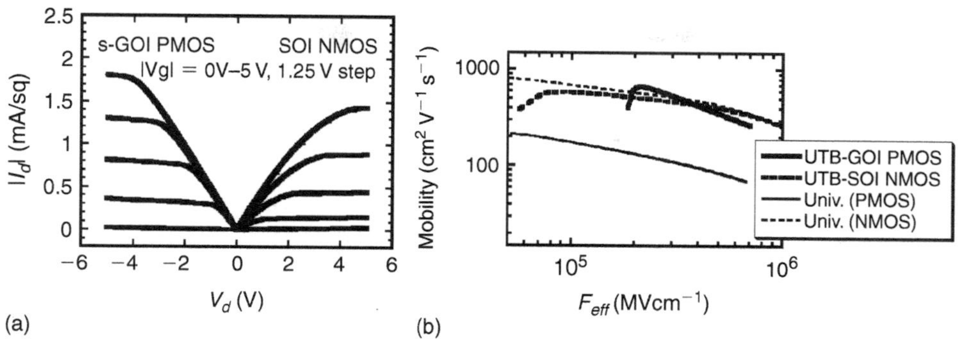

Figure 11.12. (a) Output characteristics and (b) effective mobility extracted from SOI n-channel MOSFETs and GOI p-channel MOSFETs simultaneously integrated onto the same wafer (after Tezuka *et al.* [52]).

to allow the simultaneous integration of GOI p-channel MOSFETs with SOI n-channel MOSFETs on the same substrate [52]. Starting with an SOI wafer, the $Si_{0.85}Ge_{0.15}$ layer was first selectively grown on the p-channel MOSFET areas followed by local Ge condensation. The SOI and GOI layer thickness was 10 nm apiece. On both SOI (n-channel) and GOI (p-channel) regions, gate dielectric thermal oxidation and polysilicon gate electrode deposition were subsequently carried out below 900°C. This relatively high temperature process did not degrade the Ge channel quality as indicated by the resultant compressive strain that remained in the GOI layer.

As extracted from the fully processed devices, the GOI p-channel MOSFETs revealed an on-state drive current comparable to (slightly higher than) that of the SOI n-channel MOSFETs (Figure 11.12(a)). The major reason for the p-channel device improvement was attributed to a 4 times hole mobility boost (Figure 11.12b) by the compressive strain incurred in the GOI layer. Besides, the electron mobility of the SOI devices retained its universal value. When the gate delay time of a CMOS inverter was calculated, a 30% reduction could be anticipated assuming these mobility enhancement factors. A 40% or more reduction would be expected by further process optimization.

11.6. Schottky Source-Drain Germanium MOSFETs

As discussed in Sections 11.4 and 11.3.2, the high thermal budget required to activate dopants in Ge challenges the stringent thermal stability requirement on novel gate stacks (with metal gate electrode and high-κ gate dielectric) during their integration. For instance, the commonly observed issues in Si systems including metal silicate phase separation [3] and metal and/or Si atoms inter-diffusion are also likely in the Ge system. The Schottky source–drain MOSFET structure has therefore been proposed to solve this problem [53,54]. This implantation-less approach not only eliminates the problem of low dopant solid solubility and fast dopant diffusion, but also guarantees low resistivity and atomically abrupt source–drain junctions. In order to achieve high MOSFET drive current, Schottky junctions with low barrier height between the source and the channel would be necessary [55]. Owing to the preferential Fermi level pinning near the valence band in most Ge Schottky junctions [56], the realization of Ge p-channel MOSFETs with Schottky source–drain is highly favorable.

The first successful demonstration is Ge p-channel MOSFETs with nickel germanide (NiGe) Schottky source–drain [53]. The starting substrates were (100) oriented n-type Ge wafers having a resistivity of 0.05 Ω-cm at room temperature. The surfaces were first cleaned in dilute nitric acid (HNO_3) solution and dipped in dilute HF solution. They were then nitrided in NH_3 at 600°C for 30 s followed by ~6 nm HfAlO deposition and *in-situ* N_2 anneal at 600°C for 1 min. Next, ~50 nm of HfN and ~100 nm of TaN were sequentially sputtered at room temperature as the metal gate electrode. After gate definition by lithography and reactive ion etching, the wafers were dipped in dilute HF solution to slightly undercut the TaN/HfN electrode stack and form spacers. Around 100 nm of Ni was sputtered thereafter and Ni germanidation reaction was performed by RTA at 600°C for 1 min. The un-reacted Ni was removed by wet etching in a mixture of ammonium hydroxide (NH_4OH), hydrogen peroxide (H_2O_2), and water. As a final note, each MOSFET was electrical isolated from others by an individual guard ring.

Figure 11.13a shows the current–voltage characteristics measured from the NiGe–Ge Schottky diode and its cross-sectional transmission electron microscopy (TEM) image. When the forward currents were fitted with the traditional thermionic emission model, Schottky barrier height (Φ_n), ideality factor (m), and series resistance (R_S) were extracted to be 0.50 eV, 1.49, and 110 Ω, respectively. The corresponding Φ_n for holes was found to be ~0.16 eV. The relatively high reverse leakage current and large m could be attributed to the existence of a Ge-rich NiGe layer at the germanide–substrate interface, which should be eliminated to

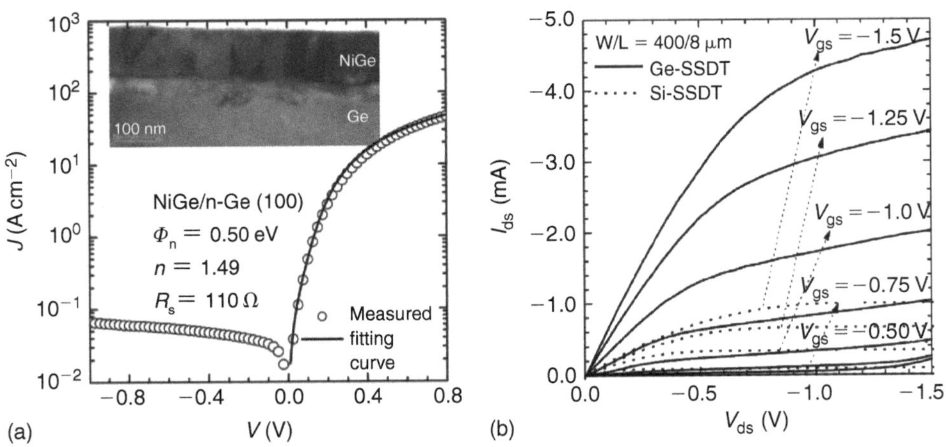

Figure 11.13. (a) Current-voltage characteristics of NiGe-Ge Schottky diode with the cross-sectional TEM image shown in the inset. (b) Output characteristics of a Ge p-channel MOSFET with NiGe Schottky source–drain and a control Si p-channel MOSFET with PtSi Schottky source-drain (after Zhu *et al.* [53]).

further improve the rectifying behavior. Figure 11.13b shows the output characteristics of Ge p-channel MOSFETs with NiGe Schottky source–drain. For comparison, control Si devices with platinum silicide (PtSi) Schottky source–drain were also fabricated using a similar process. After scaling to the same EOT, the Ge device exhibited ~5 × higher drive current than the Si control with similar V_T. Although the higher Ge hole mobility could be partially responsible for the drive current improvement, the major reason is attributed to the smaller source-to-channel Φ_n from the NiGe–Ge contact (versus the PtSi–Si contact in the Si control). On the other hand, the on-to-off current ratio of these Ge Schottky MOSFETs is substantially lower than that of the Si Schottky MOSFETs because of the lower barrier height to electrons in Ge owing to its smaller bandgap. In order to improve this ratio, advanced device structures including ultrathin body or DG FET could be employed.

The second demonstration is p-channel GOI MOSFETs with platinum germanide (PtGe) Schottky source–drain [54]. The starting substrates were (100) GOI substrates fabricated using the Ge condensation technique [46] with a net p-type background concentration of $\sim 2 \times 10^{15}\,\text{cm}^{-3}$ for accumulation mode operation. Mesa isolation of the GOI layer was first performed to define the active device area followed by source–drain Pt evaporation and liftoff. The Pt germanidation reaction was performed by annealing at 400°C for 30 min in forming gas. The resultant device structure had the Si substrate as the back gate, 210 nm buried SiO_2 as the MOS gate dielectric, 30 nm GOI layer as the channel, and 60 nm PtGe as the source–drain junctions.

From these GOI p-channel MOSFETs with PtGe Schottky source–drain, accumulation mode MOSFET operations with good linear and saturation behavior were obtained. The low parasitic series resistance, as evident from the good linearity in the output characteristics (omitted here), was due to the low PtGe sheet resistance and Φ_n. Also, well-behaved MOSFET action was observed with a respectable on-to-off current ratio of $\sim 10^4$ (Figure 11.14a). At high drain voltage, the increase in off-state leakage current was primarily owing to the injection of electrons from the drain-to-channel Schottky junction [54], which is commonly known as the ambipolar conduction in Schottky source–drain MOSFETs. Finally, the effective hole mobility

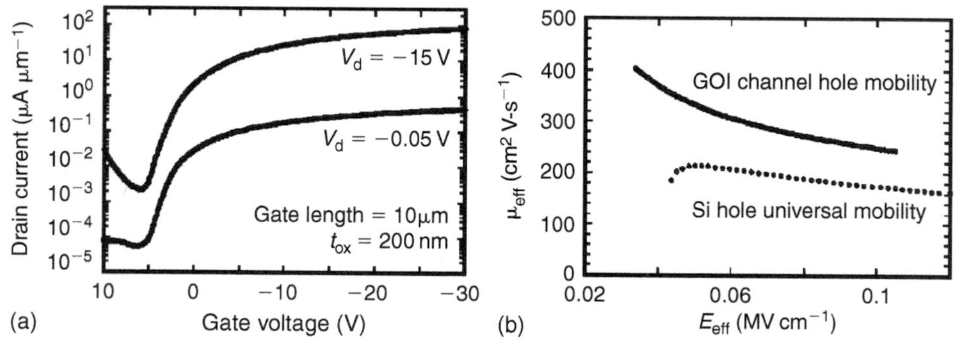

Figure 11.14. (a) Transfer characteristics and (b) extracted mobility from a GOI p-channel MOSFET with PtGe Schottky source–drain and *W*/*L* = 100/10 μm/μm (after Maeda *et al.* [54]).

was extracted from the Schottky GOI device showing approximately 40–50% enhancement against the Si universal hole mobility (Figure 11.14b).

11.7. Germanium Nanowire MOSFETs

Semiconductor nanowires and nanotubes have attracted considerable attention as one-dimensional (1-D) structures for fundamental studies and also as potential building blocks for nanoscale devices [57–59]. Having diameters of a few nanometer, the carriers traveling along these 1-D systems are quantum confined that leads to the observation of ballistic transport and conductance quantization. Given the superior hole mobility in Ge as discussed in Section 11.2.2, prototype Ge nanowire p-channel MOSFETs have been fabricated [60,61].

Ge nanowires are commonly synthesized by the vapor–liquid–solid approach in which vapor phase Ge is provided by a precursor in a CVD process. This approach is particularly attractive as it allows the precise control of reactants and dopants and thus the composition and doping of the nanowires. Besides homogeneous nanowires, this technique can also be extended to produce complex structures such as core-shell nanowires [62]. The Ge nanowires were first nucleated from gold nanoclusters at 320°C followed by their elongation at a reduced temperature of 285°C, all using germane (GeH_4) gas in hydrogen ambient [61]. The nanowire surfaces were then doped with either PH_3 or diborane without the presence of GeH_4 and finally coated with a thick undoped Ge shell.

The p- and n-channel MOSFETs were fabricated by synthesizing Ge nanowires on 60 nm ZrO_2 capped Si substrates with degenerate doping. Ti source–drain contacts were then applied to complete these back-gated Ge nanowire MOSFETs with gate length of 1.5 μm. As shown in Figure 11.15a, the p-channel MOSFETs revealed reasonable MOSFET characteristics with a subthreshold slope of 240 mV dec^{-1}, on-to-off current ratio of 10^5, and field-effect mobility of $\sim 115\,cm^2V^{-1}s^{-1}$. In contrast, the n-channel devices had inferior performances (Figure 11.15b) with a subthreshold slope of 675 mV dec^{-1}, on-to-off current

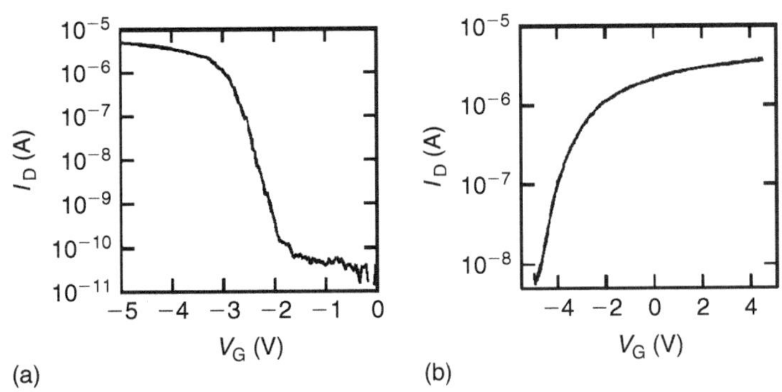

Figure 11.15. Transfer characteristics of Ge nanowire (a) p-channel and (b) n-channel MOSFETs with the Si back gate, ZrO_2 gate dielectric, Ti source–drain (after Greytak *et al.* [61]).

ratio of 10^3, and field-effect mobility of $\sim 20\,cm^2\,V\text{-}s^{-1}$. It is worth noting that the effective mobility extraction from nanowire or nanotube MOSFET remains challenging at this point owing to the inability to obtain any practical MOS capacitance–voltage characteristics. In any event, a high yield of above 85% on both device types was achieved showing great promises. Even though the Ge nanowire MOSFETs processed to date are far from optimized, the low thermal budget involved indeed opens up the possibility of 3-D integration of electronic devices on substrates with pre-fabricated IC.

11.8. Conclusions

During only the last few years, a wide variety of advanced Ge MOSFETs has been proposed and demonstrated to replace conventional Si transistors in future technology nodes. Surface channel bulk Ge MOSFETs with advanced gate stacks had been the initiator of all, which were then evolved into the strained epitaxial channel versions. While mobility enhanced Ge p-channel MOSFETs incorporating metal gate electrode and high-κ gate dielectric have been readily attained, the performance of their n-channel counterparts are still unsatisfactory. Two possible reasons are the inferior dielectric–substrate interface quality and the fundamental difficulty in n-type dopant incorporation into Ge. Any future research to remedy the n-channel MOSFET pathology should go into the direction of buried channel operation and/or implantation-less source–drain architecture.

In order to fully exploit the excellent transport properties in Ge whereas eliminating the respectable off-state leakage, ultrathin body device structures have been suggested and intensely investigated. The two most promising embodiments until now include the GOI MOSFET and Ge nanowire MOSFET, where the improved electrostatics, reduced junction leakage, and reduced parasitic capacitance have been well justified. As the next step, one should take a serious look at the ultrathin body non-planar Ge channel MOSFET design that combines with all the aforementioned benign features such as a strained buried channel with Schottky source–drain.

References

[1] *The International Technology Roadmap for Semiconductors*, Semiconductor Industry Association, 2004 Update. (http://public.itrs.net/)

[2] P.J. Wright and K.C. Saraswat, *IEEE Trans. Electron Dev.* **37** (1990) 1884.

[3] G.D. Wilk, R.M. Wallace and J.M. Anthony, *J. Appl. Phys.* **89** (2001) 5243.

[4] J.D. Plummer and P.B. Griffin, *Proc. IEEE* **89** (2001) 240.

[5] A. Asenov, *IEEE Trans. Electron Dev.* **45** (1998) 2505.

[6] E.J. Nowak, *IBM J. Res. & Dev.* **46** (2002) 169.

[7] M. Lundstrom, *IEEE Electron Dev. Lett.* **18** (1997) 361.

[8] M. Lundstrom and Z. Ren, *IEEE Trans. Electron Dev.* **49** (2002) 133.

[9] A. Lochtefeld and D.A. Antoniadis, *IEEE Electron Dev.* Lett. **22** (2001) 95.

[10] S. Takagi, *VLSI Symp. Tech. Dig.* (2003) 115.

[11] R. Chau, S. Datta, M. Doczy, B. Doyle, B. Jin, J. Kavalieros, A. Majumdar, M. Metz and M. Radosavljevic, *IEEE Trans. Nanotech.* **4** (2005) 153.

[12] C.O. Chui, A.K. Okyay and K.C. Saraswat, *IEEE Photon. Technol. Lett.* **15** (2003) 1585.

[13] L.L. Chang and H.N. Yu, *Proc. IEEE* **53** (1965) 316.

[14] K.L. Wang and P.V. Gray, *IEEE Trans. Electron Dev.* **22** (1975) 353.

[15] S. Iwauchi and T. Tanaka, *Jpn. J. Appl. Phys.* **10** (1971) 260.

[16] E.E. Crisman, J.I. Lee, P.J. Stiles and O.J. Gregory, *Electron. Lett.* **23** (1987) 8.

[17] J.J. Rosenberg and S.C. Martin, *IEEE Electron Dev. Lett.* **9** (1988) 639.

[18] M. Randolph and L.G. Meiners, *J. Electrochem. Soc.* **136** (1989) 2699.

[19] D.G. Leupp and G.M. Oleszek, *Symposium on Low Temperature Electronic Device Operation Tech. Dig.* (1991) 141.

[20] C.M. Ransom, T.N. Jackson and J.F. DeGelormo, *IEEE Trans. Electron Dev.* **38** (1991) 2695.

[21] A.V. Rzhanov and I.G. Neizvestny, *Thin Solid Films* **58** (1979) 37.

[22] S.C. Martin, L.M. Hitt and J.J. Rosenberg, *IEEE Electron Dev. Lett.* **10** (1989) 325.

[23] S.C. Martin, L.M. Hitt and J.J. Rosenberg, *IEEE Trans. Electron Dev.* **36** (1989) 2629.

[24] T.N. Jackson, C.M. Ransom and J.F. DeGelormo, *IEEE Electron Dev. Lett.* **12** (1991) 605.

[25] H. Shang, H. Okorn-Schmidt, K.K. Chan, M. Copel, J.A. Ott, P.M. Kozlowski, S.E. Steen, S.A. Cordes, H.-S.P. Wong, E.C. Jones and W.E. Haensch, *IEEE IEDM Tech. Dig.* (2002) 441.

[26] C.O. Chui, H. Kim, D. Chi, B.B. Triplett, P.C. McIntyre and K.C. Saraswat, *IEEE IEDM Tech. Dig.* (2002) 437.

[27] K. Chen, H.C. Wann, P.K. Ko and C. Hu, *IEEE Electron Dev. Lett.* **17** (1996) 202.

[28] S.J. Lee, H.F. Luan, C.H. Lee, T.S. Jeon, W.P. Bai, Y. Senzaki, D. Roberts and D.L. Kwong, *VLSI Symp. Tech. Dig.* (2001) 133.

[29] Y. Kamata, Y. Kamimuta, T. Ino, R. Iijima, M. Koyama and A. Nishiyama, *IEEE IEDM Tech. Dig.* (2005) 441.

[30] B. De Jaeger, M. Houssa, A. Satta, S. Kubicek, P. Verheyen, J. Van Steenbergen, J. Croon, B. Kaczer, S. Van Elshocht, A. Delabie, E. Kunnen, E. Sleeckx, I. Teerlinck, R. Lindsay, T. Schram, T. Chiarella, R. Degraeve, T. Conard, J. Poortmans, G. Winderickx, W. Boullart, M. Schaekers, P.W. Mertens, M. Caymax, W. Vandervorst, E. Van Moorhem, S. Biesemans, K. De Meyer, L. Ragnarsson, S. Lee, G. Kota,

G. Raskin, P. Mijlemans, J.-L. Autran, V. Afanas'ev, A. Stesmans, M. Meuris, and M. Heyns, *Proc. ESSDERC* (2004) 189.

[31] N. Wu, Q. Zhang, C. Zhu, D.S.H. Chan, A. Du, N. Balasubramanian, M.F. Li, A. Chin, J.K.O. Sin and D.-L. Kwong, *IEEE Electron Dev. Lett.* **25** (2004) 631.

[32] S.J. Whang, S.J. Lee, F. Gao, N. Wu, C.X. Zhu, J.S. Pan, L.J. Tang and D.L. Kwong, *IEEE IEDM Tech. Dig.* (2004) 307.

[33] H.Y. Yu, J.F. Kang, J.D. Chen, C. Ren, Y.T. Hou, S.J. Whang, M.-F. Li, D.S.H. Chan, K.L. Bera, C.H. Tung, A. Du and D.-L. Kwong, *IEEE IEDM Tech. Dig.* (2003) 99.

[34] H. Shang, K.-L. Lee, P. Kozlowski, C. D'Emic, I. Babich, E. Sikorski, M. Ieong, H.-S.P. Wong, K. Guarini and W. Haensch, *IEEE Electron Dev. Lett.* **25** (2004) 135.

[35] C.O. Chui, H. Kim, P.C. McIntyre and K.C. Saraswat, *IEEE IEDM Tech. Dig.* (2003) 437.

[36] C.O. Chui, L. Kulig, J. Moran, W. Tsai and K.C. Saraswat, *Appl. Phys. Lett.* **87** (2005) 091909.

[37] M.V. Fischetti and S.E. Laux, *J. Appl. Phys.* **80** (1996) 2234.

[38] T. Irisawa, S. Tokumitsu, T. Hattori, K. Nakagawa, S. Koh and Y. Shiraki, *Appl. Phys. Lett.* **81** (2002) 847.

[39] A. Ritenour, S. Yu, M.L. Lee, N. Lu, W. Bai, A. Pitera, E.A. Fitzgerald, D.L. Kwong and D.A. Antoniadis, *IEEE IEDM Tech. Dig.* (2003) 433.

[40] M.L. Lee, C.W. Leitz, Z. Cheng, A.J. Pitera, T. Langdo, M.T. Currie, G. Taraschi, E.A. Fitzgerald and D.A. Antoniadis, *Appl. Phys. Lett.* **79** (2001) 3344.

[41] H. Shang, J.O. Chu, X. Wang, P.M. Mooney, K. Lee, J. Ott, K. Rim, K. Chan, K. Guarini and M. Ieong, *VLSI Symp. Tech. Dig.* (2004) 204.

[42] H. Shang, J.O. Chu, S. Bedell, E.P. Gusev, P. Jamison, Y. Zhang, J.A. Ott, M. Copel, D. Sadana, K.W. Guarini and M. Ieong, *IEEE IEDM Tech. Dig.* (2004) 157.

[43] O. Weber, Y. Bogumilowicz, T. Ernst, J.-M. Hartmann, F. Ducroquet, F. Andrieu, C. Dupré, L. Clavelier, C. Le Royer, N. Cherkashin, M. Hytch, D. Rouchon, H. Dansas, A.-M. Papon, V. Carron, C. Tabone and S. Deleonibus, *IEEE IEDM Tech. Dig.* (2005) 143.

[44] T. Krishnamohan, Z. Krivokapic, K. Uchida, Y. Nishi, and K.C. Saraswat, *VLSI Symp. Tech. Dig.* (2005) 82.

[45] C.C. Yeo, B.J. Cho, F. Gao, S.J. Lee, M.H. Lee, C.-Y. Yu, C.W. Liu, L.J. Tang and T.W. Lee, *IEEE Electron Dev. Lett.* **26** (2005) 761.

[46] S. Nakaharai, T. Tezuka, N. Sugiyama, Y. Moriyama and S.-I. Takagi, *Appl. Phys. Lett.* **83** (2003) 3516.

[47] E. Pop, C.O. Chui, S. Sinha, R. Dutton and K. Goodson, *IEEE IEDM Tech. Dig.* (2004) 411.

[48] C.H. Huang, D.S. Yu, A. Chin, C.H. Wu, W.J. Chen, C. Zhu, M.F. Li, B.J. Cho and D.-L. Kwong, *IEEE IEDM Tech. Dig.* (2003) 319.

[49] D.S. Yu, A. Chin, C.C. Liao, C.F. Lee, C.F. Cheng, M.F. Li, W.J. Yoo, and S.P. McAlister, *IEEE Electron Dev. Lett.* **26** (2005) 118.

[50] N.A. Bojarczuk, M. Copel, S. Guha, V. Narayanan, E.J. Preisler, F.M. Ross and H. Shang, *Appl. Phys. Lett.* **83** (2003) 5443.

[51] Y. Liu, K. Gopalakrishnan, P.B. Griffin, K. Ma, M.D. Deal and J.D. Plummer, *IEEE IEDM Tech. Dig.* (2004) 1001.

[52] T. Tezuka, S. Nakaharai, Y. Moriyama, N. Hirashita, E. Toyoda, N. Sugiyama, T. Mizuno and S.-I. Takagi, *VLSI Symp. Tech. Dig.* (2005) 80.

[53] S. Zhu, R. Li, S.J. Lee, M.F. Li, A. Du, J. Singh, C. Zhu, A. Chin and D.L. Kwong, *IEEE Electron Dev. Lett.* **26** (2005) 81.

[54] T. Maeda, K. Ikeda, S. Nakaharai, T. Tezuka, N. Sugiyama, Y. Moriyama and S. Takagi, *IEEE Electron Dev. Lett.* **26** (2005) 102.

[55] J. Guo and M.S. Lundstrom, *IEEE Electron Dev. Lett.* **49** (2002) 1897.

[56] E.D. Marshall, C.S. Wu, C.S. Pai, D.M. Scott and S.S. Lau, *MRS Symp. Proc.* **47** (1985) 161.

[57] P.L. McEuen, M.S. Fuhrer and H. Park, *IEEE Trans. Nanotechnol.* **1** (2002) 78.

[58] Y. Xia, P. Yang, Y. Sun, Y. Wu, B. Mayers, B. Gates, Y. Yin, F. Kim and H. Yan, *Adv. Mater.* **15** (2003) 353.

[59] C.M. Lieber, *IEEE IEDM Tech. Dig.* (2004) 521.

[60] D. Wang, Q. Wang, A. Javey, R. Tu, H. Dai, H. Kim, P.C. McIntyre, T. Krishnamohan and K.C. Saraswat, *Appl. Phys. Lett.* **83** (2003) 2432.

[61] A.B. Greytak, L.J. Lauhon, M.S. Gudiksen and C.M. Lieber, *Appl. Phys. Lett.* **84** (2004) 4176.

[62] L.J. Lauhon, M.S. Gudiksen, D.L. Wang and C.M. Lieber, *Nature* **402** (2002) 57.

CHAPTER 12

Alternative Ge Applications

Erich Kasper

12.1. Introduction

Historically, the first bipolar transistors were made with germanium as semiconductor material. At the same time and even earlier [1] the photo response properties of germanium were discovered. For a theoretical description of the early Ge photodiode properties, see e.g. [2–3]. The rather high saturation currents of the low band gap Ge made theory and comparison with experiments easier, but led later to the replacement by the larger band gap semiconductor silicon. Smaller devices, new insulation schemes and heterojunction growth are now available which allow to reconsider the use of Ge as components in silicon-based heterojunction circuits [4].

We will discuss the attractive properties of Ge for alternative applications, especially for the integration of optoelectronics and microelectronics, but also for solar cells, quantum dot (QD) applications, alternative field effect transistors (FETs) and spintronics. A key technological issue is the adjustment of lattice constants and strain, because Ge is 4.2% larger than Si. The adjustment technique suitable for integration is called virtual substrate.

12.2. Attractive Properties for Alternative Applications

The material properties of frequently used semiconductors are collected in book series [5–6]. In the group IV material sequence C (diamond), SiC (cubic), Si, Ge, Sn (semiconducting) the atomic weight is increasing, while the band gap is decreasing. The lattice

Germanium-Based Technologies: From Materials to Devices. C. Claeys and E. Simoen (Editors)

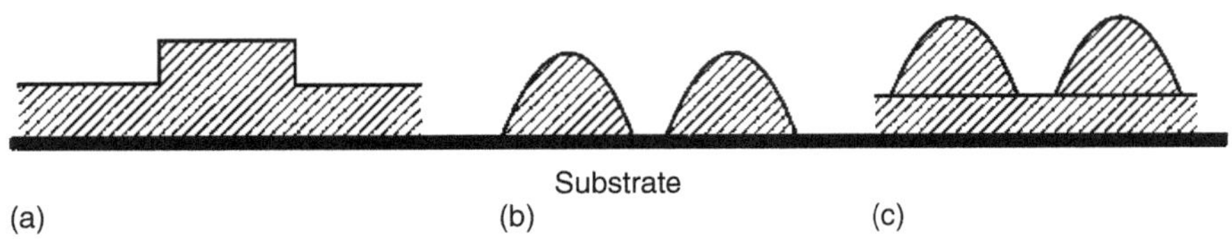

Figure 12.1. GOI fabrication growth modes in heteroepitaxy. (a) Two-dimensional (van de Merwe) growth, (b) Three-dimensional (Volmer–Weber) growth and (c) Mixed (Stranski–Krastanov) growth.

mismatch between Ge and Si is about 4.2%. The elastic stiffness of Ge is about 75% of that of silicon.

12.2.1. Growth modes

The morphology of a heterostructure depends strongly on the interplay of surface, interface and strain energies. Three basic types are described for the nucleation and thin layer behavior (Figure 12.1).

Without lattice mismatch and with low surface energy of the film a smooth surface (two-dimensional nucleation with monatomic steps) appear. With increasing lattice mismatch the nucleation tends to get three dimensional (3D), either directly on the substrate (Volmer–Weber growth mode, e.g. Si on Ge) or on top of a wetting layer (Stranski–Krastanov growth mode). Smooth van de Merwe films or wetting layers are preferred for a planar semiconductor technology, but in the meanwhile a tremendous increase in interest on 3D nucleation emerged because, topics as self-organization or self-assembling of islands, quantization in wires or dots or stress manipulation enrich the material and device design. The growth mode may be influenced by modifications of the surface energy by surfactants. Antimony (Sb) on Ge acts as such a surfactant [7].

12.2.2. Strain influence on electronic alignment

The lattice mismatch $\Delta a/a$ is accommodated in thin layers by strain (Figure 12.2) and in thick layers by misfit dislocations (MD). In uniform layers the in-plane biaxial stress causes a tetragonal distortion of the unit cell (100 orientation of the substrate). The total strain can be considered as composed of a hydrostatic and a uniaxial component, respectively. The band edges (Figure 12.3) are shifted in energy by the hydrostatic component whereas the uniaxial component splits degenerate band edges.

The band splitting concerns heavy and light holes in the valence band (and this valence band splitting is similar in direct and indirect semiconductors) and the sixfold (Si) or eightfold (Ge) degenerated electrons in the conduction band (the conduction band splitting is only seen in indirect semiconductors). Therefore, especially for indirect semiconductors, the band gaps and the band ordering at interfaces are strongly influenced by the strain status. As an example the band ordering for a compressively strained (100) SiGe is given in Figure 12.4.

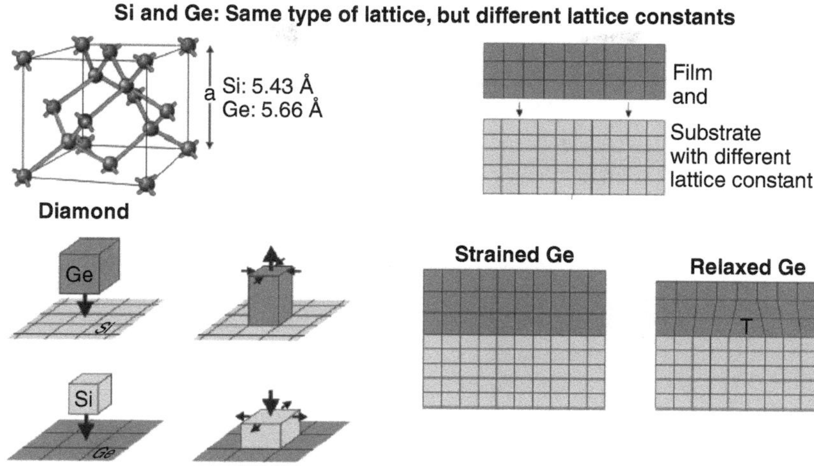

Figure 12.2. Biaxially stressed Ge (tetragonal distortion).

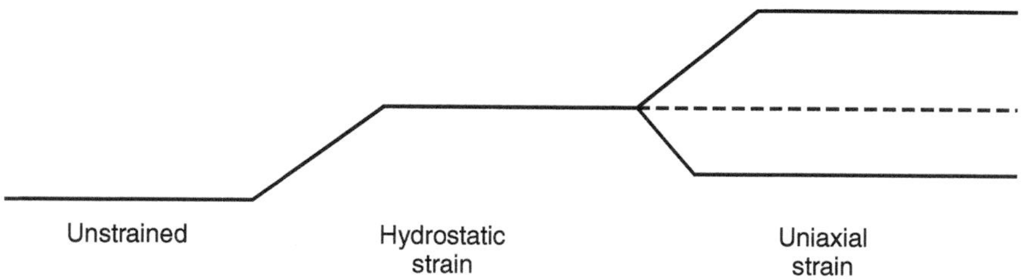

Figure 12.3. Strain effects on the band structure. Energy shift by hydrostatic components, splitting by uniaxial components.

The L-band (dominant in unstrained Ge) is not influenced by the (100) biaxial strain (its degenerate momentum vectors lie along [111]). The in plane Δ band (four fold degenerate, 4 g) is split downwards below the L-band. Therefore, in compressively strained Ge the silicon like 4 g Δ electrons and the heavy holes (v_2) define the band gap. Examples of band ordering with tensile strain are given in the sections on modulators and spintronics.

12.2.3. Wave guiding

Optical waveguides need a high index of refraction core layer. At low energies (below the band gap) the index of refraction tends towards $n = (\varepsilon_{\mathrm{rel}})^{1/2}$ with $\varepsilon_{\mathrm{rel}}$ the relative permittivity ($\varepsilon_{\mathrm{rel}} = 16$ for Ge). The common insulator materials as SiO_2 and Si have lower index of refraction as Ge. Ge is therefore wave guiding in a SiO_2 or Si surrounding. Table 12.1 gives the index of refraction for Ge n_{Ge} and the difference $\Delta n_{\mathrm{Si/Ge}}$ to silicon.

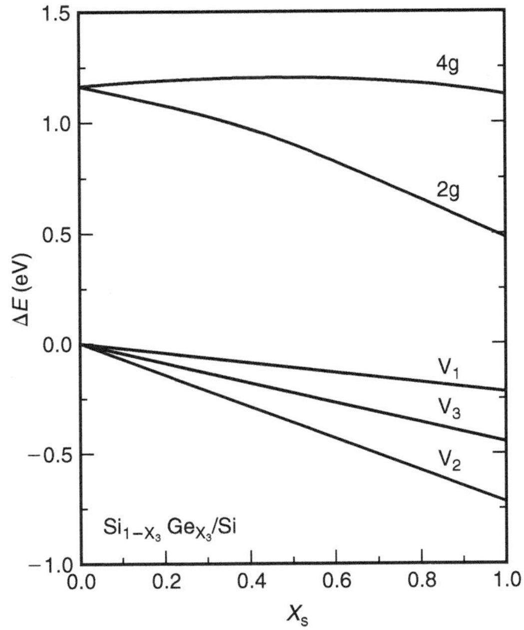

Figure 12.4. Band structure for compressive strained Ge (and SiGe).

Table 12.1. Energy dependence of index of refraction n_{Ge} of Ge and index difference $\Delta n_{Si/Ge}$ to Si, i.e. $n_{Si} = n_{Ge} - \Delta n_{Si/Ge}$.

Energy (eV)	0.5	0.7	0.9	1.1
n_{Ge}	4.06	4.12	4.21	4.32
$\Delta n_{Si/Ge}$	0.62	0.65	0.69	0.77

Table 12.2. Mobility μ of Ge, Si and GaAs (for comparison) in m^2/V-s.

	Ge	**Si**	**GaAs**
Electrons (μ_n)	0.39	0.15	0.85
Holes (μ_p)	0.19	0.045	0.04

12.2.4. Transport properties

The transport properties of Ge are very favorable compared with other semiconductors because of their high hole mobility (Table 12.2).

The comparison shows that Ge has the highest hole mobility (0.19 m^2/V-s = 1900 cm^2/V-s) of this three semiconductors, and that is true also with other semiconductors of reasonable band gap. The electron mobility is also high (more than twice the silicon value, half of GaAs). By applying of strain these values can even be improved. The most interesting configuration is the strain adjusted Si/Ge heterostructure, which means Ge is compressively strained and Si is tensile strained. In this configuration the band ordering is of the type II, which means, that electrons jump on the wide band gap (Si) side and the holes on the small band gap (Ge) side. This is the ideal material configuration for symmetric complementary FET (same mobilities for electrons and holes).

12.2.5. Brillouin zone folding

The first Brillouin zone in an energy E versus momentum k plot is limited to π/a (a is the lattice constant) in the principal axes of a primitive cubic lattice. In the diamond lattice cell (eight atoms per cell) the Brillouin zone edge is given by $2\pi/a$. By applying a superlattice (SL) period in direction z the first Brillouin zone in k_z is reduced to π/L. In the reduced Brillouin zone minibands are created by folding. The minibands should allow stronger band-to-band transitions, subband transitions (infrared, IR to Terahertz frequencies) and Bloch oscillations. Ge is an excellent candidate for lattice matched superlattices (SL), e.g. Si/GaAs/Si. . . . and for strained layer superlattices (SLS), e.g. Ge/GaAs/Ge. . . .

12.3. Optoelectronics

Photonics and optoelectronics play an essential role in many areas of communication and information technology. This key importance and the application in mass products drive the need to integrate different types of optoelectronic devices on silicon – the base material for microelectronics. Prominent examples of large scale integration are charge-coupled device (CCD) – and CMOS receiver arrays for cameras. Silicon optoelectronic devices suffer from negligible responsivity in the near infrared (NIR) – the preferred wavelength for fiber optical systems – and from the self-absorption in waveguide structures. These locks can be overcome by silicon-based heterostructures with a low band gap material. Within the now available technologies SiGe and Ge are the natural choice for those heterostructures. Low band gap materials shift the absorption edge toward the IR, these IR wavelengths are not absorbed in silicon or silicon oxide-based waveguides, and the index of refraction increases which forces the IR light to be focused into the heterostructure.

12.3.1. Integration aspects

Integration of microelectronics and optoelectronics would be much easier on the same substrate material – silicon. Figure 12.5 shows a schematic view on such an integrated circuit (IC) with an optoelectronic receiver array as part of it. We prefer to name this type of ICs as MOIC (micro-/optoelectronic integrated circuits) to underline that the already existing ultra-large scale integration (ULSI) of microelectronics on silicon dictates the conditions under which the integration of optoelectronics is possible. The principally also correct name OEIC (optoelectronic integrated circuit), which is mainly used in the III/V material area, focuses on optoelectronic devices which are supported by small scale electronics, e.g. amplifiers.

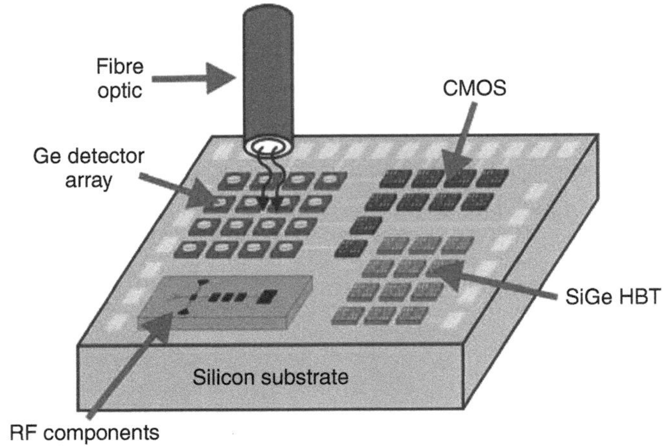

Figure 12.5. MOIC with CMOS digital electronics, high frequency analog circuitry (HBT, SIMMWIC) and optoelectronics – in the given example a receiver array for vertical incidence of light.

For MOICs not only the choice of silicon as substrate but also the very stringent requirements of optical submicron lithography are mandatory. These demand a surface with only small height differences because of the small depth of focus of lithography systems. The heteroepitaxy parts of the IC should therefore be not higher than about 1 μm above the silicon substrate surface level to allow for surface planarization techniques. One possible route could be a selective growth of the thin heterostructures in the predefined areas or a uniform growth on the silicon surface covered by a patterned oxide. The latter growth is called differential epitaxy [8–9], because it produces epitaxial islands in the oxide windows surrounded by polycrystalline layers on the oxide. The poly layers can either be removed by preferential etching or chemical mechanical polishing (CMP) or they can be used for resistors, absorbers. The limited thickness of the heterostructures has an often overlooked influence on the technical route to accommodate the lattice mismatch in the silicon/heterostructure compound. The thick strain relaxed buffers (SRB) in artificial substrates (described in Section 12.4 on solar cells) are not applicable but they have to be replaced by thin and ultra-thin buffer layers. This concept is named virtual substrate because it provides a silicon with a thin surface layer of a different lattice constant (described in more detail in Section 12.8).

12.3.2. Detectors for the visible to the NIR

For a high quantum efficiency (QE) of a given detector the absorption coefficient should be high enough that the majority of light quanta is absorbed in the active detector region. This is rather easily obtained for direct semiconductors and light with energies above the band gap. See e.g. Figure 12.6 where GaAs and InGaAs are given as examples of direct band gap materials. Above the band gap (or below the corresponding cut-off wavelength) the absorption increases steeply. An absorption depth of 1 μm is obtained for wavelengths below 800 nm (GaAs) or 1400 nm (InGaAs), the absorption depth of 10 μm is 855 nm and

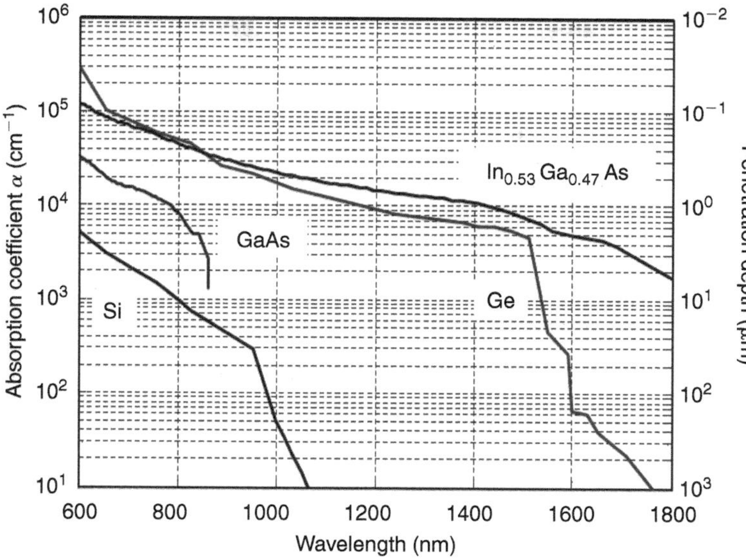

Figure 12.6. Absorption versus wavelength for two direct band gap (GaAs, InGaAs) and two indirect semiconductors (Si, Ge).

1850 nm, respectively. Both, Si and Ge, are indirect semiconductors with a much weaker absorption directly above the band gap. The typical behavior is shown by the silicon curve with a cut-off wavelength of 1100 nm. The 10 μm absorption depth is only reached in the red (800 nm) and the 1 μm depth deep in the visible (550 nm). Despite this, silicon detectors with good QE in the red and very near infrared are fabricated because of a large detector region with diffusion controlled carrier movement, but these detectors are inherently slow. Although Ge is also an indirect semiconductor its absorption curve looks very different. Down from the cut-off at about 1800 nm a slight increase in absorption is seen as typical for the indirect band gap but down from 1600 nm a steep increase follows. This is caused by the lowest direct transition which is in Ge only slightly above (0.2 eV) the indirect transition whereas in Si the lowest direct transition (3.4 eV) is far above the indirect one (1.12 eV).

The absorption of the SiGe alloy (Figure 12.7) is also strongly influenced by this transition (which is always above the indirect one). Even up to a high Ge content the absorption curve looks similar to Si with only the cut-off wavelength shifted to the infrared.

One can distinguish three regions if we compare the absorption of silicon and germanium. From the visible up to the very NIR (600–950 nm) the absorption in Ge is about 50 times stronger than in Si (direct versus indirect transitions). That means even in the visible is Ge a superior detector material compared to silicon.

In the NIR (950–1500 nm) the absorption in Si fades out (band edge at 1100 nm) and therefore the absorption in Ge is many orders of magnitude larger. Beyond 1500–1800 nm the

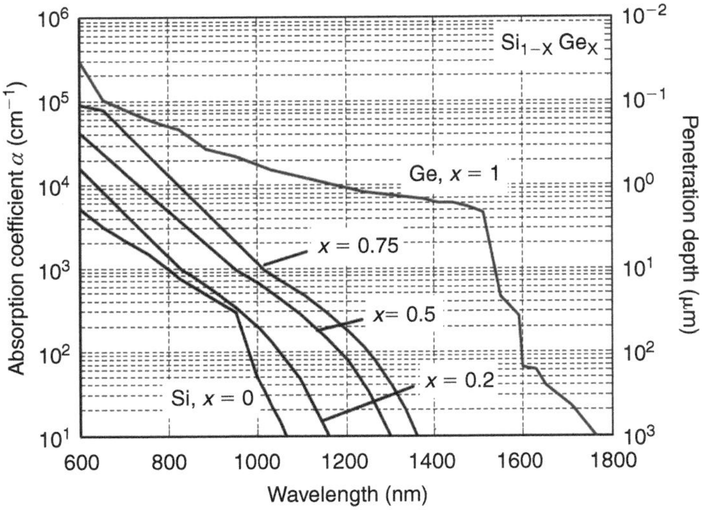

Figure 12.7. Absorption of the SiGe alloys with different Ge contents x ($x = 0, 0.2, 0.5, 0.75, 1.0$).

absorption of Ge fades out (transition from direct to indirect transition) which requires specific absorption geometries (resonant cavity, waveguide detector).

The absorption process creates an electron-hole pair (fundamental absorption). In doped semiconductors a carrier could be increased in energy by absorption of a light quantum (free carrier absorption). Free carrier absorption is rather weak and can only be observed for energies below the fundamental absorption (band gap). The electron-hole pair increases the conductivity (photo conductor) and – more important – it creates a dark current in an unbiased (zero bias operation) or reverse biased pn-junction. The basic structure of a pin photodiode consists of highly doped p^+ and n^+ contacts on both sides of a low doped intrinsic (i) region. The doping of the i-region is preferentially that low that the depletion layer reaches through. In a Ge/Si pin diode (Figure 12.8) the absorption region may be either strain relaxed (Ge or SiGe layer above the critical thickness) or pseudomorphic as in a Ge (QDs)/Si overlayer SL (right side of Figure 12.8). We will concentrate here on the relaxed Ge/Si pin-photodiode. The QDs are treated in a later section. The p/n-junction may be replaced by a reverse biased Schottky contact which also has a depletion layer.

The cross section of a vertical incidence photodetector is shown in Figure 12.9. For integration purposes the back-side contract is led to the front side by a buried layer, in this case a p^+layer but the inverse structure may also be used. With a thin intrinsic Ge layer (300 nm) compatible with integration technology requiring planarization measures for lithography a QE of 10–25% could be obtained between 650 and 1500 nm wavelength [10]. The photo response was nearly independent on the applied voltage and allowed zero bias operation (Figure 12.10). The dark current is enhanced (80 nA at −1 V) because of threading dislocations from the MD network. Zero bias operation is therefore recommended for low light

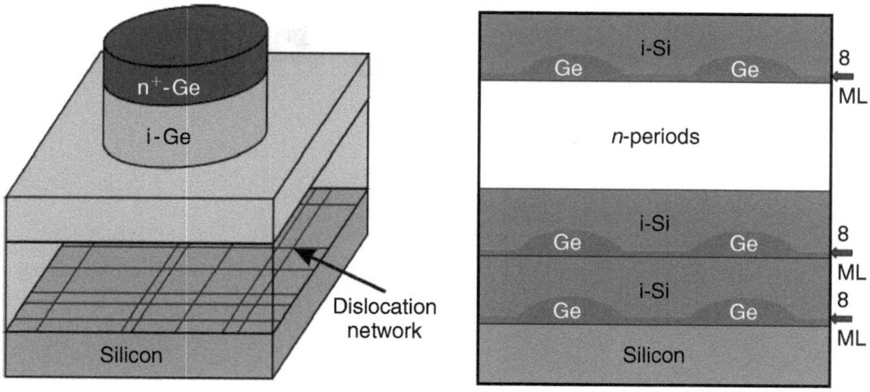

Figure 12.8. Intrinsic absorption region in a Ge/Si pin diode. Left side: Pure Ge or SiGe alloys on top of a Si substrate. Because of the lattice mismatch between Si and Ge a MD network is positioned at the interface. Right Side: Ge QD embedded in silicon without MD network.

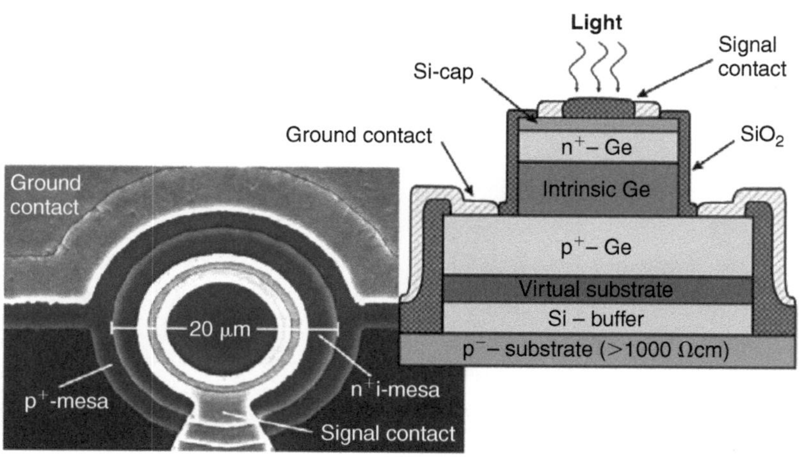

Figure 12.9. Scheme of a vertical incidence Ge/Si photodiode.

levels. Fast operation of the photodetector requires extraction of the electron-hole pairs from the depletion layer (carrier speed equals saturation velocity) but negligible contributions from slowly diffusing species from the surrounding regions (which would improve QE but on the cost of speed reduction). The relative contribution of carriers from the depletion layer can be assessed by the voltage dependence of the photocurrent I_{ph} at the forward biased regime when the intrinsic layer is only partially depleted. If only the carriers from the depletion layer contribute to the photocurrent, then the current increases with increasing depletion layer width. For absorption lengths larger than the intrinsic width W_i the photocurrent I_{ph} can be written as

$$(I_{ph}/I_{pho})^2 = (2/q)(n_i W_i^2)^{-1}(V_{bi} - V) \qquad (12.1)$$

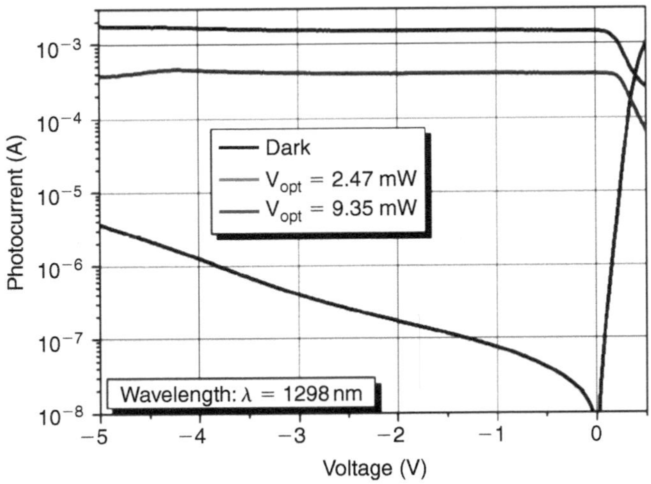

Figure 12.10. Photocurrent (at 1.3 μm wavelength) of the vertical GeSi photodiode versus voltage.

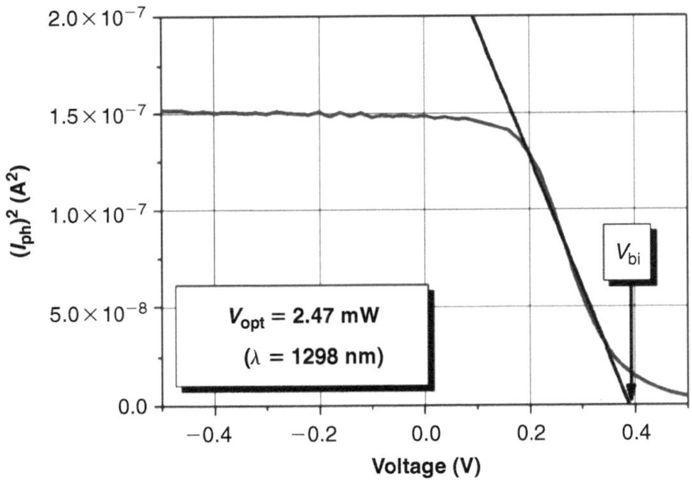

Figure 12.11. Square of the photocurrent I_{ph}^2 versus voltage. The linear slope in the forward region marks the partial depletion of the intrinsic region.

with I_{pho} the saturated current, n_I intrinsic doping, V_{bi} the built-in voltage. From a plot I_{ph}^2 versus V the validity of the assumption can be tested (Figure 12.11). Indeed the photodetectors turned out to be very fast. The bandwidth was 25 GHz for zero bias operation (Figure 12.12) and 40 GHz for a moderate 2 V reverse bias. Presently these are the fastest Si-based IR detectors. This type of detectors has the potential for 100 Gb s^{-1} communication links when *RC* limitations [11] are reduced. The reduced diffusion is obtained by the dislocation network, the abrupt junctions and the high doping levels in p$^+$ and n$^+$ contact layers which

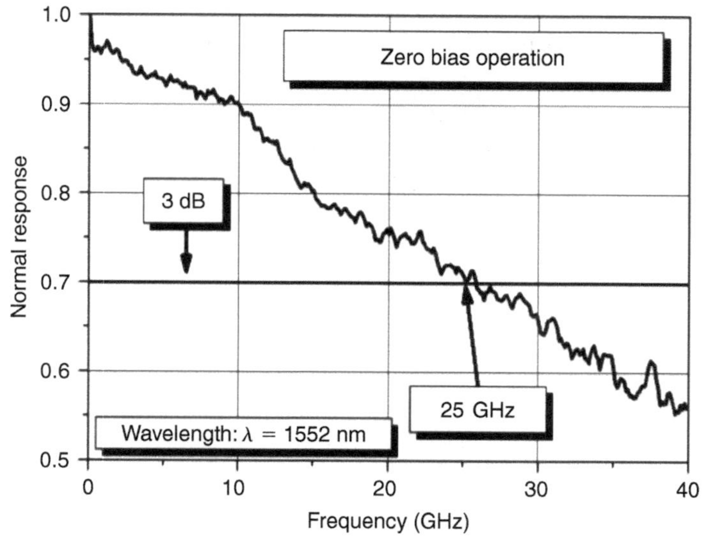

Figure 12.12. Photo response versus frequency. A 3 dB bandwidth of 25 GHz at 1.55 μm wavelength was obtained with zero bias operation.

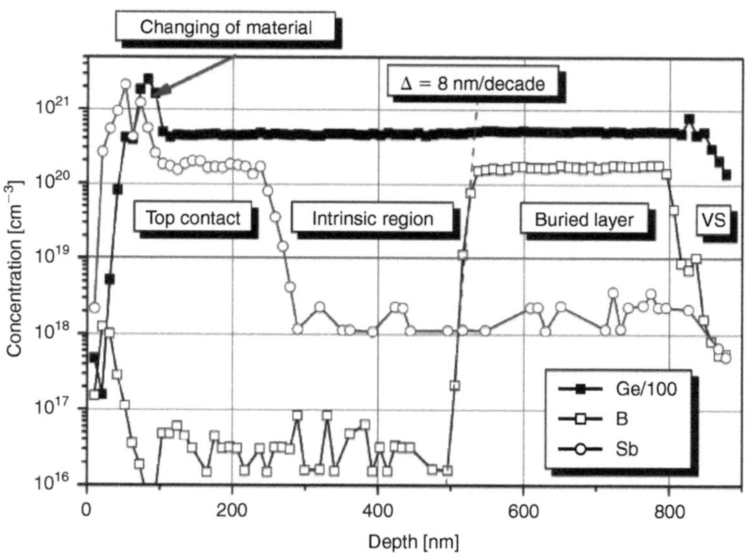

Figure 12.13. SIMS measurement of the vertical photodetector doping profile grown by molecular beam epitaxy (MBE). B(p) and Sb (n) doping. The intrinsic doping is not visible because of the background level (B: 10^{16} cm^{-3}, Sb: 10^{18} cm^{-3}) of the SIMS method.

reduce minority carrier lifetime in the surrounding regions by dislocation recombination and Auger recombination. The SIMS profile (Figure 12.13) demonstrates the high doping levels and the abrupt transitions. The doping of the intrinsic layer (10^{15} cm^{-3} range) is below the detection limit.

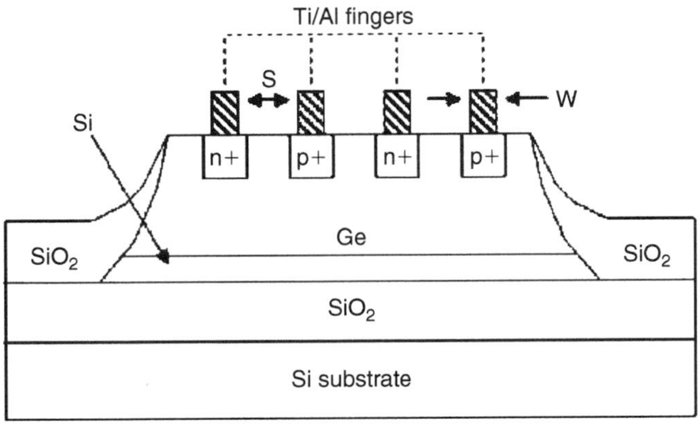

Figure 12.14. Cross section of the lateral pin design of a Ge on SOI photodiode (after Dehlinger *et al.* [12]).

High speed operation at 850 nm was also demonstrated with lateral Ge pin photodiodes on SOI substrates [12]. The design of the lateral pin is given in Figure 12.14. In the nominal intrinsic Ge on SOI (in the given reference the real doping was p^- type, $10^{16}\,cm^{-3}$) alternative doping fingers $n^+ip^+in^+ \ldots$ were implanted and contacted. The measurements were made at $\lambda = 850$ nm with good quantum efficiencies (34–46%). The QE fade out up to 1360 nm (1.3 μm : 3.4% QE). The best frequency bandwidth value was 29 GHz with a bandwidth – efficiency product of 13.2 GHz. Low voltage operation (V = −4–0 V) was demonstrated.

Low voltage operation was also reported for polycrystalline Ge on Si detectors with responsivities of $16\,mAW^{-1}$ at 1.3 μm wavelength [13]. These photodetectors were integrated as linear array with a readout electronics. The responsivity dropped for this first test to $0.8\,mAW^{-1}$ when the array was fabricated on the CMOS circuit with a low temperature post-processing step. Integration and digital readout of the generated photocurrents at the standard TV-line frequency were demonstrated.

All these experiments have shown that with vertical GeSi photodetectors high speed can be obtained in the wavelength range from 850 to 1550 nm, but only at 850 nm good efficiencies and at 1300 nm acceptable efficiencies and band width – efficiency products can be obtained. At 1550 nm the efficiency of fast photodetectors drops to a few percent. A well-known method in the III/V area is enhancement of detection (or emission) by a resonant cavity (Fabry–Perot-resonator). Resonant cavity enhanced (RCE) photodetectors were realized in the Si/Ge system by two methods [14–15]. In one approach [14] a double SOI substrate was used as back reflector (Figure 12.15). This special substrate was obtained by a repeated SOI process. With 300 nm Si, 250 nm SiO_2 layer thicknesses a reflectivity of more than 90% could be obtained for the back reflector. The highly reflecting Ge layer itself was used as front reflector. The measured quantum efficiencies were 40% and 14% at 1.3 and 1.55 μm wavelength, respectively. At zero bias 25% and 8.6% efficiencies were

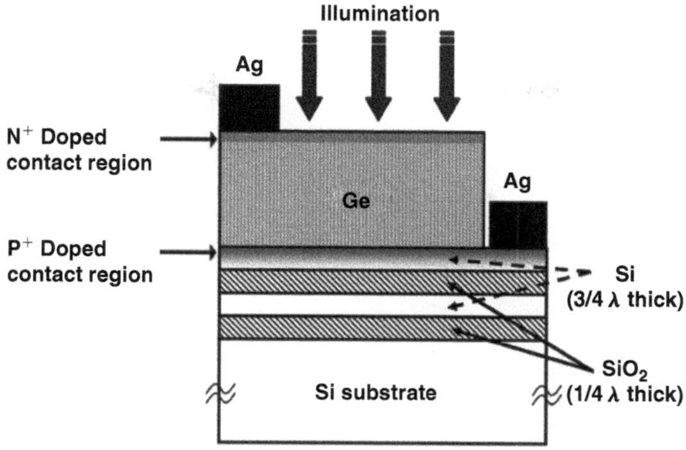

Figure 12.15. RCE photodetector on a double SOI substrate (after Dosunmu *et al.* [14]).

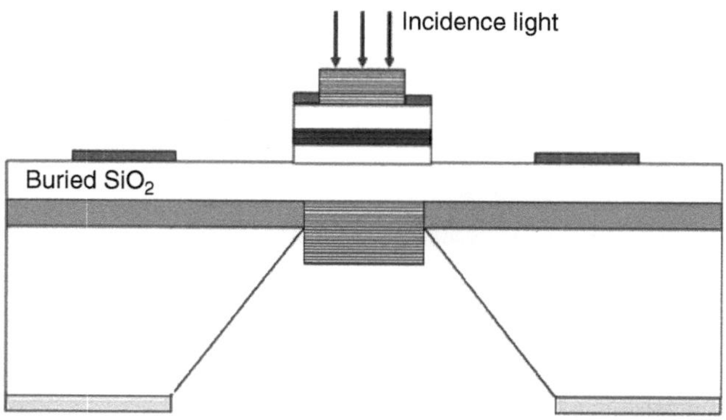

Figure 12.16. Structure of the RCE photodiode with DBR mirrors front side (upside) and back side (bottom). The bottom DBR mirror was produced after thinning the SOI handling wafer (after Yu *et al.* [15]).

obtained at these two wavelengths. For an optimized structure up to 80% efficiency is predicted. In the other approach [15] dielectric Bragg reflectors (DBR) were used as back- and front-side mirrors (Figure 12.16). These gave very narrow responses. At 1.55 μm a bandwidth as small as 5 nm was realized.

The most elegant way to overcome the limited thickness of the absorber is to use waveguide detectors (Figure 12.17). These detectors were used for SiGe alloys [16] but their demonstration with pure Ge layers is outstanding.

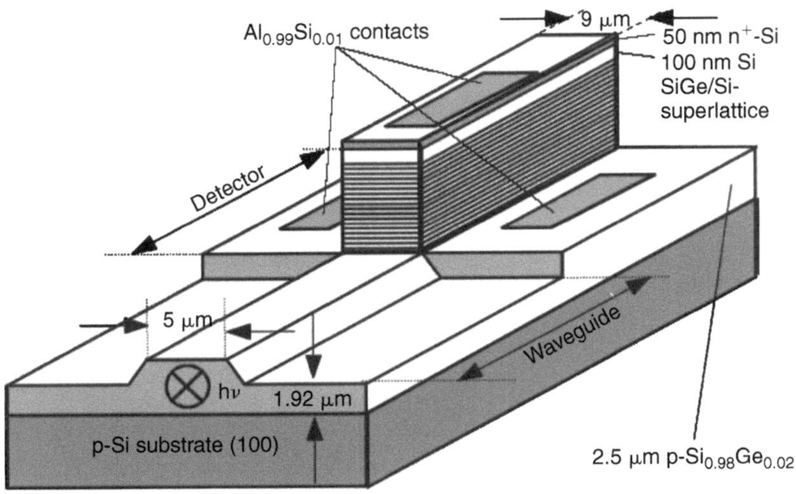

Figure 12.17. Scheme of a waveguide detector (SiGe on Si) for 1.3 µm detection wavelength (after Splett *et al.* [16]).

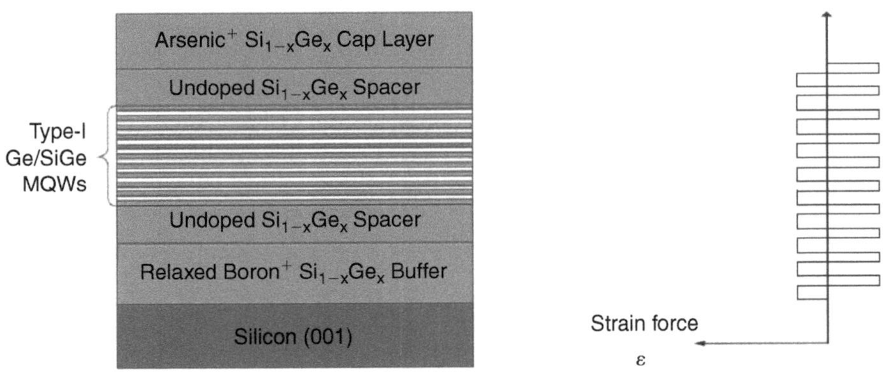

Figure 12.18. Ge/SiGe quantum well modulator. Cross section of the pin structure on a virtual substrate (Si + relaxed SiGe) (after Kuo *et al.* [17]).

12.3.3. Modulators

Modulators are an essential element of optoelectronic systems. Silicon-based optical modulators have recently been successfully demonstrated but because the light modulation mechanisms in silicon are relatively weak, several mm-long devices or sophisticated high-quality factor resonators have been necessary (see [17] and references therein). The quantum-confined Stark effect (QCSE) mechanism which is much stronger would allow modulator structures with only micrometers of optical path length. Since the break through work in 2005 [17] it is known that strong QCSE can be obtained in Ge quantum wells on Ge rich virtual substrates. The multi-quantum well (MQW) structure is in the intrinsic zone of a pin diode on a virtual substrate (Figure 12.18). The direct transitions at the zone center (Brillouin zone) are used as

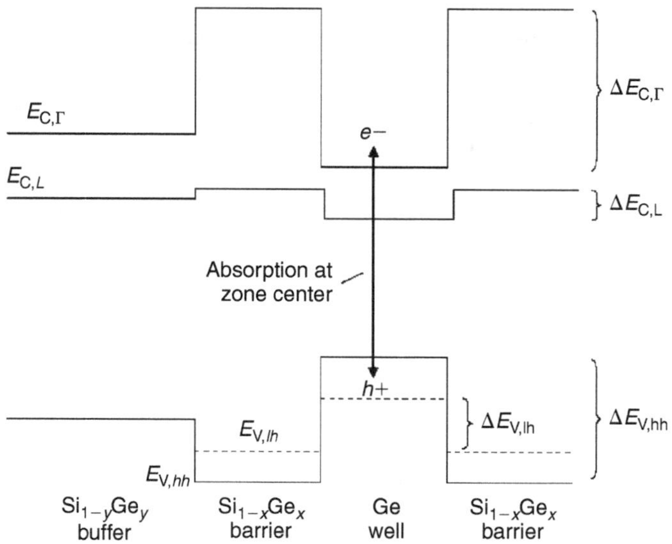

Figure 12.19. The band gap structures of a Ge/SiGe quantum well. The valence band in thin wells and barriers is split into heavy and light holes (hh, lh). The conduction band is shown for indirect (L) and direct transitions (Γ) (after Kuo *et al.* [17]).

shown in Figure 12.19. A strain symmetrized MQW is grown on the relaxed SiGe buffer of the virtual substrate. The direct conduction band (Γ) is only slightly above the indirect one in Ge but much higher in SiGe creating a type I Ge well for direct absorption. A relaxed SiGe ($x = 0.9$) below the Ge/SiGe ($x = 0.85$) MQW leads to band discontinuities of about 101 meV, 47 meV, 400 meV for heavy holes, light holes and Γ electrons.

When an electric field is applied a clear QCSE shift of the absorption edge to lower photon energies is seen in Figure 12.20. Consider first the zero bias (0 V) absorption curve. Two exciton peaks at 0.88 and 0.91 eV correspond to the heavy hole and light hole to Γ electron transitions. The strong electric field provided by the applied voltage forms a so called Stark ladder [18] which shifts the absorption edge to lower energies (Figure 12.20). The strong electro-absorption comparable to that of QCSE III/V QW structures are very promising for electro-absorption modulators operating at high speed, low power, low operating voltage and with small device areas.

12.3.4. Waveguides

The basic structure (Figure 12.21) of an optical waveguide is given by a core layer embedded between cladding layers whose index of refraction n_2 is smaller than that of the core layer (n_1).

The numerical aperture NA of the waveguide is given by [19]

$$NA = (n_1^2 - n_2^2)^{1/2} \tag{12.2}$$

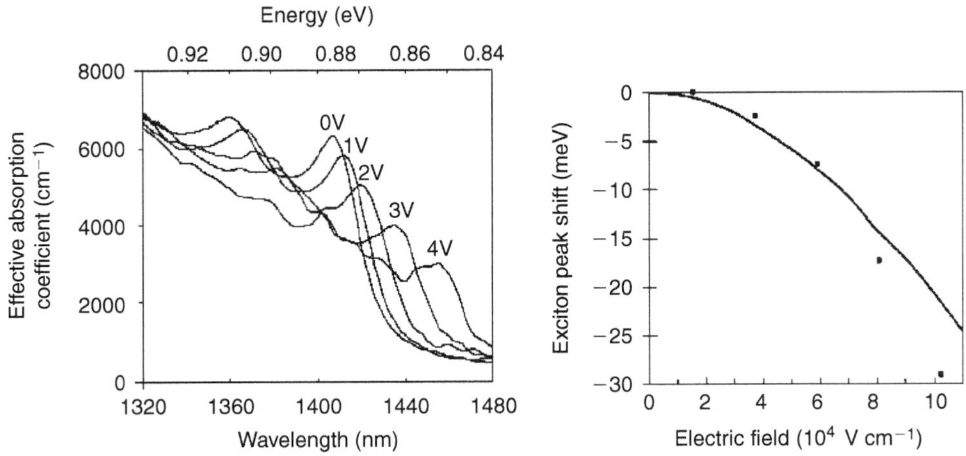

Figure 12.20. QCSE. Effective absorption coefficient versus wavelength for different applied reverse voltages.

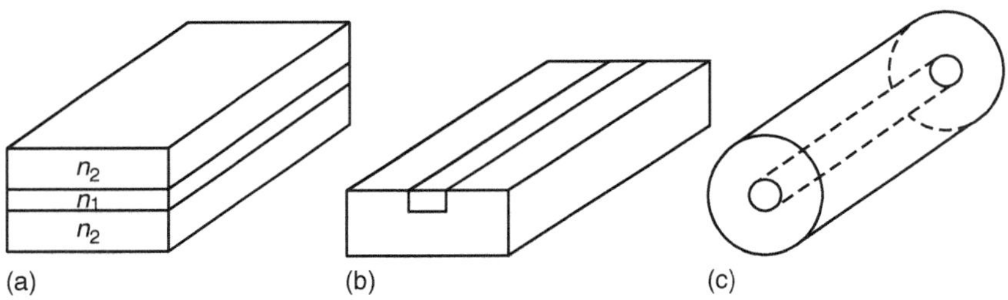

Figure 12.21. Optical waveguide structures with (a) layer, (b) stripe and (c) fiber geometries [19].

Within the high index core layer discrete waveguide modes are able to propagate. The mode number is increasing with increasing thickness d of the core layer. To each mode number $m = 0, 1, 2 \ldots$ a certain angle of total reflection ϕ_m and a certain phase angle φ_m belongs and this angle ϕ_m is described by:

$$2n_1 d\cos\theta_m - (\lambda/\pi)\varphi_m = m\lambda, m = 1, 2, 3\ldots \quad (12.3)$$

The phase angle φ_m at the reflecting interface is given by:

$$\tan(\varphi_m/2) = \frac{1}{\cos\theta_m}\sqrt{\sin^2\theta_m - \left(\frac{n_1}{n_2}\right)^2} \quad (12.4)$$

for transverse electrical (TE) waves:

$$\tan(\varphi_m/2) = \frac{1}{\cos\theta_m}\left(\frac{n_1}{n_2}\right)^2 \sqrt{\sin^2\theta_m - \left(\frac{n_1}{n_2}\right)^2} \tag{12.5}$$

for transverse magnetic (TM) waves.

The maximum number (TE and TM modes) of modes M is determined by the numerical aperture NA and the thickness to wavelength ratio d/λ:

$$M = 2[2(d/\lambda)\,\mathrm{NA} + 1] \tag{12.6}$$

Monomode operation (strictly speaking one TE and one TM mode) will be obtained when the thickness d is small:

$$d < \lambda/(2\mathrm{NA}) \quad \text{monomode operation} \tag{12.7}$$

Waveguides may be designed either by specific substrates (SOI or germanium on insulator, (GOI) substrates), by cover layers (SiO_2, SiON layers) or the device structure itself. SOI or GOI substrates benefit from the low refractive index n_2 of the oxide layer, the cover waveguide structures utilize the refractive index differences between silicon oxide, nitride and air. These waveguides are used as in silicon technology [20]. Specifically interesting for Ge devices is the high index of refraction (see Section 12.2) which means that optical waves are guided by the Ge device structure, a distinctive advantage of Ge receiver or emitter structure, e.g. in a Si surrounding.

12.3.5. Optical emitter

With direct semiconductors the band-to-band transition turned out to be the most useful for light emitting diodes (LED) or lasers. With indirect semiconductors this band-to-band transition is very useful for characterization (photoluminescence, PL), but is weak to be utilized in electroluminescence (EL) devices. Instead, transitions to localized defect levels (impurities, point defects (PD), dislocations), atomic transitions in rare earth incorporations [21] are used. A systematic approach to overcome the indirect nature of the band-to-band transitions follows the confinement of carriers in a quantum structure [22]. A one-dimensional SL confines the electrons or holes within a reduced Brillouin zone (Figure 12.22). By proper choice of the SL period a quasi-direct band structure (minibands) is created [23].

The increased transition probability from the minibands was confirmed in PL and EL experiments. More recent attempts try to further increase the confinement by formation of Ge QDs. The work on dots is described in Section 12.5.

12.4. Solar Cells

The basic structure of a solar cell is that of a zero biased pin junction. The built-in field in the depletion layer separates electron/hole pairs generated by the fundamental absorption of

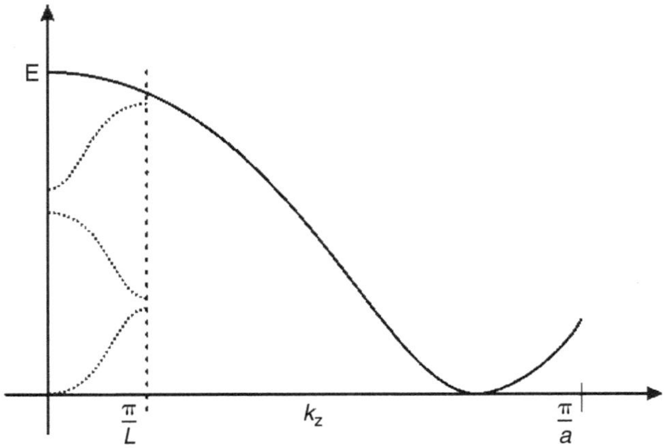

Figure 12.22. Zone folding concept of a SL with period length L. The first Brillouin zone is folded from (π/a) in a primitive lattice to (π/L) in the z-direction of the SL period L.

light. For the fundamental absorption a minimum energy E_G of the light quantum is required:

$$hf = hc/\lambda \geq E_G \tag{12.8}$$

Given the wavelength λ in μm and the band gap E_G in eV the condition above reads:

$$\lambda(\mu m) < 1.24/E_G\ (eV) \tag{12.9}$$

Light with lower energy either transmits or is absorbed by free carriers which are inefficient for energy conversion. The reason is the rapid energy relaxation of hot carriers (Figure 12.23) which also limits the useable energy extraction from fundamental absorption to the band gap energy. A part of the generated carriers is lost by recombination processes where radiative recombination is the one which cannot be suppressed by thermodynamic reasons. This was shown in a fundamental article by Queisser and Shockley [24].

The above described three mechanisms limit the efficiency of solar cells to about 30%. The efficiency is slightly dependant on the solar spectrum which can be either approximated by a blackbody radiation (BB, T_s = 6000 K) or by the sun outside the earth (AMO) or by the sun spectrum (AM 1.5 G) below the atmosphere (1.5 atmospheric mass). For the AM 1.5 G spectrum the efficiency of an ideal pn-junction would be 33%. Si solar cells realized about 25% efficiency (research results on small area cells).

12.4.1. Tandem cells

The efficiency of the solar cell may be improved by combining two semiconductor p/n-junctions with different band gaps. For a one band gap cell the optimum efficiency is obtained for band gaps between 1.1 eV (Si) and 1.45 eV (GaAs). Figure 12.24 shows the theoretical efficiencies of a single cell as function of the band gap (AM 1.5 G sun spectrum). For comparison also the efficiency for a 5800 K blackbody radiation is given. With

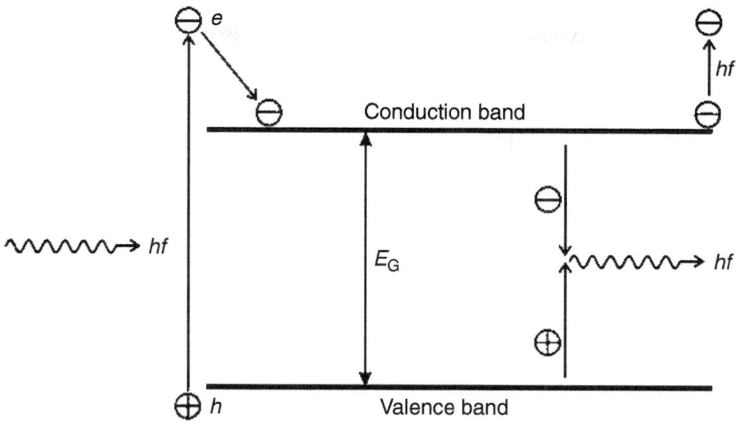

Figure 12.23. Absorption/emission in a semiconductor. Left: Fundamental absorption generates an electron/hole pair which looses its excess energy (hf-E_G) rapidly by energy relaxation of hot electrons (holes); Right: Free carrier absorption generates a hot carrier which looses its complete energy $h\nu$ by relaxation via phonons; Middle: Radiative recombination of electron/hole pairs is necessary even in perfect semiconductors for thermodynamic reasons.

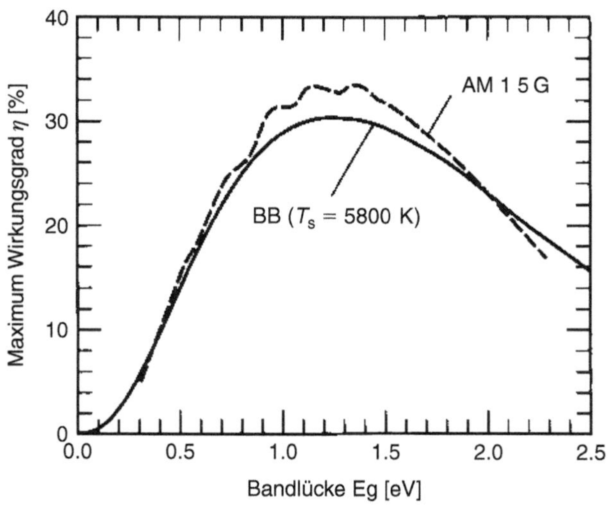

Figure 12.24. Maximum efficiency of a single solar cell as function of the band gap E_g (after Werner [25]).

stacking of different semiconductors the efficiency may be substantially increased (theoretically by a factor of about two). The practically achieved improvement is not that impressive but in either case tandem cells with three stacked layers (GaInP/GaAs/Ge) deliver the record values [26] now obtained for solar cells, (33% efficiency and 35.2% efficiency for unconcentrated and concentrated sun light, respectively). The Ge layers filter out the low energy spectrum of the sun which would be lost by a single cell structure.

12.4.2. Artificial substrates for group III/V solar cells

There is a lack of good material combinations for silicon-based solar cells. An alternative approach benefits from the good absorption of direct band gap III/V materials and the broad choice of band gaps available within III/V heterostructures whereas the main body of the silicon substrate assures light weight (density of silicon is 2320 kgm^{-3}), good cost efficiency and large area substrates. This III/V on Si approach requires in most cases a buffer layer which accommodates the different lattice constants. For GaAs-based heterostructures the lattice accommodation is best done by a rather thick (10 μm) Ge buffer layer whose lattice constant is very similar to that of GaAs. Silicon as a potential substrate for III/V – compound photovoltaics has many advantages [27] over traditional substrates that include higher thermal conductivity, lower weight, lower material costs and the potential to leverage the extensive manufacturing base of the Si industry. Such a technology would result in space solar cells that possess high specific power ($W\,kg^{-1}$) and high power density ($W\,m^{-2}$). For terrestrial solar cells the use of SiGe graded interlayers could provide for high conversion efficiency in concentrator cells. Relaxed, compositionally graded SiGe buffer layers, where the substrate lattice constant is effectively tuned from Si to that of Ge have shown great promise as artificial substrates for subsequent GaAs overlayers with a close lattice match. Typical growth conditions for low threading dislocation densities (10^5–$10^6\,cm^{-2}$) are given by a grading rate of 10% Ge μm^{-1} (Figure 12.25). The growth is interrupted at the 50% Ge layer [27] to perform a **CMP** step to remove the deepest "cross hatch" features from dislocation glide. The formation of antiphase domains (APD) is eliminated by growth on (100) substrates with a 6° off cut toward the {111} plane. Solar cell properties comparable with that of control cells grown on GaAs substrates were found.

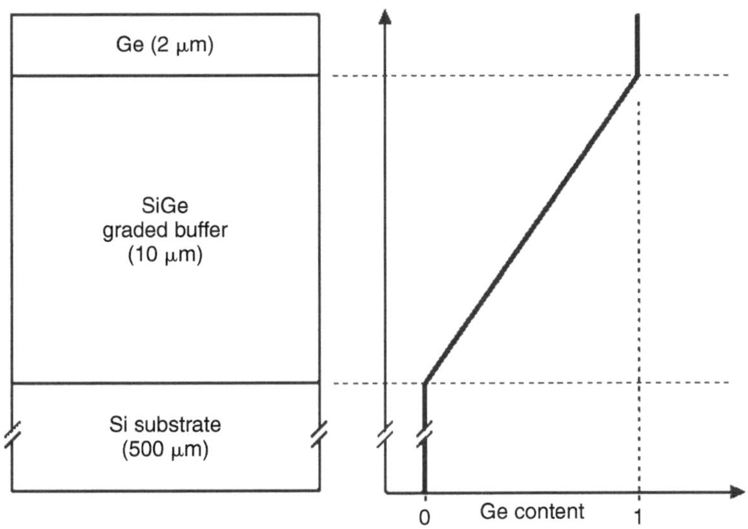

Figure 12.25. Artificial Ge substrate consisting of a silicon wafer, a thick (typical 10 μm) graded SiGe buffer layer and a relaxed Ge (1–2 μm) top layer.

The use of artificial substrates in solar cells could promote the more general integration of III/V – electronics and optoelectronics on Si substrates. As an example the fabrication of an InGaP/GaAs heterojunction bipolar transistor is mentioned [28]. Here the artificial substrate consisted of a GOI structure with a silicon handling wafer, an oxide layer and on top a Ge layer. The GOI wafer was fabricated by wafer bonding of a Si wafer and a wafer with graded SiGe buffer, a process which is described in more detail in Chapter 1. The low threading dislocation density was confirmed by good characteristics of the minority carrier device.

12.5. QD Applications

Ge QDs will be fabricated by mainly two ways. Single crystalline and oriented Ge dots are obtained by Stranski–Krastanov growth (see Section 12.2.1) of lattice mismatched Ge (preferably on Si). With epitaxy (e.g. MBE or advanced chemical vapor deposition, CVD methods) within a couple of seconds a dense array of islands (densities range from 10^9–$10^{12}\,cm^{-2}$) is created. The island dimensions are given by a rather large base (10–100 nm) and a small height (1–20 nm). From the viewpoint of the different confinement energies it is more appropriate to consider these dots as localized two-dimensional carriers. The main efforts in growth are directed to improve ordering and size uniformity [29]. Three-dimensional dot structures may be grown by vertical stackings of dot layer/cover, layer/dot layer . . . (QD SL). Ordering and size uniformity is improved strongly by growth on pre-patterned substrates.

The other method produces crystalline Ge QDs in an amorphous matrix, mainly SiO_2 or GeO_2. The dots are therefore not oriented and the electrical addressing is limited. The sizes of these QD can be small enough (3–7 nm) to show strong confinement in all three directions. Ge dots in GeO_2 are prepared [30] by decomposition of unstable solid germanium monoxide $2GeO \rightarrow GeO_2 + Ge$. Room temperature luminescence and the shift of the optical gap toward the visible (2.0–2.5 eV) proved the carrier confinement in the 1–3 nm large Ge dots. For a review, see Ref. [31].

12.5.1. Stressors

The Ge dots embedded in a surrounding mismatched matrix are stressed and stress also the surrounding material. If the surrounding material is silicon then the Ge is compressively strained and the silicon on top of the dot is tensile strained (Figure 12.26).

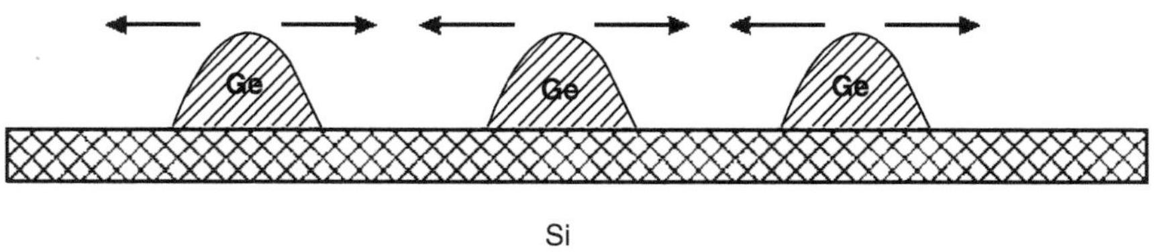

Figure 12.26. A buried Ge QD layer as Stressor for the semiconductor surface. Above the Ge QD tensile strain appears in the surface layer.

Therefore, a buried layer of Ge QDs may be used to produce strain on the surface layer which modifies the electronic properties, e.g. tensile strain increases the electron mobility considerably and it improves also the hole mobility if above a certain level.

12.5.2. Memories

Flash memories use a floating gate to store the information. The floating gate is loaded by a tunneling current (high voltage pulse). Within several years the floating gate is deloaded by low, but steady tunneling currents. Using instead a uniform floating gate a QD gate could improve loading and reduce failures.

12.5.3. Tunneling

QDs have discrete energy levels. They may be addressed by tunneling from doped semiconductors or metal contacts, often the excitation of carriers is done by a photo absorption process and the transport is done by tunneling [32–33]. The energy scheme of a photodetector is shown in Figure 12.27.

Excitation of NIR light as well as sub-band excitation of far IR light ($\lambda = 8$–$20\,\mu\text{m}$) was proven and used in photoconductive detectors.

12.6. Field Effect Transistors (Other than MOS)

For all FETs it is now accepted that Ge improves the transport properties and strain could be additionally helpful. The reader is referred to Chapter 11 for the details. Here we want to direct your look to the prospects of two different FET structures. The modulation doped FET (MODFET) is the dominant FET structure in advanced III/V – materials, more or less only with n-channels because in the typical III/V – materials the hole mobility is by a factor 20 (not only 3 as in Si) worse than the electron mobility. We will show the adoption of the concept for the Ge/Si system and will hint to the only symmetric complementary FET structure. The DotFET concept was specifically developed for Si MOS technology where a buried Ge dot array provides local strain to the surface channel. The strain mechanisms were explained in Section 12.5.

12.6.1. MODFET

The MODFET (a good selection of the first original articles is found in Ref. [34]) is based on carrier transport in a quantum well (2D electron or hole gas, 2DEG or 2DHG). The mobility of the confined carriers is additionally enhanced by a doping outside the channel (called modulation doping). Later also doping inside the channel was allowed to increase carrier density and on current although loosing much of the mobility gain. The classical adoption to the Ge/Si system would give the following vertical structure:

1. Virtual substrate (Si or SOI, relaxed SiGe buffer).
2. Strained Si (s-Si) channel (tensile strain), typical 5–10 nm thick.

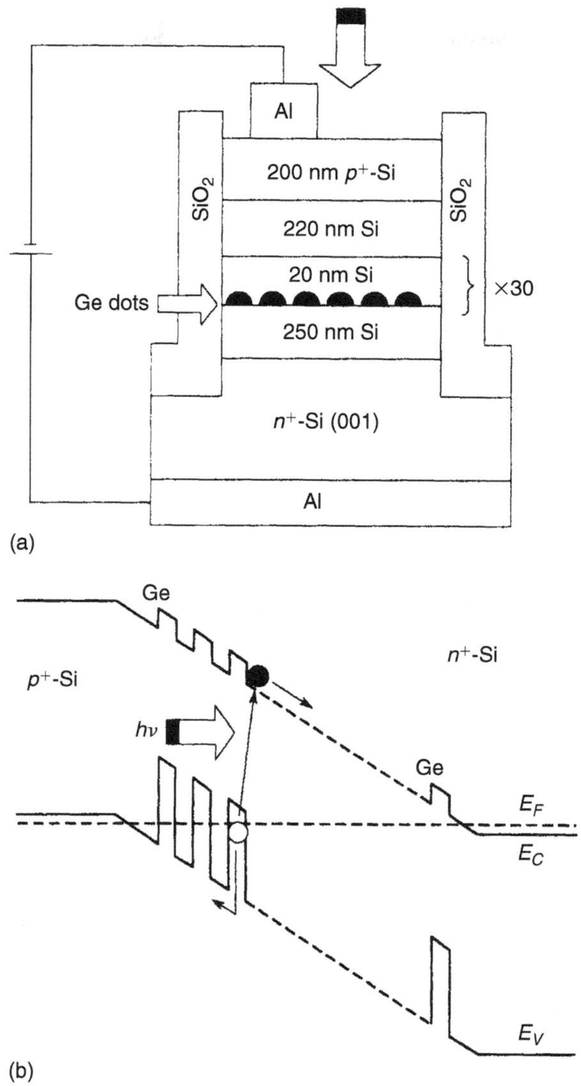

Figure 12.27. Cross-section and energy diagram of a multi-layer Ge QD detector.

3. SiGe barrier (spacer), undoped (1–4 nm thick).
4. SiGe doping layer (δ-doping, 1–5 nm).
5. SiGe/Si cap layer to provide a Si surface for the processing.

This n-MODFET (with a buried s-Si channel) proved to be a high speed, low noise transistor with relaxed lithography demands. Many of the technical problems were solved for moderate Ge content (30%).

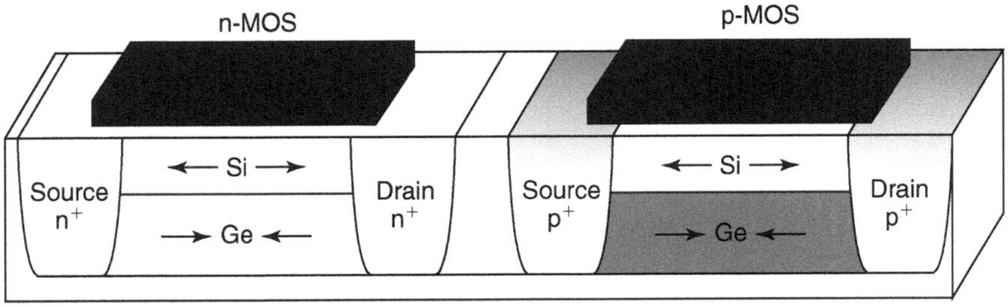

Figure 12.28. Complementary FET based on a strain symmetrized Ge/Si channel structure (after Kasper *et al.* [35–36]).

More radical suggestions [35–36] target to a unique Ge/Si solution for a high performance symmetric complementary FET (Figure 12.28).

The suggestion is based on the type II interface properties of a symmetrically strained Ge/Si heterostructure. The strain symmetrization is obtained by a vertical substrate (relaxed SiGe with about 50% Ge content). Electrons jump into the surface Si channel, while holes jump into the buried Ge channel. The n-MOS and p-MOS transistors are defined by the source/drain (S/D) doping. A modulation doping of the structure is only done for adjustment of the threshold voltage. In this structure the high hole mobility in Ge would allow to have symmetric channel properties. The technical challenge of the structure is given by the high strain level in Si and Ge, and the different temperature regimes of Si and Ge processing.

In the traditional III/V MODFET a Schottky gate is used, for Si/Ge this can be used but also replaced by a MIS gate. The layer sequence in Figure 12.28 was chosen to obtain a Si surface but also the inverted structure is principally possible.

12.6.2. DotFET

Strain in recent MOS generations is introduced locally by process steps. An elegant method [37] to introduce strain to the surface is given by using Ge dots as stressors (see Section 12.5). Above the dots tensile strain is produced. In order to get a pseudomorphic lattice spacing the Si between the dots has to be compressed. This delivers a local strain distribution depending on the positioning of dots.

Technical challenges to be solved in the future concern the positioning of dots, the self-alignment procedure and the process stability.

12.7. Spintronics

All electronics is now based on the discrete charge flow of moving electrons. But all electrons carry a spin which is a combination of two basic quantum states spin-up and spin-down. Spintronics uses also the movement of electrons but now monitoring the spin

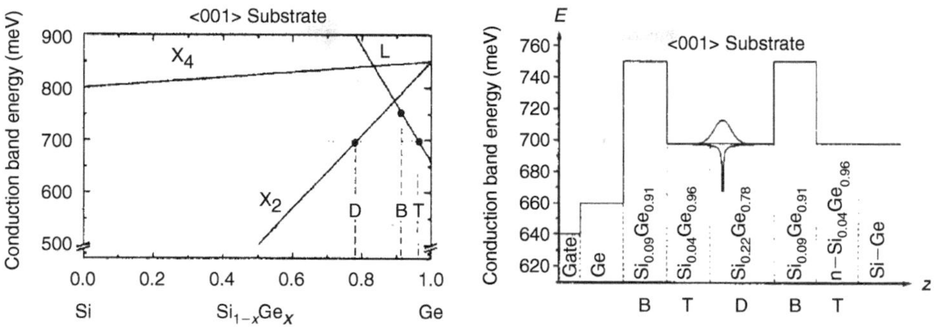

Figure 12.29. Quantum well structure with a split well equal in energy to L and X2 electrons. (after Vrijen *et al.* [39]).

properties of the crowd. A fascinating application could be quantum computing with semiconductors [38]. When monitoring the electron spin the interaction of electron spin with the nuclear spin of the atoms turns out to be a problem. Fortunately, Si and Ge have got in their natural isotope mixture some of them without any nuclear spin. This makes the semiconductors Si and Ge (from isotopes without nuclear spin) attractive for spintronics.

GeSi band gap engineering [39] may be used to bring the electron state into and out of resonance with an applied radio-frequency field (electron spin resonance). In tensile strained SiGe on Ge is the conduction band determined by the competition between the Ge like L-band and the Si like X2 band (twofold degenerate X band, electrons move perpendicular to the layer) as shown by Figure 12.29. For very high Ge contents ($x = 0.9$–1.0) the L-band is lowest, whereas for low Ge contents ($x < 0.9$) the X2 band dominates. Two Ge values deliver the same energy (one L, one X2). For the given example these Ge values are marked with D, T. A quantum well composed of one layer with $x = 0.96$ (T) and one with $x = 0.78$ (D) can be either populated with L-electrons or X2 electrons when a gate voltage is applied (Figure 12.29, right side). The g-factor of L and X2 electrons is different and so the radio frequencies for electron spin resonance are different.

12.8. Virtual Substrates

A strain relaxed SiGe as substrate would allow to adjust the strain in following layers. A strain relaxed SiGe buffer on a Si substrate (or SOI substrate) is called virtual substrate (Figure 12.30) which is an essential component in all SiGe-systems with strain adjustment.

12.8.1. Strain adjustment

The SiGe buffer is relaxed by the generation of a MD network at the interface. The dislocation network (distance p between dislocations) must be dense otherwise the buffer layer

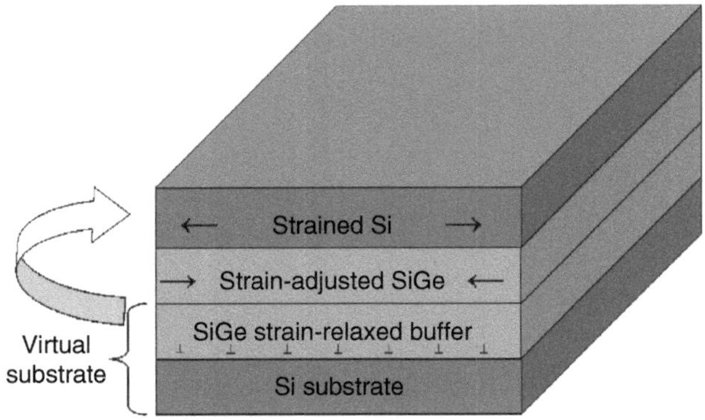

Figure 12.30. Virtual substrate (SRB on Si) with SiGe/Si heterostructure.

is only partially relaxed with a residual strain ε. The relation between p and ε reads:

$$\Delta a/a + \varepsilon = \frac{b_{\text{eff}}}{p} \tag{12.10}$$

With $\Delta a/a$ the lattice mismatch and b_{eff} the effective Burgers vector length of the dislocation, depending on type and geometry of the dislocation. For 60° dislocations (typical glide dislocation in the diamond lattice) on (100) substrates the effective Burgers vector amounts to 0.19 nm. The in-plane lattice constant of the buffer layer is given by:

$$a_{\text{b}} = a_{\text{b0}}\,(1 + \varepsilon) \tag{12.11}$$

This in-plane lattice constant is the reference mesh for all following coherent layers and it defines the strain ε_i in the ith layer by:

$$\varepsilon_i = \frac{a_i - a_b}{a_i} \tag{12.12}$$

12.8.2. Thin virtual substrates

For monolithic integration the relaxed buffer layer should be very thin. But full relaxation and good quality of the buffer are obtained with rather thick buffer layers (up to 15 μm for Ge/Si). In order to meet integration demands (and also to avoid thermal constraints) methods delivering lower buffer thicknesses (thin virtual substrates were investigated [40–42]). Our group [43] lets intentionally introduced PD nucleate to dislocation loops and we obtained high degrees of relaxation with ultra-thin (60–80 nm) buffer layers. The PD supersaturations were generated by growth at very low temperatures (VLT growth) below 200°C (Figure 12.31).

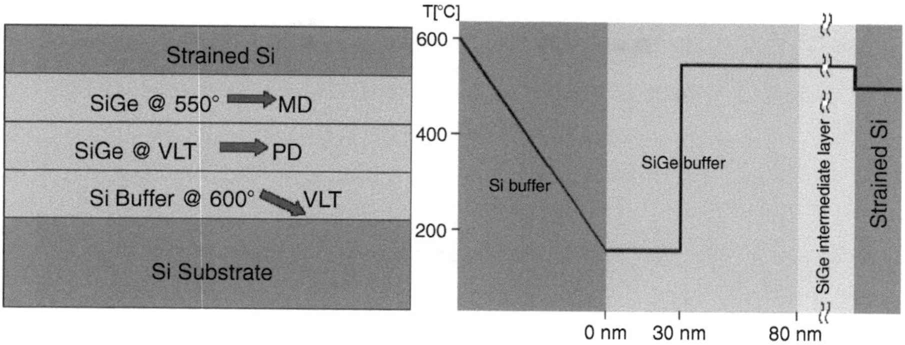

Figure 12.31. Growth sequence for a ultra-thin virtual substrate. PD are generated during a VLT growth stage. They nucleate at usual growth to MD (after Kasper *et al.* [43]).

12.9. Conclusion

Germanium is an indirect semiconductor like silicon but with a smaller band gap and a direct band gap transition slightly above the indirect one. This properties makes Ge superior to Si for optical applications from the visible to the NIR (1.6 μm). Fast IR detectors approaching the mm-wave frequency band and effective wave guiding were already demonstrated. With carrier confinement in quantum wells and QDs quantum confined stark shift (QCSE) modulators are on the way. The question of light sources is an open one, because of efficiency reasons.

With strain adjustment in Ge/Si heterostructures routes to symmetric complementary FET structures and spintronics will be opened if the material and processing challenges of highly mismatched heterodevices can be solved.

References

[1] S. Benzer, *Phys. Rev.* **70** (1946) 105.

[2] D.E. Sawyer and R.H. Rediker, *Proc. IRE* **46** (1958) 1122.

[3] H.U. Harten and W. Schultz, *Z. Phys.* **141** (1955) 319.

[4] E. Kasper and D.J. Paul, *Silicon Integrated Circuits*, Springer-Verlag, Berlin, 2005.

[5] Landolt-Börnstein, New Series **17**, Springer-Verlag, Berlin, 1982.

[6] EMIS Datareviews **24**, *IEE INSPEC*, London, 2000.

[7] T.F. Wietler, E. Bugiel and K.R. Hofmann, *Thin Solid Films* **508** (2006) 6.

[8] E. Kasper, H.J. Herzog and K. Wörner, *J. Cryst. Growth* **81** (1987) 458.

[9] H.J. Herzog and E. Kasper, *J. Electrochem. Soc.* **132** (1985) 2227.

[10] M. Jutzi, M. Berroth, G. Wöhl, M. Oehme and E. Kasper, *IEEE Photonics Techn. Lett.* **17** (2005) 1510.

[11] M. Oehme, J. Werner, E. Kasper, M. Jutzi and M. Berroth, *Appl. Phys. Lett.* **89** (2006) 071117.

[12] G. Dehlinger, S.J. Koester, J.D. Schaub, J.O. Chu, Q.C. Ouyang and A. Grill, *IEEE Photonics Techn. Lett.* **16** (2004) 2447.

[13] G. Masini, V. Cencelli, L. Colace and G. Assanto, *IEEE J. Quantum Electr.* **10** (2004) 811.

[14] O.I. Dosunmu, D.C. Cannon, M.K. Emsley, B. Ghyselen, J. Liu, L. Kimerling and M.S. Ünlü, *IEEE J. Quantum Electr.* **10** (2004) 694.

[15] J. Yu, E. Kasper and M. Oehme, *Thin Solid Films* **508** (2006) 396.

[16] A. Splett, T. Zinke, K. Petermann, E. Kasper, H. Kibbel, H.J. Herzog and H. Presting, *IEEE Photonics Technol. Lett.* **6** (1994) 59.

[17] Y.H. Kuo, Y.K. Lee, Y. Ge, S. Ren, J.E. Roth, T. Kamins, D. Miller and J. Harris, *Nature* **437** (2005) 1334.

[18] H.G. Grimmeiss, V. Nagesh, H. Presting, H. Kibbel and E. Kasper, *Phys. Rev. B* **45** (1992) 1236.

[19] D. Kuehlke, OptikVerlag Harri Deutsch, Frankfurt (2004).

[20] R.A. Soref, J. Schmidtchen and K. Petermann, *IEEE J. Quantum Electr.* **27** (1991) 1971.

[21] M. Mosbach, Private communication.

[22] E. Kasper and C. Klingshrin, Group IV Quantum Structures, subvolume C34/3 in Landolt-Börnstein, New Series (Series Editor: K. Mattiensen), Springer Verlag, 2007.

[23] R. Zachai, K. Eberl, G. Abstreiter, E. Kasper and H. Kibbel, *Phys. Rev. Lett.* **64** (1990) 1055.

[24] W. Shockley and H. J. Queisser, *J. Appl. Phys.* **32** (1961) 510.

[25] J.H. Werner, *Nova Acta Leopoldina* **339** (2004) 65.

[26] R.R. King, C.M. Fetzer, P.C. Colter, K.M. Edmondson, A.A.P. Stavrides, G.S. Kinsey, H.L. Cotal, J.H. Emmer, R.A. Sherif and N.H. Karam, *Proc. 3rd World Conf. Photovoltaics*, 2003, 622.

[27] M.R. Lueck, C.L. Andre, A.J. Pitera, M.L. Lee, E.A. Fitzgerald and S.A. Ringel, *IEEE Electr. Dev. Lett.* **27** (2006) 142.

[28] G. Thomas, E.S. Johnson, C. Tracy, P. Maniar, X. Li, B. Roof, Q. Hartmann and D. Ahmari, *IEEE Electr. Dev. Lett.* **26** (2005) 438.

[29] C. Huang, J. Ju and Q. Wang, *Prog. Nat. Sci.* **14** (2004) 388.

[30] E.B. Gorokhov, D.A. Martin, D.A. Orekhov, A.G. Cherkov, A.K. Hutakovskii, V.A. Shvets, A.G. Borisov and M.D. Efrimov, *Semiconductors* **39** (2005) 1210.

[31] Y. Masumoto, T. Tokahare (ed), *Semiconductor Quantum Dots: Physics, Spectroscopy, Applications*, Springer-Verlag, Berlin, 2002.

[32] A.I. Yakimov, A.V. Dvurechenskii, A.I. Nikiforov, S.V. Chaikovskii and S.A. Tiis, *Semiconductors* **37** (2003) 1345.

[33] A.I. Yakimov, A.V. Dvurechenskii, A.A. Nikiforov and Yu. Proskuryakov, *J. Appl. Phys.* **89** (2001) 5676.

[34] Modulation-Doped Field-Effect Transistors, edited by H. Daembkes, IEEE Press, New York, 1991.

[35] E. Kasper and G. Reitemann, Future Trends in Microelectronics – The Road ahead, edited by S. Luryi, J. Xu, A. Zaslavsky, John Wiley, New Jersey, New York, 1999, pp. 125–132.

[36] E. Kasper, *ULSI Process Integration II*, edited by C. Claeys, F. Gonzales, J. Murota and K. Saraswat, *Electrochem. Society Proceeding.* **2001–2**, 2001, p. 143.

[37] G. Schmidt and K. Eberl, *IEEE Trans. Electr. Dev.* **48** (2001) 1175.

[38] I. Shlimak, Private communication.

[39] R. Vrijen, E. Yablonovich and K. Wang, *Phys. Rev. A* **62** (2000) 012306.

[40] H. Chen, L.W. Guo, Q. Cui, Q. Hu, Q. Huang and J.M. Zhou, *J. Appl. Phys.* **79** (1969) 1167.

[41] H. Trinkaus, B. Holländer, S. Mantl, H.J. Herzog, H. Kibbel and T. Hackbarth, *Appl. Phys. Lett.* **76** (2000) 3552.

[42] O. Pchelyakov, Private communication.

[43] E. Kasper, K. Lyutovich, M. Bauer and M. Oehme, *Thin Solid Films* **336** (1998) 319.

CHAPTER 13

Trends and Outlook

Eddy Simoen and Cor Claeys

13.1. Introduction

Chapters 8 till 11 clearly demonstrated that so far encouraging results have been obtained regarding the development of metal-oxide-semiconductor field-effect transistors (MOSFETs) on germanium. Particularly for pMOSFETs the feasibility of achieving higher drive currents than on strained Si or SiGe channels has been shown for long-channel transistors. The performance improvement on short devices is yet to be established, whereby the parasitic series resistance, associated with the contact and lowly-doped drain (LDD) junctions, is one of the major obstacles to be overcome. The situation for n-channel devices on germanium is more problematic, owing to the observed high density of acceptor-like interface traps near the conduction band, pinning the Fermi level and inhibiting the formation of an inversion layer. An additional constraint is the low activation and high diffusion of n-type dopants like As and P, rendering the fabrication of low resistance (ultra)-shallow junctions difficult if not impossible. One approach to tackle this technological barrier is the use of ultra-short annealing schemes, like laser [1,2] or flash-lamp annealing [3,4]. Another possibility to control the concentration-enhanced diffusion is by point defect engineering, whereby for the case of germanium the supply of vacancies should be suppressed e.g. by co-doping with neutral (C, F) or non-standard dopants.

Even if Ge holds promise for the sub-22 nm complementary MOS (CMOS) node, one should first of all consider the materials issue. As shown in Chapter 1, Ultra-large scale integration (ULSI) compatible 200-mm Cz Ge bulk substrates can be grown essentially dislocation free and the feasibility of 300-mm growth and wafering has been demonstrated.

Germanium-Based Technologies: From Materials to Devices. C. Claeys and E. Simoen (Editors)

Nevertheless, it is clear that from a viewpoint of materials cost and supply and considering practical processing using standard silicon compatible equipment (lower mechanical strength of Ge compared with Si), realistic Ge CMOS technologies will be developed on thin channel layers, e.g. employing Ge-on-Insulator (GOI) or thin epitaxial germanium on a silicon substrate. In Chapter 1, the fabrication of GOI based on the so-called smart-cut technique has been described in some detail. Here, in a first section we will focus on an alternative approach, namely, the Ge condensation technique, which is currently receiving quite some research efforts. Besides, the epitaxial growth of thin germanium layers on silicon is a competing technology, although one has to deal with the large 4% lattice mismatch, generating strain relaxation by extended defect formation. Several schemes for controlling the threading dislocation density (TDD) will be discussed in Section 13.2 together with some alternative techniques to grow thin crystalline Ge layers on a Si or SiO_2 template.

The second section will focus more on alternative device schemes, which may alleviate some of the problems stated above, like Schottky barrier Ge MOSFETs. In order to enhance the on current, one may consider the use of alternative channel materials like GaAs or other III–V binary and ternary compounds. Finally, the potential of 1D and 0D device structures based on Ge nanowires and quantum dots (QDs), respectively, will be briefly discussed.

13.2. GOI and Epitaxial Germanium Substrates

As shown in Chapter 1, good progress has been made in the fabrication of GOI substrates by wafer bonding techniques. For a recent overview, the readers are also referred to [5]. Here alternative techniques to fabricate a thin germanium channel layer on silicon or SiO_2 will be briefly outlined.

13.2.1. Ge condensation technique

The principle of the Ge condensation technique is schematically represented in Figure 13.1 [6,7] and relies on the ejection of Ge from high-temperature oxidized $Si_{1-x}Ge_x$. In order to obtain a GOI layer, one starts from an Si-on-insulator (SOI) wafer, where a thin $Si_{1-x}Ge_x$ layer is grown by epitaxy, the volume fraction of Ge(x) usually being in the range 15–25%. This is followed by a high-temperature dry oxidation (>700°C), resulting in an enrichment of the SiGe layer in between the growing top oxide and the buried oxide (BOX). At full completion of the oxidation process a Ge rich layer is obtained, where x can be close to 100%. Recently, models have been developed to describe this enrichment in function of the oxidation time and thickness [8–10].

The advantage of this technique is that it is compatible with standard silicon processing and that it can also be applied locally, i.e. in selected parts of the wafers defined by Shallow Trench Isolation (STI) or mesa etched regions. Evidence has been advanced that the Ge enriched layer formed by the condensation technique is partially relaxed, whereby the BOX underneath helps to relax the strain without significant dislocation formation [5]. Nevertheless, TDDs in the range of $\sim 10^7\,cm^{-2}$ indicate that the relaxation is also assisted by dislocation nucleation. In order to improve the stress relaxation, it has been shown that medium-dose ($5 \times 10^{15}\,cm^{-2}$) H^+ implantation followed by a two-step annealing helps in

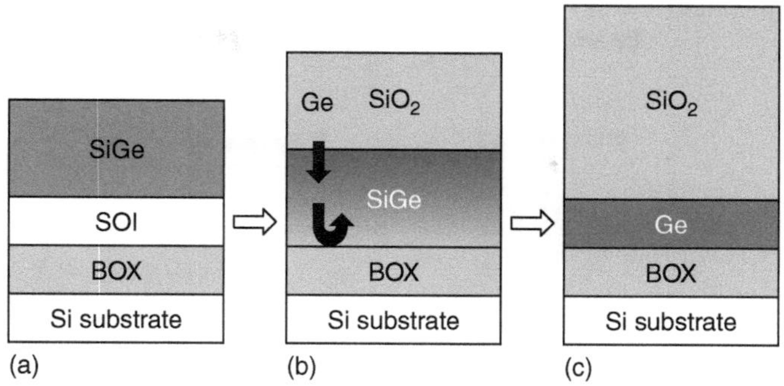

Figure 13.1. GeOI fabrication process by the Ge condensation technique. (a) SiGe layer is grown epitaxially on an SOI wafer. (b) Oxidation of SiGe-on-insulator (SGOI). (c) Ge is condensed completely.

relaxing the strain while keeping the TDD at values below $10^6 cm^{-2}$ [11]. It is believed that this improvement is related to the weakening by the ion implantation of the SiGe/SiO_2 bonding at the BOX interface, facilitating the glide of the SiGe layer over the BOX during the high-temperature oxidation.

Recently, the germanium condensation technique on vertical compliant structures has been demonstrated, which opens the door for the development of multiple-gate field-effect transistors (FETs) (or FinFETs) on germanium fins [12].

More exotic techniques for the growth of crystalline germanium on SiO_2 are currently also under study. Rapid melt growth has been applied using a so-called SiO_2 microcrucible [13]. In this case, amorphous germanium is sputtered on a silicon substrate capped by Si_3N_4 or SiO_2 and next encapsulated by a low-temperature SiO_2 layer. Good recrystallization with low defect density is obtained by using the defect necking technique [13]. Hetero-epitaxy of good quality germanium layers on silicon can also be obtained by nanoscale Ge seeds grown through a thin SiO_2 layer [14]. While promising results have been obtained it remains to be seen which of these techniques will survive, the main criterion will be compatibility with standard ULSI processing and cost.

A final remark with respect to GOI materials is that as for the front interface, one should pay attention to the passivation of the buried interface. This is related to the fact that GOI devices will probably operate in the ultra-thin film, fully-depleted mode, whereby the characteristics are equally determined by the interface states at the back interface. Recently, research has been oriented to optimize the passivation of the BOX relying on plasma oxidation and nitridation [15,16].

13.2.2. Germanium epitaxial growth on silicon

The growth of pure Ge on silicon remains a tremendous challenge due to the large lattice mismatch. It has been shown that up to three atomic layers of Ge can be grown layer by layer on

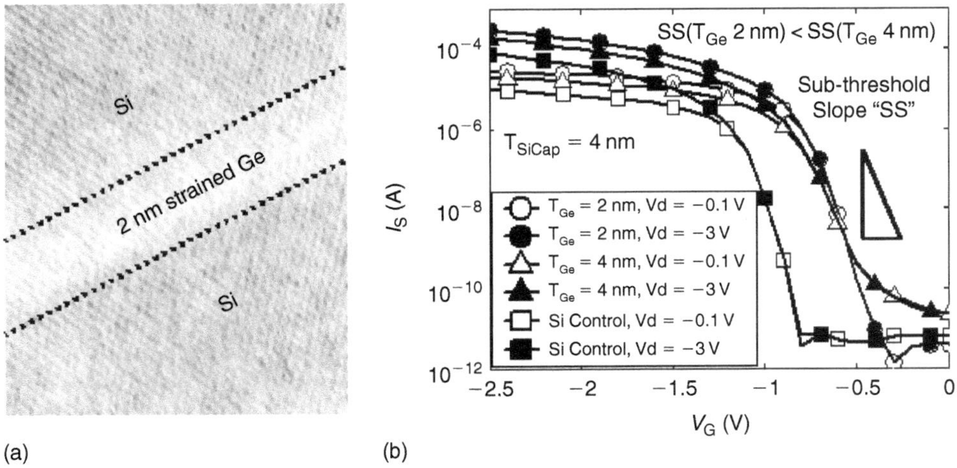

Figure 13.2. (a) Transmission Electron Micrograph (TEM) of defect free ultra-thin strained Ge on relaxed Si. (b) Source current versus gate voltage for the strained Ge (on bulk Si) pMOSFETs (after T. Krishnamohan *et al.* [19]).

a Si(100) substrate (2D growth) [17] before 3D Stranski–Krastanov growth sets in, resulting in the formation of Ge islands. This is related to the fact that the formation of a 3D structure is energetically favored by the large increase in the strain energy upon growing thicker layers. However, it has recently been proposed that such thin Ge layers (~2 nm) could be beneficially exploited to reduce the band-to-band tunneling issue in Ge, which is compromising the off-state current of a transistor [18–20]. Figure 13.2a shows the thin strained Ge layer covered by a silicon cap layer, while Figure 13.2b depicts the improved off-state characteristics of pMOSFETs fabricated in this channel stack. It is clear that both the Ge and Si cap layer thickness need to be carefully optimized to achieve the best I_{on}–V_{off} performance.

Although the use of ultra-thin strained Ge layers sounds attractive, it may be extremely demanding from a processing viewpoint. Moreover, as will be argumented below, a future high-performance integration scheme could involve III–V epitaxial layers, deposited on a relaxed Ge template. For that purpose thicker germanium epi layers are required.

Several methods have been developed in order to deposit rather thick epitaxial Ge layers on silicon. A first approach relies on the deposition of a thick, relaxed compositionally graded $Si_{1-x}Ge_x$ virtual substrate [21–28], also called strain-relaxed buffer (SRB) layer. Figure 13.3 illustrates the concept. Usually, a chemical mechanical polishing (CMP) step is implemented in order to reduce the surface roughness of the epitaxial layer, which is related to the so-called cross-hatch pattern associated with the misfit dislocations at the Si–$Si_{1-x}Ge_x$ interface. Other drawbacks of the use of a thick virtual substrate is the high material's cost and the lower thermal conductivity of Ge and SiGe compared with silicon, which leads to self-heating of the transistors under large drain bias operation. As the mobility and drain current are reduced at higher temperatures due to phonon scattering, this will deteriorate the overall device performance.

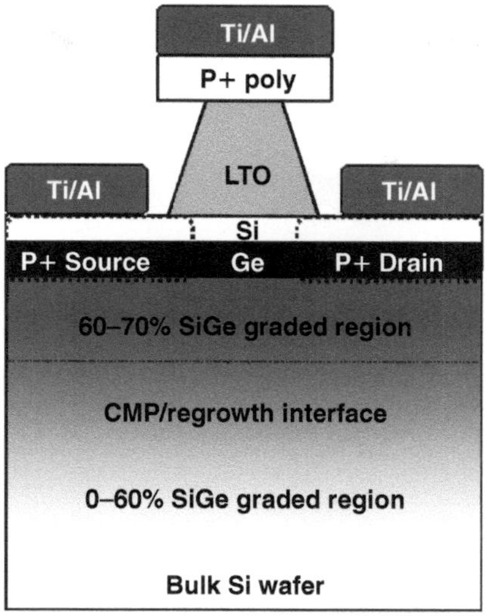

Figure 13.3. Schematic of a Ge channel pMOSFETs fabricated on a thick relaxed compositionally graded SiGe virtual substrate (after Lee *et al.* [24]).

Attempts have been made to grow directly thick germanium layers on silicon by a two-step deposition approach followed by cyclic thermal annealing [29]. The two-step approach is composed of a low-temperature (350°C) deposition of a thin Ge buffer, after which the substrate temperature was raised to e.g. 600°C for the deposition of a 1 μm thick Ge layer. Finally, the wafers were cyclic annealed between a high and a low annealing temperature to reduce the TDD. The combination of selective area growth with cyclic thermal annealing enabled an average TDD as low as a few $10^6 cm^{-2}$ [29]. This method has mainly been developed by the MIT Group for the integration of Ge p-i-n photodetectors on silicon [30,31].

Recently, it has been proposed to apply a hydrogen annealing step post-epitaxial Ge deposition by chemical vapor deposition (CVD) [32,33]. As sketched in Figure 13.4, the initially rough Ge surface, characterized by islanding is smoothened considerably by the hydrogen anneal. When performed at 825°C, a reduction of the roughness by 90% has been obtained [32]. On this smooth layer, a second epitaxial growth can be performed, followed by another hydrogen anneal until the desired epi layer thickness is reached. This method has been termed multiple hydrogen annealing for hetero-epitaxy (MHAH), whereby the final Ge layer is defect free. It has been shown that on these layers W/GeO_xN_y capacitors with good *C–V* characteristics are obtained [32].

The use of surfactants for the hetero-epitaxial growth of germanium on silicon by molecular beam epitaxy (MBE) is also well documented in the literature. The main role of a surfactant is to lower the surface energy and promote the layer by layer growth. For that purpose, one mono-layer of certain impurities is deposited on a clean silicon substrate before the deposition of germanium. The role of the surfactant is twofold: the adatoms passivate the surface dangling bonds, thereby controlling the energetics of the epi growth and they change the kinetics of the process by affecting the diffusion barrier for the deposited

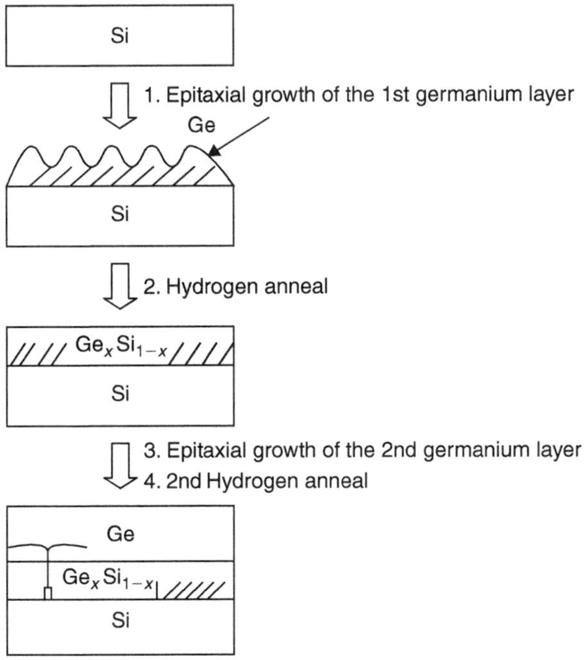

Figure 13.4. Schematic process flow of the MHAH method used to grow hetero-epitaxial Ge layers on Si (after Saraswat *et al.* [33]).

Ge atoms [34]. Ideally, the surfactant continues to lower the surface energy by diffusing up to the surface without being incorporated in the layer. As a result a smooth interface is obtained. A popular element for surfactant mediated epi (SME) growth is Sb, which can terminate a clean Si surface by a 1x1 reconstruction, perfectly suited for subsequent epitaxial growth [35–42]. High electron mobilities ($3100\,cm^2V^{-1}s^{-1}$ at 300 K) have been obtained on 1 μm thick Ge epitaxial layers on (111) Si by Sb SME [43]. The technique has also been applied to the growth of graded SiGe buffers for high-quality Ge films on Si [44].

Other candidate surfactants for SME of germanium on silicon comprise Bi [45,46], Ga [47–49] and Te [50–52]. One drawback is the fact that part of the surfactant atoms do get incorporated in the germanium layer, giving rise to a significant p- or n-type doping density when Ga, on the one hand, or Te, Bi or Sb, on the other are employed. This problem can be overcome by implementing hydrogen as a surfactant [34,53–57]. A more serious handicap is the fact that the method is mainly used in ultra-high vacuum (UHV) epitaxial deposition systems and is not so easily transferable to more industrial low- or reduced pressure CVD tools.

Some progress has also been made by using Ar-plasma-enhanced GeH_4 reaction for the CVD of smooth Ge films on Ge(100) [58]. The crystallinity of the films can be improved by avoiding the incorporation of Ar through a reduction of the Ar ion energy and of the substrate temperature.

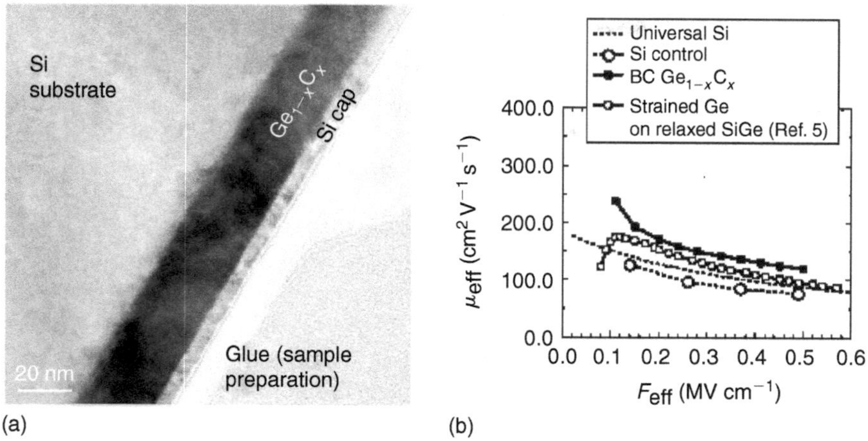

Figure 13.5. (a) High-resolution cross-section transmission electron microscopy (XTEM) image of a 30-nm $Ge_{1-x}C_x$ layer grown directly on a Si(100) substrate, with a 6-nm thick cap layer, showing good crystalline quality with no visible defects. (b) Effective hole mobility (μ_{eff}) of a buried-channel $Ge_{1-x}C_x$ pMOSFET as a function of the vertical effective field F_{eff}. Mobility data for surface-channel strained Ge on relaxed $Si_{1-x}Ge_x$ from Ref. [26] are included for comparison (after Kelly *et al.* [64]).

A final trend in the epitaxial deposition of Ge-based layers on silicon is the formation of metastable $Ge_{1-x}C_x$, with $x \sim 1\%$. It should be reminded that, as shown in Chapter 2, C has a very low solubility in germanium and no stable crystalline GeC phases have been identified because of the thermodynamic instability of the Ge–C bond. Initial experiments were based on MBE growth [59–61]. Later, the feasibility of UHV–CVD deposition has been demonstrated, based on the reaction of GeH_4 with methylgermanes at 470°C [62,63]. This resulted in the growth of $x = 0.015$–0.03 alloys on silicon with a diamond-cubic lattice and high crystallinity, with defects mainly confined to the epitaxial interface [63]. This is illustrated in Figure 13.5a showing a transmission electron microscopy (TEM) cross section [64] of a 30-nm $Ge_{1-x}C_x$ layer with 1 at % of carbon covered by a 6-nm thick Si cap layer to passivate the Ge surface for improved interfacial quality. Thin layers with an etch pit density (EPD) of $<3 \times 10^5 cm^{-2}$ can be achieved which is much lower than for pure Ge epi layers. It is believed that the stress compensation exerted by the small substitutional C atoms lowers the overall stress in the layer [63–65], resulting in a lower defect density and a 2D growth. There is still compressive strain in the final films, which opens good perspectives for the development of p-channel devices [65]. This is illustrated in Figure 13.5b, where the highest low-field mobility is found for the buried-channel pMOSFETs fabricated in $Ge_{1-x}C_x$ layers [64]. The high mobility indicates that alloy scattering for $x = 1\%$ does not play a significant role. Other potential advantage is the slightly larger band gap of the alloy layer, increasing with x. It has, finally, been demonstrated that direct deposition of a HfO_2/TaN gate stack on a 30-nm $Ge_{1-x}C_x$ film yields a favorable density of interface traps ($\sim 4.8 \times 10^{11} eV^{-1} cm^{-2}$), as derived from C–V measurements on the resulting MOS structure [65].

13.3. Alternative Ge-based Device Concepts

One of the key issues in the development of CMOS on germanium is the poor activation of the n^+ source/drain (S/D) junctions, contributing to the underperformance of the resulting n-channel transistors. As pointed out in Chapter 11, one way out is to replace an implanted junction by a metal (germanide) Schottky contact. However, since all germanides give acceptable barrier heights on n-type material only, this merely provides an alternative for pMOSFETs and does not alleviate the n^+ S/D issue. While it becomes more and more obvious that for the ultimate p-channel performance, stressed germanium is the favorite choice, this is less obvious for nMOSFETs. From the diagram in Figure 13.6, one can derive in a straightforward manner that InSb (and III–V compounds in general) offer much higher low-field electron mobilities compared with silicon and germanium. This raises the idea of exploiting this potential for the development of high-performance scaled devices.

13.3.1. GaAs and III–V on germanium FETs

The same arguments motivating the current interest in metal-insulator-semiconductor (MIS) devices on germanium equally hold for III–Vs: a higher mobility, combined with the implementation of a high-κ gate dielectric. Again, it is clear that the technology will not be developed on III–V substrate materials but on epitaxial layers, preferable deposited on a silicon template. This enables the co-integration of more standard silicon devices with high-performance III–V components and eventually on a longer term with III–V photonics. However, due to the large lattice mismatch, hetero-epitaxy of III–V compounds on silicon is problematic, suffering from strain relaxation and extended defect formation. On the other hand, there is a nearly perfect match between the lattice parameters of Ge and GaAs e.g. leading to a possible scenario depicted in Figure 13.7 for ultimate performance CMOS.

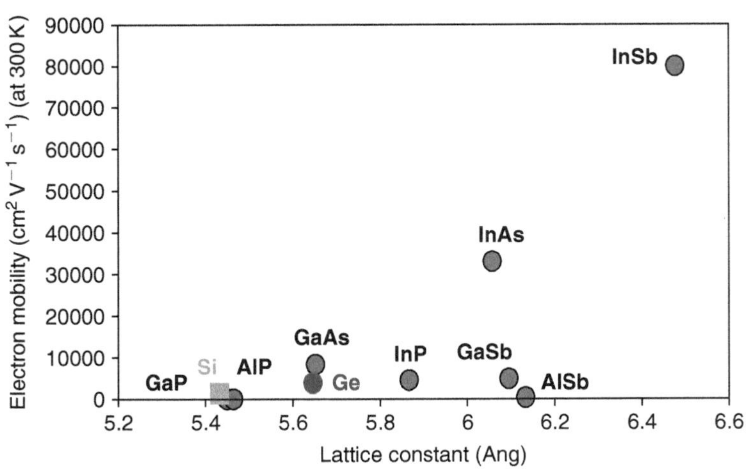

Figure 13.6. Electron mobility versus lattice constant for Si, Ge and several III–V binary compounds at 300 K.

While most of the foregoing Chapters were dealing with developing CMOS on a single channel material, in this case GOI, it could turn out advantageous to perform the exercise on a GOI platform (Figure 13.7b) whereby a Ge p-channel is utilized, while nMOSFETs are fabricated in a III–V layer selectively deposited on a germanium template.

While the idea sounds logic and attractive, many problems will have to be solved before becoming industrial reality. Basically the same issues have to be tackled as for the case of germanium: namely, surface and interface passivation and dopant activation. This explains the growing number of papers published in the last year on MIS capacitors on III–V semiconductors [66–72]. The same range of dielectrics as for high-κ applications on Si and Ge are being explored: ZrO_2, HfO, MgO, Al_2O_3, Si_3N_4, SiO_2, . . . [73]. The ultimate goal is the development of high-performance MOS III–V or heterostructure FETs as depicted in Figure 13.8 [74–81]. As can be seen there, no junctions but metal contacts are being employed, eliminating the issue of junction formation.

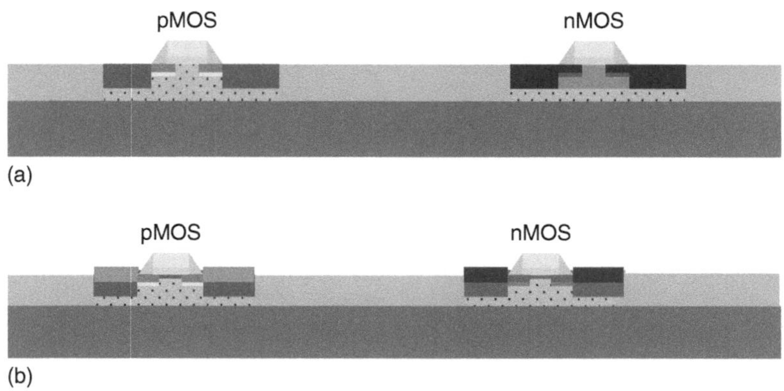

Figure 13.7. Schematic representation (a) of GOI CMOS and (b) of future mixed Ge/III–V CMOS.

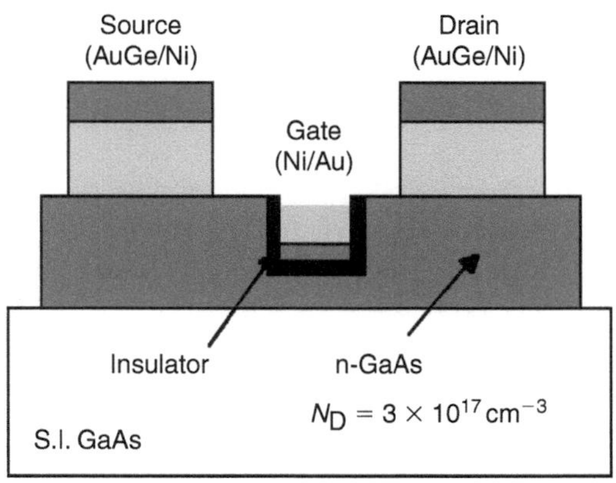

Figure 13.8. Cross-sectional view of a GaAs MOSFET (after Fujino *et al.* [81]).

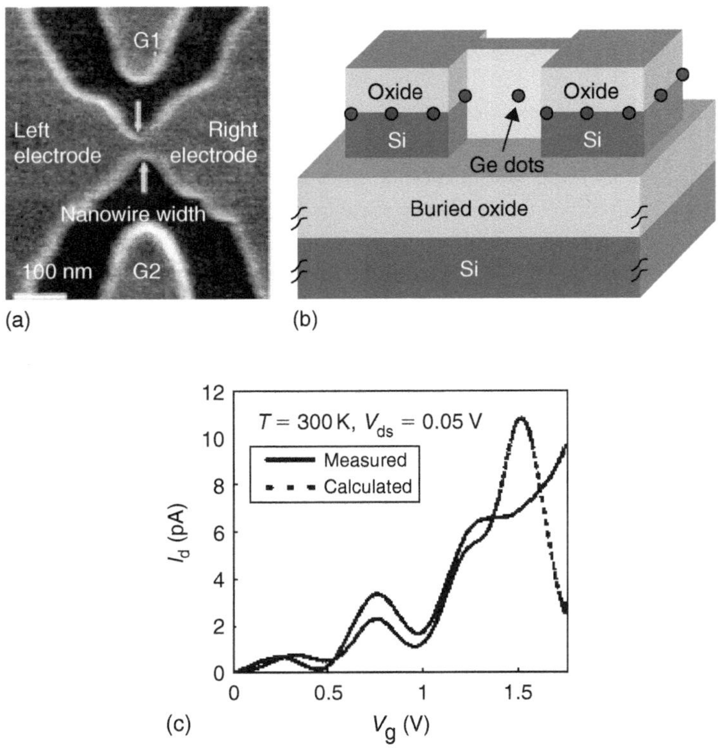

Figure 13.9. (a) Plan-view scanning electron micrograph of a quantum point contact structure. (b) Schematic of Ge QDs formed after thermal oxidation of SiGe/Si-on-Insulator (after Liao *et al.* [87]). (c) Drain current as a function of gate voltage with a drain voltage of 50 mV at 300 K (after Li *et al.* [86]).

13.3.2. Germanium nanowire and QD devices

Looking further ahead past the International Technology Roadmap for Semiconductory (ITRS) 1D and 0D germanium-based device structures may also play a significant role. In Chapter 11, the fabrication and operation of Ge nanowire FETs has been briefly described. Such nanowires are typically formed using organic mono-layer coated gold nanocrystals, acting as catalysts during a vapor–liquid–solid synthesis in a CVD process [82]. Once more, key issue for successful device operation is the surface passivation [83], whereby Cl-termination appears to be more stable in air (HCl treatment) than H-termination from an HF-treated surface. Well behaved nanowire FETs have been realized in the meantime, implementing an HfO_2 [84] or ZrO_2 [85] gate dielectric.

At the same time, the operation of germanium QD single-electron transistors, represented in Figure 13.9 has been demonstrated [86,87]. They are formed by the selective oxidation of $Si_{0.95}Ge_{0.05}$/Si wires on an SOI substrate [86]. Similarly as for the condensation technique, Ge atoms segregate out of the growing oxide and condense into Ge nanocrystals. They were observed to have a large Coulomb barrier, allowing functioning at 300 K as shown in Figure 13.9b [86,87].

13.4. Conclusions

It is evident from the foregoing that the development of CMOS on germanium is not just a simple translation of what worked for silicon to germanium. While both semiconductors belong to the same Group IV column and are similar in many aspects, in some others, which are technological relevant they are so different. One is related to the susceptibility of the germanium surface/interface to form interface states – intuitively one can reason that due to the smaller band gap and considering the same number of broken bond at a Si or Ge surface (in the worst case, one per surface atom) they have to be distributed over a 40% smaller energy interval. Anyhow, this higher density of interface traps has to be dealt with in order to preserve the potentially higher Ge mobility in a deep submicron MOSFET. A bit ironically maybe, but the reason for the abandoning of germanium as the leading semiconductor material is still a threat for its use in future deep submicron CMOS. The success of the large research efforts currently undertaken world wide will largely depend on a breakthrough in the adequate passivation of surface states. Apparently, the recipes that work for silicon, like hydrogen passivation by a forming gas anneal, do not for Ge. Fundamentally, this could point to a different nature of the interface states, which probably are not dangling bonds. This has led people to implement a thin epitaxial silicon layer on top of germanium to replace the Ge–oxide interface by a much better behaving silicon–oxide interface. The question here is whether this is scalable to an Equivalent Oxide Thickness (EOT) of 0.5 nm required for high-performance sub-22 nm transistors. Anyhow, a clear conclusion coming out of this chapter is that epitaxy will play a more and more prominent role in future CMOS technologies.

The role of epitaxial processing will be even more pronounced if the semiconductor industry decides to turn to III–Vs for n-channel transistors. This is a formidable task as the possible options are wide open given the wide scope of candidate III–V binary and ternary compounds. While intense materials investigations are certainly necessary, device simulations may help in pointing out fruitful directions. In fact, besides the choice of channel material, one should also be open for alternative architectures, whereby Schottky contacts may replace the traditional junctions. At the same time, it appears to be advantageous to operate the channel in a buried mode, away from the interface or scattering centers, as is also done in a high electron mobility transistor, although this could compromise the scalability of the device through short-channel effects.

Finally, even if it may turn out that germanium CMOS is only but a nice dream, current and future research efforts generate a lot of lacking information on basic materials parameters and engineering, which may turn into new exciting applications.

References

[1] J. Huang, N. Wu, Q. Zhang, C. Zhu, A.A.O. Tay, G. Chen and M. Hong, *Appl. Phys. Lett.* **87** (2005) 173507.

[2] S. Heo, S. Baek, D. Lee, M. Hasan, H. Jung, J. Lee and H. Hwang, *Electrochem. Solid-St. Lett.* **9** (2006) G136.

[3] E. Simoen, A. Satta, A. D'Amore, T. Janssens, T. Clarysse, K. Martens, B. De Jaeger, A. Benedetti, I. Hoflijk, B. Brijs, M. Meuris and W. Vandervorst, *Mat. Sci. in Semicond.* Proc. **9** (2006) 716.

[4] A. Satta, A. D'Amore, E. Simoen, T. Janssens, T. Clarysse, W. Anwand and W. Skorupa, *Nucl. Instrum. Meth. In Phys. Res. B.* (in press).

[5] G. Taraschi, A.J. Pitera and E.A. Fitzgerald, *Solid-State Electron.* **48** (2004) 1297.

[6] T. Tezuka, N. Sugiyama and S. Takagi, *Appl. Phys. Lett.* **79** (2001) 1798.

[7] S. Nakaharai, T. Tezuka, N. Sugiyama, Y. Moriyama and S. Takagi, *Appl. Phys. Lett.* **83** (2003) 3516.

[8] P.E. Hellberg, S.-L. Zhang, F.M. d'Heurle and C.S. Petersson, *J. Appl. Phys.* **82** (1997) 5773, 5779.

[9] M.A. Rabie, Y.M. Haddara and J. Carette, *J. Appl. Phys.* **98** (2005) 074904.

[10] M.K. Bera, S. Chakraborty, R. Das, G.K. Dalapati, S. Chattopadhyay, S.K. Samanta, W.J. Yoo, A.K. Chakraborty, Y. Butenko, L. Šiller, M.R.C. Hunt, S. Saha and C.K. Maiti, *J. Vac. Sci. Technol.* A **24** (2006) 84.

[11] M. Tanaka, I. Tsunoda, T. Sadoh, T. Enokida, M. Ninomiya, M. Nakamae and M. Miyao, *Jpn. J. Appl. Phys.* **45** (2006) 3147.

[12] T.-Y. Liow, K.-M. Tan, Y.-C. Yeo, A. Agarwal, A. Du, C.-H. Tung and N. Balasubramanian, *Appl. Phys. Lett.* **87** (2005) 262104.

[13] S. Balakumar, M.M. Roy, B. Ramamurthy, C.H. Tung, G. Fei, S. Tripathy, C. Dongzhi, R. Kumar, N. Balasubramanian and D.L. Kwong, *Electrochem. Solid State Lett.* **9** (2006) G158.

[14] Q. Li, Y.-B. Jiang, H. Xu, S. Hersee and S.M. Han, *Appl. Phys. Lett.* **85** (2004) 1928.

[15] T. Signamarcheix, T. Chevolleau, O. Renault, C. Deguet, T. Akatsu and F. Rieutord, *Proceedings of the EUROSOI 2006*, Grenoble (2006) 41.

[16] H. Wadsworth, S. Bhattacharya, D.W. McNeill, F. Ruddell, B.M. Armstrong and H.S. Gamble, *Proceedings of the EUROSOI 2006*, Grenoble (2006) 43.

[17] J. Tersoff, *Phys. Rev. B* **43** (1991) 9377.

[18] T. Krishnamohan, Z. Krivokapic, K. Uchida, Y. Nishi and K.C. Saraswat, *VLSI Techn. Dig. The IEEE* (Piscataway, New York) 2005, p. 82.

[19] T. Krishnamohan, Z. Krivokapic, K. Uchida, Y. Nishi and K.C. Saraswat, *IEEE Trans. Electron Dev.* **53** (2006) 990.

[20] T. Krishnamohan, D. Kim, C.D. Nguyen, C. Jungemann, Y. Nishi and K.C. Saraswat, *IEEE Trans. Electron Dev.* **53** (2006) 1000.

[21] J. Tersoff, *Appl. Phys. Lett.* **62** (1993) 693.

[22] S.B. Samavedam and E.A. Fitzgerald, *J. Appl. Phys.* **81** (1997) 3108.

[23] S.B. Samavedam, M.T. Currie, T.A. Langdo and E.A. Fitzgerald, *Appl. Phys. Lett.* **73** (1998) 2125.

[24] M.L. Lee, C.W. Leitz, Z. Cheng, A.J. Pitera, T. Langdo, M.T. Currie, G. Taraschi, E.A. Fitzgerald and D.A. Antoniadis, *Appl. Phys. Lett.* **79** (2001) 3344.

[25] R.E. Jones, S.G. Thomas, S. Bharatan, R. Thoma, C. Jasper, T. Zirkle, N.V. Edwards, R. Liu, X.D. Wang, Q. Xie, C. Rosenblad, J. Ramm, G. Isella, H. von Känel, J. Oh and J.C. Campbell, *2002 IEDM Techn. Dig. The IEEE* (Piscataway, New York) 2002, p. 793.

[26] A. Ritenour, S. Yu, M.L. Lee, N. Lu, W. Bai, A. Pitera, E.A. Fitzgerald, D.L. Kwong and D.A. Antoniadis, *2003 IEDM Technical Digest, The IEEE* (Piscataway, New York) 2002, p. 433.

[27] G. Luo, T.-H. Yang, E.Y. Chang, C.-Y. Chang and K.-A. Chao, *Jpn. J. Appl. Phys.* **42** (2003) L517.

[28] K. Sawano, Y. Abe, H. Satoh, K. Nakagawa and Y. Shiraki, *Jpn. J. Appl. Phys.* **44** (2005) L1320.

[29] H.-C. Luan, D.R. Lim, K.K. Lee, K.M. Chen, J.G. Sandland, K. Wada and L.C. Kimerling, *Appl. Phys. Lett.* **75** (1999) 2909.

[30] J. Liu, J. Michel, W. Giziewicz, D. Pan, K. Wada, D.D. Cannon, S. Jongthammanurak, D.T. Danielson, L.C. Kimerling, J. Chen, F.Ö Ilday, F.X. Kärtner and J. Yasaitis, *Appl. Phys. Lett.* **87** (2005) 103501.

[31] Y. Ishikawa, K. Wada, J. Liu, D.D. Cannon, H.-C. Luan, J. Michel and L.C. Kimerling, *J. Appl. Phys.* **98** (2005) 013501.

[32] A. Nayfeh, C.O. Chui, K.C. Saraswat and T. Yonehara, *Appl. Phys. Lett.* **85** (2004) 2815.

[33] K.C. Saraswat, C.O. Chui, T. Krishnamohan, A. Nayfeh and P. McIntyre, *Microelectron Eng.* **80** (2005) 15.

[34] S.-J. Kahng, J.Y. Park, K.H. Booth, J. Lee, Y. Khang and Y. Kuk, *J. Vac. Sci. Technol.* A **15** (1997) 927.

[35] J.M.C. Thornton, A.A. Williams, J.E. McDonald, R.G. van Silfhout, J.F. van der Veen, M. Finney and C. Norris, *J. Vac. Sci. Technol. B* **9** (1991) 2146.

[36] M. Horn-von Hoegen, F.K. LeGoues, M. Copel, M.C. Reuter and R.M. Tromp, *Phys. Rev. Lett.* **67** (1991) 1130.

[37] H.J. Osten, J. Klatt, G. Lippert, B. Dietrich and E. Bugiel, *Phys. Rev. Lett.* **69** (1992) 450.

[38] G. Meyer, B. Voigtländer and N.M. Amer, *Surf. Sci. Lett.* **274** (1992) L541.

[39] H.J. Osten, J. Klatt, G. Lippert, E. Bugiel and S. Higuchi, *J. Appl. Phys.* **74** (1993) 2507.

[40] M. Katayama, T. Nakayama, M. Aono and C.F. McConville, *Phys. Rev. B* **54** (1996) 8600.

[41] Th. Schmidt, R. Kröger, T. Clausen, J. Falta, A. Janzen, M. Kammler, P. Kury, P. Zahl and M. Horn-von Hoegen, *Appl. Phys. Lett.* **86** (2005) 111910.

[42] T.F. Wietler, A. Ott, E. Bugiel and K.R. Hofmann, *Mater. Sci. Semicond. Process* **8** (2005) 73.

[43] D. Reinking, M. Kammler, M. Horn-von Hoegen and K.R. Hofmann, *Jpn. J. Appl. Phys.* **36** (1997) L1082.

[44] J.L. Liu, S. Tong, Y.H. Luo, J. Wan and K.L. Wang, *Appl. Phys. Lett.* **79** (2001) 3431.

[45] M. Horn von-Hoegen, F.J. Meyer zu Heringdorf, M. Kammler, C. Schaeffer, D. Reinking and K.R. Hofmann, *Thin Solid Films* **343–344** (1999) 579.

[46] T. Schmidt, J. Falta, G. Materlik, J. Zeysing, G. Falkenberg and R.L. Johnson, *Appl. Phys. Lett.* **74** (1999) 1391.

[47] J. Falta, M. Copel, F.K. LeGoues and R.M. Tromp, *Appl. Phys. Lett.* **62** (1993) 2962.

[48] J. Falta, T. Schmidt, A. Hille and G. Materlik, *Phys. Rev.* B **54** (1996) R17288.

[49] S. Maruno, S. Fujita, H. Watanabe, Y. Kusumi and M. Ichikawa, *Appl. Phys. Lett.* **68** (1996) 2213.

[50] S. Higuchi and Y. Nakanishi, *Surf. Sci. Lett.* **254** (1991) L465.

[51] S. Higuchi and Y. Nakanishi, *J. Appl. Phys.* **71** (1992) 4277.

[52] M.R. Bennett, C.J. Dunscombe, A.A. Cafolla, J.W. Cairns, J.E. Macdonald and R.H. Williams, *Surf. Sci.* **380** (1997) 178.

[53] A. Sakai and T. Tatsumi, *Appl. Phys. Lett.* **64** (1994) 52.

[54] A. Sakai, T. Tatsumi and K. Aoyama, *Appl. Phys. Lett.* **71** (1997) 3510.

[55] S.-J. Kahng, Y.H. Ha, J.-Y. Park, S. Kim, D.W. Moon and Y. Kuk, *Phys. Rev. Lett.* **80** (1998) 4931.

[56] D. Dentel, J.L. Bischoff, T. Angot and L. Kubler, *Surf. Sci.* **402–404** (1998) 211.

[57] S.-J. Kahng, Y.H. Ha, D.W. Moon and Y. Kuk, *Appl. Phys. Lett.* **77** (2000) 981.

[58] K. Sugawara, M. Sakuraba and J. Murota, *Mater. Sci. Semicond. Process* **8** (2005) 69.

[59] H.J. Osten, E. Bugiel and P. Zaumseil, *J. Cryst. Growth* **142** (1994) 322.

[60] H.J. Osten and J. Klatt, *Appl. Phys. Lett.* **65** (1994) 630.

[61] J. Kolodzey, P.A. O'Neil, S. Zhang, B.A. Orner, K. Roe, K.M. Unruh, C.P. Swann, M.M. Waite and S.I. Shah, *Appl. Phys. Lett.* **67** (1995) 1865.

[62] M. Todd, J. McMurran, J. Kouvetakis and D.J. Smith, *Chem. Mater.* **8** (1996) 2491.

[63] M. Todd, J. Kouvetakis and D.J. Smith, *Appl. Phys. Lett.* **68** (1996) 2407.

[64] D.Q. Kelley, J.P. Donnelly, S. Dey, S.V. Joshi, D.I. García Gutiérrez, M.J. Yacamán and S.K. Banerjee, *IEEE Electron Dev. Lett.* **27** (2006) 265.

[65] D.Q. Kelly, I. Wiedmann, J.P. Donnelly, S.V. Joshi, S. Dey, S.K. Banerjee, D.I. García-Gutiérrez and M.J. Yacamán, *Appl. Phys. Lett.* **88** (2006) 152101.

[66] R. Droopad, M. Passlack, N. England, K. Rajagopalan, J. Abrokwah and A. Kummel, *Microelectron. Eng.* **80** (2005) 138.

[67] M. Passlack, *J. Vac. Sci. Technol.* B **23** (2005) 1773.

[68] C.J. Kao, M.C. Chen, C.J. Tun, G.C. Chi, J.K. Sheu, W.C. Lai, M.L. Lee, F. Ren and S.J. Pearton, *J. Appl. Phys.* **98** (2005) 064506.

[69] H.C. Lin, P.D. Ye and G.D. Wilk, *Appl. Phys. Lett.* **87** (2005) 182904.

[70] I. Ok, H. Kim, M. Zhang, C.-Y. Kang, S.J. Rhee, C. Choi, S.A. Krishnan, T. Lee, F. Zhu, G. Thareja and J.C. Lee, *IEEE Electron Dev. Lett.* **27** (2006) 145.

[71] S. Koveshnikov, W. Tsai, I. Ok, J.C. Lee, V. Torkanov, M. Yakimov and S. Oktyabrsky, *Appl. Phys. Lett.* **88** (2006) 022106.

[72] Y.-Z. Chiou, *Jpn. J. Appl. Phys.* **45** (2006) 3045.

[73] S. Rai, V. Adivarahan, N. Tipirneni, A. Koudymov, J. Yang, G. Simin and M. Asif Khan, *Jpn. J. Appl. Phys.* **45** (2006) 4985.

[74] M. Passlack, R. Droopad, K. Rajagopalan, J. Abrokwah, R. Gregory and D. Nguyen, *IEEE Electron Device Lett.* **26** (2005) 713.

[75] N. Maeda, C. Wang, T. Enoki, T. Makimoto and T. Tawara, *Appl. Phys. Lett.* **87** (2005) 073504.

[76] C. Wang, N. Maeda, M. Hiroki, H. Yokoyama, N. Watanabe, T. Makimoto, T. Enoki and T. Kobayashi, *Jpn. J. Appl. Phys.* **45** (2006) 40.

[77] C. Liu, E.F. Chor and L.S. Tan, *Appl. Phys. Lett.* **88** (2006) 173504.

[78] Y. Dora, S. Han, D. Klenov, P.J. Hansen, K. No, U.K. Mishra, S. Stemmer and J.S. Speck, *J. Vac. Sci. Technol.* B **24** (2006) 575.

[79] Y. Cao, X. Li, J. Zhang, P. Fay, T.H. Kosel and D.C. Hall, *IEEE Electron Dev. Lett.* **27** (2006) 317.

[80] M. Marso, G. Heidelberger, K.M. Indlekofer, J. Bernát, A. Fox, P. Kordoš and H. Lüth, *IEEE Trans. Electron Dev.* **53** (2006) 1517.

[81] Y. Fujino, H. Seto, M. Takebe, M. Tametou, N.C. Paul, K. Iiyama and S. Takamiya, *Jpn. J. Appl. Phys.* **45** (2006) 2417.

[82] T. Hanrath and B.A. Korgel, *J. Am. Chem. Soc.* **124** (2002) 1424.

[83] H. Adhikari, P.C. McIntyre, S. Sun, P. Pianetta and C.E.D. Chidsey, *Appl. Phys. Lett.* **87** (2005) 263109.

[84] D. Wang, Q. Wang, A. Javey, R. Tu, H. Dai, H. Kim, P.C. McIntyre, T. Krishnamohan and K.C. Saraswat, *Appl. Phys. Lett.* **83** (2003) 2432.

[85] A.B. Greytack, L.J. Lauhon, M.S. Gudiksen and C.M. Lieber, *Appl. Phys. Lett.* **84** (2004) 4176.

[86] P.W. Li, W.M. Liao, D.M.T. Kuo, S.W. Lin, P.S. Chen, S.C. Lu and M.-J. Tsai, *Appl. Phys. Lett.* **85** (2004) 1532.

[87] W.M. Liao, P.W. Li, D.M.T. Kuo and W.T. Lai, *Appl. Phys. Lett.* **88** (2006) 182109.

Appendix

This appendix gives more detailed information on important parameters of metals in Ge. The different cross-section data is given in the following tables.

Table A1. Summary of capture cross-section data available in the literature for the different charge states of Cu_s.

Cu charge state	Capture cross section (cm^2)	Temperature range (K)	Measurement technique	References
$Cu_s^{(0/-)}$	$\sigma_p = 8.1 \times 10^{-14}$	20	GR noise	[3]
	$\sigma_p = 10^{-13}$	24	GR noise	[8]
	$\sigma_p = 1.1 \times 10^{-13}$	24	GR noise	[8]
	$\sigma_p \sim T^{-1.1}$	5–60	GR noise	[8]
	$\sigma_p = 2.5 \times 10^{-12}$	20	Photo-Hall/lifetime	[2]
	$\sigma_p = 2 \times 10^{-11}$	4.2	Photo-Hall/lifetime	[2]
	$\sigma_p = 3.3 \times 10^{-13} \exp(0.0033/kT)$		DLTS[a]	[7]
	$\sigma_p = 2.2 \times 10^{-12}$	20	DLTS[a]	[7]
	$\sigma_n = 10^{-16}$	85	PC[b] versus T	[9]
	$\sigma_n = 2.4 \times 10^{-17}$	295	PC[b] versus T	[9]
$Cu_s^{(-/2-)}$	$\sigma_p = 1.5 \times 10^{-16}$		Constant T-dependent PC[b]	[1]
	$\sigma_p = 2.86 \times 10^{-14} \exp(0.015/kT)$		DLTS[a]	[7]
	$\sigma_p = 9.5 \times 10^{-14}$	145	DLTS[a]	[7]
	$\sigma_p = 1.05 \times 10^{-5} T^{-3.16}$	125–150	DLTS[a]	[5]
	$\sigma_p = 2.5 \times 10^{-12}$ – 1.4×10^{-12}	125–150	DLTS[a]	[5]
	$\sigma_p = 1 \times 10^{-20}$	125	Extrinsic PC[b]	[10]
	$\sigma_p = 10^{-17}$	85	Extrinsic PC[b]	[11]
	$\sigma_n = \sigma_{e0} \exp(-\Delta E/kT)$ $\Delta E > 0.11$ eV		T-dependent PC[b]	[1]

(continued)

Germanium-Based Technologies: From Materials to Devices. C. Claeys and E. Simoen (Editors)

Table A1 (Continued)

Cu charge state	Capture cross section (cm^2)	Temperature range (K)	Measurement technique	References
	$\sigma_n = 10^{-17}$	300	*T*-dependent PC[b]	[1]
	$\sigma_n \sim 1/T^2 = 1.5 \times 10^{-17}$	300	Lifetime (PC)[b]	[12]
	$\sigma_n = 1 \times 10^{-17}$	125	Extrinsic PC[b]	[10]
$Cu_s^{(2-/3-)}$	$\sigma_p = 10^{-16}$	120–300	Recombination lifetime	[17]
	$\sigma_p \sim 10^{-16}$	120–300	Recombination lifetime	[6]
	$\sigma_p = 1.8 \times 10^{-16}$	300	p–n lifetime	[13]
	$\sigma_p = 1 \times 10^{-16}$	125	Extrinsic PC[b]	[16]
	$\sigma_p = 1 \times 10^{-16} (\sim T^{-3})$	300	PGM effect[c]	[6]
	$\sigma_n = 3.4 \times 10^{-17}$ $\exp(-0.65/kT)$	165–185	DLTS[a]	[4]
	$\sigma_n = 4.4 \times 10^{-17}$ $\exp(-0.01/kT)$	115–130	DLTS[a]	[5]
	$\sigma_n = 1.2 \times 10^{-17}$	115–130	DLTS[a]	[5]
	$\sigma_n = 2 \times 10^{-19}$	125	Extrinsic PC[b]	[10]
	$\sigma_n = 6 \times 10^{-20}$	85	Extrinsic PC[b]	[11]
	$\sigma_n = 1.3 \times 10^{-17}$	300	p–n lifetime	[13]
	$\sigma_n = 0.9 \times 10^{-20}$	77	Extrinsic PC[b]	[14]
	$\sigma_n \sim \exp(-T_0/T)^{1/3}$[d]	60–235	Extrinsic PC[b]	[15]
	$\sigma_n = 10^{-22}$	30	Photocurrent	[16]
	$\sigma_n \sim T^{1/2}\exp(-T_0/T)^{1/3}$[d]	20–250	Photocurrent	[16]

[a]Variable pulse duration method at fixed peak temperature.
[b]PC: photoconductivity.
[c]PGM: photogalvanomagnetic effect.
[d]Such a cross section is interpreted as being due to field-assisted tunneling through a repulsive potential. $T_0 = 1.3 \times 10^5$ K.

References

[1] R.G. Shulman and B.J. Wyluda, *Phys. Rev.* **102** (1956) 1455.

[2] P. Norton and H. Levinstein, *Phys. Rev. B* **6** (1972) 489.

[3] B.V. Rollin and J.P. Russell, *Proc. Phys. Soc.* **81** (1963) 578.

[4] P. Clauws, G. Huylebroeck, E. Simoen, P. Vermaercke, F. De Smet and J. Vennik, *Semicond. Sci. Technol.* **4** (1989) 910.

[5] I.M. Kotina, V.V. Kuryatkov, S.R. Novikov and T.I. Pirozhkova, *Sov. Phys. Semicond.* **21** (1987) 635.

[6] R.A. Paramonova and A.V. Rzhanov, *Sov. Phys. Solid State* **4** (1963) 1335.

[7] F.X. Zach, H. Grimmeiss and E.E. Haller, *Mat. Sci. Forum* **83–87** (1992) 245.

[8] V.A. Besfamil'naya and V.V. Ostroborodova, *Sov. Phys. Semicond.* **3** (1969) 15.

[9] B.H. Schulz, *Philips Res. Repts* **16** (1961) 182.

[10] R.A. Paramonova and A.F. Plotnikow, *Sov. Phys. Solid State* **4** (1963) 2077.

[11] B.V. Kornilov, *Sov. Phys. Solid State* **4** (1963) 1771.

[12] G.S. Kalashnikov and A.I. Morozov, *Sov. Phys. Solid State* **1** (1959) 1182.

[13] M.I. Iglitsyn and Yu.A. Kontsevoi, *Sov. Phys. Solid State* **2** (1960) 1039.

[14] Yu.L. Ivanov, *Sov. Phys. Solid State* **4** (1963) 1665.

[15] N.G. Zhdanova and S.G. Kalashnikov, *Sov. Phys. Solid State* **6** (1964) 350.

[16] S.G. Kalashnikov, *J. Phys. Chem. Solid.* **8** (1959) 52.

[17] N.G. Zhdanova, S.G. Kalashnikov and A.I. Morozov, *Sov. Phys. Solid State* **1** (1959) 481.

Table A2. Electron and hole capture cross section for the four levels of Au_s in Ge.

Charge state	Capture cross section (cm^2)	*T* (K)	Method	References
σ_n^{2-}	10^{-14}	300	Carrier lifetime	[4]
	$(1-3)\times10^{-18}$	37	DLTS	[2]
	1×10^{-19}	25	PC[a] kinetics	[18]
	$\sim\exp(-0.017/kT)$			[18]
σ_p^{2-}	1.5×10^{-13}	300	PC[a] effect/PEM[b]	[6]
	1×10^{-14}		PC[a] effect/PEM[b]	[11]
σ_n^{-}	2×10^{-16}	300	Carrier lifetime	[4]
	2×10^{-16}	Independent	PC[a] effect/PEM[b]	[5]
	1×10^{-16}	300	PC[a] effect/PEM[b]	[6]
	$\sim\exp(-0.01/kT)$	(60–150 K)		[6]
	2×10^{-16}	90	PC[a] effect/PEM[b]	[11]
	1.2×10^{-17}	77	PC[a] decay	[13]
	$\sim\exp(-T_0/T)^{1/3}$	(55–170)		[13]
	6×10^{-18}		DLTS	[1]
	$(1-2)\times10^{-17}$	125	DLTS	[2]
	1.5×10^{-16}	77–90	DLTS	[3]
	$\sim10^{-17}$	80	PC[a]	[17]
	$10^{-16}\exp(-0.018/kT)$			[17]
	6×10^{-19}	25	PC[a]	[18]
	1.5×10^{-17}	77	PC[a] decay	[19]
	5×10^{-18}	100	PC[a] decay	[20]
	3×10^{-17}	200	PC[a] decay	[20]
σ_p^{-}	1.5×10^{-12}	90–100	GR noise	[8,9]
	7×10^{-13}	100	Pulsed field effect	[10]
	$7\times10^{-13}\,(T/100)^{-n}$ with $0.5<n<1$ (50–100)			[10]
	$(1-1.6)\times10^{-14}$	77	GR noise[c]	[12]
	4×10^{-13}	30	PC[a] decay	[15]
	$\sim T^{-3.5}$ (7–35 K)			[15]
	10^{-15}		DLTS	[1]
	6.2×10^{-14}	77–95	DLTS	[3]
	1×10^{-13}	80	PC[a]	[17]
σ_n^{0}	5×10^{-17}	Independent	PC[a] effect/PEM[b]	[5]
	$\sim10^{-16}$	300	PC[a] effect/PEM[b]	[7]
	6×10^{-14}	80	PC[a]	[17]
σ_p^{0}	3.7×10^{-18}	90–100	GR noise[c]	[8,9]
	6.7×10^{-19}		Field-effect study	[9,10]
	$(1-2)\times10^{-14}$	24	GR noise[c]	[14]
	1.2×10^{-14}	20	GR noise[c]	[16]
	$\sim1/T$			

[a] PC: photoconductivity.
[b] PEM: photoelectromagnetic effect.
[c] GR noise: generation-recombination noise.

References

[1] S.J. Pearton, *Solid-State Electron* **25** (1982) 305.

[2] E. Simoen, P. Clauws, G. Huylebroeck and J. Vennik, *Semicond. Sci. Technol.* **2** (1987) 507.

[3] I.M. Kotina, V.V. Kuryatkov, S.R. Novikov and T.I. Pirozhkova, *Sov. Phys. Semicond.* **21** (1987) 635.

[4] K.D. Glinchuk, E.G. Miselyuk and N.N. Fortunatova, *Sov. Phys. Solid State* **1** (1959) 1234.

[5] V.G. Alekseeva, I.V. Karpova and S.G. Kalashnikov, *Sov. Phys. Solid State* **3** (1961) 699.

[6] R.L. Williams, *J. Phys. Chem. Solid.* **22** (1961) 261.

[7] V.G. Alekseeva, S.G. Kalashnikov, I.V. Karpova and E.G. Landsberg, *J. Phys. Chem. Solid.* **22** (1961) 45.

[8] L.J. Neuringer and W. Bernard, *J. Phys. Chem. Solid.* **22** (1961) 385.

[9] L.J. Neuringer and W. Bernard, *Phys. Rev. Lett.* **6** (1961) 455.

[10] G. Rupprecht, *J. Phys. Chem. Solid.* **22** 255 (1961).

[11] I.V. Karpova, V.G. Alekseeva and S.G. Kalashnikov, *Sov. Phys. Solid State* **4** (1962) 461.

[12] E.V. Buryak, S.A. Kaufman and K.M. Kulikov, *Sov. Phys. Solid State* **5** (1963) 249.

[13] N.G. Zhdanova and V.G. Alekseeva, *Sov. Phys. Solid State* **5** (1963) 397.

[14] V.A. Besfamil'naya and V.V. Ostroborodova, *Sov. Phys. Solid State* **6** (1965) 3005.

[15] I.A. Kurova and V.V. Ostroborodova, *Sov. Phys. Solid State* **7** (1965) 547.

[16] V.A. Besfamil'naya and V.V. Ostroborodova, *Sov. Phys. Semicond.* **3** (1969) 15.

[17] L. Johnson and H. Levinstein, *Phys. Rev.* **117** (1960) 1191.

[18] I.A. Kursova, S.G. Kalashnikov and N.D. Tyapkina, *Sov. Phys. Solid State* **4** (1962) 1104.

[19] S.A. Kaufman, K.M. Kulikov, N.P. Likhtman and N.Sh. Khaikin, *Sov. Phys. Solid State* **7** (1965) 668.

[20] K.D. Glinchuk and E.G. Miselyuk, *Sov. Phys. Solid State* **4** (1962) 2684.

Table A3. Electron and hole capture cross section for the four levels of Ag_s in Ge.

Charge state	Capture cross section (cm^2)	*T* (K)	Method	References
σ_n^{2-}	10^{-18}	100 $\sim \exp(T_0/T)^{1/3}$	PC decay	[4]
	1.1×10^{-17}	80	DLTS	[1]
	$10^{-15}\exp(-0.031/kT)$			[1]
σ_p^{2-}	5×10^{-15}	300	PEM and PC	[5]
σ_n^{-}	10^{-16}	300	PEM and PC	[5]
	5.7×10^{-17}		DLTS	[3]
	$1.2 \times 10^{-16}\exp(-0.01/kT)$	120–140		[3]
	2×10^{-16}		DLTS	[2]
	1.6×10^{-17}	192	DLTS	[1]
	$1.9 \times 10^{-16}\exp(-0.041/kT)$			[1]
	5×10^{-19}	100	PC decay	[6]
	3×10^{-18}	200	PC decay	[6]
	0.75×10^{-16}	100–300	PCD lifetime*	[7]
σ_p^{-}	$1.5 \times 10^{-13} - 6.8 \times 10^{-14}$		DLTS	[3]
	$1.2 \times 10^{-7}\,T^{-3.38}$	55–70		[3]
σ_n^{0}	1.5×10^{-16}	100–300	PCD lifetime*	[7]

*PCD: photoconductive decay.

References

[1] G. Huylebroeck, P. Clauws, E. Simoen, E. Rotsaert and J. Vennik, *Semicond. Sci. Technol.* **4** (1989) 529.

[2] S.J. Pearton, *Solid-State Electron* **25** (1982) 305.

[3] I.M. Kotina, V.V. Kuryatkov, S.R. Novikov and T.I. Pirozhkova, *Sov. Phys. Semicond.* **21** (1987) 635.

[4] I.G. Aksyanov, V.G. Ivanov, S.K. Novoselov, M.A. Reshchikov and Yu.S. Smetannikova, *Sov. Phys. Semicond.* **21** (1987) 1223.

[5] K.D. Glinchuk, E.G. Miselyuk and N.N. Fortunatova, *Sov. Phys. Solid State* **1** (1959) 1234.

[6] K.D. Glinchuk and E.G. Miselyuk, *Sov. Phys. Solid State* **4** (1962) 2684.

[7] W.K. Schoenmaekers and R. Henck, *IEEE Trans. Nucl. Sci.* **19** (1972) 329.

Table A4. Capture cross sections for the two nickel levels in germanium.

Level	Cross section (cm^2)	*T* dependence (K)	Method	References
σ_p^{2-}	2×10^{-14}	150–300 constant with *T*	PEM and PC	[6]
	$\geq 10^{-13}$	300	GR noise	[7]
	$\geq 5 \times 10^{-14}$	300	GR noise	[8]
	$(1-4) \times 10^{-14}$	Lowers with *T*	"Lifetime"	[11]
σ_n^-	2×10^{-15}	150–300 constant with *T*	PEM and PC	[6]
	2.3×10^{-16}	300	GR noise	[6,7]
	3.0×10^{-16}	300 independent	PEM, PC and PC decay	[9]
	1×10^{-18}	100 moderately dependent	PC modulation	[10]
	3×10^{-18}	200~exp(−0.15 eV/*kT*)	PC decay	[4]
	5×10^{-20}	80–100~exp(−0.02/*kT*)	PC decay	[4]
	3×10^{-16}		DLTS	[1]
	4×10^{-18}	165	DLTS	[5]
	1.8×10^{-14}	105–125	DLTS	[2]
	5×10^{-16}	Independent	"Lifetime"	[11]
σ_p^-	3.5×10^{-14}	200~ $T^{-1.5}$	GR noise	[7]
	8×10^{-13}	100	Pulsed field effect	[3]
	8.8×10^{-13}	95–110	DLTS	[2]
	10^{-13}	77	"Lifetime"	[11]
σ_n^0	1.7×10^{-15}		GR noise	[7]
	0.5×10^{-16}	300 independent	PEM, PC and PC decay	[9]
	0.96×10^{-16}	Independent	"Lifetime"	[11]
	10^{-16}	85	PC	[12]

References

[1] S.J. Pearton, *Solid-State Electron* **25** (1982) 305.

[2] I.M. Kotina, V.V. Kuryatkov, S.R. Novikov and T.I. Pirozhkova, *Sov. Phys. Semicond.* **21** (1987) 635.

[3] G. Rupprecht, *J. Phys. Chem. Solid.* **22** (1961) 255.

[4] K.D. Glinchuk and E.G. Miselyuk, *Sov. Phys. Solid State* **4** (1962) 2684.

[5] G. Huylebroeck, P. Clauws, E. Simoen and J. Vennik, *Solid State Commun.* **82** (1992) 367.

[6] S.G. Kalashnikov and K.P. Tissen, *Sov. Phys. Solid State* **1** (1959) 1603.

[7] F.M. Klaassen, J. Blok and H.C. Booy, *Physica* **27** (1961) 48.

[8] F.M. Klaassen, K.M. Van Vliet and J.R. Fassett, *J. Phys. Chem. Solid.* **22** (1961) 391.

[9] P.G. Eliseev and S.G. Kalashnikov, *Sov. Phys. Solid State* **5** (1963) 233.

[10] A.D. Belyaev and E.G. Miselyuk, *Sov. Phys. Solid State* **6** (1965) 2101.

[11] G.K. Wertheim, *Phys. Rev.* **115** (1959) 37.

[12] B.H. Schulz, *Philips Res. Repts* **16** (1961) 182.

Index

FSC
www.fsc.org
MIX
Papier | Fördert
gute Waldnutzung
FSC® C083411

Kolibri 360 GmbH
Ferdinand-Jühlke-Straße 7
99095 Erfurt, Deutschland
produktsicherheit@kolibri360.de